Environment

ENVIRONMENT

PHOTO BY GEORG GERSTER/RAPHO GUILLUMETTE

Resources, Pollution and Society

SECOND EDITION

William W. Murdoch, Editor
University of California, Santa Barbara

SINAUER ASSOCIATES, INC. PUBLISHERS
Sunderland, Massachusetts

THE COVER

The photo on the front cover depicts solar furnaces near Odeillo, France. Giant panels reflect sunlight into a large parabolic reflector (not shown), which focuses it on furnaces mounted atop derrick-like masts. Arrays of such furnaces may be used in the future to generate electric power. The photo on the back cover depicts the shoreline of Fire Island, New York, as taken by a fisheye lens. (Photos: front, Georg Gerster/Rapho Guillumette; back, Dick Hyman.)

ENVIRONMENT, Second Edition
First printing
© 1975 by Sinauer Associates, Inc.
Sunderland, Massachusetts, 01375

Manufactured in the U. S. A.

Library of Congress Catalog Card Number 74–24361.

ISBN: 0–87893–503–7

Contents

Preface

Quite suddenly in the mid-1960's the general public became aware that there were environmental problems of truly crisis proportions. In the ensuing decade there has been a great outpouring of heavy rhetoric on the environment. The result frequently has been to bewilder the beginning student: Are all of these problems real? Is our "survival" really threatened from so many sides? Should one believe the advertising by the major polluters, claiming that they will solve the problem by technological fixes? Because of this bewilderment, and also because of inflation, recession, and bizarre politics, by the mid-1970's the environment has receded somewhat in the public consciousness. But make no mistake—the environmental crisis has not gone away; indeed, in some respects it has become worse. However, in one important respect we are now much better equipped to deal with environmental issues: Environmental Studies has matured into a firm, solidly based discipline. True, the problems remain complex, value-laden, and multidisciplinary. But we are pretty sure what the problems are (and what they are *not*) and we have gained tremendously in our understanding of them.

It is possible to select a few problems that stand out in importance: food supply and population in poor nations; energy, its production, our demand for it, and its effects on human health and our environment; the ubiquitous distribution of chemical pollutants in air, water and food supply that threaten human health over long periods of chronic exposure; and, in the intermediate and long run, the possibility that we will seriously interfere with complex feed-back mechanisms that determine global climate. A central point of this book is that, while technological approaches to these problems are vital, solutions to the problems almost always depend upon economic, legal and political factors. This interplay of science, technology, and human values and institutions is what makes environment not only interdisciplinary but also fascinating.

This completely revised and updated Second Edition provides a solid basis for the study and discussion of environmental problems. Its great strength is the fact that each of the wide array of disciplines covered has been treated by an authority in the field. These disciplines encompass ecology, demography, economics, law, sociology, health, resource management, pollution in its various forms, the environmental effects of pesticides, pest control, and meteorology. Although there are many authors, the book is nonetheless a unified whole. The authors themselves, by carefully writing for the beginning student in the light of our experience with the First Edition, have accomplished much of this. To add further unity I have carefully edited the chapters, modified some topics to fit the overall plan (especially the new chapter on health), and added editorial comment throughout. In all of this I have been aided by advice and comments from many of the instructors who have used the book.

My own experience in teaching environmental subjects is that the book can be fully used almost regardless of which environmental issue is central to the course. For example, in teaching about world food and agriculture, the following sequence was useful:

	Chapters
Malnutrition and the food supply	3,7
Demand for food	2,3
Agriculture and demand for other resources	5,6,(4)
Food production and climate	15
Agriculture and pollution	8–11,13,15
Pests and food	14
Obstacles to the adoption of alternative pest control techniques	14,16,17,19
Ecology and the design of agriculture	1

The book is intended for undergraduates in beginning and intermediate courses dealing with Man and Environment, particularly interdisciplinary courses. It is written so that a student with no particular background in any of the sciences, law or economics can understand it. It does not contain any material that a non-science student will find difficult. It contains some arithmetic, but no mathematics. However, because some very large (and very small) numbers are used, the student will often see a number such as 5×10^{11}. This particular number is 500 billion—the number of anchovies caught by Peru in 1972 (Chapter 7). The notation 5×10^{11} is simply a shorthand. To find out how large the number is, write down the 5 (or, a 1 if the number is simply 10^{11}) and the 11 zeroes. Two commonly used numbers are 10^6 (one million, 1,000,000) and 10^9 (one billion, 1,000,000,000). One can also write numbers such as 10^{-6}, which is $1/10^6$ (i.e., a one-millionth), or 10^{-9}, a billionth.

Finally, I should like to stress that the major thing a student needs to develop in learning about the environment is an attitude, a way of viewing and analyzing problems. The facts, the numbers, the arguments are all important; but the way you set them in perspective is crucial. Ecological systems (and our own industrial system is no exception) are made up of interconnected components, and actions taken at one place reverberate and resonate throughout the system. Environmental decisions will affect other resources and other pollution problems, and will have economic and social implications. In addition, they are made on the basis of judgments that imply economic, ecological and social values, and an awareness of these underlying values is a central part of an ecological attitude.

I am deeply grateful to Sharon Avery for her assistance throughout the preparation of this book.

William W. Murdoch
Santa Barbara, California

Cypress trees rise from the Okefenokee Swamp in Georgia. Photo by Grant Heilman.

1

Ecological Systems

William W. Murdoch *is Professor of Biology at the University of California, Santa Barbara. After graduating from the University of Glasgow (Scotland) in 1960, he did research on insect population dynamics for three years with Charles S. Elton at the Bureau of Animal Population, Oxford. In 1963 he earned his doctorate from the University of Oxford, and then did post-doctoral research at the University of Michigan.*

Murdoch's research centers on predator-prey relationships, an area that provides some of the theory underlying biological pest control. Recently he served on a National Academy of Sciences study of pest control problems. Murdoch has written on a number of other environmental issues and is particularly interested in the extent to which ideas from the biological discipline of ecology can be applied to environmental problems.

Like any other organism, man is inextricably dependent upon ecosystems. More than any other organism, we depend on a wide range of ecosystems, both natural and heavily managed by man, and we affect almost every ecosystem that exists. Certainly very few (perhaps no) ecosystems exist that have not been changed by man; the deep oceans and some areas of tropical rain forests are examples. But even the majority of natural ecosystems get traces of pollutants, and most ecosystems are of some use to us.

The diversity of types of ecosystems is great, but we can order them according to the extent to which we manage and use them. At one end of the spectrum are ecosystems that we hardly manage at all, such as marine communities that support enormous fisheries, and some forests. Intermediate are such systems as grazing lands, where we do minor herbicide spraying and manage cattle; most forests; and some orchards and other long-lived crop types. At the other end of the spectrum are heavily managed ecosystems, such as short-lived crops, which are often grown in monoculture and cultivated, sprayed, and fertilized. Modern agriculture is so dependent on industry for energy, machinery, fertilizers, and the like, that we really ought to consider the agroecosystem as including the city, but it is more useful for us to keep these components separate.

In the rest of this chapter we shall explore the most important processes that occur in these ecosystems and the principles and rules that govern their operation and constrain the kinds of things that we can do to ecosystems and the kinds of ecosystems that we can design — for we cannot create and sustain all the sorts of ecosystems that we might like to have. Furthermore, although some principles apply to all ecosystems, it is important to understand that there are also differences among systems regarding basic considerations. For example, in designing agricultural systems we might be better off to forget about what happens in natural communities, because the two differ in important ways. An understanding of ecosystems is crucial, for at the core of our environmental crises is the problem of how much we can alter ecosystems, both intentionally and by accident, and still rely upon them to sustain us into the indefinite future.

In this chapter we shall look at biological communities in terms of the flow of energy and materials, food webs, populations, and communities; and we shall try to see what has been discovered by each of these approaches.

A word about ecology before continuing. Ecology is an exciting discipline that has changed rapidly in recent years. New theory is developing, much more field evidence is available, and ecologists are thinking about the relevance of their work to environmental problems. As a consequence, today's conventional wisdom is likely to be discarded tomorrow. Thus, some of the material in this chapter (e.g., that on diversity and stability) may not be in agreement with most ecology textbooks. Where this happens I shall note the disagreement and try to explain it.

Two processes impose a gross pattern and organization on ecological systems. These are the flow of energy and the cycling of materials.

Energy and Matter on a Global Scale

ENERGY

Energy sources on earth include tidal energy, nuclear energy, the heat of the earth's core, and so on. But for living systems the

2

unique energy source is solar radiation. Biological systems depend on the flow of this energy from the sun, through the system, and back into outer space in the form of heat. The earth is in a steady state, so over any appreciably long period, the amount of energy that enters from the source (the sun) is equal to the amount lost as heat to the "sink" (outer space). The existence of both source and sink is essential, because if the earth did not maintain this energy balance, it would heat up or cool down, and in either state, biological systems would not persist (Chapter 15). The earth is thus an open system with respect to energy.

Energy flows and in so doing produces material cycles, the water cycle (Chapter 6) being an example. Part of the solar energy eventually absorbed by the atmosphere and oceans produces temperature differences; these cause convection currents in the air and produce winds. The heat energy absorbed by water also produces evaporation, and water vapor is then convected high in the atmosphere. When the water vapor precipitates over land, the heat energy absorbed during evaporation is released. The cycle is completed when the water returns to sea level via runoff, rivers, and lakes, the potential energy of the water being dissipated as heat. Man concentrates the potential energy by building dams and converting the energy to hydroelectric power.

Other energy sources that play a part in cycles include the thermal energy of the earth, which produces volcanoes and mountain uplifting; and the solar energy trapped by plants, which drives biological cycles. The latter energy is transferred as food from one part of a biological system to another. Each transfer involves biochemical rearrangements during which heat is produced, and eventually all the energy first trapped by photosynthesizing plants leaves the system as heat.

MATERIALS

In contrast to the situation with regard to energy, the earth is essentially a closed system with respect to matter. The amount of matter is fixed, so it merely passes from one state to another. For some elements, or some portion of them, the natural cycle is fairly simple. For others, for example the 30 to 40 elements used by biological systems, the cycle is more complex. And, of course, man now uses almost all the elements, complicating and disrupting cycles.

The most perfect or complete cycles tend to be those in which there is a *gaseous phase*, for example water, nitrogen, and oxygen. These, in particular, are balanced cycles, in that the amount in any one phase (such as oxygen in the air) tends to be constant. Small disturbances are corrected for naturally, and changes are only temporary. However, as Hutchinson[1] has noted, gaseous cycles may not be able to return to the steady state following large disturbances, which may cause severe disruption of the cycle.

From man's point of view, small, gradual changes may also have profound environmental effects. The carbon dioxide (CO_2) in the atmosphere forms an important reservoir in the carbon cycle, and the oceans and the world's plants together act to regulate its quantity. However, in burning fossil fuels, we produce enough CO_2 to exceed the capacity of the regulators, so the CO_2 content of the atmosphere is increasing (Chapter 15).

A second type of cycle involves movement of materials from land to sea and back again. Elements going through such *sedimentary cycles* generally take millions of years to cycle, and the cycles tend to be less perfect than those with a gaseous phase, in that some of the element may get "stuck" in one phase of the cycle. In sedimentary cycles, materials are leached or eroded from rocks on land and carried by rivers to the oceans. Most of the element is deposited on the bottom of the shallow inshore areas of the oceans and is returned to the land eventually, when these sediments are raised into mountains by uplifting. However, there is some slight "loss" of calcium and phosphorus (important elements for biological systems), as small amounts are deposited in deep ocean sediments. Such sediments are only very slowly incorporated into continental rocks by spreading of the ocean mantle.

Biological systems, of course, are important parts of the cycles in many elements. One example involves the guano birds off the Peruvian coast, which eat ocean fish and produce concentrations of nitrates and phosphates on land (Chapter 7). This process has probably occurred in other times at other places. Of major importance also is the part

3

played by organisms in the movement of carbon dioxide, as noted above.

A second example is that of the phosphorus cycle, illustrated in Figure 1. Phosphorus is a necessary constituent of protoplasm. Note in the figure that phosphorus continually cycles to some extent within biological communities, going through plants and sometimes animals to soil bacteria and back through plants again. There is always a tendency, however, to lose phosphorus in the form of soluble phosphates, in runoff, down streams, and eventually to the ocean. However, rocks and other phosphate sources act as a reservoir, and the phosphorus leaving biological communities is partly replaced from these sources by leaching.

It has been estimated[1] that about 20 million tons of phosphorus is leached from land to the oceans each year. In addition, man quarries

phosphorite and makes from it superphosphate for use as fertilizer, most of which eventually reaches the sea. About 7 million tons of phosphorus are quarried each year, and much of that binds to the soil when it is applied, so man adds 1 to 2 million tons annually to the loss of phosphorus from the continents. The major pathway for returning phosphorus to land is the uplifting of marine sediments, but this is a geologically intermittent process. The other major and more continuous process is the deposition of guano by seabirds, which may return several hundred thousand tons each year. Some thousands of tons are returned by marine fishing, but it is clear that at present the rate of loss to the oceans is much greater than the rate of return.

Man can now use enormous amounts of energy to interfere in a massive way with the

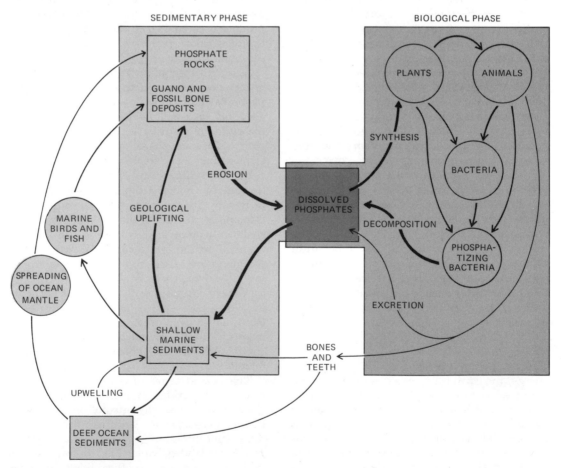

1   **Biogeochemical Cycle of Phosphorus.**

cycling of materials, as just described for phosphorus. In particular, in mining and drilling operations elements are removed from more-or-less concentrated forms and dispersed on the earth's surface, a process that clearly cannot persist indefinitely for any element (Chapter 4). This process of dispersion is, for our purposes, a permanent disruption of the cycling process. Thus copper, once mined and dispersed, will not recycle naturally to an extent useful to human populations. Of course, we could recycle it artificially within our industrial system, and we do to some extent.

Perhaps the most astounding fact illustrating our influence on the environment is that we are now element movers on a scale approaching that of natural worldwide forces. Table I lists the rate at which we remove, by mining, various common elements. Note that man's rate actually *exceeds* natural rates in all these examples! Remember, also, that the average rate of increase of commercial mineral production is about 5 percent, so the amount moved will double roughly every 14 years; furthermore, for each ton of useful yield, we need to move progressively more parent material as we mine progressively lower grades of ore.

Man disrupts material cycles in exploiting resources. The other side of this coin is that disruption of the cycles leads to pollution. Here it is important to note that we do not consume materials, we merely move them around. Pollution frequently involves the breakdown of the natural cyclic processes so that materials accumulate where we do not want them. This results in part simply from the fact that, because we extract materials from the earth at very high rates, the natural cyclic processes cannot cope with the increased rate at which materials are returned to the system. Possibly the most graphic illustration of this is eutrophication (Chapter 10). Wastes are spewed into rivers as sewage and as effluent from industries such as pulp mills. Such organic and inorganic materials (e.g., phosphates) occur naturally, as described above, but the stream organisms normally responsible for cycling them can handle only a limited rate of input. In eutrophication the rate of input is too high and the cycling is continued mainly by a different set of organisms, often lacking oxygen. This can result in the loss of fish and recreation. It is instructive that this change in

quantity of input results in a change in the *quality* of the biological situation. This illustrates the point that biological systems frequently have *thresholds*, so that responses to rate changes are not simply linear.

Another example in which part of the problem is simply an increase in the rate of production of materials is the solid waste problem — what to do with the junk heaps of metal and other materials that are littering the landscape. But the solid waste problem also illustrates a second aspect of pollution: that technology produces new combinations of materials for which no natural breakdown processes (or poorly developed processes) have been evolved. Plastics are a good example. Various persistent pesticides and phenols from the petroleum industry are also in this category. Some of these materials are a particular problem, because they are toxic (Chapter 8). Some naturally occurring materials when in high concentrations are also toxic; in this case it is our producing them at high rates and in concentrated form that provides the toxic effect; lead from gasoline additives and mercury from pesticides and paper mills are examples.

A peculiar addition to some material cycles is that of radioactive isotopes from nuclear explosions and nuclear reactors. For example, strontium cycles with calcium in the sedimentary cycle and passes through biological systems. As is well known, radioactive strontium

I Effect of Man on the Movement of Materials

Element	Rate of element movement (thousands of metric tons/year)	
	Geological (in rivers)	Man-induced (mining)
Iron	25,000	319,000
Nitrogen	8,500	9,800
Manganese	440	1,600
Copper	375	4,460
Zinc	370	3,930
Nickel	300	358
Lead	180	2,330
Molybdenum	13	57
Silver	5	7
Mercury	3	7
Tin	1.5	166
Antimony	1.3	40

The numbers were estimated in 1967. *Source:* Reference 2.

is now involved in this cycle. Ecologists have taken advantage of the fact that radioactive isotopes are easily monitored in organisms to trace the cycling of materials in the system. An example is discussed later in this chapter (see Figure 4).

So much for the broad-brush picture of the flow of material and energy. We turn now to materials and energy in biological systems and to the processes that occur in the continuous cycling of inorganic matter to biological production, to inorganic matter, and back to biological production.

Production in Ecosystems

TROPHIC LEVELS AND FOOD WEBS

All organisms are factories for taking in, rearranging, and passing on materials, and, like all factories, they need energy for the work involved. Plants get their materials (from soil, air, and water) and energy (from the sun) separately, but other organisms (animals, fungi, bacteria) get their materials and energy in the single package that is their food.

In the oceans, energy is trapped and used to synthesize organic tissue, mainly by minute plants — the phytoplankton (algae, diatoms, etc.). The phytoplankton are fed upon by very small invertebrates (the zooplankton). These, in turn, are eaten by larger invertebrates and small fish, which are attacked and eaten by large fish, seals, and so on; and these may finally be eaten by killer whales. Each link in this food chain can be viewed as a trophic level, a group of organisms feeding on similar food sources. Typically, the organisms in each trophic level are larger than those in the lower trophic levels. Food chains, or, more accurately, food webs, are thus the pathways along which energy and nutrients move, and trophic levels are transfer stations along the way at which nutrients are first broken down and used or reassembled into new tissue, and where energy is dissipated.

A word of caution is required regarding trophic levels. Many organisms regularly or at different stages in their life history are in different trophic levels. In insects, for example, more than half the species may spend time in more than one trophic level, and even seed-eating birds feed small animals to their young.

To this extent trophic levels are an abstraction, although they are useful in thinking about ecosystems and in the study of energy flow.

For any biological system to persist, there has to be a basic trophic level that traps solar energy and synthesizes food. This function is universally performed by green plants, and this first level is called the *producer level*. Just as necessarily, nutrient cycles must be completed and organisms must be present that can break down the complex organic materials and molecules in plant tissues. This job is finally done by microorganisms (bacteria and fungi). In general, intermediate activities take place in which the plants are eaten (live or dead) and large chunks of plants and animals are broken down into smaller parts.

There are two more-or-less clear series of trophic levels, one (the *herbivore chain*) using live plant materials as a base and containing plants, herbivores (which eat green plants), predators (which eat other animals, mainly herbivores), and parasites; the other using mainly dead materials as a base (the *detritus* or *decomposer chain*). The distinction between the two types of food chains is not complete; for example, snails eat both live and dead plant material. In a perfect, idealized ecosystem the links in these chains form a closed circle, so that biological material is being continually synthesized by plants from simple inorganic molecules (e.g., phosphates in the soil) and passed along the circle until bacteria finally complete the decomposition process by breaking down complex organic molecules to inorganic compounds that plants can reuse.

BIOLOGICAL PRODUCTIVITY

Roughly speaking, the productivity of a trophic level is the amount of material that it passes on as food to other organisms. If the first trophic level is to stay approximately constant, this food should be "interest" rather than "principal." Biological productivity and the rules that govern and constrain it are clearly of great importance. Thus, where we enter into a fairly natural community, such as the oceans, we want to divert as much productivity as we can through our own population without digging too severely into the "principal." This problem is discussed later in this chapter and in Chapter 7. We are especially interested in what determines the productivity of the plants in the system, because, in turn, this determines the productivity of the whole community.

Discussions of productivity can be confusing because different authors use different units to measure it; further, although we are interested in the amount of material that is produced, productivity is often expressed as the amount of energy (calories) produced. However, because a given weight of a particular organic molecule always contains exactly the same amount of energy, we can always calculate how much energy there is in a given amount of fat, carbohydrate, or protein; or we can measure it directly by burning the material.

Plants, the producers, synthesize organic molecules (carbohydrates, fats, proteins) from inorganic molecules in the environment (CO_2, sulfates, nitrates, etc.) and use solar energy for the purpose. This energy is trapped during photosynthesis, during which CO_2 combines with water in the presence of radiant energy and enzymes associated with chlorophyll, to form glucose (oxygen is given off). By far the largest proportion of plant biomass is made up of such carbohydrates.

The basic reaction of photosynthesis joins six water molecules (taken up by the roots) with six molecules of CO_2 absorbed through the leaves to yield one molecule of glucose, which contains chemical energy, plus six molecules of oxygen, which is released from the leaves:

$$\downarrow 6CO_2 + H_2O \xrightarrow[\text{energy}]{\text{light}} \begin{array}{c} C_6H_{12}O_6 + 6O_2 \uparrow \\ \text{(contains} \\ \text{energy)} \end{array}$$

This chemical reaction has caused widespread confusion about plants and our oxygen supply: because oxygen is produced during photosynthesis, many people believe that as we kill plants, we reduce our oxygen supply. This notion is false. The photosynthetic reaction is a balanced one, in that every time a molecule of glucose is broken down to release energy for use, six oxygen molecules are absorbed from the atmosphere and six CO_2 molecules are given off; the gases exchanged during photosynthesis are thus in balance. For all practical purposes, the earth's total plant biomass is in equilibrium; for each plant molecule that is created by photosynthesis, a molecule is broken down by respiration, the process opposite to photosynthesis. The molecules are respired either by the plant, or by the organism that eats the plant, or by the organism that eats the organism that eats the plant. . . . If, for example, we killed off all the phytoplankton in the ocean, many organisms would starve, but the amount of oxygen in the atmosphere would not change.

The apparent paradox is that the oxygen in the atmosphere was produced by plants during the past 2 billion years. This is because at various times plant material was *not* decomposed but accumulated in the earth's crust. Thus we now also have fossil fuels, which are concentrations of old organic material that was not decomposed. The accessible fossil fuels are but a tiny fraction of the buried organic material, and their combustion (which is what respiration is) will not significantly affect the oxygen supply. However, CO_2 is rather rare in our atmosphere, so the CO_2 produced by burning fossil fuels *does* affect the concentration in the atmosphere and hence our climate (Chapter 15).

The amount of solar energy actually available to plants, on the average, is about 5×10^8 calories per year per square meter of the earth's surface, and plants manage to "fix" about 1 to 2 percent of this energy. All organisms need energy, the ability to do work, for their maintenance, growth, and reproduction, and they get it in the form of potential energy, that is, chemical energy stored as food. Thus, the total amount of energy fixed by plants sets an absolute upper limit to biological production and activity.

Solar energy is not trapped uniformly over the earth's surface, nor is biomass produced uniformly. The surface of the open oceans is by and large a biological desert — as are the deserts on land — and in these areas, solar heat is almost entirely lost via radiation to outer space, without flowing through a biological community. In contrast, a few places are very high in their annual production. In particular, in the wet tropics natural communities and intense cultivation can produce 50 to 75 metric tons per hectare (2.5 acres) in a year, and estuaries, coral reefs, and rich alluvial farmland are also very productive. Some idea of the world distribution of production can be gained from Table II.

The main determinants of productivity are: solar radiation and temperature, rainfall (on land), and nutrients. Thus, the desert can be made more productive by irrigating it, and water bodies become eutrophic by getting fertilizer and feedlot runoff. Productivity generally is high in the tropics and decreases toward the poles. (Production is defined as the amount of organic matter elaborated over a specified time period, whether or not it all survives to the end of that time.)

The primary plant production is eaten and passes through various trophic levels. At each transfer, the great proportion of the energy contained in the production of one trophic level is dissipated as heat during the process of forming the production of the next higher trophic level. This follows from thermodynamic and biochemical rules.[5] First, every time that energy is transformed naturally from one form to another, some portion always goes into heat and is lost. For example, when an animal releases the potential energy in glucose (by oxidation during respiration), about two-thirds of the energy is available for growth and activity (work) and one-third is lost to the environment as heat. Second, not all the food eaten can be assimilated or used,

and a varying proportion is lost as feces, which passes to decomposers. Third, organisms need energy to maintain themselves, to move about, and to grow and reproduce. They get this energy by breaking down large food molecules into smaller, simpler molecules, which are then further degraded to produce energy plus waste products in the process of respiration. An example of maintenance is that every day almost 8 percent of our body protein becomes degraded and is re-synthesized. New biomass (growth or reproduction) is synthesized from broken-down food molecules.

In addition to these losses, there are more ecological losses. For example, herbivores do not usually eat all the plant production nor carnivores all the herbivore production. Thus, the herbivore food chain is continually "leaking" large quantities of energy to the decomposer food chain in the form of dead organisms and feces. These losses are summarized in Figure 2.

From this discussion it can be concluded that the amount of energy available to a trophic level and its production of biomass over a period of time must decrease very sharply as we move up trophic levels.

Plant community	Dry material (metric tons/hectare/year)
Terrestrial communities	
Desert scrub	0.7–1.0
Tundra	1.4
Temperate grassland	5.0
Savanna	6.0
Temperate deciduous forest	12–13
Tropical rain forest	20–50
Aquatic communities	
Open ocean	1.25
Continental shelf	3.5
Attached plants (freshwater)	6.0
Estuaries	20–32
Temperate freshwater reedswamp	43
Tropical freshwater reedswamp	75
Agricultural communities	
Annual crops (temperate)	22
Annual crops (tropics)	30
Perennial crops (temperate)	30
Perennial crops (tropics)	75

II Net Primary Production by Various Plant Communities

Source: estimates based on References 3 and 4.

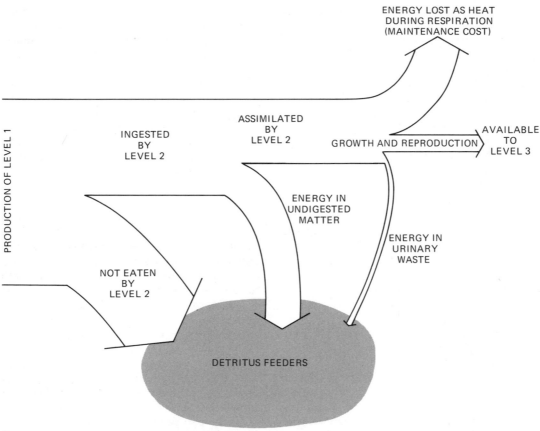

2 **Energy Losses in a Food Chain.** Energy leaves trophic level 1 and passes through trophic level 2.

In the herbivore food chain a special constraint applies in converting material from plants to animals. The chemical composition of plants is very different from that of animals. In particular, a gram of plant biomass has a much lower energy (calorific) content than a gram of animal biomass. This is because plants consist mainly of carbohydrate, with very little protein and fat, whereas animals are very high in protein and fat but low in carbohydrate. The energy in carbohydrate is about 3.7 kilocalories per gram dry weight, that of protein is 5.8, and for fat the figure is over 9.0. Therefore, a great deal of plant biomass must be degraded to turn the energy into the concentrated form found in animals. Clearly, since the food of carnivores is similar to their tissue, carnivores can convert assimilated food into biomass more efficiently than can herbivores. These

relationships explain why grain production must increase so rapidly as the *quality* of people's diets increases, even if they are not eating much more in total (Chapter 3). To go from a grain to a meat diet, we must greatly increase the amount of grain.

Since energy is always lost as material moves up the food chain, it is important to know how much is lost at each step and whether in some ecosystems the loss is less than in others. Such movement of energy is called *energy flow*. Some data are available for the efficiency of transfer between trophic levels, but it is not clear how generally applicable they are.[6] This efficiency is the product of (1) the percentage of the production of one trophic level taken by the next higher level, and (2) the percentage of this food that is turned into new biomass (production). It is difficult to get such data for major portions of communities, and the data available are approximate. Plants themselves use some of the energy they fix; about 20 to 50 percent is eventually lost as heat during respiration, and 50 to 80 percent of the total energy fixed over a year is available to the herbivore food chain or the detritus food chain as potential energy. This amount is what has been defined as net *primary production*. Herbivores (primary consumers) do not eat all this plant production (in fact, in forests they may eat only 10 percent or less). Of the plant production actually consumed by herbivores, about 10 percent is turned into new herbivore biomass (i.e., herbivore production). Thus, for every 100 grams of land-plant production in natural systems, generally we only get 1 or 2 grams of herbivore production!

Secondary consumers (predators feeding on herbivores) tend to take a large percentage of the production of herbivores (at least 30 percent and possibly sometimes nearly 100 percent). The efficiency of predators in converting food to new biomass was previously believed to be about 10 percent, as in herbivores, but it may be as high as 30 percent or more in cold-blooded predators such as invertebrates, and less than 10 percent in warm-blooded vertebrates. If predators in general were 30 percent efficient, predator production would be about 10 percent (30 percent times 30 percent) of herbivore production and possibly as high as 30 percent. Estimates for ocean herbivores average 15 percent, and for ocean fish predators reach 20 percent.[7] In the case of modern animal husbandry, where highly selected corn is fed to cattle, pigs, and other animals that have also been highly selected for efficiency and do not move around much, conversion is a very high 20 percent; that is, 5 pounds of corn can produce 1 pound of meat.

It is generally stated (although more data are needed) that in aquatic systems based on phytoplankton, almost all the primary production passes through the herbivores. In such aquatic systems, where the herbivores consume nearly all the plant production, for every 100 grams of plant production there will be almost 10 grams of herbivore production. As on land, predators may take a high proportion of this herbivore production, so that one might get 20 grams of primary carnivores for each 100 grams of herbivores. Thus, a conversion factor of 1 to 2 percent between the first and third trophic levels may be common in aquatic systems, compared with 0.1 to 0.3 percent for the land.

This kind of information can be summarized in a "production pyramid." Figure 3 shows the estimated amount of new biomass produced over a year by each trophic level in all the world's oceans. Herbivores were assumed to convert about 10 percent of the plant productivity into herbivore productivity, and at all higher trophic levels the conversion rate between levels was assumed to be 15 percent. Since interest centered on the potential fish production of the oceans, a small fraction of production was omitted at carnivore levels to take account of organisms, such as jellyfish, corals, and barnacles, not in the trophic path to fish.

The basic measurements are necessarily rather rough, but the significance of energy losses between levels is strikingly obvious. Man exploits the oceans at about trophic level 4 (mollusks and some fish are at a lower level, some fish even higher). The 300 million tons of production at this level is slightly higher than some other recent estimates, and of this productivity man probably can harvest rather less than half, that is, not much more than double the present 60 million metric tons of harvest (Chapter 7). By comparison, on land man can exploit the second trophic level for protein. We have to come in at higher trophic levels in the sea because the organisms in the first two or three levels are very small.

Presumably the drastic decreases in production as one moves up trophic levels is a major reason why ecological food chains rarely have more than five links from producer to top predator. There probably are other reasons, of course. For example, predators have to be larger than their prey generally, so individual animals are large and population sizes become very small in top predators.

By contrast with the herbivore food chain, on the average and over long periods the decomposer food chain consumes all the incoming production and completes the dissipation of biologically fixed energy in the form of heat. If this were not the case, most ecosystems would be chest-deep in a growing pile of dead and decaying organisms and detritus. In fact,

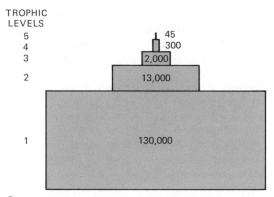

TROPHIC LEVELS

5	45
4	300
3	2,000
2	13,000
1	130,000

3 **New Biomass Production Pyramid.** The estimated total annual production at different trophic levels in the world's oceans, in millions of metric tons. The areas of the rectangles are proportional to the estimated production. The estimate of primary production (mainly by phytoplankton) is 130 billion tons. *Source:* Reference 8.

temporary exceptions (such as bogs and lakes) do occur and some organic material does accumulate, as we shall see when the process of *succession* is discussed. Furthermore, important exceptions in the geologic past permit us now to burn fossil fuels (Chapter 5).

The detritus food chain is crucially important in nutrient cycling. Here the same thermodynamic and biochemical principles apply as in the rest of the ecosystem. A system of trophic levels can be outlined, although perhaps less clearly. Although many of the species are small and inconspicuous by habit, tremendous complexity exists here also.

There are two rather different parts of the decomposition process, and both are necessary for its completion. In the first place, large pieces of detritus (dead animals, leaves, etc.) are fragmented and partly decomposed, generally by invertebrates and by fungi. Second, these coarse organic particles are decomposed into simple inorganic compounds by simple organisms — bacteria and other fungi. The part played by the different kinds of organisms varies with soil conditions.

Throughout this discussion emphasis has been on the quantity of energy transfer and loss between trophic levels. However, it must be stressed that energy considerations alone may not explain the limits of production at some or all trophic levels. Thus, primary production by plants in an area is rarely limited by

the solar energy available, because nutrients of some sort usually set a lower limit. This is particularly well illustrated in lakes and in the ocean. In lakes, algal "blooms" of production occur seasonally when the water stratification breaks down and nutrients are brought to the lighted surface waters from the bottom of the lake. In the ocean, regions of upwelling are highly productive; in these regions, nutrients in deep cold water are brought up to the lighted surface and are used by the phytoplankton (Chapter 7).

At higher trophic levels food *quality* may be important in determining production. For example, aphids feed on plant sap, and it has been shown[9] that the limits to the transformation of this primary production into aphid biomass is set, not by energetics, but by the concentration of nitrogen (an essential ingredient of proteins) in the plant sap. In fact, aphids go into diapause (a sort of suspended animation) and shut off production when nitrogen levels are too low, even though the sap is still an adequate source of energy.

In general, much of what looks like food for herbivores — the leaves in a forest, for example — may, in fact, not be available, for plants have been selected over millions of years to avoid being obliterated by their herbivore enemies. During this evolution, plants have developed physical and chemical defenses; for example, oak leaves in England are high in tannin and are very tough, and this serves as a defense against caterpillars[10]; cotton, to take another example, has gossypoll glands that act to deter insects. In the selection of crops for agricultural purposes we have removed many of the defense mechanisms, often to make the crops palatable, and have, as a result, made them more susceptible to attack by pests (Chapters 13 and 14).

ENERGY FLOW AND SYSTEMS ANALYSIS

As yet, little direct use has been made of energy-flow analyses for the practical management of ecosystems. However, several major community types (e.g., temperate forest, tundra, grassland, desert) are presently being studied in various parts of the world to try to describe the major pathways for nutrients and energy and to determine the important limiting processes involved. These studies, done under the auspices of the International Biological Program (IBP), are unique in ecology in that they involve large teams of people whose separate projects are synthesized in large systems-analysis models of the community. These models, which are manipulated in computers, can involve hundreds of variables that workers try to measure in the field. The ultimate aim of, for example, the desert program, which is directed by David Goodall of Utah State University, is to build a model that will be able to predict the effect upon the ecosystem of various patterns and intensities of grazing, so that ranching can be managed more successfully.

Systems analysis is particularly well suited to these complex problems and also to those involving the management of populations. For example, systems analysis is beginning to be used to manage fishery stock (Chapter 7) and complex pest problems (Chapter 14). This is a promising approach because systems analysis is suited to the peculiar nature of ecological systems. These are composed of very large numbers of different components, which interact in a large variety of ways. Each biological component is affected by a large number of physical variables, and all variables change not only in time but from place to place, because the environment is heterogeneous. Distant components interact, and one component affects all others in one way or another. Hence "the complexity of the system of interlocking cause–effect pathways confronts us with a superficially baffling problem."[11] Systems analysis is designed precisely to handle such situations. Basically, the system is analyzed in terms of simple components. The variables and processes affecting each component are analyzed and described so that changes through time can be described and predicted. The way in which the various components fit together is also analyzed, and finally the units are reassembled in a model of the whole system. Almost invariably this involves the use of a computer to simulate the consequences of changes in the variables that control the system.

In the case of real ecological systems, one does not try to describe every interaction and the relationships among all the variables. No simulation can be complete. Indeed, the art, as in any science, is to draw a caricature of the system, etching in the really crucial lines which describe the major features of the system. This is why systems analysis will be particularly useful to practical ecologists who have to make decisions on an incomplete description of the system. Systems analysis is a method of allowing one to make decisions and to make the best decision on the available data and at different times as the system changes.

BIOLOGICAL CONCENTRATION

Before leaving the subject of trophic levels and transfer of nutrients, it is important to note that in many cases materials become concentrated as they move up food chains. This is a characteristic of considerable practical importance, as the widespread occurrence of DDT illustrates so dramatically (Chapter 13).

Heavy metals and some radioactive elements also become concentrated as they move up food chains. An example of the latter occurs in Perch Lake in Ontario and is illustrated in Figure 4. Perch Lake is fed by a stream that contains radioactive material which seeps from a nearby liquid disposal area of Atomic Energy of Canada Limited. Strontium 90 is one of these radionuclides. Over the 5-year study period the amount of strontium 90 entering the lake each year was fairly constant. Some of the isotope is removed from the water and the substrate by plants and by organisms in the bottom sediment (where most of the strontium 90 occurs) and also directly by animals swimming in the water. Figure 4 shows concentration factors for radiostrontium in selected organisms in the lake. The concentration in yellow perch, a top fish predator, is 3,000 times that in the water.

Populations

For some purposes we are interested only in particular components of the ecosystem, such as the population of a single species. A population is a group of organisms that belong to the same species and live in the same area, so that there is a possibility of interaction among its members. Populations thus present us with less of a problem of delineation than do trophic levels. They have birth rates, death rates, age distribution, genetic composition, and so on (Chapter 2). For many purposes the ecosystem can be seen as being organized at

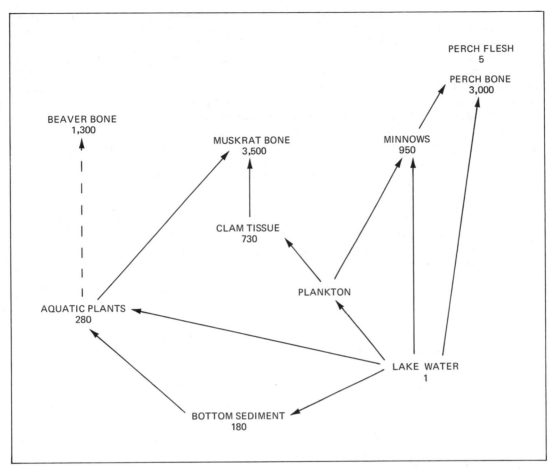

4 **Biological Concentration of Strontium 90.** The average concentration factors for strontium 90 in the Perch Lake food web (from I. L. Ophel, 1963. The fate of radiostrontium in a freshwater community). *Source:* Reference 12.

the population level and we can examine interactions within and among populations. Frequently, we are interested in this level of organization from a very practical point of view. We may wish to maximize the yield from the population of a crop or fishery, and we often wish to reduce the numbers of a pest population. We are also very much interested in the problem of the human population. The general field of study that deals with this level is population biology; that dealing with population abundance is population dynamics. A population, of course, cannot be studied to the exclusion of all other parts of the ecosystem. Rather, the population forms the focus and other components are studied as they appear to influence it.

Perhaps most important from a practical point of view (as well as being interesting theoretically) is the question of how the numbers of a population are determined, what causes them to fluctuate, and what helps to stabilize them if they exhibit stability. It is important to get clear that there is more to finding out about populations than merely listing all the things that kill organisms. Animals fall over cliffs, are struck by lightning, die of disease, and perish in many other ways, yet many causes of mortality may not be "important" in the sense that they provide no explanation for the particular question asked of the population. It is sometimes difficult for the beginner in population biology to drag his attention away from the specific events that occur to individual organisms and to learn to look at, say, the effect of mortality rates upon populations as they vary through time and with population density. For example, in spite of the fact that there may be many causes of death in a population, it may be possible to explain the fluctuations in the numbers on the basis of changes in the death rate caused by only one or a few factors. This has been shown to be the case in many of the long-term studies of insects, where it has been found (1) that there is one crucial stage in the life history which has a particularly variable survival period from year to year, and (2) that much of the annual variation in survival can be attributed to a single cause or factor. Thus, although dozens of processes affect the birth rate and death rate of the population, one *key factor* can explain (and serve to predict) most of the annual fluctuations. Some easily measured aspect of the weather often has been the key factor, although in other cases it has been the amount of parasitism (by insects). This finding has obvious practical importance in any situation where we want to predict population fluctuations or where we want to control them.

The numbers of organisms in a population are determined by three processes: birth rate (natality), death rate (mortality), and movement. Movement is sometimes important in determining the numbers of a species as a whole. Thus, the British starling increased its numbers by invading and spreading throughout North America in this century. Emigration from a population which results in lowering the local density must in general be a net loss to the species in the form of eventual mortality, because most species cannot be continually expanding their range in a finite world.

Dispersal is a crucial process in many species, for example in organisms such as plants and barnacles, where the adult is sedentary. Migration is important in many bird species and, for example, in the spectacular case of the California gray whale, which migrates between the Arctic and Baja California. Also, organisms that live in temporary habitats (such as some carrion feeders and crustaceans living in temporary ponds) have to have well-developed powers of dispersal.

In managed systems, movement can sometimes be important.[13] Many pest insects, such as aphids and the fruit fly, travel large distances as "plankton" in air currents. They descend upon crops, and frequently the factors that control such invasions determine the size of the pest problem. It has been suggested, for example, that the prevailing wind pattern determines which areas in the Middle East are infested by cereal pests.

For simplicity, we will now ignore population movements and discuss population growth and limitation in terms of birth and death rates. Populations increase in density when births exceed deaths and decrease when the reverse is true. For populations that are approximately constant numerically, the birth rate must equal the death rate. Growing populations, therefore, become limited in numbers either by an increase in the death rate or a decrease in the birth rate or both.

Since populations grow by multiplication, they have the capacity to expand exponentially — that is, the numbers present can increase at a constant rate (the numbers can double at a constant interval) even though the numbers get larger at each doubling. This potential for exponential growth is, of course, rarely realized in nature, for a variety of reasons. Populations are usually already at a high density with respect to factors that limit them, although some may frequently decline catastrophically and build up again almost exponentially.

In its pure form, exponential growth is like money in the bank, growing at a constant interest rate that is compounded, say, daily. Table III shows the relationship between the yearly rate of increase (interest) of a popula-

III Relationship Between Annual Growth Rates and Number of Years That a Quantity Takes to Double

Annual increase (%)	Doubling time (years)
0.1	693
0.5	139
0.8	87
1.0	69
2.0	35
3.0	23
4.0	17
5.0	14
6.0	12
8.0	9
10.0	7

tion and the time it takes to double. A simple rule of thumb is that the doubling time is 70 years divided by the interest rate (a rule that derives from the formula for exponential growth).

However, some populations do exhibit approximately exponential growth, especially when they invade a new environment (Figure 5). Strict exponential growth arises when birth and death rates are constant, and you can imagine that even in a new environment, the death rate sooner or later begins to increase so that the growth rate declines. Sometimes, if the decrease occurs smoothly, this results in the sigmoid, or S-shaped form of growth curve, which is also illustrated in Figure 5. In most real populations the rate of increase does not change smoothly and the sigmoid curve is only roughly approximated.

The startling thing about the human population is that it has grown even faster than exponentially until the present! The "interest rate" has itself been increasing because, as the population has become larger, the death rate has decreased while the birth rate has remained the same or decreased more slowly (Chapter 2). It is patently obvious that positive exponential population growth cannot continue indefinitely in *any* population, because all populations face finite resources. You will hear more of this point throughout this book; the point at issue, really, is not whether infinite population is possible in a finite world, but *what* the finite limits are, when we shall encounter them, and whether we ought to set limits ourselves before then.

Two final details about exponential growth. First, exponential growth can be *negative*, so that the population decreases. For example, if you start out with 1,000 test tubes and they "die" accidentally through use but are not replaced (zero birth rate), the population of test tubes will decrease at a constant rate, halving their number in each constant time period. In a radioactive substance, which is described as having a constant "half-life," this is called *exponential decay*. The rules are exactly the same as for positive exponential growth, but in this case the death rate exceeds the birth rate. Usually when populations are described as growing exponentially, we take it to mean positive growth. Second, exponential growth is sometimes called *logarithmic growth*, because

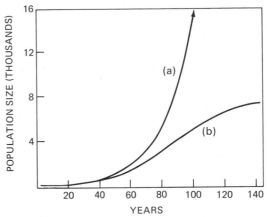

5 **Comparison of Exponential (a) and Limited (b) Growth.** Two populations were started in year zero at a size of 100. Both had an initial yearly rate of increase of 5 percent, which remained constant in (a). Population (b), which is "logistic," has a limit of 8,000. After 140 years population (a) exceeds 109,000.

if density is transformed to logarithms, a plot of the numbers versus time gives a straight line.

FACTORS THAT LIMIT POPULATION

Populations do not increase indefinitely; in fact, they generally increase for only short periods. This is because all populations are constrained by limiting factors of one sort or another. The following list includes a number of things that can set upper limits to populations or can be important in population dynamics in some other way.

1. Sources of Mortality
   a. Enemies, for example predators (including insect parasites), disease
   b. Competitors
   c. Food supply (quality and quantity)
   d. "A place to live," for example nesting site, shelter
   e. The weather
   f. Factors internal to the population, for example cannibalism, changes in the quality of organisms, stress from crowding, waste products

2. Factors that Affect Natality
   a. Food supply (quality and quantity)
   b. Oviposition sites
   c. The weather
   d. Interference among members of the

population, for example territorial behavior, stress, social hierarchy

It is probably the case that many populations in nature are kept at low densities through the action of one or more *enemy species*. This idea is certainly often implicit in discussions of "natural balance." In fact, there is little direct evidence in the way of detailed population studies that many populations are limited by enemies, but this is probably because there are few thorough studies.

Some of the most convincing examples of the limitation of populations by enemies come from studies on the rocky seashore.[14] Examples are also available from the field of biological control (Figure 6 and Chapter 14). But we do not have good direct evidence from complex natural ecosystems, such as forests, although such systems strike us as excellent examples of natural balance. A basic problem is that few such studies have been made.

However, it is probably a good assumption that many herbivore species are kept at low densities by their enemies. The indirect evidence that is available comes mainly from comparing managed ecological systems with natural systems. Thus pest populations occur where natural enemies generally are missing. The trouble with such evidence is that in managed ecosystems many factors other than the absence of natural enemies have been altered from the natural system and that we are therefore not presented with clear, controlled experiments.

One example of population control by an enemy is particularly illustrative of the difficulties of doing ecological research and of the fact that a historical perspective is necessary in field studies.[15] The St.-John's-wort (*Hypericum*) is an introduced plant in California. Generally, it is not very abundant in pastures there. Also, if one searches very diligently, an occasional chrysomelid beetle (*Chrysolina*) can be found, but an analysis of the present scene would rate this insect as unimportant. In actual fact, the plant was once a very important weed in California pastures and it was controlled by introducing the beetle. The beetle has been extremely effective and now the weed is rather rare and patchy and is kept that way by the beetle, which of course is also rare.

The *food supply* must set the upper limit to the potential population size in most organisms, although it is not clear how often populations reach this limit or if they are usually held below it by other factors. It is again difficult to get clearcut examples of populations being limited in a simple, direct fashion by the quantity of food available to them. Few instances are known of populations simply eating out their food supply and being limited by mass mortality through starvation, although food limitation need not involve such occurrences. The numbers of animals feeding on plants in lake plankton seem in many cases to be limited by food supply. This can be shown by the increase in their numbers following an increase in the phytoplankton caused by adding nutrients.[16]

There are several good examples of the quality of the food, or the quantity of the right quality of food, being limiting. In aphids the quality of the sap can help limit numbers.[9] It seems quite likely that there is a complex interaction between food quality, the behavior of the aphids, and aphid density. The food quality and the frequency of contact with other aphids may interact to determine when reproduction is shut off.

Although I have presented food and enemies as separate limiting factors, it is probable that the density of many species is determined at the same time by two or even more factors operating together and interacting. For example, L. B. Slobodkin showed that the density of the water flea *Daphnia* in the laboratory was determined both by the food supply and the predation rate. In nature, well-fed animals probably can escape predators better than poorly fed animals, so the effect of predation will depend on the food level, and the effect of a given food supply will depend upon the number of predators. This is analogous to the phenomenon of *synergy* that sometimes occurs in pollution and toxicology problems, in which the action of a pollutant or toxin is influenced by the level of another pollutant or toxin (Chapter 8).

There is growing evidence that *self-regulation* occurs in some populations. That is, numbers may be determined largely by the qualities of the organisms in the population and by interactions among them. Perhaps the clearest example is territoriality in birds, where only a portion of the population is allowed to breed,

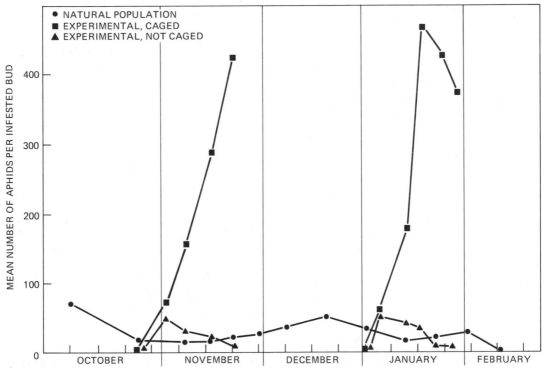

6 **Limitation of Population by an Enemy.** The numbers of aphids on rose buds are normally kept low and stable by predatory insects (ladybugs, syrphid flies, and lacewings), as shown by the curve for the natural population. This was demonstrated by placing aphid colonies on buds and then excluding predators from some of these colonies with cages and letting the predators attack colonies that were not caged. Numbers in protected colonies then increased enormously. This work was done by Derek Maelzer, Adelaide University, Australia.

others being excluded by the aggressive behavior of the territory holders. The density in many such cases seems to be geared to the food supply. Self-regulation is also brought about by social hierarchy or aggressiveness in some small mammals, which allows some individuals to feed and reproduce better than others. Clearly a *sine qua non* for the functioning of such mechanisms is the existence in the population of animals with different qualities.

Other mechanisms include cannibalism, which is very common in insects, and prevention of reproduction via social hierarchy and strong selectivity by the females, which appears to operate in wolves. We must hope that our own self-regulatory mechanisms will operate soon and that they will consist of the more bearable mechanism of social pressure against high birth rates rather than some of the nastier methods found in other animals.

To summarize, all populations become lim-

ited, and food supply or space usually sets the potential upper limit. Other processes, such as predation, prevent many populations from reaching this limit. But it is not clear how frequently the various kinds of limiting factors operate, and ecologists are not yet in a position to make generalizations on this subject.

The factors that limit populations are of great importance in many applied areas. For example, they determine the yield we get from fisheries or wildlife species. A major reason why we have pests is that we have selectively removed plant defenses, and predators are becoming more and more important in pest control (Chapter 14).

POPULATION STABILITY

For reasons of clarity I have kept separate the discussion of limiting factors and the mechanisms that act to keep populations stable. The question of population stability (used here synonymously with "regulation") is an interesting and vexing one. It seems likely that some local populations of species are very unstable and frequently become extinct and are reconstituted by immigration. Others are also unstable numerically, although they seem to persist for quite a long time (see, e.g., Figure 7). Unstable populations are common in simple agricultural systems, where pest outbreaks either occur or are continually threatening but are prevented artificially, but some natural populations also erupt when conditions are right. For example, in balsam fir forests the Canadian spruce budworm (a moth caterpillar) historically seems to have remained at low densities for 30 or 40 years at a time and then increased enormously when a combination of climatic and tree-"stand" factors were favorable.[18]

On the other hand, large numbers of populations seem to jog along maintaining fairly

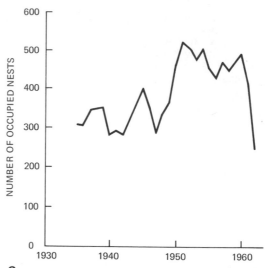

8   **Stable Bird Population.** The number of breeding pairs of heron on the Thames drainage area, England. Note that abundance is plotted on an arithmetic scale and that fluctuations over the entire period were about twofold. *Source:* Reference 19.

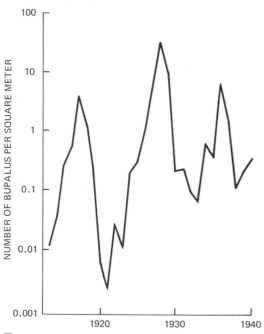

7   **Unstable Moth Population.** The number of hibernating pupae of *Bupalus* per square meter of forest floor at Letzlingen, Germany. *Bupalus* is a moth whose larvae are pests that eat the needles of pine trees. The population peaks were severe outbreaks that caused extensive damage. Note that abundance is plotted on a logarithmic scale and increased more than 10,000-fold between 1920 and 1930. *Source:* Reference 17.

constant numbers, although different populations fluctuate around different density levels. It may be that most populations in most natural systems are stable, or "regulated," to this extent. Indeed, the fact that populations persist for long periods suggests that there are mechanisms, either in the populations themselves or in the communities they live in, which tend to keep them stable. An example of a stable bird population that has been studied over a long time is shown in Figure 8.

Other evidence is the contrast between the pest problem in our agricultural systems and the continued existence in natural communities of grass, shrubs, trees, and so on, which rarely are destroyed or defoliated, which suggests that herbivores in natural systems rarely become enormously and erratically abundant. We would prefer our ecosystems to be dependable and stable, so gaining an understanding of natural stability is of great value to us.

Figure 8 shows a stable population, and the fact that territorial birds generally seem to be stable suggests that one important set of stabilizing mechanisms comes under the general heading of self-regulation. These mechanisms aid populations in buffering themselves against some vagaries in the environment and probably occur in many kinds of animals. In fact, populations that ecologists consider stable do fluctuate, but the fluctuations are relatively small and the average density over several generations in the future will be the same as it was for several generations in the past.

In trying to explain why a given population is stable, we need to discover the mechanisms that tend to make the population decrease when it becomes higher than average and increase when it is below average. To illustrate how such mechanisms work, let us take a simplified example. Suppose that we are interested in the numbers of adults in an animal population in May of each year, and we observe that every year there are 10 adults per square meter. In this example we have perfect regulation of the adult numbers. However, each year the birth rate (number of eggs laid) is a density-disturbing process; it tends to be highly variable, so that in some years there may be only 20 eggs per square meter but 500 in other years. Clearly, mortality as a whole is regulatory. We assume that adults do not live a second year. Table IV shows what happens each year.

Notice, not only does the number killed increase with increasing initial density, so does the *percentage* killed. Thus, regulating factors are those which cause an increasing percentage mortality as density increases. Of course, natality can also be regulatory; one hopes that it will be so in human populations. Such regulatory factors are said to be *density-dependent* in operation.

Populations that are stable, then, are, by definition stabilized by density-dependent processes. To explain such stability we must search for factors that operate in this way. Since the effect of a factor acting upon a population can result from changes in both the factor and in the individuals in the population, in principle any factor might operate in this way. In general, however, ecologists tend to look to such factors as food, enemies, nesting sites, and interactions among organisms in the population for density-dependent effects.

The importance of the concepts of density dependence, regulation, and stability goes beyond pure ecology into the realms of pest control and the management of resource species. Pests are species that are highly unstable and prone to outbreak with great rapidity. The ideal solution is to keep such populations regulated at densities so low that they cause little harm. Ideally, then, pests should be controlled by density-dependent mechanisms, as discussed later in this chapter and in Chapter 14.

The concept of density dependence is also important in resource mangement.[20] Although the problem of getting the optimum yield from an exploited natural population is complex, at the simplest level, for naturally stable stocks, to guarantee the persistence of the population, exploitation should be density-dependent over the long run. Disregard of this obvious point has resulted in the past in the near-extinction of the blue whale, the sperm whale, and the bison, and is now leading to overfishing in some of the world's fishing stocks. Similarly, in the management of game-bird stocks the aim is to shoot the surplus each year, and this is simply to act as a density-dependent mortality factor.

On the other hand, the great ability of some populations to compensate (i.e., regulate) for large perturbations is extremely important. It explains why pesticides often fail to control pests (Chapter 14), and it allows us to fish some populations very heavily and still have them persist (Chapter 7).

GENETICS AND EVOLUTION

One final important characteristic of populations is that they are variable. Because individuals are different and some of the differ-

IV A Mortality-Regulated Population

Initial no. eggs laid in May, year 0	No. adults produced in May, year 1	No. dying	Mortality (%)
20	10	10	50
40	10	30	75
60	10	50	83
100	10	90	90
200	10	190	95
300	10	290	97
400	10	390	98
500	10	490	98

ences are inherited, the population can adapt to new circumstances by *evolving*. Environmental decision makers have often overlooked the ability of species to evolve, sometimes to our disadvantage.

An individual's characteristics are determined by its genes, or genotype, although they are more or less modified by environment. For example, the tendency to be tall is inherited, but a poor diet during childhood will keep a genetically tall person small. Within natural populations there is great genetic variation among individuals, some genes being reduced in frequency by selection each generation, while other genes are added by mutation and new genotypes are added as new genetic combinations occur through sexual reproduction.

Many members of a population have similar attributes, but there is also an almost infinite variety in a large population, and there are usually a few rare individuals or genes that would be at an advantage in a different environment. It is this largely inconspicuous variation that allows nature to keep surprising us. When the English industrial midlands became blackened, black moths became abundant because those rare black individuals with appropriate genes were missed by birds that ate the normal light-colored moths. And when we have sprayed DDT or one of dozens of other insecticides, there have been rare insect genotypes resistant to them that have now become the norm (Chapter 14). However, not all species can adapt to the massive and rapid changes that we create, and thousands become extinct. Those that survive near man tend to be preadapted to living in temporary and frequently disrupted habitats — so we have selected "weeds" as our main companions.

In spite of the examples provided by insects and insecticides, we have not really learned the lesson of evolution. For example, the "third-generation" pesticides have been hailed as a panacea. These include hormones (e.g., insect juvenile hormone) that you can spray on an insect to prevent it from becoming an adult or — worse — cause it to become part adult and part larva. (These hormones occur naturally in the insect but are shut off before it molts to the adult stage.) No doubt, though, there are some tricky insects who will be able to ignore the particular biochemical message we spray

and will go ahead and become complete adults anyway, and the pest will again become abundant. In fact, geneticists predict that populations can make such a change in 10 generations or less if the right mutations are available. Thus it is foolish to put all our hope in such apparent panaceas.

Most weed experts, in a similar vein, claim that weeds do not become resistant to herbicides because the action of herbicides is "too basic." But a more likely explanation for the lack of resistant weeds is that selection operates very slowly in weeds because they have a long generation time, many weed seeds lie deep or dormant in the soil and are missed, herbicides are changed frequently, and so on. One can predict that weeds will eventually produce the same headache of resistance as do insect pests, unless we keep changing herbicides, and indeed there have been some reports of resistant weeds in recent years.

Perhaps even more intriguing is the question of why very few predatory insects are resistant to pesticides. Maybe there are more than we have found, but again it seems likely that for ecological reasons, which we cannot go into here, selection is occurring very slowly in these insects. If we could find resistant strains, and if we could raise them in large numbers, perhaps we could use them in pest control in areas where continued spraying is necessary. Some work is being done on this problem, but there are great technical difficulties.

Although the most dramatic selection we see is caused by man, in natural communities species are selected upon by other species. In nature, interacting populations have a long-shared evolutionary history and are coevolved, or *coadapted*. We noted earlier an example of this in the defenses that plants have against the insects that attack them. This again is a point that is often missed. For example, in order to preserve genetic variation in crops, we keep thousands of distinct genetic strains in bags of seeds. But "out there," where the wild strains of crops are growing as weeds, their genetics are constantly changing in response to diseases and herbivores. Out there may well be the most important seed bank for future crops, but almost no effort is being made to make sure that natural patches of communities with wild crop species are preserved.

20

Thus, the environmentalist who is interested in biological systems has to add a whole new way of thinking about them: he or she has to appreciate the variability of populations and the dynamic, ever-adapting nature of the variation.

Ecosystems

We can learn a good deal that is useful by studying the ecology of individuals and populations. But a population does not exist in a vacuum; it depends on and interacts with its physical environment and with populations of other species (i.e. in an ecosystem), and as we manipulate nature and our own ecological systems, we have to deal with whole communities, not just one species. Growing "monocultures" has dramatically shown us that we actually cannot have just one species! Indeed, I think this is an important point often missed by conservationists concerned with rare or interesting species: outside zoos, we cannot preserve endangered species except by *preserving the communities they depend on.* Raptorial birds need small mammals or small birds, and they in turn need a variety of seeds and insects, and they in turn require a variety of plant species. The birds may need one kind of habitat to nest in and another to feed in and these may need to be close together, and the insects may have different things they eat at different times of their lives. Furthermore, it is extremely unlikely that we can preserve solely the web of species that directly supports the raptorial bird and its prey species; since we cannot usually predict the consequences of major changes in a community, we usually have to preserve the community as it comes — with perhaps some management. Thus, we need to understand the workings of communities, how much abuse they can stand, and what happens if we alter them. In the remainder of this chapter we shall look at ecosystems from several points of view, and in each case we shall find ideas of great practical importance.

SUCCESSION

Succession is the change in communities over long periods of time, usually decades or even centuries, and two types are recognized:

primary succession and secondary succession. In *secondary succession* we want to know what happens after an existing community, say a woodland, has been removed. This type of succession is the major one of interest to us, because we are continually removing communities and may want back a version of the one we destroyed. *Primary succession* is the creation of new communities, for example the transformation of bare rock to soil that supports a forest. First I shall describe the standard ideas on succession and then present another point of view.

The succession of terrestrial communities in various circumstances has been fairly well described and seems to occur in an orderly way. The initial stages of development may take a long time — for example, filling in a lake or bog, or breaking down a bare rock surface by wind, rain, and the action of small plants, such as lichens, which produce a top soil. Once soil and some plants are present, succession then tends to proceed faster. Each stage or community tends to modify the environment, making it more suitable for some other group of organisms, so that there is an interaction between the biological and physical parts of the system.

Secondary succession on abandoned fields is fairly well studied. In the southeastern United States the order often is grasses → grasses plus forbs → mainly shrubs → pine forest → oak–hickory "climax" forest. This history may take 150 years or more, with each successive stage lasting longer than the preceding one. Various features about the process are rather general. (1) Low, structurally simple communities which persist for only a short period give way to communities with larger plants, a more complex structure which persists longer. (2) The number of species present tends to increase with time (although there may be a slight dip in the number of species in the final, climax community), and food webs become more complex. (3) The later stages are more stable in the sense that the individuals tend to be longer-lived, the community tends to stabilize its internal environment (e.g., temperature and wind speed), and later stages last longer and continually replace themselves. Thus, diversity and stability seem to go hand in hand, a claim that we discuss later.

Other changes occur in succession.[21] The

21

production of plant biomass in early successional stages is high. Typically, this is associated with a low standing crop or biomass (the production/biomass ratio is high). From one point of view, such early successional stages are highly efficient, in that the rate of production of plant material is greater than the rate of respiratory loss of energy by the system, so biomass tends to accumulate throughout succession. As this occurs and we move to a larger, older community with greater biomass, the relative rate of biomass production declines; that is, the production/biomass ratio tends to decrease. Climax communities then tend to be balanced, in that the energy fixed by photosynthesis is equal to that respired in maintaining the community, so biomass no longer increases but remains constant. This low production/biomass ratio might be considered inefficient. From another point of view, though, such communities are highly efficient, in that the energy fixed is used to maintain a large, diverse, and very stable community which has gained a great deal of control over its physical environment. The relative importance of the two basic types of energy pathways also changes during succession. In early successional stages a high proportion of the energy passes through the herbivores, whereas in later stages most of the energy passes through the decomposers.[21]

The conventional view outlined above is that succession is directed, orderly, and predictable; that each stage modifies the environment, making it more suitable for later stages; and that *each stage is a necessary precursor of the succeeding stage*. It is the last point that is particularly in some doubt, and which is important. An alternative explanation for the sequential events is simply that the early stages — grasses and forbs, for example — are species that are quickly dispersed and grow quickly, whereas the later stages (such as tree species) tend to disperse more slowly and grow more slowly. Furthermore, in the alternative view (expressed by Joseph Connell of the University of California at Santa Barbara), a given set of vegetation may grow up by a variety of stages, not necessarily predictable. Most important, the alternative view holds that the intermediate stages, say forbs and shrubs, are *not* necessary for the growth of the later tree community.

This difference of opinion is of more than academic importance. Suppose, for example, that you were in charge of strip mining for coal in a huge tract of natural forest. One way of saving the forest and still mining the coal would be to scrape off sections of forest, leave other tracts untouched, and then regrade the hole and cover it with soil, on the assumption that secondary succession will occur and that the intact surrounding forest will serve as a source of propagules (colonizers). Using this scheme, and with the correct strategy, it might be possible to mine the whole area over a period of 100 years and *still have viable forest over much of the area* at the end of that period. We would need to know how big a bare area could be recolonized naturally from surrounding forest, after mining, and how long it would take it to be recolonized and for the forest to grow back. Are intermediate successional stages necessary, or can we accelerate succession by judicious seeding — and still get a replica of the natural community? This question is very important, because if the answer is yes, we might be able to cut the recovery period by many years. In fact, we do not know the answer to such questions, because succession has rarely been studied experimentally. Clearly, this is work that needs to be done.

Even more important than getting back the original community following the destruction of its vegetation is ensuring that the soil and its nutrients are not lost by erosion and leaching while the area is bare. Here, again, succession is important. It has generally been thought that there is a trend during succession from an open to a more closed system with respect to nutrients. Climax communities tend to retain the nutrients they have and continually recycle them (although no ecosystem, except the biosphere as a whole, is completely closed with respect to matter).

Erosion and loss of topsoil rich in nutrients can be severe when natural vegetation is removed, and it has been claimed that nutrient loss is increased an average of eight- to ninefold when agriculture replaces a natural community.

These ideas concerning community function have been examined recently in a pioneering study done in New Hampshire by F. Herbert Bormann, Gene Likens, and a group of colleagues.[22,23] They have measured the input and output of nutrients into several forest *ecosystem units*, each ecosystem unit being a small complete watershed (i.e., drainage or catchment area). They are also measuring the pathways for nutrients within the ecosystems and the rates at which nutrients cycle from inorganic to biological and back to inorganic. Most important, however, they are actually doing experiments with entire ecosystems!

The forest sits on a granite base that is watertight, so all nutrient loss can be measured in the streams leaving the forest. The most significant finding about the natural systems is that, although there are huge amounts of nutrients cycling within the ecosystem, very little is lost to the streams by leaching and erosion. Indeed, the gain of important nutrients such as nitrate and phosphorus in precipitation is greater than the natural losses to streams, so the natural system is actually conserving these nutrients; whereas for calcium, magnesium, and a few other nutrients, there are very small losses each year that are made up by the weathering of rocks. Bormann and Likens then showed that man's activities can fundamentally alter this stable situation. In 1965 an entire watershed was clear-cut: every tree and shrub was cut. No roads were built or logs removed, so the experiment was less drastic in this respect than normal logging. However, to prevent regrowth of the vegetation, the area was sprayed with herbicides for 3 years. An adjacent watershed was left undisturbed.

The first major shift observed was in runoff, which increased 40 percent. Most of that water would have been lost by transpiring vegetation, so the serious problem was not water loss but the damage done by increased runoff. Nitrate concentration in the streams (leached there from the soil) increased eightyfold, so instead of gaining nitrates, the ecosystem lost 97 to 142 kilograms of nitrate-nitrogen per hectare per year. The average yearly loss of nitrate-nitrogen was double the amount taken up each year by the undisturbed ecosystem, and it would take about 100 years to replace the nitrogen lost in the 3-year experiment if precipitation input were the only source. Calcium losses increased tenfold, potassium about twentyfold.

The project also examined the effects of clear-cutting with no herbicide treatment. The pattern of response was the same as before, although the nutrient loss was not quite as great. Vegetation (pin cherry in this case) covered the area again in about 2 years after cutting, although nutrient loss and erosion continued for some time. The encouraging aspect of this experiment is that loss rates came back to about normal in most systems in 3 to 5 years, although in at least one commercially logged area 10 years was not enough for recovery to normal. A summary of events in the absence of herbicide is therefore as follows. After cutting there is a very rapid, serious loss of nutrients (especially nitrates), caused by leaching. Erosion increases somewhat later and reaches a peak about 3 years after cutting. The nutrient pool for the whole ecosystem is thus reduced. However, there is cause for cautious optimism in that the rate of loss goes back to normal after 3 to 5 years but takes longer in some systems. This recovery is caused by early-colonizing plants, so to get nutrient conservation, we do *not* need to wait for the return of the original climax community.

The undisturbed systems are able to hold on to their nutrients for both physical and chemical reasons. Runoff is slowed down because much water is transpired. Living and dead vegetation lying on the ground and on stream banks keeps the soil in place, preventing erosion. The nitrogen cycle in the undisturbed forest produces ammonium, which is taken up by microorganisms and vegetation; but in the felled watershed, ammonium was converted to nitrate, which was quickly leached from the soil into the streams.

Deforestation also caused water pollution. The increased nitrification made the streams more acid. Indeed, the nitrate content almost always exceeded drinking standards. Increased stream nutrients plus sunlight on the exposed streams caused significant eutrophication (a good example of interaction between parts — the terrestrial vegetation and the streams — of an ecosystem). Results from commercially logged areas in New England show the same pattern of nutrient loss and eutrophication of drainage waters. In these areas logging roads cause increased erosion.

INTERACTIONS IN AND BETWEEN ECOSYSTEMS

From our point of view it is of great importance that ecosystems are dynamic, not only in that they change in space and time but in that enormous numbers of interactions are going on within them all the time. Perhaps feeding interactions are the most significant. On an old field in Michigan where about 1,800 species of insects were found, 31 categories were needed to classify all the insects simply into general feeding types (predators on insects, parasites, plant feeders, scavengers, etc.).[24] More than half the insect species, mainly parasites, feed on other insects, providing great potential for interactions among species. This point is illustrated by the food web in Figure 9, which is based on only one species of plant (broccoli). This shows interactions among only the 50 most abundant insect species of the 200 in an agricultural community.

Apparently distant components in the ecosystem, as well as adjacent ones, interact and are interdependent. For example, the adult stage of insects that parasitize other insects feed on flowers; thus, the time of flowering of plants might indirectly affect the parasitism rate observed in a caterpillar population later in the year. This has given Kenneth Hagen, of the University of California at Berkeley, the idea of providing to the natural enemies (parasites and predators) of pests an artificial diet that could be applied to fields as a substitute for flowers, thus keeping high the numbers of adult enemies when flowers are scarce.

If interactions within communities are of interest to us, interactions between communities are even more important in a practical sense. The fact is that ecosystems are open systems, and our activities in one area inevitably have effects elsewhere — the DDT story (Chapter 13) should be enough to convince us of this.

The openness of ecosystems means that they are linked by a flow of materials and energy, and therefore they are not independent of one another. Freshwater bodies, in particular, are open systems. Thus streams gain nutrients in the runoff from surrounding terrestrial communities and in the leaves and other debris that fall in from the riparian vegetation, as we discussed above in relation to Bormann and Likens's work. There is a constant downstream drift of nutrients in the form of detritus and organisms, although this is

sometimes reversed, for example when migrating salmon come back as nutrients from the ocean. Lakes receive inputs of potential energy and nutrients from upstream, and although the rate of flow is slowed within lakes, they still pass on nutrients downstream to the estuaries, continental shelf, ocean plankton, and the deep ocean floor. Sometimes the passage of materials occurs on a hugh scale. The annual flooding of the Nile, the Tigris and Euphrates, and other great river systems of the world provided rich alluvial farmland with which great civilizations were associated. It is estimated that every year the Mississippi carries 730 million tons of soil — 38 thousand acres 3 feet deep — into the Gulf of Mexico.

Perhaps the most dramatic examples of interactions between ecosystems can be found in systems managed by man. The well-intentioned but often poorly conceived building of dams for reservoirs illustrates this. For example, the Aswan High Dam on the Upper Nile was built to provide hydroelectric power. It has had three major bad effects on surrounding systems.[26]

The catch of sardines from the eastern end of the Mediterranean, once 18,000 tons per year, dropped to 500 tons per year. It is not clear whether this is a consequence of cutting off the nutrients that formerly came into the sea during annual flooding and produced large blooms of phytoplankton, or if it is due to changes in the salinity regime of the area. In either case, the dam is the cause of the reduced catch.

Another ecological effect of the dam has been the replacement of an intermittent flowing river with a permanent stable lake. This has allowed aquatic snails to maintain large populations, whereas before the dam was built they had been reduced each year during the dry season. Because irrigation supports large human populations, there are now many more people living close to these stable bodies of water. The problem here is that the snails serve as intermediate hosts of the larvae of a blood fluke. The larvae leave the snail and bore into humans, infecting the liver and other organs. This causes the disease schistosomiasis. The species of snail that lives in stable water harbors a more virulent species of fluke than that found in another species of

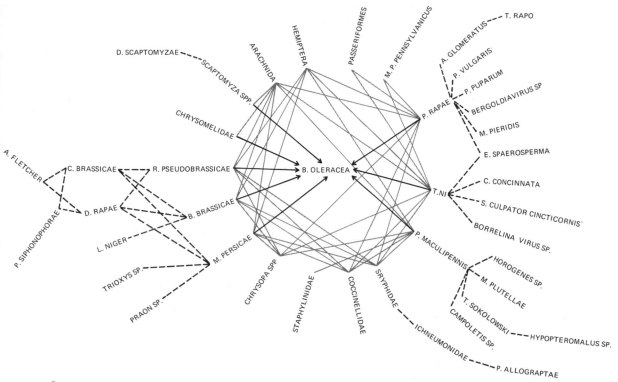

9   **Food Web for Broccoli.** The relationships herbivorous (solid black line), parasitic (dashed black line), and predaceous (shaded line) between a number of the more abundant insect species associated with the plant *Brassica oleracea* (Crucifera). *Source:* Reference 25.

snail in running water. Thus, the lake behind the Aswan Dam has increased both the incidence and virulence of schistosomiasis among the people of the Upper Nile.

Finally, the flooding of the river historically produced rich alluvial farmlands. Below where the dam now sits, these soils are no longer being replenished, and the fertility of the farm soils is decreasing.

Another example of interactions over large distances is the problem of acid rain in Europe, illustrated in Figure 10. The problem here is that the precipitation falling on the areas shown has become more and more acid over the years, owing to increased concentrations of sulfur and nitrogen oxides, which can be converted in the air into strong acids. Although the acidity is partly determined by natural sources, it is estimated that over 75 percent of the sulfur in the rain over Scandinavia is produced by man. This example is peculiarly ironic from Sweden's point of view, because Sweden is a recipient of acid rain but is itself not highly industrialized and indeed has been a leader in environmental matters. In fact, most of Sweden's acid rain probably originates from the industrial areas of England and the Ruhr Valley, a beautiful example not only of interaction between distant ecosystems, but of the efficacy of tall chimney stacks in getting pollution out of your own back yard, and of the difficulties of internalizing environmental costs (Chapter 16).

Acid precipitation may at times be quite widespread.[28] Its effects on ecosystems are not well studied but probably include: increased leaching of nutrients from vegetation and soil, acidification of lakes and streams with unexamined effects on the metabolism of the organisms there, and the corrosion of man-made structures.

The chapters in this book that deal with

environmental degradation are replete with examples of interactions between distant components of the environment.

Complexity, Stability, and Agriculture

One of the most interesting ideas in ecology is that complex communities are more stable than simple communities. In particular, it is now a widely held belief that communities with many species (high diversity) are more stable than those with few species. If true, this idea is profoundly important, for several reasons. First, the trend in agriculture, our basic life-sustaining ecosystem, is toward more and larger monocultures. If the belief is valid,

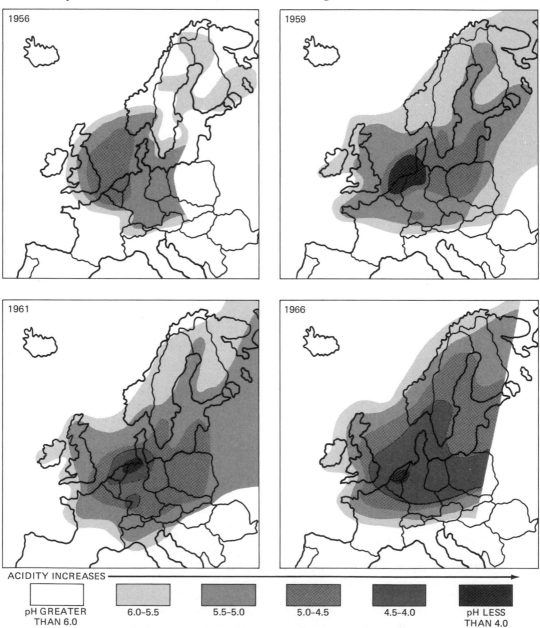

ACIDITY INCREASES ⟶

| pH GREATER THAN 6.0 | 6.0–5.5 | 5.5–5.0 | 5.0–4.5 | 4.5–4.0 | pH LESS THAN 4.0 |

10   **Acid Precipitation in Europe.** Changes in the pH of precipitation over northern Europe from 1956 to 1966. *Source:* Reference 27.

this is a dangerous trend: agriculture will become less and less stable and we will not have a dependable food supply. Second, pollution of all sorts is turning out to be a great accidental and undirected simplifier. It does this by killing off and reducing sensitive species and, in the case of excess nutrients, by greatly increasing the abundance of some species that come to dominate the community. Remarkably, many kinds of pollution have this effect, including chronic radiation, air pollution, and herbicides.[29] These also tend to cause structural simplification, because taller types of vegetation seem to be more sensitive.

Although we do not farm the sea to any extent (except to culture shellfish), we are certainly reducing its biological complexity. In particular, J. H. Connell has pointed out that we tend to remove predators in the ocean — frequently top predators with wide-ranging food habits. We have done this for a long time by harvesting whales and fish, although for most fish species we do not yet seem to have had a large effect on their numbers. More recently, we have started to make serious inroads into marine predators by pollution. Pesticides and heavy metals are concentrated in predators, as discussed earlier, and we are now seeing breeding failures in ospreys and pelicans — both at the top of marine food chains. Crude-oil spills have most effect on diving birds, and, of all marine birds, these are the main predators of fish.

Thus, our crop systems are becoming simpler by design, and natural systems are being made simpler, often by accident. It is therefore urgent that we find out more about the relationship between diversity and stability before these trends have resulted in disastrously irreversible changes. In the rest of this section we shall examine this problem. First I will present the conventional view, that diversity causes stability, then an alternative view, and

then the consequences for agricultural and natural systems.

The evidence for the claim that species diversity leads to stability is of three primary types:

1. Natural systems, which are also diverse, are more stable than artificial systems, such as crops or populations in the laboratory, which are also simpler than natural communities (Table V).
2. The tropics, especially tropical rain forests, are more diverse and more stable than simpler temperate communities, and simple arctic ecosystems, exemplified by lemming populations, are less stable than more complex temperate communities.
3. It makes good intuitive sense that a system with many links, or multiple fail-safes, is more stable than one with few links or feedback loops. For example, if a herbivore is attacked by several predatory species, the loss of any one of these species will be less likely to allow the herbivore to erupt than if only one predator species were present. There seem to be obvious parallels in human society: pluralistic societies may tend to be socially stable; economic systems with multiple resources, diverse manufacturing processes, and export markets are supposedly less vulnerable to fluctuations in one part of the system; and so on.

The message of this analysis for crop systems is clear. Monocultures are dangerously unstable, because their pests have few species of enemies, and we should so design crop systems (agroecosystems) that their diversity is increased. Thus, instead of one crop we should have a mosaic of several crops, and, as far as possible, we should allow weeds, trees, and hedgerows to survive because they add diversity to the system. In these ways, one

V	Three Major Differences Between Natural Ecosystems and Agroecosystems	
	Natural communities	Crop systems
1.	Component species are coevolved	Many component species not coevolved
2.	Not simplified (diversity variable)	Greatly simplified
3.	Massive disruption infrequent	Massive disruption frequent, especially in short-lived crops

hopes to reintroduce some of the stability properties of natural communities to crop systems.

A nice example of this philosophy at work is provided by the control of a leafhopper pest of vines in the San Joaquin Valley in California (Chapter 14). The biological-control workers from the University of California at Berkeley noticed that in some vineyards a small parasitic insect arrived early in the season and controlled the leafhopper. It turned out that these vineyards were near patches of blackberry bushes that had another species of leafhopper. This second leafhopper served as a winter host for the parasitic insect, which was therefore abundant by spring, when it migrated into the vineyard. The leafhopper problem has thus been solved by planting blackberry bushes near vineyards.

So much for the standard view. However, before we sally forth to diversify crop systems, we must ask whether the evidence to support the basic idea, diversity → stability, is adequate. The idea, if valid, could be a tremendous help in pest control in the future, but at the same time its implementation would require a loss in efficiency in the use of machinery and an increased use of labor. So we need to subject it to careful analysis.

EVIDENCE AGAINST THE IDEA

Let us take the three points made above, in reverse order. Unhappily, our intuition seems to be wrong as far as multiple fail-safes go. Recent mathematical theory[30] shows that, for a wide range of models of ecosystems, adding more species results in *less* stability! Note that this says nothing about *real* communities; it simply says that we cannot trust our intuition on this matter and must turn to evidence from the field.

In the field, however, our evidence is equally shaky, especially for point 2, which claims that diverse *natural* communities are more stable than simple natural communities. The fact is that there is no good evidence for this. Usually there is no evidence either way — we do not have enough measurements from natural communities. However, some recent observers have claimed that populations in the tropics seem to fluctuate about as much as they do elsewhere, and Charles Krebs of the University of British Columbia has shown that

mice in Indiana and California fluctuate about as much as do their lemming relatives in the Arctic, whereas Arctic ground squirrels and prairie dogs seem to be quite stable. The real problem here is that many variables, other than species diversity, make the Arctic different from the temperate zone and the temperate zone different from the tropics — the weather, to name but one. What we would really like to know is: Are fluctuations greater in a simple community (say a *Sequoia* forest) than in a more complex one (say a mixed hardwood forest) when all other characteristics of the community are the same? But, of course, all other characteristics never are the same. However, some recent work in three very similar old fields in Michigan shows that the least diverse was the most stable, and vice versa.[31] Thus, in natural communities there is no compelling evidence that more diverse communities are more stable.

Turning to species diversity in agriculture, we are able to say that for cotton, which is the only example that has been fully analyzed, there is no correlation between diversity and stability (by stability here I mean severity of pest outbreaks). The evidence available for cotton illustrates an important point already made by the leafhopper example above: diversity in the form of mixed cropping (e.g., planting maize next to cotton) adds stability *when it helps the natural enemies more than it helps the pest*. Mixed cropping decreases stability when it helps the pest (e.g., by providing it with an alternative food source) without increasing the performance of the enemies. This point is worth emphasizing: when natural enemies are effective, the way to get stability is to ensure continuity in the enemy populations. We shall return to this point later.

We must therefore reject the general claim, both in natural systems and in agroecosystems, that diversity leads to stability, except that the first piece of evidence remains: diverse natural systems are more stable than simple artificial systems. This is an interesting fact. Let us examine it more fully.

NATURAL VERSUS ARTIFICIAL SYSTEMS

Table V lists three major differences between natural systems and crop systems. The fact that man actively disrupts crops does add to their instability, but let us concentrate on the other two points. It is my hypothesis (and it is no more than hypothesis) that a major cause of instability in the form of pest outbreaks in crops is the fact that the interacting species, unlike those in natural communities, do not have a long-shared coevolutionary history. In natural systems this means that the plant species that we see today are those that have evolved defense mechanisms (thick tough leaves, spines, chemical defenses, etc.) against the herbivores that have attacked them for thousands of generations. So some plants survive in each generation, and rather severe limits are set on how abundant the herbivores can become. Plants species that did not evolve such defenses are no longer with us. Similarly, existing herbivores have "kept up" well enough so that each generation they get enough of a harvest to survive. In turn, the herbivores and their predators have evolved in a similar way. The process is dynamic and ongoing, but relative gains in one generation by any species must be slight because species evolve by very small steps.

Contrast this situation with a crop system. The dominant plant species is thrust willy-nilly into an often alien landscape. It collects a more-or-less haphazard and frequently disrupted set of species that, on average, must have had a relatively brief shared history. Furthermore, the crops have undergone radical selection, often losing their genetic defense mechanisms, and no doubt the major herbivores have also undergone rapid evolution, in response to rapidly changing crops, so that *their* enemies now face a different kind of prey. In addition, the selective regime, from irrigation to pesticides, has changed and is continuously and rapidly changing so that the species come to be more and more a haphazard assortment of genotypes — haphazard to the extent that the interactions between species are no longer such dominant or consistent selective forces.

Finally, crop systems are simplified (they are often not simple, although usually they have fewer species than natural communities). For example, many of the species that the pest interacted with in natural systems are gone, and in this way yet more of the coevolutionary links are gone. Thus, what is important in this respect is that crops are *simplified* communities rather than simple communities.

This comparison of natural versus artificial communities suggests that ecologists have seized upon the wrong variable (number of species) in explaining the difference in their stability properties. Rather, the *kinds* of species and interactions are crucial.

The analysis leads to an additional conclusion. If the sorts of interactions that occur in crop communities are fundamentally different from those in natural coevolved communities, and if coadaptation is central in explaining the stability of natural systems, we can expect important differences in the rules that govern the two types of systems. Thus, *natural communities provide the wrong model for crop systems and we must look elsewhere for principles*. (However, where the crop is a lightly managed forest, and thus is similar to a natural community, this will be less true than in a cotton field or vegetable patch.)

THE DESIGN OF CROP SYSTEMS

Before suggesting an alternative model to the diversity → stability one, we need to address another question: Is stability something that we should aim for in agriculture, in pest control in particular? The farmer wants stability in that he wants dependable, consistent high yields and as few pest outbreaks as possible. But, in fact, *instability* has been perhaps the major tool that the farmer has used in pest control. Indeed, until recently, much pest control has been based upon what we might call a "catastrophe strategy"! The farmer has tried to eradicate or frequently drive the pest to low numbers by applying poisons, burning or plowing stubble, rotating crops, and so on — hardly a strategy based on stability. However, I would argue that this has been changing in some crops recently and will change in many more as pesticides become less and less effective and as substitutes become less available (Chapters 14 and 19). The catastrophe strategy may continue to work in some crops, but more and more, pest control is likely to become *integrated control* (Chapter 14). Here, the fundamental strategy is to rely on natural enemies to control actual and potential pests, and this is a *stability* strategy. The object is to maintain *continuity* in the enemy populations, even though the crop is disrupted, and in so doing to maintain pests at low, fairly stable densities.

To maximize the chance that integrated control will work, we need to know how to design crop systems so that, by means of the maintenance of natural enemies and other techniques, pest problems are prevented, thus avoiding the disruption that occurs when we have to cure pest problems. This is perhaps the most interesting and important challenge that faces applied ecology. Surprisingly, almost no work is being done on the problem. Nowhere is there large-scale experimentation with different designs of crop systems. Even worse, we really have no established principles to guide such large-scale experimental work.

Recent work with laboratory populations and mathematical models, together with some evidence from forests, suggests a simple approach. Pest problems are likely to be minimized if crops are broken up into discrete units, perhaps with barriers between units, rather than having large continuous stands of a crop. At the simplest level, this would make it more difficult for a pest species to spread throughout the crop. For example, in Canadian balsam fir forests, barriers to pest dispersal, in the form of discrete stands of trees, nonsusceptible trees of other species, or trees of the appropriate age, cut down the spread and infestation rate of the spruce budworm. At a more complex level, a number of mathematical models developed recently by various authors[32] suggest that the pest–enemy interaction will be stabilized if the crop system consisted of separate subunits.

These are rather vague suggestions, and we would have to do experiments with crops to find out how large our units should be in different crops, what sort of barriers might be useful, and so on. The suggestions I have made are not a blueprint for agriculture but an indication of the type of approach that might be worth following.

Before leaving crop systems I must clarify two points. First, the above analysis is *not* a recommendation to grow monocultures. Given a policy of using integrated control, we need to add enough plant species to the system to ensure continuity in the natural enemies of the insect pests. However, there is no point in adding diversity blindly and for its own sake. Second, we *do* need to maintain genetic diversity within the crop. The Green Revolution in particular is potentially vulnerable to new

pests and diseases because it uses so few genetic strains which are planted over wide areas.

I should remind you that the above analysis of the diversity → stability concept is not the conventional one, nor has it by any means proved to be correct. This is only one example of an environmental issue on which there is as yet no final single correct answer.

PRESERVING NATURAL SYSTEMS

If diversity does not necessarily lead to stability, there is an important corollary for natural communities. These communities are stable because their component species are coevolved; therefore, simplified natural communities, which have coevolutionary links removed, can be expected to be less stable than undisturbed communities. Thus, the coevolution argument stresses the point made earlier that if there are species in nature that we wish to preserve, the best way to do this is to preserve as undisturbed as possible the natural communities in which they live.

But should we make any effort to preserve the species (i.e., the genetic diversity) that now exists? Millions of species have become extinct through geological time, and in our own history we have accounted for tens of thousands. In the last three centuries we have driven to extinction about 100 bird species and over 50 species of mammals (perhaps as many as 80). Furthermore, the rate of extinction is accelerating alarmingly: 25 of the 47 species of wildlife lost in the United States in the past 270 years became extinct in the last 50 years. There are now, for example, 59 endangered species of artiodactyls (antelope, deer, etc.), 45 endangered species of carnivores (cats, bears, weasels, etc.), and 35 endangered species of primates.[33]

In spite of all this, we seem to be surviving pretty well; it seems that, after all, we do not need the dodo. Couldn't we do just as well without the condor, bald eagle, African big game, and all these bees, butterflies, centipedes, protozoans, and the myriad of other creepy-crawlies? The answer undoubtedly is that we could survive pretty well without some of them, but it is *not* clear what the minimum number is that constitutes a viable biosphere, or which species they are; and, more difficult, it is not clear which species are now or might someday be useful, or which species are a necessary part of an ecosystem that contains species that might be useful at a future time. Each species gone is an option foreclosed, and it is impossible for us to tell which options the human race might someday want to take up. Thus, the conservative procedure is to preserve as many types of ecosystems as we can, in each case preserving at least the minimum number of areas necessary for the community to maintain itself indefinitely. Since we know so little about the factors that are important in determining the long-term survival of ecosystems, we have to make generous estimates of the minimum needed.

Estuaries on the California coast provide an example of rare ecosystems that probably are dangerously close to the minimum. These are few and far apart, and, except for a very few, they are small. Yet they are extremely important in supporting a large number of bird species, many of which migrate along the coast. Furthermore, the organisms in the estuaries cast their young into the ocean and "seed" other estuaries, so that estuaries depend upon other estuaries for recruitment. As estuaries become fewer and smaller, we therefore run the risk of losing not only bird species but resident estuarine species also.

Natural or semiwild ecosystems, which we call *unmanaged systems* because we do not need to spend much effort maintaining them, do help us in a number of ways.

First, many ecosystems are of obvious direct benefit. For example, forests on hills and along stream and river courses protect the soil from erosion; they trap rain and slow its runoff rate, which helps the area to operate as a watershed. They also lower wind speed and water movement, thus slowing erosion in adjacent agricultural land. The same purpose is served by trees and other vegetation in agricultural areas. Forests also cause the local climate to be wetter, and their removal in semiarid areas such as the shores of the Mediterranean has helped to produce deserts in formerly agricultural areas. Many natural systems also directly provide food, game, fishing, and wildlife. Estuaries are extremely important as nurseries for many species of fish and for the prey species that the fish depend on. Perhaps as much as 30 percent of the world fish catch is in some way dependent on estuarine waters, and this percentage is at least twice as high in the United States. Inshore ocean communities themselves are the world's major source of fish protein. The significance of these ecosystems is enhanced because the seashore, in particular the estuaries, are precisely the regions that have the fastest-growing populations and that attract industry. Thus, estuaries are not only exceptionally useful areas to man, they are also under the greatest pressure from development and are "sinks" for our pollutants.

Second, unmanaged communities serve a number of indirect functions through which they aid crop systems, the two obvious functions being pollination and the maintenance of a store of natural enemies for pest control. More than half of our crops are pollinated by insects, a service that we generally receive at no cost. It is true that for some crops we could, and sometimes do, transport honeybees to the fields to help pollination. But this cannot be done with all crops; it is an extra cost, and it would be dangerous to depend completely on our ability to maintain this substitute into the indefinite future. The problem of substituting for natural enemies that move back and forth between crops and wild vegetation is even more complicated and fraught with difficulties, as the history of pesticides illustrates.

Third, natural communities are a storehouse of genetic information. Each species is a unique experiment that we can never repeat once the species becomes extinct. It is impossible to predict where and for what reason we might want to go to that storehouse — for new crops for food or fiber, for natural enemies, to purify an organic molecule so that we can make its analogue, and so on.

Fourth, we need natural systems as outdoor ecological laboratories. How is it that natural systems are self-maintaining and relatively stable? What determines where species occur and which species occur together? What is the reason for the great diversity of the tropics, and is diversity related to stability? How do predators operate in nature, and can we get ideas from natural systems to use in crop systems? These are just some of the questions that we need to ask about natural systems.

Finally, but just as important as the others, a

large fraction of the population seems to feel a need to visit natural communities now and then, and perhaps for many more there is comfort in the knowledge that such communities exist. Furthermore, as we become richer and have more leisure, we can expect this need to expand. This is illustrated by the fact that visits to National Parks have been increasing by 10 percent each year while the U.S. population increases only at roughly 1 percent a year. The fact that it is hard to "prove" that we need some contact with nature does not make it less important to preserve the opportunity for such contact. We have never done the experiment of preventing an entire society from having access to nature, so we cannot predict the consequences.

Theory, observations, and experiments from the discipline of geographical ecology provide guidelines for preserving natural ecosystems.[34] Preserved areas are islands in the ocean of the human landscape. Fewer species are found both on smaller islands (because extinction rates are higher there) and on islands far from a source of colonists. So we should make our preserves as large as possible and should cluster similar preserves together so that they can recolonize each other.

The above list of reasons for preserving wild ecosystems is based on considerations of what we may need in the future; it is an attempt to provide a rationale for keeping at least a minimum amount. The reason that we have to be concerned that even this minimum will be preserved is our growing population and our growing standard of living, which requires the exploitation of more and more resources. Yet it is obvious that a maximum population size is not necessarily an optimum population size, and maximizing the standard of living probably does not optimize the quality of our lives. Clearly, the optimum population will not easily be defined, but it surely will be one that can afford to maintain enormous ecological diversity as an aesthetic pleasure, rather than one that has a minimum of diversity forced upon it as a necessary cost of survival.

Conclusions

Certain important features of ecological systems, and of studies of these systems, can be summarized as follows: First, limits are ubiquitous (Chapters 3 through 7). Not only are there limits to resources, there are limits to the rate at which the environment can receive wastes and return them to the system in usable form, and to its capacity for storing them in innocuous form (Chapters 8 through 15).

Second, ecosystems are made up of interacting and interdependent components, and ecosystems are open and linked to each other. As a result, events at one place in the environment are bound to have repercussions in other places and at other times. Because of the interconnectedness and complexity of the environment, some of the consequences are bound to be unpredictable.

Third, actions that are massive enough, drastic enough, or simply of the right sort will cause environmental changes that are irreversible. This is partly because the genetic material of extinct species cannot be reconstituted. Furthermore, some changes are ecologically irreversible. For example, typically when tropical forests are removed and the soil exposed, the mineral nutrients (already poor) are leached by the rain. The soil usually becomes hardened also and therefore the forest will not grow back again, nor can crops be grown. Such irreversible changes will almost always produce a simplification of the environment.

Fourth, simplified communities tend to be unstable, and almost without exception man's activities result in simplification. Further, when man's simple ecosystems develop instability, the actions taken (e.g., pest control) tend to be inherently destabilizing in the long run. In particular, in artificial simple systems the instability is enhanced because the organisms do not have a shared evolutionary past.

In dealing with systems with the above characteristics, the present limitations of methods of study must be remembered. First, the complexity of ecosystems makes them very difficult to study. Systems analysis can help, but it is only beginning to be used for the management or study of whole ecosystems, where we are interested in the behavior of many variables in the output as well as in the input. Second, it is not clear that removing portions of the problem to the laboratory for experimentation is an appropriate technique. Possibly the necessary simplification that this involves removes exactly the elements from the system that determine how it functions. Yet field experiments are difficult to do and usually difficult to interpret. Third, ecology is almost the only field of biology that does not boil down to physics and chemistry, and it seems certain that a direct extension of physics and chemistry will not solve ecological problems. Yet exactly such an extension has proved very helpful for the rest of biology. Finally, there is the problem that each ecological situation is different and has a unique history. Therefore, we do not have, and may not have in the future, broad and profound generalizations as a basis for action. Each problem requires particular analysis, and there will generally be a sizable time lag between posing the question and receiving the ecological answer. This does not make all environmental problems hopeless of solution, merely difficult and interesting. However, it does suggest that above all we need to be conservative in our management of biological systems.

References

1. Hutchinson, G. E. 1948. On living in the biosphere. *Sci. Monthly 67*: 393–397.
2. Study of Critical Environmental Problems (SCEP). 1970. *Man's Impact on the Global Environment*. The MIT Press, Cambridge, Mass.
3. Westlake, D. F. 1963. Comparisons of plant productivity. *Biol. Rev. 38*: 385–425.
4. Whittaker, R. H. 1970. *Communities and Ecosystems*. Macmillan Publishing Co., Inc., New York.
5. Morowitz, H. J. 1968. *Energy Flow in Biology*. Academic Press, New York.
6. Ricklefs, R. E. 1973. *Ecology*. Chiron Press, Newton, Mass.
7. Ricker, W. E. 1969. Food from the sea. In *Resources and Man* (Preston Cloud, ed.). W. H. Freeman and Company, San Francisco.
8. Committee on Resources and Man of the Division of Earth Sciences, National Academy of Sciences–National Research Council. 1969. *Resources and Man: A Study and Recommendations*. W. H. Freeman and Company, San Francisco.
9. Dixon, A. F. G. 1970. Quality and availability of food for a sycamore aphid population. In *Animal Populations in Relation to Their Food Resources*. Blackwell Scientific Publications Ltd., Oxford, England.
10. Feeny, P. 1970. Seasonal changes in oak leaf tannins and nutrients as a cause of spring feeding by winter moth caterpillars. *Ecology 51*: 565–581.
11. Watt, K. E. F. 1968. *Ecology and Resource Management*. McGraw-Hill Book Company. New York.
12. Ophel, I. L. 1963. The fate of radiostrontium in a freshwater community. In *Radioecology* (V. Schultz and A. W. Klement, Jr., eds.), Reinhold Publishing Company, New York.
13. Southwood, T. R. E., and Way, M. J. 1970. Ecological background to pest management. In *Concepts of Pest Management* (R. L. Rabb and F. E. Guthrie, eds.). North Carolina State University Press, Raleigh, N.C.
14. Connell, J. H. 1972. Community interactions on marine rocky intertidal shores. *Ann. Rev. Ecol. Systematics 3*: 169–192.
15. Harper, J. L. 1969. The role of predation in vegetational diversity. In *Diversity and Stability in Ecological Systems*. Brookhaven Symposia in Biology 22.
16. Hamilton, D. H., Jr. 1969. Nutrient limitation of phytoplankton in Cayuga Lake. *Limnol. Oceanog. 14*: 579–590.
17. Varley, G. C., 1949. Population changes in German forest pests. *J. Animal Ecol. 18*: 117–122.
18. Morris, R. F. (ed.). 1963. The dynamics of epidemic spruce budworm populations. *Mem. Entomol. Soc. Canada 31*.
19. Lack, D., 1966. *Population Studies of Birds*. The Clarendon Press, Oxford, England.
20. Watt, K. E. F. 1968. *Ecology and Resource Management*. McGraw-Hill Book Company, New York.
21. Odum, E. P. 1969. The strategy of ecosystem development. *Science 164*: 262–270.
22. Likens, G. E., and Bormann, F. H. 1972. Biogeochemical cycles. *The Science Teacher 39*.
23. Pierce, R. S., et al. 1972. Nutrient loss from clearcuttings in New Hampshire. In *Watersheds in Transition*. American Water Resources Association.

24. Evans, F. C., and Murdoch, W. W. 1968. Taxonomic composition, trophic structure and seasonal occurrence in a grassland insect community. *J. Animal Ecol. 37:* 259–273.

25. Pimentel, D. 1961. Competition and the species-per-genus structure of communities. *Ann. Entomol. Soc. Amer. 54:* 323–333.

26. McCaull, J. 1969. Conference on the ecological aspects of international development. *Nature and Resources (UNESCO) 2:5–12.*

27. Odén, S. 1971. Nederbördens försurning-ett generellt hot mot ekosystemem. In *Forurensning og biologisk miljovern* (I. Mysterud, ed.). Universitetsforlaget, Oslo. pp. 63–98.

28. Likens, G. E., Bormann, F. H., and Johnson, N. M. 1972. Acid rain. *Environment 14:* 33–40.

29. Woodwell, G. M. 1970. Effects of pollution on the structure and physiology of ecosystems. *Science 168:* 429–433.

30. May, R. M. 1973. *Stability and Complexity in Model Ecosystems.* Princeton University Press, Princeton, N.J.

31. Murdoch, W. W., Evans, F. C., and Peterson, C. H. 1972. Diversity and pattern in plants and insects. *Ecology 53:* 819–829.

32. Murdoch, W. W., and Oaten, A. 1975. Predation and population stability. *Advan. Ecol. Res. 9:* 1–131.

33. Miller, R. S., and Botkin, D. B. 1974. Endangered species: models and predictions. *Amer. Scientist 62:* 172–181.

34. Wilson, E. O., and Willis, E. O. 1974. Applied biogeography. In *Memorial to R. H. MacArthur.* Harvard University Press, Cambridge, Mass.

Further Reading

Kormondy, E. J. 1969. *Concepts of Ecology.* Prentice-Hall, Inc., Englewood Cliffs, N.J. 209 pp.

Krebs, C. J. 1972. *Ecology: The Experimental Analysis of Distribution and Abundance.* Harper & Row, Inc., New York. 694 pp.

Odum, E. P. 1971. *Fundamentals of Ecology*, 3rd ed. W. B. Saunders Company, Philadelphia. 574 pp.

Population and Resources

PHOTO BY GEORG GERSTER/RAPHO GUILLUMETTE

Editor's Commentary—

Chapter 2

The U. N. World Population Conference (August 1974) in Bucharest was a disappointment to those from the rich countries who had hoped to see some firm resolution towards population control on the part of poor nations with rapidly expanding populations. Instead, these poor nations attacked the rich countries on the subject of economic imperialism and on the difference in development rates between rich and poor nations. The final, compromise, statement on population merely said that countries that feel that their population growth interferes with their "goals of promoting human welfare are invited . . . to consider adopting population policies . . . which are consistent with basic human rights and national goals and values".

The most forthright statement was reserved for world economics: "Recognizing that *per capita* use of world resources is much higher in the more developed than in the developing countries, the developed countries are urged to adopt appropriate policies in population, consumption and investment, bearing in mind the need for fundamental improvement in international equity."

Before condemning as improvident such half-hearted support of population control it is worth noting some comments made by Lester Brown (co-author of Chapter 3) in his recent book *In the Human Interest*. Brown emphasizes that social improvements are needed to bring down birth rates even in the short term; these include improved education and health care, income redistribution, fuller employment and a secure food supply. He notes further that such programs are not necessarily impossibly expensive. He estimates, for example, that to teach everyone in the developing countries to read, and to provide contraceptives and simple health care for every woman in these countries, would cost less than $6 billion per year for the next five years. (He excludes China since it is well on its way to achieving these goals, yet China is a poor nation.) Six billion dollars is less than the amount of money the U. S. Department of Defense spends in a month.

Thus, the point of view of the poor countries, and their priorities, are not as mistaken as they might appear. It is in fact well within the capacities of the rich nations to contribute enough funds to make a huge difference to population growth in the poor countries. Of course, political and economic reform within many of these countries is also needed; it is tragic that, as a general rule, the poorer nations also have the most inequitable distribution of wealth and income, which serves to worsen their population problems.

In this chapter, Professor Keyfitz points out the relationships among birth rates, economics, and a sense of security, and he also points to one feature of population change that provides some hope: attitudes and values that are important in population control can in fact change very rapidly, as recent developments in the United States concerning the legality of contraception and abortion have shown. Such changes in poor countries need to be stimulated by economic improvement and political reform, but the evidence is that, once the stimulus is there, the changes can occur quickly.

In fact, the recent history of declining birth rates in some poor countries is encouraging. At least ten such nations have experienced declines in birth rates of 1.2 to 1.8 per 1000 per year over a period of 7 to 16 years. Thus, for example, Taiwan's birth rate fell from around 45 or 50 births per 1000 population per year in the 1950s to a current 24 per 1000. Other nations in the group include Egypt, Chile, and South Korea. Their annual per capita incomes range from $210 (Egypt) to $970 (Hong Kong), and they represent a broad range of political, cultural and religious systems. In addition to these nations, China (annual per capita income $160), Sri Lanka (per capita income $110), and Cuba (per capita income $530) have reduced their birth rates to 30/1000 or less. The common factor in these countries seems to be that the *majority* of the people have experienced significant increase in social welfare (literacy, rudimentary health care, low infant mortality rates, better diets), in spite of the fact that most of the countries are very poor. If such rates of decline were to spread to the rest of the developing world and were to continue for a few decades, the world population would grow less than the U.N.'s optimistic predictions and could stabilize around 6 billion.

Two Indonesian girls spread rice on mats to dry in the sun. FAO photo by Jack Ling.

2

Population Growth: Causes and Consequences

Nathan Keyfitz *is Andelot Professor of Sociology and Demography at Harvard University. Born in Montreal in 1913, he did his undergraduate work at McGill University and after an early career as a statistician and census taker, earned his doctorate from the University of Chicago in 1952. He has held appointments as Professor of Sociology at the universities of Toronto and Chicago. Before assuming his present position in 1972, he was Professor of Demography at the University of California, Berkeley, and taught demography in Buenos Aires, Santiago, Mexico and Indonesia.*

An Inhabited Satellite and Its Crew

Before this century it did not occur to anyone that the world of human habitation is effectively finite. Discovery of new features on the earth's surface was still in progress, and the maps contained white spaces of unknown topography. Names had indeed been given to the continents and regions within them, and seaports were established around their edges, but men had not seen, let alone possessed, all their interior space. Today every part of every continent is mapped, and most parts are claimed by some national state; each square mile is under daily photographic and electronic surveillance. World population censuses have been attempted in 1950, 1960, and 1970. We have the outline of precise knowledge about man and his habitat, with enough detail at least to be sure that no new fertile lands will be turned up by a Columbus of the twenty-first century.[1,2]

But is not our world infinite in the sense that new knowledge will make it ever more productive? Science and invention have accelerated and their future is open. They seem a modern form of magic, slaves at our elbows to grant all our wishes. Yet just as the wishes in fairy tales are granted, but each with some unexpected rebound that brings calamity on the wisher, so ecologists are finding that the magic of science can bring an assortment of disasters. From the DDT that kills insect pests but also poisons useful animals, to the Green Revolution that produces mountains of food but encourages an increase of population that defeats economic growth, this book contains many instances of what science can do on the rebound. At least pending the resolution of some major points of ignorance, we have to think of the frontier of science as ultimately closed, just as the land area of our planet is closed. Finiteness in both senses is a theme of this book.

Is finiteness of the food supply the ultimate limit on growth, as Malthus, living in a simpler age, believed, or is that limit set by poisoning of the atmosphere or exhaustion of fuels? About 125 million babies will be born this year, and their first-year consumption will be small; their demands will grow as they age; a world that can support the babies cannot necessarily support the school children and adults that they become. Every birth represents a commitment of resources for the following half-century or more, not to food only, but to jobs, housing, clothing, transport. Perhaps the environment can stand damage up to a certain point, at which time it suddenly becomes unlivable — because of smog, for instance. Of all the human problems, that of population makes the heaviest demands on foresight. We may be climbing to a position of instability in the habitat analogous to the position in which a physical object topples over. If so, the world's capacity for sustaining population could at some future time drop sharply, perhaps far below the present level.

With separate nations and identifiable races, the intergroup competition has through history taken a demographic form; any groups that were not pronatalist disappeared under the universal high mortality. Once mortality was brought under control, outbreeding one's national, cultural, or racial rivals acquired planetary implications that it did not have before. Even governments, in both poor and rich countries, have become officially aware that competitive breeding is destructive for individual countries and for the planet, and they are trying to bring their populations under control. More drastic measures are likely to be necessary in many areas before runaway populations are brought to a halt, and some changes in the value system that would permit drastic measures are appearing. These are the principal topics to be discussed in this chapter.

Numbers of People

Table I shows how the world's population has grown since 1930. The increase of less than 50 percent for the rich countries contrasts with almost 100 percent for the poor ones. Income also has grown in both rich and poor countries, but the effect of this on the world average income is offset in considerable part by the present heavier weight of the poor in the average. Indications for the 1970s are that the population differential is increasing, with the rich countries becoming stationary while the poor ones substantially maintain their rate of population increase.

Within the poor group Latin America is increasing fastest, followed by Africa and then Asia. There are signs that this order may change; Latin America could well slow its births while African mortality falls, so that by the 1980s Africa may be growing fastest. But this

could be proved wrong, for example by the spread of famine in Africa beyond the areas affected in the past year or two.

PEOPLE AND NATIONS

That part of the earth's crust not under salt water is mostly divided among national states, and humanity is allocated among these. International movement for residence purposes is sharply restricted. Eighteenth- and nineteenth-century notions of the state and citizenship, after bringing to an end the unity of the European Middle Ages, have now triumphed around the world.

Whatever its other consequences, nationalism has certainly promoted censuses and statistical counting. Present statistical knowledge is an offshoot of the national state, as the word "statistics" itself reminds us. Let us briefly review the status of this knowledge for the eight states with the most people.

China is the largest of the national entities in terms of population and the one on which information is most conspicuously lacking. Much of what we know pivots around the total of 583 million reported by Peking in 1953. That this total was about 100 million persons more than the regime itself had thought were present up to that time lends a degree of credibility to the result, and this level of evidence is most of what we have to go on. Professionals in the field have accepted the

I Population and the Land Area of the Earth

	Population (millions)					Area (1,000 km²)	Hectares per person, 1970
	1930	1940	1950	1960	1970		
Planet	2,070	2,295	2,517	3,005	3,632	135,767	3.74
Poor countries	1,328	1,494	1,683	2,058	2,580	78,040	3.02
Africa	164	191	222	278	344	30,313	8.81
Asia (except Japan)	1,056	1,173	1,298	1,567	1,953	27,162	1.39
Latin America	108	130	163	213	283	20,565	7.27
Rich countries	742	801	834	947	1,055	57,727	5.47
Europe	355	380	392	425	462	4,929	1.07
Japan	64	71	83	93	103	370	0.36
Northern America	134	144	166	199	228	21,515	9.44
Oceania	10	11	13	16	19	8,511	44.79
USSR	179	195	180	214	243	22,402	9.22

Source: Reference 3.

41

figure, and compilations of world population raised their totals by about 100 million at the time of the announcement.

Even less certain is our knowledge of the rate of increase of the Chinese population. It could be well below 2 percent, or 15 million per year, and the total still have passed the 800 million estimated by the United Nations for the present time.

Because of extensive Chinese record keeping in the past, we know a good deal about the population of China over the last two millennia, perhaps more than we know about it today.[4] For other developing countries, current information is more reliable than historical records.

India's tradition of census taking was established in British times, and the record up to 1951 has been pulled together and analyzed by Davis.[5] We are reasonably sure that the 1971 population was nearly 548 million. If the births are 42.8 per thousand and the deaths 16.7 per thousand, a net of 26.1 per thousand or 2.61 percent, as the United Nations tells us in the *Vital Statistics Report* (1973)[6] and other sources, then India has passed 600 million by 1975 and is gaining at the rate of 15 million per year.

Third is the USSR, which counted 241.7 million people in its census of January 1970.[7] No information is provided on the accuracy of this census, of which it can be said that it agreed with the precensal estimate of 241 million. Like the United States, the Soviet Union has seen a steady fall in its birth rate over the past decade, down to a 1972 level of 18.0 per thousand; its net natural increase of 2.3 million per year compares with 1.3 million for the United States.

The United States is the fourth country in respect of population. Its latest census was taken in April 1970, using very modern techniques, and counted 203 million[8] with about 2 percent omissions, as judged from intensive analysis made by the U.S. Bureau of the Census. In a period of exceptional leadership the Bureau developed important new census and sampling techniques which are now being diffused throughout the world.

The fifth country is Indonesia, whose 1971 population was counted at very nearly 120 million. Omission as high as 4 percent or more has been claimed on the basis of earlier figures. On the other hand, the count agrees very

closely with a precensal estimate by Widjojo N.[9] Current estimates suggest an increase of up to 3 million per year.

Sixth is Japan, with about 107 million at the present time; its drop from fifth in population has coincided with its rise to third industrial power. Brazil is seventh in population, now passing 100 million and rising faster than any other country in its size class; before the 1970s are over it will pass Japan, on present indications. Bangladesh is eighth, with an estimated 83 million.

These eight countries account at the present time for over 2 billion people, over half of humanity. They do not include any of Western Europe, whose statistics are complete enough that we can do what has not been possible for any other continent: compile a grand total of a solid international unit. A compilation[10] shows the aggregate of Europe from Ireland to the boundaries of the USSR to have been 442 million in 1965, and probably about 460 million by 1970, some 12 percent of the world's total.

Estimaters of error are inevitably subjective in those countries whose censuses include no sample check. We probably know China's population within 10 percent, that of Europe and the United States within 2 or 3 percent, and most countries of South America and Asia with accuracy intermediate to these. Adding together such evidence for national units suggests that we know the world total to within about 5 percent and can assert that by 1975 it will be between 3.8 and 4.2 billion.

WORLD CARRYING CAPACITY AND THE OPTIMUM

The optimal or best population cannot be discussed without first deciding some contentious questions of value. Is the best population the largest possible? Is it the one that has the greatest diversity of culture? The greatest degree of freedom? The optimum depends on what is to be maximized. Three possibilities are numbers, diversity, and freedom, and we shall see that they suggest quite different optima.

The 4 billion people who inhabit the earth in the mid-1970s are a biomass of 200 million tons, increasing at about 4 million tons per year. This increase is about one-twenty-fifth of the annual world production of beef and pork, and about one-fifteenth of the annual catch of the world's oceans. It is less than 1 percent of the combined annual output of wheat, rice, corn, potatoes, and other crops raised as food for men and animals.

The human biomass that the earth is capable of maintaining is limited by inevitable losses in a trophic chain. Five pounds of corn are required to produce 1 pound of pork, and man has not been selected, as has the pig, for efficiency in conversion. Moreover, the turnover of the human biomass is 30 to 70 years, much slower than that of farm animals and crops. Suppose that it takes 40 pounds of cereal to produce 1 pound of person — during the growing ages a child might gain 10 pounds per year and eat 400 pounds of cereal. Suppose also that the child grows for 15 years, and thereafter remains an adult of unchanging weight for 15 to 55 years, during which time he or she still requires the same 400 pounds of cereal per year. Then regardless of the mix of children and adults, or of the death rate, 1 billion tons of cereal and potatoes per year can maintain 5 billion people. World production of cereals and potatoes in 1970 was safely above this, at 1.3 billion tons, so the present population is easily sustainable.

If productivity on existing agricultural land were doubled, and if an equal additional amount of land could be brought into use, the population could exceed 20 billion. This depends on higher-yielding varieties, which have the merit that they can feed into the human biomass the stored-up energy of fossil fuels made available through fertilizers. The process would be aided by the shortening of trophic chains — that is, if meat eating declined and men fed themselves directly on the corn now consumed by pigs, cows, and chickens. All this could come about with substantially existent agricultural technology. Whether agricultural systems can sustain such high productivity indefinitely is another question[11,12] (see Chapters 1, 3, 13, and 14).

With the use of algae and other more efficient primary converters of the sun's energy, and with some quite feasible scientific advances, another doubling, to 40 billion, could well be possible. The globe's carrying capacity might thus be tenfold the present population. Such figures have been given in the literature and have also been widely criticized. Let us accept them here, nonetheless, as suggestive of what might be possible if science and technology were to be devoted to increasing human subsistence.

But the total effort to maximize the human biomass can only arise from seeing man as a species of domestic animal, and represents a thoroughly inhumane notion of what people are on the earth for. Let us suppose instead that the object is to attain the greatest volume and duration of civilization — not maximization of biomass, but deepening of culture. The question then is not how many we can have, but how many we need.

Would a world population of 1 million suffice? This is far more than the 5,040 free citizens that Plato thought ideal for carrying the culture and doing the work of the city, but then we have more specialities than Plato knew about, and even he wanted to duplicate cities. A world population consisting of 1 million educated and skilled persons could continue the main branches of knowledge and of art for one culture. To provide as well for a wide range of industrial products and consumer goods, the skills and division of labor in a population of the order of 100 million would be required. To go further and preserve a reasonable selection of the world's diverse languages and cultures might require half a billion. Remember that even with our 4 billion people, local cultures and specialized traditional knowledge and occupations are being lost, and good citizens worry about the loss.

A conscious policy of preserving cultures should be able to do better with half a billion than we now do with 4 billion. From the viewpoint of cultural variety a relatively homogeneous country like France would have less claim than one like India, which has many hundreds of languages. On the other hand, an industrial country has a claim in respect of its specialized occupations, but industrial countries do duplicate one another. The extent to which a man duplicates his neighbor culturally and occupationally, or to which one nation duplicates another, would be the basis for discouraging them from having children, if it was decided that world population should be brought down in the course of a few generations to the minimum number sufficient to carry the diversity of knowledge and skills. Those of us with cultures that are widely duplicated, say middle-class Americans, would be

43

offered incentives to have only one child per couple; those with rare and unique cultures, like some of the hill peoples of South Asia, would be encouraged to have two children.

The number of people in the world that would maximize freedom is even harder to assess than the number that would maximize diversity of culture. We know that in most matters more people require more regulation than fewer people. The first automobiles required no traffic lights; regulations on waste disposal would have been out of place on the frontier; in a modern community both traffic lights and waste disposal are indispensable. As each increase of population requires an increase of regulation, it entails a decline of liberty. If freedom were the sole objective, it would make for a very small population, much smaller than maximizing civilization or diversity of culture.

Incentives might well be devised that would bring the world total down to the 500 or so million that would provide a diverse life and a reasonable degree of freedom on the planet over a long period of time. But there is little prospect — and little danger — that population policy will become an instrument for the preservation of peoples and cultures. Considering how hard it has been to attain simpler objectives in international negotiation, we cannot expect much for population. Furthermore, demographers point out that the age distribution of a group that had to drop to one child per couple for several generations would be very peculiar, and the transitional generations would have to carry the burden of large numbers at the oldest ages. The economics of contraction cannot draw the same enthusiasm as the economics of growth.

Given the difficulty of deciding how many people there ought to be in the world, it is no wonder that virtually everyone who has thought about the matter is willing to settle for voluntary control of childbearing. To provide the facilities that will permit people to have the number of children that they want is the highest goal of policy so far reached. Those few who go beyond this ask only that the world population stay within the limits of its subsistence base, or that individual countries not increase at so fast a rate that they spend their subsistence raising children rather than raising

capital. The perspective is that of the next generation or two at most. I repeat that in view of the very deep issues, unresolvable on any present knowledge, not much better can be done than to provide means of contraception and possibly abortion, and to try to induce people to use them sufficiently that the world will attain stationarity at not too much above present levels.

The one country where this is not considered good enough is China. Local communities are instructed to assemble their members and decide collectively how many children are to be born in the next five years. This number is then allocated to the married couples, with first priority to those just married, second priority to those with fewer than three children, and third to those whose last child is 5 years old. Evidently different ideals of privacy and the right of individual decision making prevail than in the United States. No information is available on how widespread this population control is — in particular, how far it has penetrated areas distant from the large cities.

THE NATURE OF THE CEILING

In times of rapid technical change few useful statements can be made about the ceiling that the environment sets on population. Malthus went out of style partly because his food ceiling on population was repeatedly raised by the discovery of new lands and new techniques. If the ceiling were completely rigid, no sudden tragedy could ever occur. A rigid ceiling would exist if there were a clear and patent division of space, food, and other resources such that the moment when the last sustainable person was born would be clearly recognized. The absence of support for one more person would be unmistakably signaled, and that person would not be brought into the world. At all times mankind would be compelled to stay within its food supply. From the moment the limit was reached we could add no further population, except that as individuals died, Nature would permit us to replace them.

There being no such overt ceiling, the population problem requires calculation and foresight; Nature is like a moneylender who allows — even encourages — us to borrow beyond our means and beyond our needs, and then calls the debt when we are most extended. The Malthusian ceiling was not rigid; it demanded foresight but, at least under a regime of private property, not an impossible amount of foresight.

Now the calculability of the Malthusian ceiling may be reduced by a new instability (which ecologists have come to fear) which has resulted from man–environment interaction in our technological age. Instead of living with a ceiling over our heads, which is bad enough, we may be on a melting iceberg that can turn over at any moment. This raises the amount of foresight demanded to much higher levels. Let us look into some aspects of the potential instability under which we may now actually be living.

Many examples throughout this book prove that the analogy of our world to an iceberg that could turn over is not farfetched. To add one more illustration of the mechanism, at least on a local scale, we can cite the Sumatran or Laotian hill country, where slash-and-burn agriculture has been practiced for centuries. Each year a family burns the trees on about an acre of land, plants with digging stick its cassava, tobacco, and corn or other cereal, and after a year or two abandons the site and makes another clearing in the forest. This neolithic agriculture is perfectly stable as long as people are few.

Above a certain population density, however, each family has to return to the same plot before the forest has had time to reestablish itself. When population increase compels such a premature return, the landscape changes in the course of a few cycles, and instead of forest a tough grass, called alang-alang in Sumatra, comes to prevail. It cannot be cut with primitive implements, and it effectively puts the land out of use indefinitely. The land could have sustained 20 people or so per square kilometer under shifting agriculture, but the attempt to make it sustain 40 has reduced its carrying capacity to virtually zero. This story can be repeated in respect of overgrazing, overfishing, and other attempts to exceed the (unknown, but nonetheless existent) limit imposed by the environment. Fortunately, the cases have so far all been local, and the people involved could move somewhere else or use other food sources, and so

have had another chance. No second chance would be possible, however, for violation of the population ceiling on the planet as a whole.

That populations can fall tends to be overlooked during a century of rapid and nearly universal increase. And yet many well-documented instances exist — in Europe with the Black Death, in the Americas after the advent of the Spaniards, in some Pacific islands. Schmitt[13] brings together available figures for Hawaii and concludes that at the arrival of James Cook in 1778 some 300,000 persons were living in the islands, after which the number fell gradually to the 56,897 counted at the end of 1872. The 1930 census was the first that showed more people than were present in Cook's time. The decline has been attributed to wars, measles and other diseases, disorganization of economic life, and other causes.[14] We need more study of instances of population decrease as well as of population growth, even though past declines do not have the same causes as possible future ones.

We know little about instability in local situations now and can only speculate on its operation on a larger scale in the future. When the farmer replaces horses with a tractor, he need no longer grow oats for traction and hence he increases the output of human food on his land. Based on this larger output, population comes into existence. But on what will the additional people be sustained when the fossil fuels that drive the tractor are exhausted?

More immediately threatening, the farmer eliminates species that are not of use to him, including weeds, rats, and other pests, and so increases effective yields. Population again grows on the increased production. One instrument used by the farmer to produce the increased yields is DDT, and DDT turns out to be an active poison. It concentrates in the fat of animals up food chains throughout the world. What happens to the population built up on it when its use has to be discontinued?

Aside from any specific effect of this kind, ecologists fear agricultural procedures that drastically reduce the number of species coexisting in an area. Seed banks to save genetic materials that have for the moment become commercially valueless do exist, but the fear is that they are not adequate — and genetic erosion is taking place throughout the world. In simplifying the ecosystem man can engage mechanisms of instability that are not yet thoroughly understood. He builds one device on another in a complex structure; at each stage production is increased, but always with an unknown risk that the whole structure will collapse after the means of recreating its simpler phases have been lost.[15]

Because the number of inhabitants is set by such considerations, the simple Malthusian ceiling no longer applies. We do not know whether the long-term carrying capacity of the earth is 1 or 5 or 10 billion people. Much more research is needed, including better means of forecasting technological change and of foreseeing its ecological consequences. Such research could lead us to conclude that the limit is below or above the present total of about 4 billion.

The preceding discussion has provided three images or ways of thinking about the problem of sustainable human numbers. One is in terms of a rigid ceiling by which we would know immediately when we had reached the last person the environment could support; if nature provided such a warning, the need for men to exercise responsibility in their affairs would be greatly diminished. The second is a flexible ceiling of the kind Malthus envisioned, which permits some temporary excess population. The third is a potential instability, by which a moment would be reached that would drastically reduce the population sustained, like the crash observed in lemmings and other species.

We turn now from considerations of overall carrying capacity to distribution in space, in particular to the distinction between rural and urban.

Distribution

URBAN AGGLOMERATION

Even in developed countries, half the population is rural and small town; and in the less-developed countries, four-fifths is rural and small town (Table II). But the trend everywhere is toward large urban centers. During the period 1960–1980 the increase of rural parts is expected to be 33 percent, of urban 78 percent, the urban population being defined for this purpose as that living in agglomerations of 20,000 persons or more. (The word "agglomeration" rather than "city" shows a fine awareness on the part of our source, the United Nations, of what has happened to urban life in this age of mass society.) This continues earlier trends. Between 1920 and 1960 the world total increased by just over 50 percent, while the urban population on this definition multiplied more than 2.5 times.

The differential in growth between the more and less developed regions, conspicuous in total population, is even more striking in respect to urban population. Between 1920 and 1960 the total population of the more-developed countries doubled, while the less-developed multiplied by 4; in respect of urban population the more-developed multiplied by 4.5, the less-developed by 20.

A striking change in the meaning of urbanization stands out from the U.N. figures. While urbanization was essentially a phenomenon of industrialization in 1920, so that there were 198 million people in urban agglomerates of developed countries and only 69 million in those of less-developed regions, we are now approaching the time when the absolute numbers will be the same in more- and less-developed regions, and by the year 2000, again according to the U.N.'s projection, the more-developed regions will have only 901 million, the less-developed 1,436 million, in places of over 20,000. In Latin America the projection shows a fivefold increase between 1960 and 2000, in Africa even more. Considering what the agglomerations of rich countries have now become — Manhattan is an example — we can only begin to imagine the problems that will confront even larger urban places in Asia and elsewhere, remembering that the latter will have to face most of the same difficulties with only a small fraction of the income.

When analysis reaches the level of individual metropolitan areas we find that the growth is especially concentrated in those that are already the largest. This is true for the United States as for the Soviet Union. The Soviet effort to persuade people not to move to Moscow, but rather to help develop the mineral and other resources of Siberia, has hardly prevented Moscow from growing[17]; one of the surprises of the 1970 Soviet census was finding

	1960	1970	1980	1980 as % of 1960 population
Total population				
World total	2,991	3,584	4,318	144
More-developed major areas	854	946	1,042	122
Europe	425	454	479	113
Northern America	199	227	262	132
Soviet Union	214	246	278	130
Oceania	16	19	23	144
Less-developed major areas	2,137	2,638	3,276	153
East Asia	794	911	1,041	131
South Asia	858	1,098	1,408	164
Latin America	212	283	378	178
Africa	273	346	449	164
More-developed regions	976	1,082	1,194	122
Less-developed regions	2,015	2,502	3,124	155
Agglomerated population *(localities of 20,000 inhabitants and over)*				
World total	760	1,010	1,354	178
More-developed major areas	389	472	568	146
Europe	188	214	237	126
Northern America	115	142	177	154
Soviet Union	78	105	141	181
Oceania	8	11	13	158
Less-developed major areas	371	538	786	212
East Asia	147	198	267	182
South Asia	118	176	266	226
Latin America	69	107	163	234
Africa	37	57	90	246
More-developed regions	450	546	661	147
Less-developed regions	310	464	693	223
Rural and small-town population *(localities of less than 20,000 inhabitants)*				
World total	2,231	2,574	2,964	133
More-developed major areas	465	474	474	102
Europe	237	240	242	102
Northern America	84	85	85	101
Soviet Union	136	141	137	100
Oceania	8	8	10	128
Less-developed major areas	1,766	2,100	2,490	141
East Asia	647	713	774	120
South Asia	740	922	1,142	154
Latin America	143	176	215	151
Africa	236	289	359	152
More-developed regions	526	536	533	101
Less-developed regions	1,705	2,039	2,431	143

Total, agglomerated, and rural and small-town population in the world and major areas, as projected for the period 1960–1980 (millions). *Source:* Reference 16.

400,000 more people in Moscow than the planners thought were there.

At the other extreme, little Panama's census of May 1970 shows an increase of 32 percent for the country as a whole, 53 percent for the capital city, since 1960. One-third of Argentinians live in greater Buenos Aires. The argument that a country needs people to fill its empty spaces is rendered specious by current migration from relatively empty spaces to the capital city.

That cities everywhere draw people from the countryside is by no means due to their aesthetic attractiveness. Large numbers of new residents who lack a tradition of urbanity have more than offset the efforts of planners and builders to create a physical milieu that would encourage civility, and a flight from the older parts of cities is in progress. That pride in community of ancient Athens, indeed of New York only a generation ago, seems to have left the central city and, insofar as it exists at all, moved to the suburbs. The suburban shopping center has become the acropolis of the modern community.

The movement of people to the cities in excess of the capacity of the cities to employ and house them is especially a problem in poor countries of rapid population increase. Yet the difficulties of city life may turn out to be the key to the population problem. In the countryside population growth seems to cost less; social services are nonexistent, and the contingency of crop failure and famine is the worry largely of individual families. The advantage of more people to rural landlords in particular has been emphasized by historians of Latin America. Before Ricardo it was suspected that land rents go up with population. Even in those countries with no large empty spaces, rural population growth can be contemplated with tolerance if not with enthusiasm.

In the city there is no one who benefits from increased numbers of unemployed people. They do not lower wages enough to compensate for their cost in relief, and they are a continual political threat to the civic order. It is often said that the cost of social services is greater in the city; this can only be because no services at all are provided in the countryside. The cost to the poor themselves of raising children is more clearly expressed in money terms in the city and so is more evident than the cost in the countryside.

As the urban proportion moves toward the 50 percent point in country after country of Latin America and elsewhere, the pressures toward population control, both among what are called the political classes and among the poor, become stronger. Religion cannot stand in the way, and indeed birth control becomes acceptable even to many of the clergy. Urbanization and urban slums may not be pretty, but they make the great contribution of discouraging population increase.

The importance of urbanization has attracted good minds to its study in recent years. Harris[17] has published a definitive work on the Soviet Union, Hodge and Hauser[18] on the United States, and Hauser and Schnore[19] have reviewed the state of knowledge in the field as a whole.

THE SUN BELT

Movement to cities may be seen as part of a loosening of the chains that have historically bound productive activities to particular locations. Subsistence farming is the main example of location-determined production, but there are many others. Iron-ore mining goes on in a section of the United States just west of Lake Superior and coal mining in Pennsylvania; the production of steel tends to be located along the land and water routes of the territory between these two areas. Recent discoveries of iron ore in northern Australia have brought settlement to that part. Oil production is heavy in Texas and Venezuela, where the crude oil is located; citrus fruit is produced in Florida and California, which have the soil and climate for it.

Other activities depend much less on locality. The film and later the aircraft industries did not need to locate on the West Coast. A university whose student body is national can locate anywhere on the national transportation network, which provides a good deal of choice. With better transport, automobile assembly plants, computer manufacture, and printing become loosed from the location of their raw materials. Those who determine the location of such activities may well do so merely on the basis of how pleasant the climate is, as among other things giving them an edge in attracting staff.

48

This liberation from place applies to increasing fractions of the population as the economy shifts from primary activities — agriculture and mining — to tertiary activities — service — and as the proportion on pension and similar incomes increases, both with earlier retirement and with a greater proportion of older people. The 1970 census of the United States shows the process in full motion. Nevada grew by 69 percent; Arizona by 35 percent; Florida, California, and Colorado by over 25 percent. Very few other states grew by as much as 25 percent. The sun belt, running from Florida west and somewhat north across the continent, has emerged as a population magnet. One can anticipate that the trend will continue with further affluence, with technological advances that make industries less dependent on bulky raw materials, with air transport that comes to include overnight air-cargo service through much of the country, and with the kind of product miniaturization that lessens the volume of both raw and finished goods to be shipped.

Insofar as climates are sunny, they lack rainfall and hence water for agricultural, industrial, and household use (Chapter 7). People discover the shortage only after they have moved, and they demand water supplies. Transportation of water will be an increasing expense as migration toward warm, dry areas continues; less fuel for heating will be required but more electric power for air conditioning.

We turn now to a closer look at time trends, in the hope of being able to say something useful about future population.

Growth of Population

ACCELERATION OF GROWTH

From the beginning of man's appearance to A.D. 1825, a period of 1 or 2 million years, the human race increased from a few savages to a billion people. In only a little over 100 years more, by about 1927, it increased another billion. The third billion took 33 years, until 1960; the fourth is being attained as this book comes to publication in 1975 (15 years); the fifth is expected by 1985 (10 years). (These future numbers are the U.N.'s medium variant projection, to which reference will be made again later.)

The time for adding 1 billion would steadily diminish even if the rate of increase were constant; indeed under constant increase the time would diminish according to the number of billions already here. Thus, if the second billion took 100 years, the third would take 58 years, the fourth 42 years, and so on. These intervals for attainment of successive billions under a constant rate of increase may be contrasted with the shorter intervals of the preceding paragraph (see Table III).

An alternative way of describing the acceleration is in terms of doubling time. A doubling took place between 1825 and 1927; a second doubling, from 2 to 4 billion, is in prospect by 1975, or 48 years later; at the rate of increase of 2.1 percent per year already attained, the next doubling would require only about 33 years. On the most optimistic assertions about the number of people that can live on the earth — however uncomfortably — only two or three doublings more are possible. Under the best of circumstances we are pushing against a ceiling that will be reached within the lifetimes of children already born.

The rate of increase is about 20 per thousand per year, and it has been rising. Table IV shows changes compiled by the United Nations. The birth rate of the less-developed countries declined marginally but deaths declined much more, resulting in a rise in natural increase. On the other hand, the more-developed countries showed no improvement in their crude death rate, and their births declined, to make a net decline in their rate of natural increase. The decline of the more developed does not offset the rise of the less developed; since the latter have a base of 2.5

III World Population Growth

	Actual growth	Average annual growth (%)	Doubling time (years)
1st billion by	1825		
2nd billion by	1927	0.68	102
3rd billion by	1960	1.23	57
4th billion by	1975	1.92	36
5th billion by	1985	2.23	31

Dates when successive billions were or will be reached, along with the average annual growth rates and doubling times implied by these; the sequence shows the acceleration in world population growth.

times as great, a rise in the rate of increase for the planet as a whole is shown. Just when the peak in the rate of increase will be reached, and what its level will be, are not yet known, but under slightly optimistic assumptions the peak could be reached at about 2.3 percent in the 1970s.

The basis for optimism is that decline of mortality can go only so far. If mortality to the end of childbearing were to be eradicated in the United States, and births continued year after year at 3,000,000 per year, our expectation of life might average 75 years for males and females together, so the population would rise to $75 \times 3,000,000 = 225,000,000$ and remain there. The reason that population would not be much higher than with present mortality is that about 95 percent of babies come to reproductive age, and this percentage cannot in the nature of things go above 100.

In the poor countries infant mortality is higher than in the United States, but still only about 10 percent of children fail to survive to reproduction in many countries of Asia and Latin America. Tropical Africa, on the other hand, remains more similar to earlier times in Europe and elsewhere, when mortality was so high that its decline could greatly raise the rate of increase. For example, West Cameroon, the Central African Republic, Guinea, and Togo show only 40 percent of girl babies surviving to the end of reproduction at age 50.[10] In the United States 92 percent survive to age 50. Averaging the world as a whole one can say that for the most part the decline of mortality has by the 1970s done its part in the accelera-

tion; at least the U.N.'s medium variant estimate shows a peaking and subsequent decline after 1980 in the world rate of increase.

While the end of acceleration is a true turning point in world demographic history, it by no means disposes of the population problem. The U.N. numbers show a rate of increase from 1975 to 1980 that averages 2.05 percent per year, and this drops to 2.03 percent per year by 1980–1985. With a rate of increase of 2 percent per year, the doubling time is 35 years; at this rate the nearly 5 billion of 1985 would become 10 billion by 2020, 20 billion by 2055, when some readers of this book will still be alive. The increase up to 1975 has been more than exponential; it now becomes "only" exponential.

GROWTH IN DEVELOPED COUNTRIES

The proportion of children under 15 is necessarily a consequence of birth rates of the preceding 15 years. For Europe in 1965, 25.23 percent of the population was under 15 years. The United States shows 30.67, higher because its birth rates were higher in the 1950s. But since then the U.S. birth rate has fallen below Europe's. A convergence in demographic as in other parameters is apparent among the United States, Western Europe, Japan, Canada, Australia, and the USSR.

Births in the United States in 1950 numbered 3,632,000, a rise from 2,858,000 in 1945 and 2,559,000 in 1940. The individuals in question (strictly the 96 percent of them who survived 20 years) came into working age during the 1960s. If growth of demand for labor were fixed at about 2 percent per year (we here abstract from fluctuations), then during the decade of the 1960s an increase of nearly 50 percent in young persons entering the labor force confronted an increase of less than 25 percent in jobs. If other things had been the same, drastic unemployment of young persons would have resulted, brought about simply through the rising births of the 1940s.

An indication that the demographic mechanism tends to operate in this direction is provided by the trend in past unemployment of young people. Comparing the unemployed in 1960 with 1970 we find that those 16–19 rose from 711,000 to 1,105,000 during the decade; those 20–24 from 583,000 to 864,000; those 25–44 actually fell, from 1,423,000 to

IV Natural Population Increase

Area	Birth	Death	Natural increase
Less-developed countries			
1960–1965	42.0	18.8	23.2
1965–1970	40.6	16.1	24.5
More-developed countries			
1960–1965	20.5	9.0	11.5
1965–1970	18.6	9.1	9.5
World			
1960–1965	35.1	15.7	19.4
1965–1970	33.8	14.0	19.8

Components of natural increase in less- and more-developed countries: crude rates per thousand population. *Source:* Reference 20.

1,230,000. The cohorts increasing in size, born in the 1940s and early 1950s, showed the major increases in unemployment.

The demographic relation tells us something about what is ahead, for 1950 was by no means the height of the baby boom. Births rose from 3,632,000 in 1950 to 4,258,000 in 1960. An increasing number of job seekers through the 1970s is inevitable, reaching 600,000 more about 1980 than appeared in 1970. Under almost any economic developments during the 1970s we can expect that new entrants into the labor market will have a harder time getting placed, at least relatively to those older.

The peak of the problem will be reached about 1980, after which the relative excess of youthful unemployment ought to decline. The early 1990s will see a substantial relief, corresponding to 1973 births of about 3,200,000.

As a result of the large cohorts born in the late 1940s and 1950s and the smaller ones born in the 1930s, we had few teachers and many students in the 1960s. Now this is turning around, and if other conditions are the same, the 1980s will see many teachers and few students.

Fractions of old people and young children are very different in a stationary population from those in most real populations up to this time. A typical developed country that has been increasing at about 1 percent in the past would have only about 10 percent of its population aged 65 and over. When it becomes stationary, if the life table does not change, we can expect an increase to about 15 percent at ages 65 and over. Specifically, the United States showed 8.4 percent of its males 65 and over in 1970; after a sufficient period of no growth, if the current life table continues to apply, we should have 12.4 percent of males 65 and over. For females the 1970 percentage 65 and over was 11.2, and the ultimate stationary percentage would be 17.4. The large improvement in female survival, especially since World War II, shown by a female expectation at age zero fully 7 years higher than male, is already influencing the age–sex distribution in the United States and will do so more in the future (Figure 1).

Corresponding to this, the proportion of young children will decrease with the advent of stationarity. That proportion was as high as

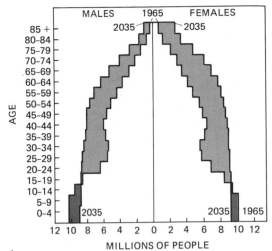

1   **Current Age Structure for the Population of the United States Compared to the Equilibrium Age Structure That We Could Reach By 2035** A.D. *Sources:* Data from T. Frejka. 1968. Reflections on the demographic conditions needed to establish a U.S. stationary population growth. *Population Studies 22:* 379–397; and N. Keyfitz and W. Flieger. 1968. *World Population, An Analysis of Vital Data.* University of Chicago Press, Chicago.

32 percent for males under 15 in the United States in 1965; it would go down to 22 percent with stationarity on the present life table. Expenditures on schools will be lower, those on medical services and other benefits for the elderly higher, in the stationary condition that is ahead.

Changes in this direction appear under the Series F projection of the U.S. Bureau of the Census, in which women are assumed to have a lifetime average of 1.8 births. According to the F projection, only 20.2 percent of the U.S. population would be under 15 by the year 2000; and 11.5 percent would be 65 and over. It takes longer, of course, for the new low birth rate to show in old people than in young.

GROWTH IN UNDERDEVELOPED COUNTRIES

We do not have to brood over subtle differences when contrasting developed and underdeveloped countries. Mexico's 1972 crude birth rate was 44.3 per thousand, possibly understated, and was nearly three times the U.S. figure; its death rate was only 8.8, and its rate of natural increase is the difference, 44.3 − 8.8

= 35.5 per thousand (3.5 percent). This rate of increase would double Mexico's present 50 million in about 20 years, quadruple them in 40 years, multiply them by eight in 60 years. If the pace continues, a child now born will not yet have retired from active life before Mexico contains 400 million people.

Among the difficulties incidental to the high rate of population increase is the large proportion of children to be cared for. Mexico's percentage under 15 in 1966 was 46.3, against that of the United States previously mentioned of 30.7 and Europe's 25.2. The effect of having many children continues into the next generation and causes population growth even after each individual family has cut its childbearing to bare replacement. With modern mortality bare replacement means that on the average fertile couples have 2.1 to 2.3 children, depending somewhat on the survivorship rate, on the fraction marrying, and on the fraction of married couples that are fertile. If Mexican average family size dropped to bare replacement immediately, this would produce a stationary population only after half a century or more, and then at 60 or 70 percent more than the present 50 million. This momentum effect, by which growth continues long after individual families are controlled at bare replacement, is due to the many children born in the years before the drop in fertility, which is to say, to the young age distribution.

Anyone aware of the difficulties of employment, schooling, and housing in Mexico today would surmise that it must be making every effort to prevent the exacerbation of these problems that is certain to follow from further population increase. Yet President Echevarria, himself the father of eight, campaigned on the slogan: "To govern is to populate." No birth control was needed; on the contrary, large families were to be encouraged. Empty parts of the country were to be colonized by pioneers engendered by the large families in the older parts. Such pioneering has an appeal to national leaders that it lacks for the intended settlers — *they* prefer to move to the capital city. This fact of population life has since acted back on politics, so that the pronatalist policy has been reversed.

Children are a primary luxury of the poor. A country has only so much available income and savings, and, willingly or not, it necessar-ily makes a choice on allocation. If it spends more on raising and educating children, it spends less on the equipment that will give them jobs when they come to working age. If it has fewer children, it will have more capital for investment (say in factories), so that when these children reach maturity, productive jobs will await them. Determination of how much expenditure should go into children and how much into factories, for a combination of the two that will be satisfactory a generation later, is not a simple problem. Its difficulty is further accentuated by the finiteness of the environment, including, for example, the pollution caused by the factories, which is not ordinarily taken into account. To determine for the right population, the right amount of capital to accumulate, the right number of people to employ, and the right degree of pollution to be created is a problem difficult to formulate, let alone solve. And even its solution would not help much if the real condition for development was inventiveness and organizational skill, and only in a lesser degree capital.

The view (presented in detail by Coale and Hoover[21]) that the harm of population growth in poor countries is its obstruction of capital accumulation and therefore future prosperity, rather than its threat of food shortage, is congenial to an age accustomed to technical progress. It would have surprised Malthus, for whom the possibilities of agricultural innovation were limited.

The capital argument and the land argument seem slowly but surely to be taking hold in the poor countries. Mexico, Turkey, and other governments have altered previous positions against birth control. The reversal accompanies urbanization and can be expected to spread with the increasing proportion living in cities. Urban living does not necessarily lower the number of children parents want, but it does make them more expensive to governments.

POPULATION PROJECTIONS

Demographers estimate future population from separate calculations of birth and death components. These estimates are often used as predictions of what will actually happen, but demographers intend them as projections, for which they have a conditional character; they tell, for example, what will happen if nothing is done to check birth rates.

Table V shows the most recent and realistic estimates which the U.N. has made for the years up to 1985 — its medium variant. In accord with the figures quoted above, it shows the world total over 4 billion in 1975, and almost 5 billion by 1985, an increase of 100 million per year against the present increase of about 75 million. Whether because prospects for population control have worsened or for other reasons, the latest estimate shows 187 million more people in 1985 than the 1985 total as assessed by the U.N. only 6 years earlier in 1963.

The difference in the rate of climb between the more- and less-developed regions — U.N. expressions for rich and poor, respectively — is dramatic. The more-developed countries grow by 23 percent in 20 years, the less-developed by 62 percent. In fact, the presently less-developed countries alone will have more people by 1985 than the world as a whole in 1970; during the 15 years they are expected to increase somewhat more than the total of the developed countries in 1970. Among continents Latin America and Africa increase by the largest fraction, with South Asia only slightly slower.

Extrapolation to the future is very dependent on what series one decides to extrapolate. If one assumes that the rates of the 20 years 1950–1970 continue, the United States will have 320 million residents by the end of the century and about 6 million annual births. If one supposes that the rate of change of the birth rate from 1957 to 1973 continues, the result is very different: the rate fell from about 25 per thousand to 15 per thousand, so its continuance downward in a straight line would mean no births at all by 1997.

No one can complain of a lack of long-range estimates of future world population. Let us consider the year 2000 and speak only of the estimates produced by one agency, the Population Division of the U.N. Secretariat.

The medium estimate for the year 2000 as made in 1963 was 6,130 million. Other estimates ranged around this, from a low variant of 5,449 million to a high variant of 6,994 million. Even this last was based on the assumption of falling fertility. If fertility of the early 1960s was assumed to persist, the 2000 figure would be 7,522 million.

About the sharpest statement that these and other materials permit is that the world population will probably be between 6 and 7 billion by the year 2000 (Figure 2).

Population, Labor, and Development

Future population numbers are notoriously difficult to forecast. Even more difficult is the forecasting of the technology and resources on which populations will be based. The following paragraphs sketch some possibilities.

AUTOMATION AND THE END OF EXPLOITATION

The United States, Western Europe, and the USSR are now moving into a new period, with the United States in the lead. The controlled chemical synthesis that enables a modern fac-

V Total Population Estimates by Major Areas, 1965–1985

Area	1965	1970	1975	1980	1985
World total	3,289	3,632	4,022	4,457	4,933
More-developed regions	1,037	1,090	1,147	1,210	1,275
Less-developed regions	2,252	2,541	2,874	3,247	3,658
East Asia	852	930	1,011	1,095	1,182
South Asia	981	1,126	1,296	1,486	1,693
Europe	445	462	479	497	515
USSR	231	243	256	271	287
Africa	303	344	395	457	530
Northern America	214	228	243	261	280
Latin America	246	283	327	377	435
Oceania	18	19	22	24	27

In millions. Medium variant. *Source:* Reference 22.

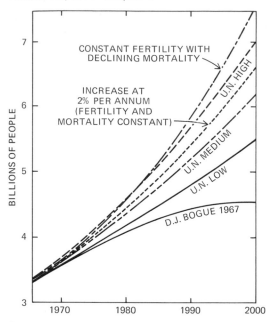

2 **Six Estimates of World Population, 1965–2000.**
Sources: Reference 23; unpublished estimate by D. J. Bogue, 1967.

tory run by a few technicians to produce the same rubber as a tropical plantation with thousands of laborers constitutes a liberation of production from human hands as well as from the natural landscape. A Dutch concern making synthetic fiber now fills much of the demand for Indonesian sisal of colonial times. But synthetics are not the only example of the new industrial power.

A few examples of the direction that industry is taking will serve. Men with pickaxes have not been important in American coal mines for some time; but even men working underground with mechanical diggers and loaders are giving place to strip mining, in which the largest earth-moving equipment ever constructed pushes away the few hundred feet of soil above the coal deposits, and then mechanical shovels take it out several tons at a time. Open-strip mining now accounts for over 20 percent of American coal production. The extraction of oil and natural gas is even cheaper, and these are moved almost without human effort through piplines to where they can be converted into electricity in nearly unmanned thermal plants.

The decline of employment in the automobile industry at the same time as more automobiles are being sold is partly due to numerical control of machine tools, which is spreading rapidly through this and other industries. Instead of craftsmen working with elaborate jigs, we have a library of control tapes (themselves made by computer from engineering drawings), and an operator who monitors whichever tape is required. Makeready time is reduced from days to minutes; inventories may practically be dispensed with when it is possible to make pieces one at a time as cheaply as several hundred at a time. Engineers, no longer held to what a craftsman can produce by hand, can develop more elaborate designs. In electronics printed circuits have reduced the man-hour content of electronic equipment in the same way that numerical control has reduced the labor content of automobiles.

Computers are the most spectacular labor savers. Whether for department store billings, for insurance company records, or for airline reservations, they produce a quality and quantity of work that makes previous methods unacceptable. Operations such as file searching or number multiplication, for which a clerk takes minutes, can be done by the computer in microseconds. It is an understatement to call a large-scale computer the equivalent of 500 people. Even based on this understatement, the 100,000 computers in existence in the United States by the mid-1970s are the equivalent of 50 million people in their capacity for nearly all varieties of clerical work. Within 25 years of the first working models — the computer era does not antedate 1950 — computers are doing as much adding and recording as could have been performed by the entire labor force of the United States in 1950.

INTERNAL LABOR DISPLACEMENT

Both economic theory and current experience show that, far from causing unemployment, the worst that automation need do is to move people from one occupation to another. The new one is usually more challenging; the work of a computer programmer is incomparably more exciting than that of a billing clerk. It is true that the shiftover cannot always take place in the same generation; many men in the transition may have to retire younger than they would like.

The major discovery has been that the level of unemployment can be kept wherever one wants by adjusting demand. If there is not enough work making automobiles, we can fill the gap by going to the moon. We could equally have filled the gap by planting flowers in the parks, but this has less appeal. We do not want the dull public works of the 1930s; our public works must be cosmic. They give employment to engineers, physicists, and technicians of all kinds. They require the expansion of the facilities for the production of such people, chiefly an expansion of the university system. And this has indeed gone forward with all possible rapidity.

EXTERNAL LABOR DISPLACEMENT

The same satisfactory observations cannot be made on the external displacements of labor caused by technical progress. Although these have not ordinarily been spoken of in the same breath, the Indonesian plantation worker dismissed because synthetic rubber limits the market for natural rubber is automated out as decisively as the Detroit automobile worker displaced by numerical control. The difference is that we neither pension the Indonesian nor offer him a way to get back into the system.

Our synthetics attack the economy of the underdeveloped country at the point where the colonial power had put its principal demand for tropical labor, the production of agricultural raw materials. Here the comparative advantage of tropical sunshine and cheap labor is greatest. If these were already being undercut in the colonial period, when Europeans owned the plantations, they will be undercut by Western factory owners even less regretfully when the plantations are in the hands of independent and sometimes unfriendly states. And while the former colonies can sell less, they want to buy more, especially the capital goods needed to industrialize themselves.

But could the ex-colony not turn to what it can now do best, relative to Western capitalized industry? Cannot India make textiles and sell them abroad, using abundant labor and local techniques, and so finance her own industrialization? Japan helped industrialize herself (without the benefit of Western good will) by selling textiles made by cheap labor. I am arguing that it may today be too late for another Asian country to do this.

Suppose Western industry were so mechanized that the competitive wages for unmechanized production would not suffice to enable the worker to buy the food that would keep him. We are actually at this point in some fields of Western production. An Indian worker with a hand loom produces about 3 yards of fabric in a long day's work; the value added in a world market whose prices are set by machine-made cloth is barely sufficient to buy him 1 pound of rice. India lacks capital to put her redundant population into factories and so make them competitive.

I have calculated that about 50,000 Calcutta clerks could do the work of a modern computer. Could they have been organized and trained by someone to take the contract for the computation of the Apollo project? Even if every other difficulty of coordination could be surmounted, it would turn out that the clerks could only be paid about 4 cents per day each (allowing a computer rental of $250 per hour, or $2,000 for an 8-hour shift), and this would not feed them, let alone feed and shelter their families.

At each point where the West learns to accomplish an industrial process at below what might be called physiological cost, that is, more cheaply than it could be done by unequipped labor willing to work for food alone, one more means of sustaining the overpopulation of the underdeveloped world is removed.

BEYOND EXPLOITATION

Much concern has been expressed on behalf of the exploited of the world. But now a lower level of the human condition has appeared: the man who is unexploitable. It is not worth anyone's while to use his labor; no one can make a profit by setting him to work, even at starvation wages.

This situation could only come about when to the technological leap of the advanced countries is added population density in the poor ones. If there were sufficient land everywhere, and each peasant could grow his own food, there would be no urgency. The peasant in the delta of the Irrawaddy can even today produce three times what he and his family need to eat; he can sell the surplus abroad and buy himself consumer goods; his government can preempt the surplus and buy machinery. The peasant could even be exploited in a factory in his spare time — since he feeds himself, no floor is set to his wages. This description does not apply to Java or to India, which are net importers of food. Increasing proportions of their people are functionless in relation to national and world production. They each come with two hands as well as a mouth, in John Stuart Mill's phrase, but without either land or capital they cannot be productive enough to feed themselves.

Notwithstanding all of this, the poor country can use its labor in conjunction with foreign capital to do those things that are relatively labor-intensive. Electronic components are made in Taiwan for assembly in Japan, and such subcontracting could expand fast in the future. If it did, the lack of capital in the poor country and the technical advance of the rich one need not prevent cooperation in production between them. The isolation of the poor described in the preceding paragraphs would be overcome, and income would rise in both rich and poor countries. Unfortunately for this constructive view, politics in many countries prevents the unlimited cooperation of poor labor and rich foreign capital. Japanese in Indonesia find themselves in trouble now that their activities are becoming widespread enough to be generally noticed, even though they are far short of the volume needed to make a noticeable difference to overall Indonesian employment. It remains to be seen whether development will be possible through this route. What cooperation does go on seems to focus on the need of the rich countries for raw materials whose extraction is capital-rather than labor-intensive.

Events of the last few years point in the direction of much higher relative prices of raw materials — oil, gold, soybeans. The result is a new division now opening up between those nonindustrialized countries that have such materials and those that lack them. The industrial countries compete with one another to equip Saudi Arabia; they are not competing to help India, which suffers simultaneously from the rise in raw material prices and the decline in foreign aid. Increased efficiency holds down prices of manufactured goods, or at least checks their rise; a shift in what is called the terms of trade is putting Third World producers of raw materials into the high-income category of the industrial countries.

ENTRY OF MILLIONS INTO HIGH CONSUMPTION

Yet despite difficulties development has in a sense already occurred in many of the poor countries of America and Asia. Automobiles are assembled, and parts for them manufactured, in Turkey, Brazil, and other countries. Indonesia has modern textile and tire plants. Some countries of small population, including Taiwan, Singapore, and Hong Kong, turn out sophisticated parts for electronic and other equipment, often subcontracting with Japanese or American assemblers. India and China may be poor on the average, but they have substantial and growing advanced sectors, with output all the way from television sets to jet aircraft. China produces thermonuclear bombs and the missiles to transport them. Both these countries have capital-goods industries and a wide range of raw materials, although apparently not enough to disregard foreign trade.

For a simplified view, consider a country whose industrial sector is expanding at 6 percent per year from its present employment of 10 percent of the labor force, and whose population is growing at 2 percent per year. Its population will double in 35 years, but in that time its industrial sector will multiply eightfold; at the end of 35 years 40 percent of its population will be employed in the industrial sector, comparable with present advanced countries.

Two difficulties, both related to population, stand in the way of this high-consumption society. The first is that within each of the large countries many nonparticipating poor people have to stand by while, in the process of economic growth, only a small fraction of the population gains much of the new wealth. Industrial wages are much higher than peasant incomes; those who are closest to the new sources of wealth take the largest part of it. Peasant communities become demoralized and their young people are drawn to the city. The demonstration effect of U.S. consumption (communicated especially through films) on demand in the poor countries is greatly reinforced by demonstration within the poor country itself. The nonparticipating population, the rural and urban citizens for whom there are as yet no jobs in industry, is impatient to divide the output of industry. Dissatisfaction may threaten civil order and have to be bought off with the premature distribution of the industrial product. Absence of a peasant sector may be one of the advantages that Singapore has over India.

The second difficulty is the expansion of the world industrial system as a whole and the resultant drawing down of reserves of nonrenewable resources. People everywhere seek to cross the threshhold from poor to middle class, and 10 to 20 million per year succeed. Most of mankind, now as in the past, eats cereals and very little meat, warms itself and cooks its food with locally cut wood, uses some fibers for clothing, and altogether has a minute impact on reserves of fossil fuels and metals. But now some families have come to drive an automobile, travel frequently by plane, live in houses fitted with electrical applicances and heated with oil, a transition in life style that takes place somewhere along the way from a per capita income of about $250 to one of about $2,500 per year.

Rather than use statistics of national income per head, averaging together rich and poor within a nation, let us take the high-consumption population of the world all together. Judging from automobile and other statistics, it may now include 90 percent of the United States population, 40 percent of that of Western Europe and Japan, 25 percent of the USSR, 0.5 percent of India, a total that may be of the order of half a billion persons. These are the people who use the 250 million motor vehicles registered in the world, at about two persons to a car. If motorcars are an index, the expansion of the modern sector has been from about 200 million persons in 1950 to 500 million in 1970.

The incorporation of more individuals into the high-consumption civilization in the USSR, Western Europe, Japan, and now in Brazil and the developing countries around the rim of Asia, could well be going forward at about 5 percent per year, say a doubling every 14 years. It is this expanded participation in high

consumption that underlies shortages of fuel and meat in the United States and elsewhere.

The expansion had to encounter resistance sooner or later, and that resistance now appears in the limits of the energy base. Americans were bound to find themselves in competition with Europeans and Japanese for the oil that powered the high-consumption style of life. With monumental lack of foresight, virtually all energy requirements were met by apparatus powered by fossil fuels — almost no other was built in the 1960s. That the crisis arrived late in 1973 was an accident of oil politics — for which we can be grateful insofar as it spurs action to prevent even more acute hardship 5 to 10 years hence. Underlying it is the cruel fact that world resources easily capable of supporting 200 million consumers, each with a modern package of automobile, electric appliances, and oil heating, are hard pressed to supply 500 million. The reader is in as good a position as I am to surmise what this means for the prospects of the remaining 3,500 million already on the planet.

Controlling Population Size

On the average more children are born to poor couples than to those better off; this is true both among countries and among individual families within countries. There is no provable genetic or inborn quality difference between rich and poor as such, but the poor command less satisfactory facilities for bringing up children. Children share the standard of living of the home into which they come, and on any definition of poverty that standard is lower for the poor. The differentials of fertility have been seen as a social problem for at least the past century, which is to say, ever since they came into existence in their modern form.

At first those differentials were most conspicuous within countries; it was noted in England and elsewhere that the more urban, the more literate, the better off generally, in short those most capable of providing amenities for children, were the ones who had fewest children. The century-old differentials of fertility are now disappearing, and differentials of income are probably less than they were a hundred years ago, although statements on this depend on the method of measurement. Within rich countries all social groups are converging to about the same birth rates — to something close to a two-child family.

However, there are few signs of convergence among countries; the birth rates continue at 50 per thousand in some parts of Africa, through 40 or so in South Asia, down to 15 or less in the United States, Europe, and Japan. For the world as a whole, the differentials in birth rates have never been higher. Some poor countries feel themselves under the demographic pressures of the past, when the problem was to produce enough children that the community would not die out under the prevailing high mortality. An individual family head wants the continuance of his line and in particular enough children surviving to take over his agricultural work and to provide him with whatever old-age security is to be had. (This association of security with large numbers of children is especially close in a regime of wage-labor; with peasant ownership or fixed tenancy, on the other hand, it is in the interest of old-age security to have *few* children.) In a regime of wage-labor where jobs for women are scarce, the interest of fathers in having children who will later support them with their wages often accords with the drive of women to motherhood as the means to securing their marriage and providing status in the community. The advent of women's careers that provide alternative sources of self-respect also turns women's thoughts away from childbearing.

The fall in mortality drastically reduces the number of sons that one needs in order to have children living into one's old age. In round numbers, eight births would be required to provide two adult sons when over 50 percent of children die before maturity; just over four births will provide the same two adult sons when most live to maturity. Thus the fall in mortality, once it is assimilated into the thinking of parents, is capable of reducing by about half the number of children they want, and four children is a figure often seen in surveys of desired family size in poor countries. The fall in mortality is now reflected in childbearing intentions and ideals in many countries, although not in all.

However, four children surviving to maturity are still twice as many as are needed for replacement, which is the level to which the United States has now dropped, and which is the inevitable long-term level for all countries. The drop in poor countries from four children to two seems to depend on the second factor mentioned above: entry of the wife into work outside the home, a substitute for traditional childbearing in providing standing in the community.

A vulgar Malthusian would think that population control in poor countries depends on high death rates, and he would be against improving health services. He would be wrong, of course; the opposite policy will be more effective. If people have confidence that a child once born will live, that if they fall sick they will have access to medical services, that some degree of security is possible — in short, if they are in a position to plan their lives — they are more likely to plan their families. Literacy is one of the factors that encourage both general planning and family planning; this we can read in Malthus. It may indeed be true that modern developments in medicine have brought the population problem upon us, but the way to solve it is not to put medicine into reverse, but to push it ahead with all speed. It is the next stages in modernization — more medicine, schooling, and economic advance — that are the best hope of solving the problems that arise in the early phases of modernization. The demographer can in good conscience recommend policies that would be desirable on general grounds as well as for their use in bringing population under control.

OFFICIAL BIRTH CONTROL PROGRAMS

The support of birth control varies among countries in the ratio of official to unofficial effort. Active official policies and programs are found in many Asian countries, including both Chinas, Pakistan, India, Ceylon, Indonesia, and Malaysia, as well as Iran and Turkey in western Asia; a few African ones (only Ghana and Kenya south of the Sahara); and virtually none in Latin America.[24]

Countries lacking formal official policies do have one degree or another of informal or private activity in the field of birth control. Chile allowed its maternal and child health clinics to provide family planning services. In Columbia and Costa Rica birth control activities are also relatively overt. Mexico has a private family planning association that is operating effectively, even though the advertising of contraceptives is illegal. In Hong Kong birth control has been successful enough to affect the official birth statistics substantially, the procedure being government subsidy of the private family planning association. Government subsidy to private medical and other family planning activity is to be found also in Korea and Taiwan.

Where contraception is being introduced officially the method varies. In India sterilization is said to be applied by just under 5 million couples out of 6.6 million using one kind of contraception or another.[24] Pakistan has favored other methods, and over 1 million women have been fitted with intrauterine devices (IUDs). In the United Arab Republic the predominant method is oral contraceptives, now used by an increasing fraction of women. Jamaica supports contraception and officially encourages emigration as a major means of population control.

Most of the countries with national family planning programs provide the necessary free supplies and a few provide some incentives. India offers 10 rupees (the equivalent of $1.30) to men who undergo sterilization, and 40 rupees to women; South Korea offers 800 won, the equivalent of $3.00, to each.[25]

No country encourages abortion, but the degree to which it is opposed in principle varies from Latin America and the Philippines at one extreme to Eastern Europe and Japan at the other. The practice of abortion varies much less than does the sternness of official injunctions against it. The United States was until the present decade a leader among those countries whose laws condemn abortion but whose women practice it. Its laws made medically safe abortion expensive, and in consequence a certain proportion of hospital beds were occupied by poor women suffering the consequences of operations by unskilled amateurs.

All this is now radically changed. The U.S. Supreme Court uncovered the right of privacy, by which the law may prohibit abortion only during the last 10 weeks of pregnancy when the fetus is presumed to be viable. The State of New York permits abortion on demand, for nonresidents as well as residents, up to the 24th week of pregnancy. More than 15 states

have liberalized their laws in one degree or another since Colorado's action in 1967. That abortion with impunity can follow by only a few years laws that provided jail sentences for the sale of contraceptives shows that the legal system can be influenced by objective conditions. The legal changes occurred about the same time that the Gallup Poll showed that the proportion of the public favoring abortion on demand during the first 3 months of pregnancy had climbed above the proportion opposed. Some of the considerable drop to below 3.2 million births in the United States in 1973, from a level that was already low, must have been due to changed laws and attitudes toward abortion.

That our values are changing rapidly in the face of population growth is also revealed in institutional action. As recently as 1967 the House of Delegates of the American Medical Association adopted a report opposing induced abortion except for therapeutic reasons and even then under restricted conditions. Only 3 years later it reversed this policy and moved to allow the decision to interrupt pregnancy to be made simply by the woman and her physician.

Until contraception attains complete reliability, abortion seems required as a backstop for its failures. Those poor who can afford children least, and whose contraception is least likely to be effective, need it most. A legal system that in effect permits abortion to the rich and withholds it from the poor hardly sets an example of fairness and responsibility.

OFFICIAL REASONS FOR BIRTH CONTROL

Governments adopting family planning policies advance varied official reasons, among which economic growth of the country is the most frequent. For the United Arab Republic, "This increase (of population) constitutes the most dangerous obstacle that faces the Egyptian people in their drive towards raising the standard of production in their country. . . ." Turkish government statements mention economic development and the capital-income ratio. Kenya emphasizes the problem of providing jobs as the adult male population increases, and the burden of child dependency. Singapore stresses the welfare of the individual family, especially the incapacity of its breadwinners to support many children.

Malaysia speaks of its population problem as engendered by lower mortality due to medical and health services and sees it as necessary to complement this health support by family planning if economic progress is not to be inhibited. Indonesia says that "the aim of family planning is in the first place to promote the welfare of the family, especially of mother and child."

Berelson[26] sums up by saying that in the entire developing world today, about 65 percent of people live in countries with policies favorable to birth control. "In Asia and Africa the movement is mainly based on the effect of population growth upon social and economic development. . . . In Latin America it is based more on medical and humanitarian concern with the prevalence of induced abortion."

The ecological view, that population out of balance with the habitat leads to disaster, is barely hinted at. The closest I see, for instance in the collection of governmental statements on family planning provided by the Population Council[27], is for India, which does refer to pressure on resources as at least a subsidiary reason for family planning in its First Five-Year Plan (1951–1956). Pakistan refers in introducing its population policy for the Third Plan (1965–1970) to the 3.5 acres held by the average cultivator, and hence must be counted as aware of the ecological problem, at least in its gross form.

Although little is said officially about the need to limit population to accord with the finiteness of the environment, much is spoken about vast natural resources. These are indeed the stock in trade of political discussion of population in Brazil and other countries of Latin America today, and in Indonesia during Sukarno's regime. Dusty notions about resources that had some substance when these two countries each contained 50 million people become obsolete as they pass the 100 million mark.

If little is said officially about the limits of national resources requiring a limiting of population, the notion of a planetary limit of resources appears not at all. That anyone should curtail his own family because the planetary environment must be protected would seem incomprehensible, if not laughable, to officials of underdeveloped countries, who face pressing day-to-day problems. Yet for the ecologists writing in this book, the finiteness of planetary resources is the ultimate reason for population restraint.

But we should not complain if countries adopt the right policies, even if for inadequate reasons. The underdeveloped world may be gradually coming to see how its own interest is served by population control. Not content to leave things to take their own course, country after country is officially promoting birth control.

VOLUNTARY BIRTH CONTROL

The dominant approach is one of providing the means to voluntary birth control, which is to say, offering cheap and effective contraceptives to those parents who feel they have enough children. Some optimists consider that this would solve the population problem — they implicitly assume that the desires of individuals for children and the carrying capacity of the earth's surface are in natural harmony. That such an assumption is gratuitous appears from many studies.[28]

To cite the most thorough of the investigations now extant, Freedman and Takeshita[29] found that for Taiwanese wives 35 to 39 years old, live births averaged 5.2, living children 4.6, and children wanted 4.2. Availability of birth control to the population under survey would reduce fertility but would still leave it very high. Some assumptions about proportions unmarried and unfertile, as well as about mortality, are required to translate a figure of 4.2 children wanted into a population rate of increase, but even with maximum adjustments for these, a mean of 4.2 births per woman comes out to an increase of over 50 percent per generation.

The groups classified as more modern want fewer children. Professionals in the Taiwan sample, for instance, stated a mean number of children wanted of 3.4, against 4.9 by farmers. Women 30 to 34, moreover, wanted slightly fewer than did women 35 to 39: 4.1 against 4.2; among wives of farmers, those 30 to 34 wanted 4.5 children, against 4.9 wanted by wives 35 to 39. That younger women, and those in the modern sector, wanted fewer children is encouraging for any country that is moving rapidly toward modernization. But the number of children wanted by even the most advanced groups is far from suggesting that a stationary condition will follow automatically from present economic trends even in Taiwan, let alone in Nepal or Bolivia.

In developed countries as well, the perfecting of contraceptive techniques by which couples will be able to have exactly the number of children they want does not by itself promise a stationary population. Reconciling the value placed on children with what the economy and ecology can stand is a worldwide problem, surmounted mostly in Japan and in Eastern Europe. But the steady fall in births, the drop to bare long-term replacement shown by the United States in 1973, and the actual excess of current deaths over current births in the same year in West Germany, suggests that stationarity may not be far away for all rich countries.

The birth control movement emphasizes the voluntary approach to family planning. Former Secretary-General of the United Nations, U Thant, in the context of a statement on overpopulation, spoke of "the right of parents to determine the numbers of their children." Is it possible that in a near future this slogan will be taken over by populationists and used to shout down those who initially devised it?

INDIVIDUAL AND COLLECTIVE GOALS

Any realistic treatment must start with the fact that most people want children. If there is any one constant of human aspiration, it is this. Nearly every culture incorporates a desire for progeny, and in most cultures children are more highly regarded than wealth. Even among Americans, who would not take the trouble to conceive 10 children if no further obligation was attached?

In fact, children are some cost and trouble to their parents everywhere, and hence most couples do not have 10 children. In industrial societies where the labor of caring for children is priced high, particularly where women have alternative opportunities of employment, the ideal family size may drop down to three or fewer. In preindustrial groups where alternative employment that could give satisfaction equal to that of childbearing is not to be had, the ideal family size is nearer to 5 children. Hundreds of surveys of ideal family size have now been carried out, each question presumably answered by each respondent on the assumption that the costs of having children and the alternatives to having children are what they are in the community of the respondent, and no large group has stated as its ideal an average as small as two children surviving to maturity. Yet this, or slightly more to allow for infertile couples, is what the long-run human average must be.

Whether seen on the national scale or on the world scale, the solution of putting safe, reliable, and easily applied contraception within the reach of every couple is only a first stage; it will enable them to have only the number of children they want. The second stage is somehow causing them to want the number of children that the crust of the earth can support. And in view of the urgency of the problem, it would be unpardonable complacency to put off facing the second stage until the first stage is disposed of.[28]

The influencing of parents is part of the ethics of population, a field developed by John Noonan,[30] whose encyclopedic work is a necessary introduction to any contemporary discussion.

MORAL DILEMMAS

If voluntary parenthood, the slogan under which contraception has made itself acceptable in wide circles, still leaves a disequilibrium, what is the next step? Must parenthood be administratively controlled? Should we require that couples fill out a form and submit it at the wicket of a government agency, wait several months, and then return to find it stamped as approved or disapproved? This is what we require of anyone adding to the national population in later life as an immigrant. Yet an immigrant entering the United States at the age of 25, his education completed and his working life just beginning, is not the burden that a newborn infant is.

Although our ethics allow officials to make choices among foreigners, we would be repelled by their making choices among natives —deciding which Americans could be parents cannot be left to administrative decision. Should we, then, allow each couple to have two children and no more? This quantitative regulation is analogous to what we now have in respect of marriage: most countries set at one the number of wives a man may have at any moment. And yet his having two wives places less strain on the community than a couple having four children. The members of the community who support laws against bigamy — presently the majority — ought to support compulsory family limitation if it is true that unrestrained reproduction does more harm than unrestrained marriage as such could do.

Or should limitation be made consistent with market freedom by giving every girl at puberty two coupons, each entitling the holder to a child?[31] Coupons are used to ration consumer goods in wartime, but here the woman could be allowed to sell the coupons, or buy others, and so the rights to children would drift into the hands of those who most wanted to be the parents of the next generation. Specifically, the coupons would drift into the hands of those who both wanted children and had the financial capacity to give them good schooling and expensive upbringing. It makes all the difference that the payment for the excess over two children would have to be *in advance* of their conception rather than after their birth. Here, as in consumer goods generally, an installment system by which one can make the decision to purchase long before he has to pay encourages improvidence. Of course, the coupon system would only be a test of foresight and saving as a prerequisite for having children if parents were forbidden to borrow to raise the amount needed to buy the coupons. We do have laws restricting purchase of stocks on margin, and such would have to be applied for children.

There are incidental benefits of the scheme. The girl who was poor would sell the two coupons at the high prices that would prevail in the market if children are desired as much as surveys of married couples show them to be. For an upwardly mobile couple, cash from the sale of the coupons could be a dowry, sufficing perhaps to set the couple up in a business.

The children would be born to those financially stronger, and therefore better able to care for them. They would be made comparable to yachts or other expensive consumer goods; those who could afford the high initial cost would be those best able to stand the upkeep. But the citizens who find it objectionable that yachts are allocated in this way would be aroused to revolutionary fury if children were so allocated, and very drastic changes in values would have to occur before such a proposal could be taken seriously.

The scheme would, in any case, require a new contraceptive technique not under the control of the couple. Guarding the nation's frontiers against illicit immigration would have its counterpart in policing the nation's wombs against unauthorized reproduction. A device would have to be invented to make women temporarily infertile and would be administered to all through the water or in some other way. Only on obtaining permission to have a child would couples be provided with a suitable antidote.

An alternative is a tax, fixed at the level that will just produce the average of two surviving children per couple needed for stability. Taxes are how we discourage the consumption of cigarettes and liquor. This would be going back to an earlier and fairly successful method of population control. When schools had to be paid for by the parents of the pupils, and free lunches were unknown, many expenditures that today are public costs had to be covered privately by those benefitting. In effect, there were quite heavy charges on children in nineteenth-century England and France, in contrast to the present day, when children are income tax exemptions. Can we restore the nineteenth-century arrangement by a tax in the nature of a user charge on schools, and so on, similar to the gasoline tax that pays for highway construction and maintenance? People being taxed differentially is not in itself abhorrent; the exemptions at the present time, which are passed down in laws inherited from an epoch of underpopulation, in effect tax couples who do *not* have children. The principle of a tax incentive relative to reproduction is well established in our laws, but heed for the environment would reverse its direction and load the tax burden on the prolific rather than the careful.

This approach of charging parents the full cost of their children has a serious disadvantage: that not the parents alone but in whole or in part the children would pay. We do not want to discourage parents from having children by means that result in poorer food and poorer education for the children that are born.

A subsidy for those who do not have children would have more appeal than a tax on those who do. Women would be invited to register for the payment, and to return to the registry office each 4 months, say, to be inspected for nonpregnancy. After a suitable number of such inspections they could claim their subsidy.[32] Since a woman may be fertile for over 30 years, and 10 years are more than enough to contribute to a dangerous overload on the environment, the subsidy per year of nonreproduction would have to be graduated upward with the length of time. Its administration would offer problems unknown in the U.S. Department of Agriculture, where farmers were long paid for leaving their soil uncultivated on a year-by-year basis.

This discussion is intended only to show the dilemma posed by the desire of individual couples to have children in the face of the incapacity of the earth's surface to contain more than a certain number of people. The balance in all previous history was maintained by bacterial diseases; now that bacteria have been substantially conquered, we need a moral equivalent to them. Those who find the above suggestions repulsive, as this writer admits he does, have to ask themselves whether they are as painful as the restoration of disease and starvation would be.

References

1. White, C. L. 1965. Geography and the world's population. In *The Population Crisis: Implications and Plans for Action* (L. K. Y. Ng and S. Mudd, eds.). Indiana University Press, Bloomington, Ind. pp. 11–20.

2. Clark, C. 1967. *Population Growth and Land Use*. Macmillan & Co. Ltd., London.

3. United Nations. 1968, 1971. *Demographic Yearbook*. U.N., New York. 1968, Table 1, p. 83; 1971, Table 1, p. 111.

4. Ho, Ping-ti. 1959. *Studies on the Population of China, 1368–1953*. Harvard University Press, Cambridge, Mass.

5. Davis, K. 1968. *The Population of India and Pakistan*. Russell & Russell, Publishers, New York. (First published by Princeton University Press, 1951.)

6. United Nations. *Vital Statistics Report*. U.N., New York. Data available as of Oct.

1, 1973, ST/STAT/SER. A/105-106, Ser. A, Vol. XX5, Nos. 3–4.

7. *New York Times*. 1970. Apr. 18, p. 3.

8. U.S. Bureau of the Census. 1969. *Statistical Abstract of the United States, 1969*, 90th ann. ed. Government Printing Office, Washington, D.C.

9. Widjojo N. 1970. *Population Trends in Indonesia*. Cornell University Press, Ithaca, N.Y.

10. Keyfitz, N., and Flieger, W. 1971. *Population: Facts and Techniques*. W. H. Freeman and Company, San Francisco.

11. Ladejinsky, W. 1970. Ironies of India's Green Revolution. *Foreign Affairs 48*: 758–768.

12. Brown, L. R. 1968. New directions in world agriculture. (Studies in Family Planning 32). The Population Council, New York.

13. Schmitt, R. C. 1968. *Demographic Statistics of Hawaii: 1778–1965*. University of Hawaii Press, Honolulu.

14. Petersen, W. 1969, *Population*, 2nd ed. Macmillan & Co. Ltd., London.

15. Ehrlich, P. R., and Ehrlich, A. H. 1970. *Population, Resources, Environment: Issues in Human Ecology*. W. H. Freeman and Company, San Francisco.

16. United Nations, Department of Economic and Social Affairs. 1969. *Growth of the World's Urban and Rural Population, 1920–2000* (Population Studies 44). U.N., New York.

17. Harris, C. D. 1970. *Cities of the Soviet Union: Studies in Their Functions, Size, Density, and Growth*. Rand McNally & Company, Chicago.

18. Hodge, P. L., and Hauser, P. M. 1968. *The Challenge of America's Metropolitan Population Outlook, 1960 to 1985*. Praeger Publishers, Inc., New York.

19. Hauser, P. M., and Schnore, Leo. 1965. *The Study of Urbanization*. John Wiley & Sons, Inc., New York.

20. United Nations. 1971. *The World Population Situation in 1970* (Population Studies 49). U.N., New York. pp. 18, 32, 41.

21. Coale, A. J., and Hoover, E. M. 1958. *Population Growth and Economic Development in Low-Income Countries*. Princeton University Press, Princeton, N.J.

22. United Nations. 1969. *New Findings on*

Population Trends (Population Newsletter 7). U.N., New York. p. 4.

23. United Nations, Department of Social and Economic Affairs. 1966. *World Population Prospects as Assessed in 1963* (Population Studies 41). U.N., New York.

24. Hughes, E. C., and Hughes, H. M. 1952. *Where Peoples Meet: Racial and Ethnic Frontiers*. The Free Press, New York.

25. Nortman, D. 1969. Population and family planning programs: a factbook. In *Reports on Population/Family Planning*. The Population Council, New York.

26. Berelson, B. 1969. National family planning programs: where we stand. In *Fertility and Family Planning: A World View* (S. J. Behrman, L. Corsa, and R. Freedman, eds.), University of Michigan Press, Ann Arbor, Mich.

27. Population Council, The. 1970. Governmental policy statements on population: an inventory. In *Reports on Population/Family Planning*. The Population Council, New York.

28. Davis, K. 1967. Population policy: will current programs succeed? *Science 158*: 730.

29. Freedman, R., and Takeshita, J. Y. 1969. *Family Planning in Taiwan: An Experiment in Social Change*. Princeton University Press, Princeton, N.J.

30. Noonan, J. T., Jr. 1966. *Contraception: A History of Its Treatment by the Catholic Theologians and Canonists*. Harvard University Press, Cambridge, Mass.

31. Boulding, K. E. 1964. *The Meaning of the Twentieth Century*. Harper & Row, Inc., New York.

32. Enke, S. 1963. *Economics for Development*. Prentice-Hall, Inc., Englewood Cliffs, N.J.

Further Reading

Hauser, P. M. (ed.). 1969. *The Population Dilemma*, 2nd ed. Prentice-Hall, Inc., Englewood Cliffs, N.J.

Keyfitz, N. 1968. *Introduction to the Mathematics of Population*. Addison-Wesley Publishing Company, Inc., Reading, Mass.

Petersen, W. 1969. *Population*, 2nd ed. Macmillan Publishing Co., Inc., New York.

——— (ed.). 1972. *Readings in Population*. Macmillan Publishing Co., Inc., New York.

Shryock, H. S., Siegel, J. S., et al. 1971. *The Methods and Materials of Demography*, 2 vols. U.S. Department of Commerce, Bureau of the Census, Washington, D.C.

Editor's Commentary—
Chapter 3

Suppose someone had arrived in the U.S. from another planet and was met by a team of scientists who recounted the fantastic achievements of 20th century science — computers, complex surgery, nuclear power, space flights. This person would surely be flabbergasted to learn that, along with all of these achievements, for almost half the world's population we have not succeeded in solving the most primitive and basic need—an adequate diet. On being given an account of the billions of dollars spent on fantastic weapons of war (called "defense" by this strange people), and on giant toys like moon-landing craft, and with the fresh outlook of a new arrival, the person would surely be interested in the ethical structure that had produced such a situation; or he or she might suppose that problems of food supply are not amenable to technological solution.

No doubt there is some truth to the second explanation; certainly the tropics often respond badly to the application of brute force technology, as is so tragically demonstrated by the violence done to the fragile agricultural system in the Sahel (the six African countries directly south of the Sahara) and the subsequent starvation there (reported in the 19 July 1974 issue of *Science*, vol. 185, pp. 234-237). But there can be little doubt that extra money diverted to research and development in tropical agriculture over the past three decades would have caused increased food production there, for as Lester Brown and Erik Eckholm point out in this chapter, the poor, tropical, nations have a large potential for increased yields, given careful agriculture and a supply of needed resources such as energy, fertilizers and irrigation water.

Unfortunately, as I write in late 1974, world agriculture has recently fared badly. This year has exposed the crucial weaknesses in the world agricultural system. Harvests are still terribly dependent upon the weather, which reduced harvests not only in India and Sahelian Africa, but in North America as well. It is important to note that, as discussed in Chapter 15, man can affect the weather and climate, and very small changes in the world average temperature can change the *pattern* of the weather, as seems to be happening now. Further, tropical agriculture, which is the most vulnerable, is highly dependent on other aspects of the world economy: The oil crisis (among other factors) has created a severe shortage of fertilizer (especially nitrogen fertilizers), resulting in doubling and in some cases tripling of the cost. The Food and Agricultural Organization of the U. N. (FAO) has estimated that in 1973-74 the developing countries were unable to purchase well over 1 million tons of fertilizers that they would have bought if the shortage had not occurred. The shortage in 1974-75 is expected to be as bad as or worse than it was in 1973-74. This "missing" fertilizer is equivalent to about 12 million tons of food. FAO foresees more shortages in the future, and also predicts that pesticides (which are also made from petroleum) will be in short supply—probably 20-30 percent short of requirements in 1974-75! As an example of these effects, in 1974 India's wheat harvest declined for the second year and was down 20 percent from the 1972 figure, owing to a combination of bad weather, inadequate fertilizers and insufficient energy for producing irrigation water.

Along with all of these difficulties, as the authors point out, world grain reserves have fallen to a very low level. Furthermore, the U. S. agricultural establishment is against setting up massive grain reserves throughout the developed world (even if excess grain were available) because this would keep down the price of U. S. grain, and thereby prevent the U. S. from using its agriculture to help solve its balance of payments problems. Since the U. S. is one of the few nations with an agriculture that can produce excess for export, the ethical dilemmas are particularly clear for Americans. Thus, this chapter illustrates a point that will be made throughout this book: environmental problems are interrelated—they are not purely physical problems, but are social and political and economic.

Chapter 7, on fisheries, examines in detail one particular aspect of world food supply. Chapters 13 and 14 look at some of the problems associated with pest control, mainly in agriculture.

Ethiopian farmer collects wheat samples from experimental plots. FAO
photo by M. Benaissa.

3

Man, Food, and Environment

Lester R. Brown   *is President of the Worldwatch Institute in Washington, D. C., a
"think tank" on emerging global problems. He has written widely on world food and population
problems and the many facets of global interdependence. He holds degrees in agriculture,
economics, and public administration, and was formerly a Senior Fellow of the Overseas
Development Council in Washington, and Administrator of the International Agricultural
Development Service. His books include* Seeds of Change: The Green Revolution and
Development in the 1970s *(1970),* World Without Borders *(1972), and* In the Human
Interest: A Strategy to Stabilize World Population *(1974).*

Erik P. Eckholm   *is an Associate Fellow of the Worldwatch Institute. He received an
M.A. from the Johns Hopkins School of Advanced International Studies, and writes on world food
and development problems. The ideas in this article are further explored in the book* By Bread
Alone *(Praeger, 1974), co-authored with Lester Brown.*

Man may have first appeared in the thin film of life that covers the earth as early as 2 million years ago, which is very recent in geologic time. For hundreds of thousands of years he hunted and gathered wild food, living as a predatory animal and a searcher for fruits, nuts, and berries. His life was largely a search for food, and starvation was a constant threat.

Then, somehow, perhaps as recently in man's existence as 10,000 years ago, he learned to domesticate animals and plants and began the great transition from hunter to tiller. Today only a small fraction of 1 percent of the human race live by hunting. The transition from hunter to tiller is virtually complete. Man has substituted the vicissitudes of weather for the uncertainty of the hunt. Man the hunter had an exceedingly limited capacity for intervening in his environment. But man the tiller developed a seemingly endless capacity for altering his environment, shaping it to his ends, and multiplying his numbers in the process. Initially quite simple and limited in scope, his interventions became successively more complex and widespread. Eventually some of the consequences of his interventions were to exceed his understanding of them, creating worrisome problems. Some of these problems are largely local, but others, the more serious ones, are global in scale.

Although still a mystery just currently being unraveled, scholars agree that the beginnings of agriculture occurred in southwestern Asia, in the hills and grassy northern plains surrounding the Fertile Crescent. Climate was hospitable and food resources relatively abundant. Wheat and barley grew wild there, as did sheep, goats, pigs, cattle, horses, and deer. To this day wild barley and two kinds of wild wheat (emmer and eikorn) flourish in the region. Available evidence indicates that the earliest agrarians were herdsmen.

The great neolithic achievements of agriculture and husbandry gave man a more abundant and secure food supply, allowing him to increase his numbers and establishing the base for civilization. Grain fields fed growing urban populations. But the problem of obtaining enough food remained; it has plagued man since his beginnings.

At the time agriculture evolved, the earth supported roughly 10 million people, no more than now live in London or Iraq. Since then a series of technological innovations have brought about an enormous expansion in the earth's food-producing capacity. Following the discovery of agriculture, food production expanded under the influence of six major technological advances: the use of irrigation, the harnessing of draft animals, the exchange of crops between the Old World and the New, the development of chemical fertilizers and pesticides, advances in genetics, and the invention of the internal combustion engine.

The earliest of these innovations was irrigation. By obstructing the flow of streams and rivers, man was able to divert water onto land under cultivation. This intervention in the hydrological cycle, closely associated with the emergence of early civilizations in the Middle East, has greatly boosted the productivity of land. Today close to one-seventh of the world's cropland is irrigated. Irrigation today is particularly prevalent in the rice-growing nations of Asia; China and India together have nearly half of the world's irrigated land.

A thousand years after learning to irrigate, man began to harness draft animals for tilling the soil. This breakthrough enabled him to convert grass and hay into a form of energy that he could use to increase his food supply. By harnessing draft animals, man supplemented his limited muscle power and raised the efficiency of his labor to the point where a small segment of the population could be spared from food-producing activities.

The third factor expanding global food production was the exchange of crops set in motion by the linkage of the Old World and the New. Few people would identify Columbus as a major figure in the effort to increase food supplies, yet his contribution in establishing such a linkage was considerable. One of the most interesting results of this exchange of crops was the discovery that some plants could be grown more successfully in their new environment than in their country of origin. For example, several times more potatoes are produced in Europe, including the Soviet Union, than in the New World, where the potato originated.

The use of agricultural chemicals both to improve soil fertility and to control pests, has been yet another crucial factor in the effort to meet the global demand for food. The early research on chemical fertilizers was undertaken in Germany by Justus von Liebig, who found that the nutrients removed from the soil by farming could be replaced artificially. Although this discovery was made in the 1840s, it did not become commercially important until well into the twentieth century, largely because the presence of frontiers made it possible to continuously bring new land under the plow. Once the frontiers began to disappear, however, farmers turned to fertilizer as the principal means of expanding production.

Then, as man's ability to alter the genetic composition of domesticated species of plants and animals progressed after the discovery of the principles of genetics, the productivity of these species increased dramatically. Advances in cereal productivity over the past generation have been particularly impressive. Many farmers in the U.S. Midwest now consistently obtain corn yields of 3 tons per acre, making corn the most productive of all the world's grains.

Advances in cereal productivity have at least been matched by those in the livestock and poultry sector. The first domesticated cow probably did not yield more than 600 pounds of milk, barely enough to support a calf until it could forage for itself. In India, milk production remains at about that level today. By contrast, the average cow in the United States last year yielded 10,000 pounds of milk. The holder of the world record is a cow in the state of Washington, "Skagvale Graceful Hattie," which produced over 44,000 pounds of milk in a single year. This productive animal could deliver 55 quarts of milk to one's doorstep daily, thus outperforming her early ancestors by a factor of more than 70 to 1.

In the case of poultry, the first domesticated hens did not lay more than about 15 eggs or one clutch per year. The average American hen produced 227 eggs in 1972. For some time U.S. hens held the world egg-laying title, but a few years ago an industrious Japanese hen set a new record by laying 365 eggs in 365 days. U.S. agricultural analysts project continuing gains in productivity per cow and hen, although the rate of increase must eventually begin tapering off as some upper biological limit is approached.

And finally, with the development of the internal combustion engine, man has been able to greatly augment the available energy resources for expanding his food supply. Today more than half the world's cropland is tilled with mechanical power; the bulk of the remainder is still tilled by animals. Three distinct forms of agriculture exist today, depending on the form of energy used — human muscle power, draft animals, or internal combustion engines. In some areas of the world all three types can be found in use within a single country. In Colombia, Andean Indians living in mountainous regions practice hand cultivation, family farmers in the lowlands employ domestic animals, and the large commercial farmers use tractors.

Man's achievements in agriculture are impressive. These principal technological advances, coupled with others not discussed here, have increased the earth's food-producing capacity several hundredfold since agriculture began. Despite these advances, hunger remains the daily lot of much of mankind.

Technological advances in agriculture and consequent increases in the food supply have permitted population to increase. Population increases in turn generate pressures for agricultural innovation. Thus trends of expanding food production and growing population have reinforced each other. Population growth continues to absorb all the increases in food production in the great majority of the poor countries, leaving little if any food for upgrading diets. The unequal distribution of income — of purchasing power with which to obtain a share of available food supplies — accentuates the problem.

The World's Food Needs

NUTRITION TODAY

There has been substantial progress in improving the nutritional state of the world over the last two decades. Hunger and malnutrition were widely prevalent in the decade following World War II. Estimates in the 1950s showed that a majority of all the people in most African, Latin American, and Asian nations (including Japan) were not consuming adequate amounts of calories or protein or both. Today, most of the countries of East Asia — Japan, South Korea, Taiwan, Hong Kong, Singapore, Malaysia — have achieved an adequate average food intake level, and have almost eliminated malnutrition. In western Asia, Israel and Lebanon are among the countries now adequately nourished. Within Latin America, Argentina, Uruguay, Chile, Brazil, and Mexico have achieved adequate *average* consumption levels, although in almost every case the distribution of food is very uneven, and the poorest fraction of the population continues to be malnourished.

Perhaps the most impressive gain among the developing countries has been China's apparent achievement of an adequate diet for its population of 800 million people. Although scores of scientists, journalists, economists, and doctors have visited China during the early 1970s, none have reported any of the obvious clinical signs of malnutrition that were prevalent in China less than a generation ago and that still exist in so many other developing countries. China's success appears to have been gained not so much by increases in per capita food production as by a more equitable distribution system and by the frugal use of available resources to produce the needed foodstuffs.

By far the most extensive remaining area of nutritional deficiency today is southern Asia — mainly India, Pakistan, and Bangladesh. In Latin America, severe remaining malnutrition is concentrated in northeastern Brazil, among the Andean Indians, among Mexico's landless laborers, and in parts of Central America. Malnutrition remains scattered in pockets throughout Africa, but it is currently most prevalent in the countries of the Sahelian zone, where outright starvation as well as severe malnutrition is now threatening the lives of many.

Probably 1 billion or more people suffer from serious hunger or malnutrition at least during part of the year. Infants, growing children, and pregnant women — the groups with the highest protein requirements — generally are the most malnourished elements of societies.

Available evidence suggests that such vitamin-deficiency diseases as rickets, scurvy, and beri-beri today remain a major problem in only a few areas of the world. However, vitamin A deficiency and iron-deficiency anemia continue to be significant nutrition problems throughout much of the world. Inadequate levels of vitamin A cause blindness or eye diseases for many millions in the developing nations. Anemia, which results in sluggishness, fatigue, and poor health, has been found to affect important portions of the population in many nations, rich and poor. Expectant mothers are particularly susceptible to anemia, which raises the chances of their death or disease from pregnancy and childbirth and also impairs the health of their newborn infants.

Man's food energy, or calorie, requirements vary with physique, with climate, and with level of physical activity. Calorie standards for an adequate daily diet range from an average of 2,300 calories per capita for the Far East to 2,700 per capita for Canada and the Soviet Union. People in most of the rich countries of North America, northern and eastern Europe, and parts of South America and Oceania consume between 3,000 and 3,200 calories daily.

Calorie intake, while a good quantitative indicator of diet adequacy, is not a good indicator of quality. Protein intake is the key indicator of diet quality in today's world. Proteins are essential to body growth and maintenance. Most people suffering from calorie malnutrition, and many with adequate caloric intake, suffer from protein malnutrition. Protein malnutrition is exacting an enormous toll in both the mental and physical development of a majority of the youngsters in the developing countries today.

This problem is not just a lack of protein per se, but a lack of protein of high quality such as that found in animal products (meat, milk, and eggs) or pulses (food legumes such as peas, beans, and soybeans). Protein quality, or usefulness to the human body, is determined by the combination of amino acids found in the food. If a key amino acid is missing, as is the case with grains, the protein is much less useful to the body. Conversely, the addition of amino acids to form the right combination, as when grains are combined with beans in a diet, results in much more usable protein than if either food is consumed in isolation.

The Joint Expert Committee of the Food and Agriculture Organization and World Health Organization has recently calculated the minimum protein requirement for an adult male weighing 65 kilograms (143 pounds) as ranging from 37 to 62 grams per day, depending on the quality of the protein consumed. Determining minimum protein needs is a very complex and controversial process, and some experts would place the minimum requirement for an active, healthy life many grams higher than these estimates.

Looking at national averages, which do not take into account internal distribution, daily protein intake ranges from more than 90 grams per day in many developed nations to about 40 in some of the poorest nations. Average consumption of animal protein, a useful indicator of the quality of available protein, varies even more, ranging from 40 to 60 grams per day in the developed nations to 6 to 20 grams in most developing nations.

THE TRAGEDY OF MALNUTRITION

In *The Nutrition Factor*[1] Alan Berg provides a graphic description of the myriad effects of malnutrition on those affected by it: "The light of curiosity absent from children's eyes. Twelve-year-olds with the physical stature of eight-year-olds. Youngsters who lack the energy to brush aside flies collecting about the sores on their faces. Agonizingly slow reflexes of adults crossing traffic. Thirty-year-old mothers who look sixty. All are common images in developing countries; all reflect inadequate nutrition; all have societal consequences."

Most children in the poor countries suffer from protein malnutrition at one time or another. Even children who received adequate protein while being breast fed often experience protein deficiencies after weaning because of what can be called the starchy food phenomenon: the transfer to a diet of easily digestible cheap starchy foods like cassava, cereals, and bananas. Advanced symptoms of protein malnutrition are well known: swollen bodies, peeling skin, reddish-brown brittle hair. What is less immediately visible is the enormous toll that protein malnutrition exacts on the development of the young.

Where death certificates are issued for preschool infants in the poor countries, death is generally attributed to measles, pneumonia, dysentery, or some other disease, when in fact these children were probably victims of malnutrition. Severely malnourished infants or children with low resistance frequently die of routine childhood diseases. The Food and Agriculture Organization states that "malnutrition is the biggest single contributor to child mortality in the developing countries." This contention is supported by the Pan American Health Organization's reports of studies in Latin America that show malnutrition to be the primary cause, or a major contributing factor, in 50 to 75 percent of the deaths of 1- to 4-year-olds.[2] In many of the poor countries of Asia, Africa, and Latin America, 50 percent of all deaths occur among children under 6 years of age. In Nigeria 180 of every 1,000 babies die before their first birthday. In India and Pakistan the number is 130 in every 1,000, and in Peru it is 110. Many others die before reaching school age and more die during early school years.

Widespread malnutrition reduces the productivity of a society. The effect of low food intake on the productivity of labor is easy to see. American construction firms operating in developing countries and employing local labor often find that they get higher returns in worker output by investing in a good company cafeteria that serves employees three meals a day.

The pervasive impact of undernourishment was dramatically illustrated during the summer of 1968, when India held its Olympic trials in New Delhi to select a track and field team to go to Mexico City. Despite its population of 525 million, not one of the contestants could meet minimum Olympic qualifying standards in any of the 32 track and field events. Outdated training techniques and the lack of public support were partly responsible, but widespread undernourishment undoubtedly contributed to this poor showing.

In contrast to India, the youth of Japan today are visible examples of the changes improved nutrition can bring. Well nourished from infancy as a result of Japan's newly acquired affluence, teenagers on the streets of Tokyo average perhaps 2 inches taller than their elders.

Protein is as important for mental development as it is for physical development. Protein shortages in the early years of life impair the growth of the brain and the central nervous system, thus preventing the realization of genetic potential and permanently reducing learning capacity. The relationship between nutrition and mental development is strikingly shown in a recent study conducted over several years in Mexico. An experimental group of 37 children who had been hospitalized for severe protein malnutrition before the age of 5 was found to average 13 points lower in intelligence quotient than a carefully selected control group that had not experienced severe malnutrition.[3]

Unfortunately, some of the effects of malnutrition in the early years of life are lasting and irreversible. No amount of compensatory feeding, education, or environmental improvement in later life can repair the damage to the central nervous system. Today protein shortages are depreciating the stock of human resources for at least a generation to come.

The problem of malnutrition is inseparable from the problem of poverty. Traditional food habits and lack of nutritional education, internal parasites and ecological constraints contribute to malnutrition; but these are in many ways simply additional manifestations of poverty. Even in the United States, malnutrition is often found where there is poverty — among migrant workers and many poor inhabitants of Appalachia, parts of the rural south, and urban slums. The connection is impossible to ignore in the poor countries. Where income is

low, cheap energy foods such as cassava and potatoes and cereals completely dominate the diet, often accounting for 60 to 80 percent of total calorie intake. The availability of protein-rich livestock products is usually quite low in these circumstances, and protein deficiencies are commonplace.

Ironically, while much of mankind is hungry, health authorities now point to obesity as a leading health problem in rich countries such as the United States. In these societies overeating has become a leading cause of heart disease and premature death. The problem is not just that most of mankind does not have enough food, but a significant and growing share of humanity has too much. If the world's rich were to reduce food expenditures 10 percent, making the resultant savings available to raise living standards among the world's poorest, the well-being of both rich and poor would improve.

The international distribution of income accentuates the problems of malnutrition in the poor countries in another way. Despite a shortage of protein in the diets of many citizens in the producing country, poor countries that produce protein-rich products such as beef, fish, or shrimp often export them to the richer countries, where consumers have the purchasing power to pay higher prices for animal products. Some developing nation economies are highly dependent on the foreign exchange brought in by protein exports, which in turn enables them to purchase foreign goods needed for development.

Thus in 1970 four South American nations — Argentina, Brazil, Paraguay, and Uruguay — exported more than 2 billion pounds of beef, nearly all of which was consumed in Western Europe and Japan. The same year, Mexico and the smaller nations of Central America supplied nearly 350 million pounds of beef to consumers in the United States. Altogether, over 90 percent of the red meats (beef and veal, pork, lamb and mutton, and horsemeat) entering world trade move into North America, Europe, and Japan.

In the 1960s Peru emerged as the world's leading fishing nation, owing to the escalating catch of anchovies off its coast, topping 10 million metric tons in some years. The anchovies are ground into high-protein fish meal, and then exported to Europe, Japan, and the United States, where they are fed to poultry and pigs. Looking at international trade in all fisheries products, including shellfish and table-grade fish as well as fish meal, the world's three poorest regions — Asia, Latin America, and Africa — all export more to the developed nations than they receive as imports.

POPULATION AND AFFLUENCE

During the 1960s and before, the world problem was generally perceived as a food-population problem, a race between food and people. At the end of each year observers anxiously compared rates of increase in food production with those of population growth to see if any progress was being made. Throughout most of the decade it was nip and tuck. During the 1970s rapid global population growth continues to generate demand for more food, but, in addition, rising affluence is emerging as a major new claimant on world food resources. Historically, there was only one important source of growth in world demand for food; there are now two.

At the global level, population growth is still the dominant cause of an increasing demand for food. With world population expanding at almost 2 percent per year, merely maintaining current per capita consumption levels will require a doubling of food production over the next generation. In demographic terms the world currently divides essentially into two groups of countries: the rich countries, which have low or declining rates of population growth, and the poor countries, most of which have rapid rates of population growth. Fully four-fifths of the annual increment in world population of an estimated 70 million occurs in the poor countries.

Some of the comparatively small poor countries add more to the world's annual population gain than the larger rich ones. Mexico, for example, now contributes more to world population growth than does the United States. The Philippines adds more people each year than does Japan. Brazil adds 2.8 million people in a year, whereas the Soviet Union adds only 2.2 million.

The effect of rising affluence on the world demand for food is perhaps best understood by examining its effect on grain requirements. Grain consumed directly provides 52 percent

of man's food energy supply. [...] rectly in the form of livest[...] provides a sizable share of [...] resource terms, grains occ[...] percent of the world's crop [...]

In the poor countries the annual a[...] of grain per person averages only about 40[...] pounds per year. Nearly all of this small amount must be consumed directly to meet minimum energy needs. Little can be spared for conversion into animal protein.

In the United States and Canada, per capita grain utilization is currently approaching 1 ton per year. Of this total only about 200 pounds are consumed directly in the form of bread, pastries, and breakfast cereals. The remainder is consumed indirectly in the form of meat, milk, and eggs. The agricultural resources — land, water, fertilizer — required to support an average North American are nearly five times those of the average Indian, Nigerian, or Colombian (see Figure 1). This is due to the fact that animals consume several pounds of grain for every pound of meat they produce. The production of a pound of beef on a feedlot requires roughly 7 pounds of grain; of pork, 4 pounds of grain; and of poultry, from 2 to 3 pounds of grain.

Throughout the world, per capita grain requirements rise with income. The amount of grain consumed directly rises until per capita income approaches $500 per year, whereupon it begins to decline, eventually leveling off at about 200 pounds. The total amount of grain consumed directly and indirectly, however, continues to rise rapidly as per capita income climbs.

The impact of rising affluence on the consumption of livestock products is evident in trends in the United States over the past generation. For example, per capita consumption of beef climbed from 55 pounds in 1940 to 116 pounds in 1972, more than doubling. Poultry consumption roughly tripled, growing from 18 pounds to 51 pounds during the same period.

There is now a northern tier of industrial countries — beginning with the United Kingdom in the West and including Scandinavia, Western Europe, Eastern Europe, the Soviet Union, and Japan — whose dietary habits more-or-less approximate those of the United States in 1940. As incomes continue to rise in this group of countries, which contain about

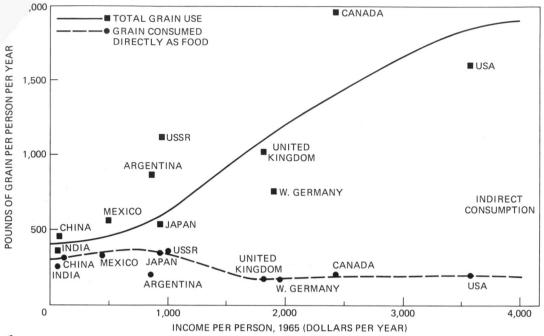

1 Direct and Indirect Grain Consumption. As incomes rise, per capita grain consumption rises rapidly. Grain consumed directly tends to be lower in the rich than in the poor countries, but the amount of grain consumed indirectly in the form of meat, milk, and eggs grows rapidly. *Source:* Reference 4.

two-thirds of a billion people, a sizable share of the additional income is being converted into demand for livestock products, particularly beef. Many of these countries, such as Japan and those in Western Europe, are densely populated. Others — the Soviet Union, for example — suffer from a scarcity of fresh water. Thus they lack the capacity to satisfy the growth in demand for livestock products entirely from indigenous resources. As a result, they are importing increasing amounts of livestock products or of feedgrains and soybeans with which to expand their livestock production.

Throughout the poor countries population growth accounts for most of the year-to-year growth in the demand for food. At best only very limited progress is being made in raising per capita consumption. In the more affluent countries, on the other hand, rising incomes account for most of the growth in the demand for food. In Japan and France, for example, where population is growing at about 1 percent annually and per capita incomes at several percent, year-to-year growth in the de-

mand for food derives principally from rising affluence.

Wherever population has stopped growing, as in West Germany, rising affluence accounts for all the growth in food consumption. In a country such as India, however, where income rises are scarcely perceptible and population continues to grow rapidly, nearly all the growth in demand derives from population growth. In Brazil, which has both rapid population growth and, in recent years at least, rapid growth in per capita income, both factors loom large in the growth in demand for food.

Expanding Food Supplies

EXPANSION OF AREA VERSUS
INCREASED YIELD

There are essentially two ways of expanding the world food supply from conventional agriculture. One is to expand the area under cultivation. The other, largely made possible by advances in the use of agricultural chemicals and in plant genetics, is to raise output on the existing cultivated area. From the beginning of agriculture until about 1950, expanding the cultivated area was the major means of increasing the world's food supply. Since mid-century, however, raising the yield on existing cultivated area has accounted for most of the increase. Intensification of cultivation has increased steadily since 1950; during the early 1970s it has accounted for an estimated four-fifths of the annual growth in world food output, far overshadowing expansion of the cultivated area.

The timing of the transition from the area-expanding method of increasing food production to the yield-raising method has been very uneven throughout the world. As population pressures built up in Japan in the late nineteenth and early twentieth centuries, the Japanese were forced to intensify cultivation and were very likely the first people to succeed in making the transition. Ironically, some of the technologies used by the Japanese in attaining a yield-per-acre takeoff, notably the use of chemical fertilizer, were introduced from Europe. Several northern European countries in which available farmland was limited were close behind and also successfully completed the transition in the early twentieth century.

In the United States the frontier had vanished just before World War I, but a yield-per-acre takeoff was not achieved until World War II, when economic conditions made the use of the accumulated technologies profitable. During the intervening period, farm output lagged, but land used to produce feed for horses was released for production when tractors began to displace horses after World War I. A second group of industrial countries, including the United Kingdom, Canada, and Australia, also made the area-to-yield transition during the early 1940s, largely as a result of the strong wartime economic incentives to utilize already available technologies to expand production. Interestingly, U.S. corn yields had remained static for nearly a century between the Civil War and World War II.

Some estimates indicate that the world's cultivated area can be doubled. But much of the potentially cultivable land is in the tropics. Experience indicates that the farming of tropical soils is often not economically feasible. The soils (which are generally not very fertile to begin with) and the protective forest cover form a fragile ecological system. Organic materials in the soil decay very rapidly in the tropical climate, and once the abundant source of new vegetative matter, the forest above, is removed, the soils often lose whatever fertility they had. Thus farming may require extremely large applications of chemical fertilizer. In addition, some tropical soils, when fully exposed to the sun and oxygen, undergo chemical changes that result in the formation of a rocklike substance called laterite, which is too hard to farm. Within a few notable exceptions, such as selected areas in Africa and the Amazon Basin in Brazil, the world's most productive farmlands are already under cultivation. At the global level, the record of the past two decades suggests that it is far cheaper and easier to expand the food supply by intensifying cultivation on the existing cropland area than by bringing new land under the plow.

During the first half of the 1960s food production in the developing countries as a group began to fall behind population growth, bringing rising food prices, growing food scarcity, and increasing dependence on food aid from the United States. The food crisis in these countries that occurred during the early and mid-1960s was brought into sharp focus by the two consecutive monsoon failures in the Indian subcontinent in 1965 and 1966, which required successive shipments to India of nearly 10 million metric tons each year. It was, however, a direct result of the fact that many of these countries had virtually exhausted the supply of new land that could be readily brought under cultivation but had not yet achieved a takeoff in yield per acre. It was in this context that the Green Revolution — the combination of new cereal technologies and production-oriented economic incentives — came into being.

THE GREEN REVOLUTION[5]

Efforts to modernize agriculture in the poor countries in the 1950s and early 1960s were consistently frustrated. When farmers in these countries attempted to use varieties of corn developed in Iowa, they often failed to produce any corn at all. Japanese rice varieties were not suited either to local cultural practices or to consumer tastes in India. When fertilizer was applied intensively to local cereal varieties, the yield response was limited and occasionally even negative.

It was against this backdrop of frustration that the high-yielding dwarf wheats were developed by the Rockefeller Foundation team in Mexico. Three unique characteristics of these wheats endeared them to farmers in many countries — their fertilizer responsiveness, lack of photoperiod sensitivity (sensitivity to day length), and early maturity.

When farmers applied more than 40 pounds of nitrogen fertilizer per acre to traditional varieties having tall, thin straw, the wheat often fell over (an occurrence called lodging), causing severe crop losses. By contrast, yields of the short, stiff-strawed dwarf varieties of Mexican wheat would continue to rise with nitrogen applications up to 120 pounds per acre. Given the necessary fertilizer and water and the appropriate management, farmers could easily double the yields of indigenous varieties.

Beyond this, the reduced sensitivity of dwarf varieties to day length permitted them to be moved around the world over a wide range of latitudes, stretching from Mexico, which lies partly in the tropics, to Turkey in the temperate zone. Because the biological clocks of the new wheats were much less sensitive than those of the traditional ones, planting dates were much more flexible.

Another advantageous characteristic of the new wheats was their early maturity. They were ready for harvest within 120 days after planting; the traditional varieties took 150 days or more. This trait, combined with reduced sensitivity to day length, created broad new opportunities for multiple cropping wherever water supplies were sufficient.

I   Plantings of High-Yielding Wheat and Rice

Year	Acres
1965	200
1966	41,000
1967	4,047,000
1968	16,660,000
1969	31,319,000
1970	43,914,000
1971	50,549,000
1972	68,000,000
1973 (prelim.)	80,200,000

Within a few years after the spectacular breakthrough with wheat in Mexico, the Ford Foundation joined the Rockefeller Foundation to establish the International Rice Research Institute (IRRI) in the Philippines. Its purpose was to attempt to breed a fertilizer-responsive, early-maturing rice capable of wide adaptation — in effect, a counterpart of the high-yielding wheats. With the wheat experience to draw upon, agricultural scientists at IRRI struck pay dirt quickly. Within a few years they released the first of the high-yielding dwarf rices, a variety known as IR-8. Since the late 1960s many dozens of high-yielding dwarf wheat and rice varieties, often bred to meet particular local growing conditions, have been developed.

The great advantage of the new seeds was that they permitted developing countries to quickly utilize agricultural research that had taken decades to complete in the United States, Japan, and elsewhere. In those areas of the developing countries where there were requisite supplies of water and fertilizer and price incentives were offered, the spread of the high-yielding varieties of wheat and rice was rapid. Farmers assumed to be bound by tradition were quick to adopt the new seeds when it was obviously profitable for them to do so.

Early in 1968, the term "Green Revolution" was coined by William Gaud, Administrator of the U.S. Agency for International Development, to describe the introduction and rapid spread of the high-yielding wheats and rices. In 1965 land planted with these new varieties in Asia totaled about 200 acres, largely trial and demonstration plots. Thereafter the acreage spread swiftly, as indicated in Table 1.

Acreage figures for Mexico are not included in Table I since the new seeds had largely displaced traditional varieties before the Green Revolution became an international phenomenon in the mid-1960s. Among the principal Asian countries to benefit from using the new seeds are India, Pakistan, Turkey, The Philippines, Indonesia, Malaysia, and Sri Lanka.

Compared with the historical record of dissemination of other major advances in agricultural technology, these figures reveal a phenomenally rapid spread of the new seeds. By 1973 high-yielding varieties accounted for 35 percent of the total wheat area, and 20 percent of the total rice area, of noncommunist

Asia. However, in each case they are planted on a much higher proportion of the lands physically suited to their use. Nevertheless, considerable opportunities for their further spread remain in many developing countries, especially in Latin America and Africa, if more progress can be made in providing adequate water supplies and control, making fertilizers and markets available to farmers, teaching farmers new techniques, and carrying out research to develop better seeds adapted closely to conditions in each locale.

During the late 1960s The Philippines was able to achieve self-sufficiency in rice, ending a half-century of dependence on imported rice. Unfortunately, this situation was not sustained because of a number of factors, including civil unrest, the susceptibility of the new rices to disease, and the failure of the government to continue the essential support of the rice program (See Table II and Figure 2).

Pakistan greatly increased its wheat production, emerging as a net exporter of grain in recent years. In India, where advances in the new varieties have been concentrated largely in wheat, progress has been encouraging. During the seven-year span from 1965 to 1972,

India expanded its wheat production from 11 million to 26 million tons, an increase in a major crop unmatched by any other country in history. Unfortunately, the production of rice, India's major crop, has not risen so dramatically, partly because the necessary control of water supply and drainage has often not been available in rice-growing regions, and because rice breeders have been less successful at developing high-yielding varieties well adapted to Indian conditions (see Table II and Figure 3).

One result of the dramatic advance in wheat production in India was the accumulation of unprecedented cereal reserves and the attainment of cereal self-sufficiency in 1972. This eliminated, at least temporarily, the need for imports into a country that only a few years before had been the principal recipient of U.S. food aid. Economic self-sufficiency in cereals — when farmers produce as much as consumers can afford at prevailing prices — is not to be confused with nutritional self-sufficiency, however, which requires much higher levels of productivity and purchasing power.

During late 1971 and in 1972, India was able to use nearly 2 million tons from its own food

II Impact of Production Using New Seeds

Year	Philippines Rice	India Wheat	India All cereals	Pakistan All cereals	Mexico All cereals
1960	88	25	164	136	202
1961	90	26	162	135	205
1962	88	27	161	137	197
1963	82	24	159	138	232
1964	83	21	160	140	251
1965	82	26	136	142	265
1966	80	21	134	127	267
1967	78	23	154	139	280
1968	81	32	162	178	274
1969	92	36	165	184	258
1970	91	37	173	190	286
1971	84	43	166	171	262
1972	78	47	149	174	226
1973 (prelim.)	88	43	165	180	272

Annual production of selected cereals in four countries using new seeds, in kilograms per person of total population. High-yielding varieties enabled major food production increases in the late 1960s, reversing what had been a deteriorating trend in food production per person in several large Asian nations. However, in each of these nations, continuing rapid population growth has to a great extent absorbed the growth in output, holding down needed gains in per capita food supplies. In India, wheat production has grown much more rapidly than population, but total grain production is struggling to keep pace with population growth.

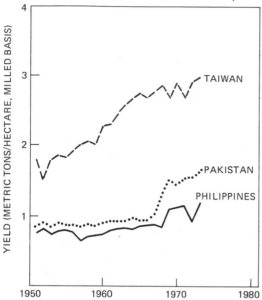

2 **Rice Yields in Taiwan, Pakistan, and The Philippines, 1951–1973.** Since the introduction of new high-yielding rice varieties, in the late 1900s, average yields per acre in Pakistan and The Philippines have risen much more rapidly than they had previously. Poor weather and flooding helped push down yields in 1972, however. The farmers of Taiwan have used higher-yielding seeds and improved agricultural practices for many decades. *Source:* FAO and U.S. Department of Agriculture.

reserves, initially to feed nearly 10 million Bengali refugees during the civil war in East Pakistan, and later as food aid for Bangladesh. A poor monsoon in 1972 temporarily forced India back into the world market as an importer of grain, but on a much smaller scale — 4 million tons — than the massive import of nearly 10 million tons that followed the 1965 monsoon failure. The shortage of fertilizer appearing in 1974 made additional Indian grain import needs a certainty.

It would be wrong to suggest that the Green Revolution has solved the world's food problems, either on a short- or a long-term basis. The droughts of 1972 clearly demonstrated that Indian agriculture is still at the mercy of the vagaries of weather. A lack of adequate rainfall resulted in reduced harvests in many nations in 1972; similarly, favorable weather conditions in most areas of the world in 1973

helped boost production substantially to record totals.

The Green Revolution can be properly assessed only when we ask what things would have been like in its absence. The grim scenario that this question calls forth lends some of the needed perspective. Increases in cereal production made possible by the new seeds did arrest the deteriorating trend in per capita food production in the developing countries. But although there have been some spectacular localized successes in raising cereal output, relatively little progress has been made in raising the per capita production of cereals among the poor countries as a whole over the past several years (see Figure 4).

Many of those involved closely with the Green Revolution have stressed from the beginning that the new technologies embodied in the Green Revolution did not represent a solution to the food problem. Rather it has been viewed as a means of buying time, perhaps an additional 15 or 20 years during

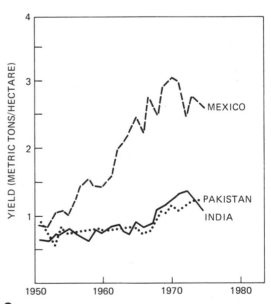

3 **Wheat Yields in Mexico, India, and Pakistan, 1951–1974.** The use of new high-yielding wheat varieties by many farmers in India and Pakistan has helped boost average wheat yields per acre in these nations. High-yielding wheats were introduced in Mexico in the early 1950s, and wheat yields have climbed substantially since then. *Source:* FAO and U.S. Department of Agriculture.

78

4 **Total and Per Capita Food Production,**
1961–1973. Total world food production (a) has increased considerably since the early 1960s. Owing to
world population growth, however, the increase in
annual production per person has increased much less
rapidly. In the economically advanced nations (b)
where populations are growing relatively slowly, there
have been substantial gains in food production per
person, allowing dietary improvements. In the less-
developed nations (c), despite a very rapid growth in
total food production, the availability of food in per
capita terms has only improved slightly, owing to the
very high average rate of population growth. *Source:*
U.S. Department of Agriculture.
[1]Excludes Communist Asia.
[2]North America, Europe, USSR, Japan, Republic of
South Africa, Australia, and New Zealand.
[3]Latin America, Asia (except Japan and Communist
Asia), and Africa (except Republic of South Africa).

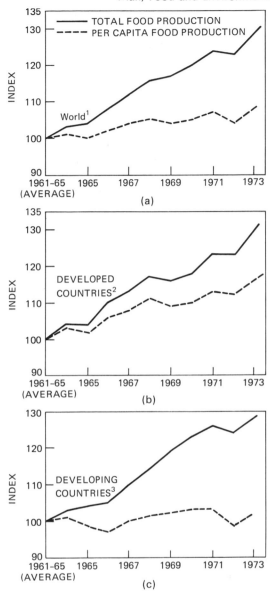

which the brakes could be applied to population growth.

Several years have passed, however, since
the Green Revolution began, but success
stories in national family planning programs in
the poor nations are all too few. Among the
population giants of Asia, China appears to be
substantially reducing its birth rates, and limited progress has been made in India, but
reductions in Indonesia, Pakistan, and Bangladesh are scarcely perceptible. Without a
sharply accelerated effort to curb birth rates,
the battle to feed Asia's massive and rapidly
growing population may yet be lost.

Another question persistently raised concerning the Green Revolution is: Who benefits
from the adoption of these new technologies?
The new seeds can be used with equal success
by both large- and small-farm owners if the
farmers have equal access to the requisite inputs and supporting services. In many countries and locales, however, the owners of large
farms have easier access to credit and to technical advisory services than the owners of
smaller ones. Where these circumstances prevail, there is a disturbing tendency for the rich
farmers to get richer and the poor ones poorer.

But many other factors determine which
farmers will benefit from the new seeds. One
is the crop they grow. Widely adapted, high-
yielding varieties exist only for wheat and rice.
Thus, in Mexico, most wheat farmers have
benefitted greatly, but corn farmers, most of

them small subsistence farmers, have gained
little from technological progress. Likewise, in
India the principal beneficiaries of the Green
Revolution are the wheat farmers, because
successful adoption of the high-yielding rices
has been very modest by comparison.

Perhaps the most important single factor
determining whether a given farmer can use
the new seeds is whether or not he has an
adequate supply of water. Wheat farmers on
the high-rainfall coastal plain of Turkey have
benefited enormously from the new seeds;

those on the arid Anatolian Plateau have scarcely been touched.

The impact of Green Revolution technology on unemployment also is often questioned. If properly managed, this technology can be highly beneficial. The new high-yielding varieties of wheat and rice require much more labor per acre than the traditional varieties they replace. Realizing the full yield potential of the new seeds requires frequent fertilization and irrigation. This, in turn, requires careful and frequent weeding lest the fertilizer and water be converted into weeds rather than food. Higher yields require more labor at harvest time.

The risk is that farmers profiting from the use of the new seeds will want to invest their profits in Western-style mechanization. This tendency may be aggravated by low, subsidized interest rates on agricultural loans for farm mechanization. Rates that are too low, which is often the case, encourage farmers to substitute machinery for labor rather than to use the maximum amount of labor.[6]

One disconcerting trend in several Asian nations over the last decade has been a decline in the acreage planted to protein-rich pulses. It is felt by some observers that improvements in the total *quantity* of food available have been accompanied by a worsening in the nutritional *quality* of food supplies in many areas, and that the availability of high-yielding grain varieties may have accelerated that trend by making it more attractive for farmers to grow grains than pulses. The evidence on this matter is not clear, however. For one thing, the trend toward declining pulse acreage appears to have begun many years prior to the Green Revolution. Second, where the increase in per capita supplies of grain has been large enough, consumers may still have ended up with more total protein in their diets than they previously had. In any case, there is clearly an urgent need for research to make pulses more productive and hence more attractive for farmers in developing nations. Pulses have a protein content double that of wheat, triple that of milled rice, and 25 times that of cassava, and even greater advantages in terms of protein quality. Any advancements in their production and consumption would have highly beneficial results in poorly nourished regions.

Future Production Prospects

CONSTRAINTS ON EXPANDING THE FOOD SUPPLY

The prospects for expanding the food supply depend on a wide range of economic, ecological, and technological factors. The traditional approach to increasing production — expanding the area under cultivation — has only limited scope for the future. Indeed, some parts of the world face a net reduction in agricultural land because of the growth in competing uses, such as industrial development, recreation, transportation, and residential development. Few countries have well-defined land-use policies that protect agricultural land from other uses. In the United States, farmland has been used indiscriminately for other purposes with little thought to the possible long-term consequences.

Some more densely populated countries, such as Japan and several in Western Europe, have been experiencing a reduction in the land used for crop production for the past few decades. This trend is continuing and may well accelerate. Other parts of the world, including particularly the Indian subcontinent, the Middle East, North and sub-Sahara Africa, the Caribbean, Central America, and the Andean countries, are losing disturbingly large acreages of cropland each year because of severe erosion.

The area of land available for food production is important, but perhaps even more important in the future will be the availability of water for agricultural purposes. In many regions of the world, fertile agricultural land is available if water can be found to make it produce. Yet most of the rivers that lend themselves to damming and to irrigation have already been developed. Future efforts to expand freshwater supplies for agricultural purposes will increasingly focus on such techniques as the diversion of rivers (as in the Soviet Union), desalting sea water, and the manipulation of rainfall patterns to increase the share of rain falling over moisture-deficient agricultural areas (Chapter 6).

During the mid-1960s many proposed that nuclear power be used to operate gigantic desalting complexes which would be sited in desert areas close to the oceans. These nuclear-powered, agro-industrial complexes, it was argued, could measurably increase the world food supply. Unfortunately, the economics have not yet proved to be attractive. With the prices of energy — the principal cost in present desalination technologies — rising rapidly, desalination on a large scale does not yet appear to be economically feasible.

One of the disturbing questions associated with future gains in agricultural production is the extent to which the trend of rising per acre yields of cereals in the more advanced countries can be sustained. In some countries increases in per acre yields are beginning to slow down, and the capital investments required for each additional increase may now start to climb sharply. With any crop, the biggest gains in yield per acre come to the lower levels of fertilizer application. When increasing amounts of fertilizer are applied, the gains in yield from each *additional* unit of fertilizer begin diminishing, until eventually an upper limit for the crop is reached and yields go no higher (see Figure 5).

In agriculturally advanced countries, such as Japan, The Netherlands, and the United States, the cost per additional increment of yield per acre for some crops is already rising. For example, raising rice yields in Japan from the current 5,000 pounds per acre to 6,000 pounds would be very costly. Raising yields of corn in the United States from 90 to 100 bushels per acre is requiring a much larger quantity of nitrogen fertilizer than was needed to raise yields from 50 to 60 bushels. By contrast, many farmers in the poor countries are using very little or no fertilizer in their fields. The introduction of moderate amounts of chemical fertilizer on these fields would generally bring substantial production gains.

Rising energy prices, and the prospects for continuing global scarcity of fossil fuels, have ominous implications for future food supplies and prices. With a substantial rise in the cost of energy, farmers already engaged in high-energy agriculture, as in the United States and Japan, will tend to use less, thus perhaps holding future production increases below current expectations. Modern food-producing systems are extremely dependent on energy from

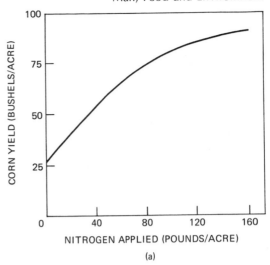

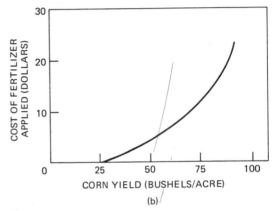

5 **Fertilizer Response Curve.** (a) When initial amounts of fertilizer are applied, yields increase rapidly, but as greater and greater amounts of fertilizer are added, the gain in yield diminishes. (b) At the same time, the amount of money a farmer must spend on fertilizer to get an additional increase in yield rises with each successive yield increase. These graphs are for corn in the United States in the early 1960s. Improvements in seeds or other farming practices may raise average yields, but for any crop at any given level of technology, the response curve resembles that of graph (a). *Source:* Reference 7.

fossil fuels for planting and harvesting crops, irrigation and the application of pesticides, and the transportation of inputs to farmers and final products to consumers. Critically needed increases in agricultural production in many developing nations, including wider use of new high-yielding varieties, will require greatly expanded inputs of energy per acre,

81

and higher energy prices will make production gains much more expensive.

Chemical fertilizers, a crucial input in modern agriculture, are a very energy-intensive product. One of the dominant costs in the process of manufacturing chemical fertilizers is energy, and in the case of the most widely used chemical nutrient, nitrogen fertilizer, natural gas or petroleum derivatives constitute the principal raw material for the production process. Fertilizer, and thus food prices, then, are closely interlinked with problems of world energy scarcity.

Dangerously complicating matters, by 1974 a worldwide shortage of chemical fertilizers had developed. The price of many important kinds of fertilizer more than doubled, and even more ominous were reports that many large Asian nations, such as China, India, Indonesia, and The Philippines, would be unable to buy needed amounts of fertilizer at any price — a development certain to cut their food production. The shortage of the mid-1970s is primarily due to an unfortunate lag in the construction of new production facilities. But even when construction catches up to demand — perhaps by the late 1970s — fertilizer prices will remain high, owing to the high cost of essential energy inputs.

It is impossible to predict precisely the level of chemical fertilizer use that will be required to meet the world's growing demand for food. Little is known about the fertilizer responsiveness of crops in the scientifically uncharted growing conditions of much of the developing world, where fertilizer use will have to grow rapidly. But in most cases the best farmland is already in production, and it seems certain that production on the increasingly marginal lands, which must be brought into production in the future, will require higher applications of fertilizer for a given yield than has been the case on more fertile lands. For many crops in the more agriculturally advanced nations, as noted previously, increasingly greater increments of fertilizer are already necessary to sustain increases in yields per acre.

If the United Nations' median population projection of 6.5 billion materializes in the year 2000, a *doubling* of current global food production would be necessary to meet the demands of population growth and modest dietary improvements. As a result, the total amount of chemical fertilizers required at century's end would be more than four times the 80 million tons being used today (Figure 6).

Assuming that the massive capital investments and energy supplies required to produce these levels of fertilizer are forthcoming, the high prices that appear inevitable for future supplies of chemical fertilizers must be regarded as an important constraint on the expansion of food production. In addition, as chemical fertilizer use grows, the eutrophication of freshwater lakes and streams will become a far more serious problem than it is at the present time, forcing societies to make very difficult choices among environmental and other social values.

CONSTRAINTS ON PROTEIN PRODUCTION

In looking ahead, one must be particularly concerned about the difficulties in expanding the world food supply of high-quality protein to meet the projected rapid growth in demand, which is now being fueled both by population growth and rising affluence. At present mankind is faced with a number of technological and other constraints in increasing the supply of three principal sources of protein.

One important source of protein is beef. Here two constraints are operative. Agricultural scientists have not been able to devise any commercially viable means of getting more than one calf per cow per year. For every animal that goes into the beef production process, one adult must be fed and otherwise maintained for a full year. There does not appear to be any prospect of an imminent breakthrough on this front. The other constraint on beef production is that the grazing capacity of much of the world's pastureland is now almost fully utilized. This is true, for example, in most of the U.S. Great Plains area, in eastern Africa, and in parts of Australia.

A second potentially serious constraint on efforts to expand supplies of high-quality protein is the inability of scientists to achieve a breakthrough in per acre yields of soybeans. Soybeans are a major source of high-quality protein for livestock and poultry throughout much of the world and are consumed directly as food by perhaps 1 billion people throughout densely populated eastern Asia. The importance of soybeans in the world food economy is indicated by the fact that they have become

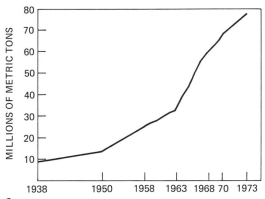

6 World Consumption of Chemical Fertilizers, 1938–1973. The use of chemical fertilizers has grown rapidly worldwide since 1950, as farmers in many countries have concentrated on raising output per acre. World fertilizer requirements will soar in the decades ahead, but the high cost of energy, which is an important input in the production of many important kinds of fertilizer, will make fertilizers more expensive to use. *Source:* Reference 8.

the leading export product of the United States. Soybeans are surpassing export sales of wheat, corn, and high-technology items such as electronic computers and jet aircraft.

In the United States, which now produces two-thirds of the world's soybean crop and supplies over 80 percent of all soybeans entering the world market, soybean yields per acre have increased by a very small amount — about 1 percent per year since 1950; corn yields, by comparison, have increased by nearly 4 percent per year. One reason soybean yields have not climbed very rapidly is that the soybean is not very responsive to nitrogen fertilizer. It belongs to the legume family, which, unlike most plant families, hosts bacteria that extract nitrogen from the air. The way the United States produces more soybeans is by planting more soybean acreage. Close to 85 percent of the dramatic fourfold increase in the U.S. soybean crop since 1950 has come from expanding the area devoted to it. As long as there was ample idle cropland available, this did not pose a problem, but if this cropland reserve continues to diminish or disappear entirely it could create serious global supply problems (see Figure 8).

The oceans are a third major source of high-quality protein. From 1950 to 1970 the world fish catch climbed steadily, tripling from 21 million tons to 70 million tons. The average annual increase in the catch of nearly 5 percent, which far exceeded the annual rate of world population growth, greatly increased the average supply of marine protein per person. Suddenly, in 1971, the long period of sustained growth was interrupted by a decline in the catch, and it has fallen for 3 straight years. Meanwhile, the amount of time and money expended to bring in the catch continue to rise every year. Many marine biologists now feel that the global catch may be at or near the maximum sustainable level. Many of the leading species of commercial-grade fish may currently be overfished; that is, stocks will not sustain even the current level of catch (see Chapter 7).

World fishery resources represent an important source of protein. The 1971 catch of 69 million tons amounted to nearly 40 pounds (live weight) per person throughout the world. Of this catch roughly 60 percent was table-grade fish, and the remainder consisted of inferior species used for manufacturing fish meal, which in turn is used in poultry and pig feed in the industrial countries. For Americans, fish are important in the diet, but direct consumption is only about 13 pounds per year, compared with about 230 pounds of beef, pork, and poultry in 1973. These data on direct consumption understate the importance of fish in the diet because a significant share of the poultry and pork consumed are produced with fish meal.

In two of the world's more populous countries, Japan and the Soviet Union, fish are much more significant as a direct source of protein in the national diet. Accordingly, as growth in the world catch of table-grade fish slows and, for some important species, begins to decline from overfishing, the Soviet and Japanese populations are particularly vulnerable. If they find themselves increasingly unable to meet protein needs from oceanic resources, they may be forced to offset this decline by substantially stepping up imports of feedgrains and soybeans to expand their indigenous livestock production.

In examining world fishery prospects, the possibility of aquaculture, or fish farming, must always be considered. There have been local successes in various forms of aquaculture in both fresh water and salt water in a number of countries throughout the world, including Norway, Germany, Israel, the Soviet Union, China, Japan, Australia, and the United States. At present, however, fish farming is not able to provide more than a small percentage, at most, of the world's fish supplies. This is not to rule out the possibility a decade or so hence of sufficient expansion in this activity to enable it to become an important global source of protein. Active research programs in aquaculture continue in a large number of countries.

A close examination of the extent of overfishing and stock depletion in many of the world's fisheries underlines the urgency of evolving a cooperative global approach to the management of oceanic fisheries. Failure to do this will result in a continuing depletion of stocks, a reduction in catch, and soaring seafood prices. Pressures on land-based protein sources will rise commensurately. It is in this context that all nations have a direct interest in the success of the Law of the Sea Conference, which was held by the United Nations in 1974. Among other things, delegates to the conference dealt with the need to devise an institutional mechanism for cooperatively managing global fishery resources.

Although there are substantial opportunities for expanding the world's protein supply, it now seems likely that the supply of animal protein will lag behind growth in demand for some time to come, resulting in significantly higher prices for livestock products during the remainder of the 1970s than prevailed during the 1960s. The world protein market is being transformed from a buyer's market to a seller's market, much as the world energy market has been transformed over the past few years.

NEW FOODS

Over the last decade many nonconventional techniques have been set forth for expanding world food supplies and improving nutrition in the poor countries. However, few have had the impact for which their proponents hoped. New foods must not only be technologically feasible, but also economically feasible, and most importantly, people must be willing to eat them.

Among past proposals was the possibility of using many species of fish, most of which are not considered edible, to produce fish protein concentrate. This protein concentrate could be added to conventional dishes of various kinds to provide a low-cost method of increasing protein intake and improving nutrition. Unfortunately, many technological problems with this approach remain, and as of 1974 it is receiving only limited attention in either the scientific or food-processing communities.

A second widely discussed possibility for augmenting human food supplies is that of using single-celled microorganisms, principally certain strains of yeast, to convert petroleum or industrial waste products into edible forms of protein known as single-cell protein (SCP). Thus far the hopes for using this process have not materialized on a significant scale. British Petroleum, the leader in this field, has been operating a small pilot plant in France, and has future plans for producing as much as 100,000 tons of protein for animal feed annually. Some SCP is being produced in the Soviet Union. However, the plans of two Japanese chemical firms to produce a total of 150,000 tons of protein yearly were recently scrapped, primarily owing to extreme resistance on the part of Japanese consumers, who felt that even the indirect consumption of protein produced in this fashion might not be medically safe.

Technical problems in the production of SCP have proved to be formidable, and clearance for marketing has been achieved only for feed-grade protein to be used in livestock feed. Serious problems must yet be overcome before this protein can be consumed directly by human beings. If world prices of high-quality protein remain high, as currently appears very likely, the production of SCP for animal feeds may grow rapidly.

The fortification of existing foods with needed vitamins, minerals, and amino acids (which determine protein quality) often represents an inexpensive way to improve nutrition. In the United States, most flour and bread sold since World War II has been enriched with vitamins and minerals. This has helped eliminate many vitamin-deficiency diseases that were prevalent in the country prior to that time.

Adding 4 pounds of the amino acid lysine to a ton of wheat costs only $4 but results in one-third more usable protein. Additions of essential amino acids to plant protein can make it equal in quality to animal protein. Government bakeries in Bombay, India, are now fortifying wheat flour with lysine, as well as with vitamins and minerals. Bread made with this flour may be the most nutritious marketed anywhere. By the early 1970s, very few developing nation governments had taken advantage of the nutritional improvement opportunities offered by fortification of such staples as flour, salt, and tea.

A principal advantage of fortification is that no changes are required in the eating habits of those consuming the enriched foods. A major disadvantage, however, is that it can only help the diets of those who purchase a high proportion of their food — mainly urban residents.

Paradoxically, in many poor countries, widespread protein hunger coexists with vast quantities of unused protein meals, largely the product of local vegetable-oil-extraction industries. In India, Nigeria, and several smaller countries, literally millions of tons of peanuts are produced for their oil, which is used in cooking. Other countries use coconut or soybean oil. After the seeds are crushed and the oil extracted, the meal remaining is largely protein. Unfortunately, little of this protein finds its way directly into foodstuffs, since most oil meal is fed to livestock or poultry, used for organic fertilizer or exported to earn foreign exchange. If the estimated 20 million tons of peanut, cottonseed, coconut, and soybean meals available each year in the poor countries could be made into attractive, commercially successful protein foods, much of the world's protein hunger would be eliminated.

Some successful new high-protein products are beginning to appear. Prominent among these are popular beverages being developed by several private firms using oilseed meals. Vitasoy, a soy-based beverage manufactured and marketed in Hong Kong, has captured one-fourth of the soft-drink market there. Other high-protein beverages are being marketed in Asia and Latin America, with mixed commercial success.

Currently pilot projects and experimentation are under way in a variety of other areas, including the recycling of animal wastes and the nutritional improvement of grain varieties. For example, corn varieties with higher lysine content, and hence higher protein quality, were developed in the early 1960s. They have not yet been widely adopted because the yields of the new seeds have often not been as

high as those of seeds already in use, but research to increase the yields continues. A more nutritional variety of sorghum, which is an important food crop in many arid regions of the world, where the plant grows well, was discovered in 1973. If the seeds prove feasible for wide use, they could help improve nutrition in such areas at a future time. Efforts are also underway to raise the protein content of wheat and rice. A 1973 breakthrough in the treatment of cotton seeds, removing a toxic element in them, is now permitting the manufacture of nutritious cotton flour for human consumption on a small scale. Research in all such areas deserves strong support.

Global Food Insecurity

DWINDLING GLOBAL RESERVES

Since World War II the world has been fortunate to have, in effect, two major food reserves. One has been in the form of grain reserves in the principal exporting countries, and the other in the form of cropland idled under farm programs, virtually all of it in the United States.

Grain reserves, including substantial quantities of foodgrains and feedgrains, are most commonly measured in terms of carryover stocks — the amount in storage at the time the new crop begins to come in. World carryover stocks are concentrated in a few of the principal exporting countries — the United States, Canada, Australia, and Argentina.

Since 1960, world grain reserves have fluctuated from a high of 155 million metric tons to a low of about 100 million metric tons. When reserves drop to 100 million tons, severe shortages and strong upward price pressures develop. Although 100 million tons appears to be an enormous quantity of grain, it represents a mere 8 percent of annual world grain consumption, a dangerously small working reserve. As world consumption expands, so should the size of working reserves, but the trend over the past decade has been for reserves to dwindle while consumption has climbed (see Figure 7).

In addition, one-seventh of U.S. cropland, or roughly 50 million of 350 million acres, has been held out of production by the government for the past dozen years to prevent the accumu-

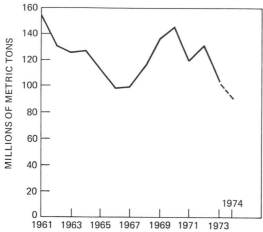

7   World Grain Reserves, 1961–1974. Includes United States, Canada, Australia, and Argentina. Reserve stocks of grain held in the principal exporting nations have provided an important safety margin in the world food economy. In 1973 and 1974 they fell to a low level, only 7 percent of annual world consumption, reflecting the shortage of grain supplies in relation to commercial demand. Efforts were under way in 1974 to organize a new system of international cooperation to build up safe levels of reserves, spread throughout both the exporting and the importing nations of the world. *Source:* U.S. Department of Agriculture.

lation of large commercial surpluses. Although not as quickly available as the grain reserves, most of this acreage was brought back into production within 12 to 18 months once the decision was made to remove controls in early 1973 (see Figure 8).

In recent years the need to draw down grain reserves and to dip into the reserve of idled cropland has occurred with increasing frequency. This first happened during the food-crisis years of 1966 and 1967, when world grain reserves were reduced to a dangerously low level and the United States brought back into production a small portion of the 50 million idle acres. Again, in 1971, as a result of the corn blight, the United States both drew down its grain reserves and brought another portion of the idle acreage back into production. In 1973, in response to growing food scarcities, world grain reserves once more declined, and the United States dipped into its idle cropland, but to a much greater degree than on either of the two previous occasions. Government decisions in early 1973 permitted all but a small fraction of the idled cropland in the United States to come back into production, and no payments to keep land idle were to be made in 1974.

From the end of World War II until 1972,

III   Index of World Food Security

Year	Reserve stocks of grain	Grain equivalent of idled U.S. cropland	Total reserves	Reserves as days of world grain consumption
	Millions of metric tons			Days
1961	154	68	222	95
1962	131	81	212	88
1963	125	70	195	77
1964	128	70	198	77
1965	113	71	184	69
1966	99	79	178	66
1967	100	51	151	55
1968	116	61	177	62
1969	136	73	209	69
1970	146	71	217	69
1971	120	41	161	51
1972	131	78	209	66
1973	106	24	130	40
1974 (proj.)	90	0	90	26

Combining reserve stocks of grain with the production potential of cropland held under farm programs in the United States, and then comparing this total reserve capacity to annual global consumption, indicates the degree of security provided by the world's reserve capacity. By 1974 reserves had fallen to the equivalent of only 26 days of world consumption. *Source:* Data from U.S. Department of Agriculture.

world prices for the principal temperate-zone farm commodities, such as wheat, feedgrains, and soybeans, were remarkably stable. Now that the reserve of idled cropland has disappeared, and is unlikely to be rebuilt substantially in the years ahead, there is every prospect of very volatile world prices for the important food commodities.

By combining global reserve stocks with the potential grain production of idled cropland, we may obtain a good indication of the actual total reserve capability in the world food economy in any given year. Taking this total as a percentage of total world grain consumption provides a rough quantitative indicator of global food security for the year. As Table III demonstrates, the world is now in an extremely vulnerable position. In 1973 and 1974, world reserve capabilities in relation to consumption needs have fallen far lower than at any time during the past quarter-century. Reserves have dwindled from the equivalent of 95 days of world grain consumption in 1961 to only 26 days in 1974, a fragile buffer against the vagaries of weather or plant disease.

THE NORTH AMERICAN BREADBASKET

The extent of global vulnerability is particularly underlined by examining the degree of global dependence on North America for exportable food supplies. Over the past generation the United States has achieved a unique position as a supplier of food to the rest of the world. Before World War II both Latin America, most importantly Argentina, and North America (United States and Canada) were major exporters of grain. During the late 1930s net grain exports from Latin America were substantially above those of North America. Since then, however, the combination of the population explosion and the slowness of most Latin American governments to reform and modernize agriculture have eliminated the net export surplus. With few exceptions, Latin American countries are now food importers.

As Table IV illustrates, over the past three decades North America, particularly the United States, which accounts for three-fourths of the continent's grain exports, has emerged as the world's breadbasket. Exports by Australia, the only other net exporter of importance, are only a fraction of North America's. The United

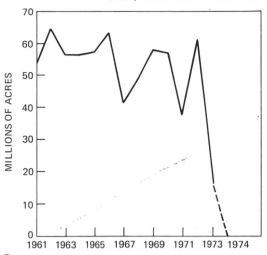

8   **U.S. Cropland Withheld from Production Under Government Programs, 1961–1974.** To prevent large commercial surpluses, the U.S. government paid farmers to keep, on the average, about 50 million acres out of production in the 1960s and early 1970s. This idled cropland has been a safety valve that could be dipped into when shortages developed. In 1974, however, in response to burgeoning global demand, no land was held idled by the government. The authors feel that the reserve of idled acreage is not likely to be rebuilt to the high levels of the past in the years ahead. *Source:* U.S. Department of Agriculture.

States not only is the world's major exporter of wheat and feedgrains, it is also now the world's leading exporter of rice. North America today controls a larger share of the world's exportable surplus of grains than the Middle East does of the world's oil exports.

Exportable supplies of the crucial soybean are even more concentrated than those of grains. Although as late as the 1930s China supplied nearly all the soybeans entering world markets, continuing population growth during the ensuing decades has gradually absorbed the exportable surplus. As of 1973, China is importing small quantities from the United States. The position of principal supplier has been taken over by the United States, which provided over 90 percent of world soybean exports in the 1960s and early 1970s. With world demand for high-quality protein surging upward, Brazil — virtually the only other nation capable of exporting soybeans on a sizable scale in the foreseeable future — has rapidly boosted its soybean production and exports. However, the United States is likely to continue supplying three-fourths or more of the world's soybean exports for many years to come.

At a time when dependence of the rest of the world on North American food exports is increasing so dramatically, there is also a growing awareness that this extreme dependence leaves the world in a very dangerous position in the event of adverse crop years in North America. Both the United States and Canada are affected by the same climatic cycles.

Some evidence indicates that North America has been subject to recurrent clusters of drought years roughly every 20 years. The cyclical drought phenomenon has now been established as far back as the Civil War, when data were first collected on rainfall. The most recent drought, occurring in the early 1950s, was rather modest. The preceding one, in the early 1930s, was particularly severe, giving rise to the dust bowl era in the United States. Drought conditions in the American Midwest in 1974 raised fears that the cycle was revisiting the region.

If the United States experiences another stretch of drought years, quite possibly during the current decade, its impact on production will not likely be as great as during the 1930s, owing to improved soil management and water conservation practices. But even a modest decline in production, given the rapid growth in global demand and extreme world dependence on North America's exportable margin of food, would create a very dangerous

situation. It would send shock waves throughout the world that would trigger intense competition for available food supplies.

With global grain reserves dwindling, most formerly idle U.S. cropland back in production, and growing world dependence on one geographic region for exportable food, the world faced a future of vulnerability by late 1973. As a result, strong efforts were underway to create a new internationally managed food reserve system. Special food reserves held by both importing and exporting nations could be built up in times of surplus and drawn down in times of scarcity. They would help stabilize food prices everywhere and ensure the availability of food when famine threatens poor nations after a crop failure, a guarantee the United States may be unable to provide throughout the coming decades. Such a plan, designed to provide "minimum food security," was proposed by FAO Director-General A. H. Boerma in 1973, and received preliminary international approval at the FAO Conference in November 1973. The plan was approved in modified form at a special U.N. World Food Conference in November 1974 in Rome with details to be worked out in 1975. The FAO plan deserves rapid implementation.

Agricultural Stresses on the Ecosystem

ECOLOGICAL UNDERMINING OF FOOD SYSTEMS

In many areas of the world, the pressure of growing demand for food is now ecologically

IV   Changing Pattern of World Grain Trade

Region	1934–1938	1948–1952	1960	1966	1973 (prelim.)
	Millions of metric tons.				
North America	+5	+23	+39	+59	+91
Latin America	+9	+1	0	+5	−3
Western Europe	−24	−22	−25	−27	−19
Eastern Europe and USSR	+5	—	0	−4	−27
Africa	+1	0	−2	−7	−5
Asia	+2	−6	−17	−34	−43
Australia and New Zealand	+3	+3	+6	+8	+6

Plus, net exports; minus, net imports. Since the 1930s, North America has emerged as the world's breadbasket, providing most of the world's exportable grain supplies. The United States alone generally accounts for about three-fourths of the continent's exportable surplus. This dependence on a single geographic region for grain exports leaves the world in a vulnerable position in the event of a poor crop year in North America.

undermining major food-producing systems. This is not a new development. What is new is the scale and acceleration of the process, which is now adversely affecting world food production prospects.

The most conspicuous examples of such ecological damage are those due to the expansion of human and livestock populations beyond the basic carrying capacity of the land, and to land mismanagement. The massive destruction of vegetable cover and the erosion of topsoil is apparent in the spreading deserts in Africa, Asia, and Latin America; in increasingly frequent and severe floods in some regions; in the silting of irrigation reservoirs and canals; and in the abandonment of millions of acres of arable land to erosion. The decline in fish catches as a result of overfishing provides another disturbing example of the loss of productive capacity as a result of excessive pressures to expand food supplies.

As livestock herds are increased to meet the growing demands for food and draft power, they graze over ever-larger areas, stripping the land of its natural cover in many parts of the world. This problem is most serious in North and sub-Sahara Africa, the Middle East, and the Indian subcontinent. And as expanding population increases the need for new agricultural land and for forest products for fuel, particularly in the poor nations, the countryside is being steadily denuded of trees.

Nature requires centuries to create the earth's thin mantle of life-sustaining topsoil; man can destroy it in a few years. Stripped of its original cover of grass and trees, much of the earth's land surface is vulnerable to erosion by wind and rain. In poor countries millions of acres of these unproductive lands are abandoned each year by rural people who are forced to go to the already overcrowded cities. Deforestation in such countries as India and Pakistan is forcing people to use cow dung for fuel, depriving the land of a natural source of fertility. Deforestation in the Himalayan foothills is also encouraging an increase in the incidence and severity of flooding in the Indian subcontinent, which in turn destroys crops, irrigation systems, and homes. In effect, deforestation is gradually undermining the subcontinent's food-producing capacity, casting a cloud over the future of its 750 million people.

As the demand for food expands, more and more land that is too steep or too dry to sustain cultivation is being brought under the plow. In the poor countries, where most of the level land is already cultivated, farmers are moving up the hillsides. Accelerated erosion is the result.

In West Pakistan, the recently completed $600 million Mangla irrigation reservoir, which originally had a life expectancy of 100 years, is now expected to be nearly filled with silt in half that time. The clearing of steep slopes for farming, progressive deforestation, and overgrazing in the reservoir's watershed are responsible for its declining life expectancy (Chapter 6).

Efforts to expand the area of farmland in one locale is reducing the water for irrigation in another. Farmers moving up the hillsides in Java are causing irrigation canals to silt at an alarming rate. Damming the Nile at Aswan expanded the irrigation area for producing cereals but largely eliminated the annual deposits of rich alluvial silt on fields in the Nile Valley, forcing farmers to rely more on chemical fertilizers. In addition, interrupting the flow of nutrients into the Nile estuary probably caused a precipitous decline in the fish catch there.

A historic example of the effects of man's abuse of the soil is all too visible in North Africa, once the fertile granary of the Roman Empire and now largely a desert or near-desert whose people are fed with the aid of food imports from the United States. Once-productive land was eroded by continuous cropping and overgrazing until much of it would no longer sustain agriculture. Irrigation systems silted, depriving land of the water needed for cultivation. Similar situations are being created in semiarid parts of Asia, such as western India, that are experiencing a rapid buildup of population during this century.

In the Sahelian zone just south of the Sahara in Africa a complex interplay of human and ecological factors is encouraging the southward spread of the desert at rates up to 30 miles per year at some points. Widespread famine in mid-1973 in Mali, Upper Volta, Niger, Chad, Mauritania, and Senegal, after several consecutive years of drought, brought international attention and emergency food aid to the region. However, there is a growing awareness that the crisis of the Sahel is not a short-term one that will end if rains begin. The pressures of overpopulation within a fragile ecosystem, and the resulting problems, such as overgrazing, lowering of water tables, and deforestation, are interacting with climatic and biological factors to gradually destroy the productive capacity of the region. The encroachment of the Sahara in Africa, if unchecked, may eventually destroy an ecosystem that currently supports about 60 million people.

An international effort to confront the long-term problems of the Sahel, led by the FAO, was launched in 1973. However, the vastness of the threatened area, and the delicate social problems involved in any attempts to alter the living habits of the proud nomadic peoples inhabiting much of the Sahel, will make large-scale progress exceedingly difficult.

While the desolating process of desertification appears most dramatic along the northern and southern interfaces of the Sahara desert in Africa, it is also occurring in parts of Asia, Latin America, and southern Africa. In most cases, a blend of human and natural factors is encouraging the spread of the deserts. Both Chile and Peru are losing arable land due to spreading deserts. The Thar desert in the Indian state of Rajasthan, which already claims half the state's area, is forcing 30,000 acres of fertile land out of cultivation each year. A massive reclamation project is being carried out by the Indian government and U.N. agencies, with plans to rechannel the waters of the Sutlej River to the desert and eventually bring over 4 million acres under irrigation.

ENVIRONMENTAL COSTS

The endangerment of food-producing systems is the most immediately serious, but by no means the only, ecological consequence of spreading agricultural activities. Even where food-producing capacity is not being adversely affected, efforts to increase the food supply — either by expanding the area under cultivation or by intensifying cultivation through the use of agricultural chemicals and irrigation — bring with them troublesome ecological consequences. Such stresses have become manifest in various ways, including the eutrophication of freshwater lakes and streams, the rising incidence of environmentally induced illnesses, and the growing number of species of wildlife

threatened with extinction. New signs of agricultural stress on the earth's ecosystem appear almost daily as the growing demand for food presses against our ecosystem's limited capacities.

One of the most costly and tragic side effects of the spread of modern irrigation in Egypt and other river valleys in Africa, Asia, and northeastern South America is the great increase in the incidence of schistosomiasis (also called bilharzia). This debilitating intestinal and urinary disease, produced by the parasitic larvae of a blood fluke that burrows into the flesh of those working in water-covered fields, affects a sizable number of the Egyptian people.

The Chinese call this disease "snail fever" and are waging an all-out campaign against it. Unfortunately, schistosomiasis is environmentally induced by conditions created by man. The incidence of the disease is rising rapidly as the world's large rivers are harnessed for irrigation. Schistosomiasis may be afflicting over 200 million people. It has surpassed malaria, the incidence of which is declining, as the world's most prevalent infectious disease.

Efforts to intensify agricultural yields also have adverse environmental consequences. If the effect of chemical fertilizer could be confined to agriculture, it would be fine, but unfortunately it cannot. The water runoff from agricultural land carries chemical fertilizer with it, raising the nutrient content of streams and lakes throughout the world, causing them to eutrophy (Chapter 10).

Thousands of freshwater lakes are threatened throughout North America and Europe, and increasingly in many poor countries in which fertilizer use is beginning to climb. Filipino villagers are finding that rising fertilizer use in rice paddies is resulting in the eutrophication of local lakes and ponds, depriving them of fish, a traditional source of animal protein. No one has calculated the cost to mankind of losing these freshwater lakes, but it is staggering.

The growing use of chemical fertilizers is causing another more localized but hazardous problem, the chemical pollution of drinking water. Nitrates are the main worry; they have risen to toxic levels in some communities in the United States. Both children and livestock have become ill and sometimes died from drinking water that contained high levels of nitrates. Excessive nitrates can cause metahemoglobanemia, a physiological disorder affecting the blood's oxygen-carrying capacity. Since the problem is generally local, it can be effectively countered by finding alternative (though usually more costly) sources of drinking water. Bottled water is being used in some California communities. Groundwater nitrates in southern Illinois rose above the FDA tolerance levels in the early 1970s, creating enough of a health threat for the state government to hold hearings on the possibility of limiting the use of nitrogen fertilizer. This proposal was eventually rejected, and the level of fertilizer applied by many Illinois farmers today continues to be high.

Solving the Food Problem

REDUCING POPULATION GROWTH

The prospect of an emerging chronic global scarcity of food as a result of growing pressures on available food resources underlies the need to reduce and eventually halt population growth in as short a period of time as possible. One can conceive of this occurring in the industrial countries as a result of current demographic trends. In the United States, attitudes toward childbearing have changed dramatically in recent years, and U.S. fertility has fallen below the replacement level of 2.1 children (Chapter 2).

Three European countries — East Germany, Luxembourg, and West Germany — have stabilized their populations within the past few years. Indeed, the growth of the German-speaking population of Europe, 90 million people who live in East Germany, Berlin, West Germany, Austria, and part of Switzerland, has very nearly ceased. Although this is a beginning toward reaching the goal of reducing or halting world population growth, it must be viewed within the context of a world containing 4.0 billion people.

Many other European countries have relatively low birth rates and appear to be moving toward a stationary population. These countries — including Hungary, Scandinavia, and the United Kingdom — plus the Soviet Union and Japan, could easily achieve zero rate of population growth within the next decade or

so, particularly if both policy makers and the people put their minds to it.

In the poor countries, however, it will be much more difficult to reduce growth rates within an acceptable time frame. For one thing, recent history suggests that birth rates do not usually decline very far in the absence of certain improvements in well-being — an assured food supply, a reduced infant mortality rate, literacy, and at least rudimentary health services.[9] Couples will not limit the number of children until infant mortality declines to the point where they are confident that most of their children will survive to adulthood. This, in turn, requires an assured food supply and jobs to earn the income to buy the needed food, a combination that removes the fear of famine and prevents malnutrition. It also requires that health services of at least a minimal nature be extended to that one-third to one-half of mankind now without any medical care whatsoever. The historical record also indicates that meaningful declines in human fertility do not normally occur among largely illiterate populations.

In short, it may well be in the self-interest of affluent societies, such as the United States, to launch an attack on global poverty, not only to narrow the economic gap between rich and poor nations, but also to help meet the basic social needs of people throughout the world in an effort to provide the preconditions for lowering birth rates. Population-induced pressures on the global food supply will continue to increase if substantial economic and social progress is not made. Populations that double every 24 years — as many are doing in poor nations — multiply eightfold in a single lifespan.

ALTERING CONSUMPTION PATTERNS

As the world shrinks under the impact of advancing communications and transportation technologies and the continuing integration of national economies into a global economy, the contrast in food-consumption levels between rich countries and poor sharpens. As indicated earlier, those living in the poor countries are sustained on 400 pounds or less of grain per year, while those in the wealthier ones require nearly 1 ton of grain. It is difficult to envisage a situation in which all of mankind could progressively increase per capita claims on the earth's food-producing resources until everyone reached the level now enjoyed by the average North American. Thus thought should be given to how diets could be simplified in the wealthy nations in order to reduce per capita claims on the earth's scarce resources of land and water. What are the possibilities of substituting less costly, more efficient forms of protein for, say, beef?

Consumption patterns in the United States suggest that there are two broad techniques to reduce per capita resource requirements for food. One is to substitute vegetable oils for animal fat; the other is to substitute vegetable protein for animal protein.

Over the past three decades vegetable oils have been extensively replacing animal fats in the American diet. In 1940, for example, the average American consumed 17 pounds of butter and 2 pounds of margarine. By 1973 the average American was consuming over 11 pounds of margarine and less than 5 pounds of butter. Lard has been almost pushed off supermarket shelves by the hydrogenated vegetable shortenings. At least 65 percent of the whipped toppings and more than 35 percent of the coffee whiteners in the United States today are of nondairy origin. This pervasive trend reduces both per capita food costs and per capita claims on agricultural resources, and it reduces the intake of saturated animal fats, now widely believed to be a factor that contributes to heart disease.

The widespread substitution of vegetable oils for animal fats in the U.S. diet over the past generation has reduced per capita claims on agricultural resources, but this has been more than offset by the simultaneous increase in beef consumption from 55 to 116 pounds. Stimulated by sharp rises in meat prices in late 1972 and early 1973, attention is now focusing on the substitution of high-quality vegetable protein for animal protein. Technology for the substitution of vegetable for animal proteins has made considerable progress, mainly in the area of soybean-based meat substitutes. The development of a technique for spinning soya protein into fibers, duplicating the spinning of synthetic textile fibers, permits the close emulation of the fibrous qualities of meat. Food technologists can now compress soya fibers into what are called textured vegetable proteins and, with the appropriate flavoring and coloring, come up with reasonable substitutes for beef, pork, and poultry. With livestock protein, particularly beef, becoming more costly, this technique is likely to gain a strong commercial foothold in the near future.

The first major meat product for which substitution is succeeding commercially is bacon. The soya-based substitute looks and tastes like bacon, and while the extent of substitution for bacon is still small, it is growing. The substitute product has the advantage of being high in protein, low in fat, and storable without refrigeration.

The greatest single area of protein substitution promises to be the use of vegetable protein to augment meat proteins in ground meats. Soya protein "extenders," as they are known, are being added to a variety of processed and ground meat products, frequently improving flavor, cooking qualities, and nutrition as well as reducing prices. Soya protein extenders are already widely used in institutions throughout the United States, and supermarket sales to the public began in 1973.

Still another way to encourage substitution of vegetable products for animal products is to introduce interesting alternatives that are part of the cuisine of other nations. In the Far East, for example, bean curd, which is high in protein quality, is an important source of protein in the daily diet. Beans, peas, or lentils — all of which lend themselves to varied preparation — are nutritious meat substitutes. A number of cookbooks recently have been published in the United States that offer nutritious and tasty recipes that do not call for livestock products.

If the substitution of high-quality vegetable protein grows, it is not inconceivable that, in the United States, per capita claims on agricultural resources could eventually begin to decline. A combination of convergent economic, ecological, moral, and health considerations could lead in this direction. This could, in turn, help make food more available and less expensive for the poorly nourished portion of mankind.

AGRONOMIC POTENTIAL OF THE POOR COUNTRIES

The world's greatest reservoir of unexploited food potential is located in the developing countries. The jump in per acre yields associated with the Green Revolution in several countries has appeared dramatic, largely because their yields traditionally have been so low relative to the potential. But today rice yields per acre in India and Nigeria still are only one-third those of Japan; corn yields in Thailand and Brazil are less than one-third those of the United States. Large increases in food supply are possible in these countries at far less resource cost than in the agriculturally advanced nations if farmers are given the necessary economic incentives and the requisite inputs.

Concentrating efforts on expanding food production in the poor countries could reduce upward pressure on world food prices, create additional employment in countries where continuously rising unemployment poses a serious threat to political stability, and raise income and improve nutrition for the poorest portion of humanity — the people living in rural areas of the developing countries. Because of the positive relationship between improved social welfare and nutrition, and lower birth rates, efforts in this direction will also be an essential element in the battle to slow the rate of population growth.

Conclusion

In the coming decade the overriding objective of a global food strategy in an increasingly interdependent world should be the elimination of hunger and malnutrition among the large segment of humanity whose food supply simply is inadequate. To be successful, such a strategy must be designed to alter existing trends in food production and population growth while seeking a more equitable distribution of food supplies both among and within societies.

Solving the food problem is a complex task — one that requires thoughtful analysis and determined action by both rich and poor nations. Assuring adequate food supplies at reasonable prices within individual countries may now be possible only through international cooperation in such areas as internationally managed food reserves, agricultural development of the poor countries, the management of oceanic fisheries, and slowing population growth. The disappearance of large surplus stocks and the return of idle cropland to production has removed the cushions that once existed as partial insurance against catastrophe.

There is little doubt that increasingly serious food supply problems will emerge unless a drastic slowdown in the growth of world demand occurs. The choice is between famine and family planning. The 10,000-year-old syndrome in which production gains have been continuously absorbed by growing populations must soon be broken if mankind is to avoid ecological and social disaster.

References

1. Berg, A. 1973. *The Nutrition Factor: Its Role in National Development.* The Brookings Institution, Washington, D.C.
2. McNamara, R. S. 1973. *One Hundred Countries, Two Billion People.* Praeger Publishers, Inc., New York. p. 56.
3. *New York Times,* 1970. Feb. 19. See also Berg, Reference 1, Chapter 2.
4. United Nations, Food and Agriculture Organization. 1971. *Food Balance Sheets, 1964–1966 Average.* FAO, Rome.
5. Brown, L. R. 1970. *Seeds of Change: The Green Revolution and Development in the 1970s.* Praeger Publishers, Inc., New York. See this book for a more extensive discussion of the origin and impact of the Green Revolution.
6. Shaw, R. d'A. 1970. *Jobs and Agricultural Development* (Monograph 3). Overseas Development Council, Washington, D.C. Discusses the employment and distributional effects of the Green Revolution and possible government policies to prevent undesirable effects.
7. Slack, A. V. 1970. *Defense Against Famine: The Role of the Fertilizer Industry.* Doubleday and Company, Inc., Garden City, N.Y. p. 57.
8. United Nations, Food and Agriculture Organization. *Production Yearbook.* FAO, Rome.
9. Rich, W. 1973. *Smaller Families Through Social and Economic Progress* (Monograph 7). Overseas Development Council, Washington, D.C. Provides further information on the interrelationship of social improvements and birth rate declines.

Further Reading

Borgstrom, G. 1972. *Too Many,* 2nd ed. Macmillan Publishing Co., Inc., New York.

Brown, L. R. 1970. Human food production as a process in the biosphere. *Sci. Amer.* 223 *(3):* 160–170.

———. 1972. *World Without Borders.* Random House, Inc., New York.

———, with Eckholm, E. P. 1974. *By Bread Alone.* Praeger Publishers, Inc., New York.

Curry-Lindahl, K. 1972. *Conservation for Survival: An Ecological Strategy.* William Morrow & Co., Inc., New York.

Ehrlich, P. R., and Ehrlich, A. H. 1972. *Population, Resources, Environment: Issues in Human Ecology,* 2nd ed., W. H. Freeman and Company, San Francisco.

United Nations, Food and Agriculture Organization. 1970. *Lives in Peril: Protein and the Child.* FAO, Rome.

———. 1973. *State of Food and Agriculture.* FAO, Rome.

Food: Readings from Scientific American. 1973. W. H. Freeman and Company, San Francisco.

Lappé, F. M. 1971. *Diet for a Small Planet.* A Friends of the Earth/Ballantine Book, New York.

Simon, P., and Simon, A. 1973. *The Politics of World Hunger.* Harper's Magazine Press, New York.

Steinhart, C. E., and Steinhart, J. S. 1974. Energy use in the U.S. food system. *Science* 184: 307–316.

Editor's Commentary—
Chapters 4-6

THE FOLLOWING COMMENTS are introductory to Chapters 4 through 6, dealing with energy, minerals, and water resources.

Resources, especially energy and mineral resources, are the keystones of modern economies and the basis of our standard of living. Whether or not the United States, the industrialized West, and the world can depend upon an adequate flow of these materials into the indefinite future (or even the next 50 years) is a question that has been debated for decades. Economists, with a few exceptions, tend to be optimists on the subject of resources. They predict that technological innovation, stimulated by the increasing costs of using old materials in old ways, will lead to expansion of existing resources and the substitution of common materials for rare ones. A clear statement of this point of view by Brooks and Andrews can be found in the 5 July 1974 issue of *Science* (vol. 185, pp. 13–19). By contrast, many earth scientists and some economists fear that for some materials the *rate* of technological innovation and expanding resources will not always keep up with the rate of increasing demand, and that shortages (and extremely high prices) may occur for some materials.

In Chapter 4 Preston Cloud points out that because of uncertainty about resources, and about such matters as the free flow of minerals from the nations that have them to the nations that need them, we cannot be sure that we will always have enough materials to satisfy our "needs". And it is surely questionable whether there will ever be enough materials, priced cheaply enough, for the needs of the poor nations of the world. Dr. Cloud shows that while there are promising possibilities for satisfying our material needs, each of these possibilities, from cheap nuclear energy to minerals from the sea, is fraught with difficulties; difficulties that are serious enough to make us conservative rather than cavalier about materials for the future. The recent oil crisis and the subsequent sky-rocketing prices of fertilizer, which has been seriously in short supply in 1974, show that these warnings are not empty. The fact that the shortages have arisen from mainly political causes and have been worsened by economic mismanagement makes them no less real—politics and economics are just as much a part of the structure of environmental problems as are, for example, geology and nuclear technology.

Dr. Cloud makes two other fundamental points. First, our hopes of satisfying the material needs for the world as a whole depend upon our putting the lid on world population. Even highly optimistic economists believe this. While the rich countries are approaching steady-state, the populations of poor countries are still increasing alarmingly (Chapter 2). The correlations among a lowered birth rate, good diet, industrial developmental, and affluence underline the absolute necessity, from an ethical point of view, of the rich countries making a real and massive contribution to the economic development of the poor countries.

Dr. Cloud's other point concerns the environment of the rich countries: it is one thing to claim that the United States can meet its needs for materials for the forseeable future—but quite another also to improve the environment while doing so. By way of illustration, inflation has been the major economic problem facing the Ford administration, and in his first few weeks in office President Ford suggested cutting $9 billion from the waste-water treatment program in order to reduce government spending. As we have seen over and over again, when the squeeze comes, environmental programs are the most vulnerable. So, not only does an increasing demand for materials put increasing pressure on the environment in a direct way, but rising prices for materials operate indirectly to weaken environmental programs.

The particular case of energy supplies (Chapter 5) illustrates dramatically the problems associated with obtaining resources, and the energy "crisis" has made us all aware of the immediacy of these problems. But if the energy problems are at this time more severe than those for many other resources, at least the issues are clear. Oil resources are just beginning to be less and less adequate for our projected "needs", and we will need to find substitutes. Substitutes do exist, for example coal, oil shale, nuclear power. But each substitute has an environmental cost associated with it, for example strip-mining and air pollution, radiation hazards and nuclear blackmail (see Table XIII in Chapter 5). The decisions we take about these alternatives are crucial, because they will lock us into patterns of energy use for decades to come. But, since we have a good understanding of the costs and benefits, it should be possible to make good decisions. Two decisions we ought to make are clear from Dr. Holdren's chapter. First, we should press now for cities, industries, transport systems, and life styles that use less energy to do a given job than do our current arrangements. Secondly, · there *are* relatively environmentally benign sources of energy (for example, solar energy and geothermal power in some areas), and these should be substituted for other resources wherever it is feasible.

The recent energy crisis, and the problems of energy supplies in general, also illustrate two points that will be made throughout this book. First, environmental problems are not only physical problems, but have important social, economic and political facets as well. Thus the energy crisis involved poor planning of refining capacity by government and the corporations, it involved international politics, it probably involved some finagling by corporations to get higher profits, and it may even reflect a real physical shortage of crude oil. As a result of the crisis, environmental constraints (such as a limit on sulfur emissions) were pushed aside, illustrating clearly the relation between resource demands and environmental quality. Secondly, the crisis showed just how interconnected environmental problems are. For example, the increased cost of oil (and natural gas) resulted directly in increased prices for products based on petroleum, including nitrogen fertilizers and pesticides, and this had the direct result of reducing recent yields from some of the crops in the Green Revolution (Chapter 3). Thus, although the energy issues are complex, they are worth studying, because energy policy is the very crux of environmental quality.

We can turn to Chapter 6, on water resources, with some relief. Although the trade-off between increasing resources and decreasing environmental quality can be seen here also, for example in the Colorado River developments, at least there seems to be no urgent overall problem of water shortages in the United States. We could certainly use our water resources more wisely—for example, by not concentrating huge populations in very arid areas like Southern California—but, generally, adequate supplies are allowing us to make mistakes without disastrous consequences. The major problem with water is its pollution, which is taken up in Chapters 10 and 11.

Strip mines have devastated
the hills of West Virginia.
Photo by Milton Rogovin/
Rapho-Guillumette.

4

Mineral Resources Today and Tomorrow

Preston Cloud   *is Professor of Biogeology and Environmental Studies at the University of California, Santa Barbara. Born in 1912, he earned his Ph.D. in 1940 at Yale. He spent three years in the Navy, worked for 16 years with the U.S. Geological Survey, did a brief stint in industry, and has taught at several universities. A member of the National Academy of Sciences and the American Academy of Arts and Sciences, he has served on many advisory panels both academic and governmental. His field work has taken him to every continent except Antarctica, and centers on problems of biogeology, sedimentology and historical geology. Biogeology is his term for the total body of evidence and theory about biological processes on the evolving Earth.*

Will mineral resources be adequate to meet the "needs" of future industrial civilizations? It is possible to give different answers to that question, depending on one's assumptions about technological and economic uncertainties, the levels of population and rates of growth assumed, the time frame chosen, how needs are defined, the way costs are counted, and other variables. Do we think of the world as a whole into a future beyond this century or of a small part of it for a shorter time? How do we estimate needs? What kind of an industrial civilization do we have in mind? Do we think of a world divided into rich and poor nations or one in which equity is a goal? How do we calculate costs? Do we include only the bare fiscal costs of removal from the ground and concentration to usable form, or do we take into consideration the larger costs of protecting and restoring mined land surfaces, of maintaining our supply of potable water and breathable air, of safeguarding public health, and of maintaining a degree of independence nationally and among nations?

We are often reminded that mankind has been in difficulty before and found technological solutions to his problems. The shortage of wood in medieval Europe was eventually solved by the introduction of coal. Malthusian food shortages have been staved off in many (but not all) parts of the world by a variety of agricultural innovations. Nineteenth-century London had serious problems of smog and pollution. If these and other problems have been solved in the past, should we lack confidence that mankind will continue to solve them in the future?

Efforts to predict the future are bound to be wrong in detail, but that is not to say that man should not exercise his uniquely human gift of foresight to anticipate and ameliorate or forestall difficulties. Technological innovation is bound to create a variety of new options that may be used for good or ill by man. Growth of populations and per capita consumption are bound to exacerbate the problem of how simultaneously to maintain a flow of raw materials, achieve a greater degree of equity in distribution, protect the environment, and achieve and maintain peace among nations.

In fact, the situation today is different from any that has ever existed in the past. More people are using more things at a greater rate, and everything is increasing exponentially at rates that, like compound interest on a loan, lead to doublings of the quantities involved within relatively short intervals. These doubling rates are very deceptive. When a given quantity of something is only one-fourth used up, it takes only two more doubling times before it is all gone. Current doubling times are about 35 years for global populations and space occupied by them, about the same time for mineral consumption on an average, and about 14 years for energy. Such trends simply cannot be sustained for very much longer. When projected, they show us elbow to elbow over the entire earth, or using energy equivalent to the earth's daily ration from the sun, or using up the entire accessible crust of the earth within a few centuries. The great danger is that quantities involved are so large and doubling times so short that general perception of impending crises may come too late to avert them.

What has brought us to such a pass and what can be done about it? What has made the world so different today is the cumulative effect of four long-range trends that have grown out of and accompanied the industrial revolution: (1) the achievements of medical technology, which have brought on the runaway imbalance between birth and death rates; (2) the hypnotic but unsustainable dream of an ever-increasing real gross national product based on obsolescence and waste of materials; (3) the finite nature of the earth, particularly its accessible mineralized crust; and (4) the increased risk of irreversible spoilation of the environment which accompanies overpopulation, overproduction, waste, and the movement of ever-larger quantities of source rock for ever-smaller proportions of useful minerals.

Granted the advantages of big technological leaps, therefore, provided they are in the right direction, I see real hope for permanent long-range solutions to our problems as beginning with the taking of long-range views of them. Put in another way, we should not tackle vast problems with half-vast concepts. We must build a platform of scientific and social comprehension while endeavoring to fill the rut of ignorance, selfishness, and complacency with knowledge, restraint, and demanding awareness on the part of an enlightened electorate. And we must not be satisfied merely with getting the United States or North America through the immediate future, critical though that will be. We must consider what effects current and proposed trends and actions will have on the world as a whole for several generations hence, and how we can best influence those trends favorably the world over. Above all, we must consider how to preserve for the yet unborn the maximum variety and flexibility of significant options consistent with resolving current and future pressures.

Nature and Geography of Resources

Man's concept of resources, to be sure, depends on his needs and wants, and thus to a great degree on his locale and place in history, on what others have, and on what he knows about what they have and what might be possible for him to obtain. Food and fiber from the land, and food and drink from the waters of the earth have always been indispensable resources. So have the human beings who have utilized these resources and created demands for others — from birch bark to beryllium, from buffalo hides to steel and plastic. It

is these other resources, the ones from which our industrial society has been created, that I address in this chapter. I refer, in particular, to the nonrenewable or wasting resources — mineral fuels which are converted into energy plus carbon; nuclear fuels; and the metals, chemicals, and industrial materials of geological origin, which to some extent can be and even are recycled but which tend to become dispersed and wasted.

Nearly all such resources, except those that are common rocks whose availability and values depend almost entirely on economic factors plus fabrication, share certain peculiarities that transcend economics and limit technology and even diplomacy. They occur in local concentrations that may exceed their average crustal abundances by thousands of times, and particular resources tend to be clustered within geochemical or metallogenic "provinces" and "epochs" (of geologic time) in which others are rare or from which they are excluded. Some parts of the earth are rich in mineral raw materials and others are poor.

No part of the earth, not even on a continent-wide basis, is self-sufficient in all critical metals. North America is relatively rich in molybdenum and poor in tin, tungsten, and manganese, for instance, whereas Asia is comparatively rich in tin, tungsten, and manganese and apparently less well supplied with molybdenum. The great bulk of the world's unmined gold appears to be in South Africa, which has relatively little silver but a good supply of platinum. Cuba and New Caledonia have well over half the world's total known reserves of nickel. The main known reserves of cobalt are in the Congo Republic, Cuba, New Caledonia, and parts of Asia. Most of the world's mercury is in Spain, Italy, and parts of China and the USSR. Industrial diamonds are still supplied mainly by the Congo. About 60 percent of the world's remaining reserves of crude petroleum (including those of the continental shelves) are located in the Middle East.

Consider tin. Over half of the world's currently recoverable reserves are in Indonesia, Malaya, and Thailand, and much of the rest is in Bolivia and the Congo. Known North American reserves are negligible. For the United States, loss of access to extracontinental sources of tin is not likely to be offset by economic factors or technological changes that would permit an increase in potential North American production, even if present production could be increased by an order of magnitude (i.e., multiplied by 10). It is equally obvious that other peculiarities in the geographical distribution of the world's geological resources will continue to encourage interest both in trading with some ideologically remote nations and in seeking alternative sources of supply — perhaps even in military adventures or threats.

MINERAL RESOURCES AND THE ECONOMY

Taking the United States as an example of a modern industrial society, one sees an intricately dependent relationship between mineral resources and the economy. Figure 1 suggests this relationship, although without numbers. Some numbers are interesting. The unrefined raw materials (including energy raw materials) upon extraction from the earth presently account for only about 3.2 percent of the total gross national product; but they have a disproportionately large effect on the economy as a whole. Their relation to the health of the economy is like vitamins to the health of a person. By the time they have been refined and processed to machines, products, structures, and tools, consuming energy in the process, they account for about 40 percent of the GNP; and, in various ways, they underpin the other 60 percent as well. There would literally be no industrial economy without those material elements. And the nonenergy materials would be unobtainable without the constant large flow of energy that drives the entire materials cycle. Thus mineral raw materials, although only a small fraction of the GNP, are the foundation from which all the rest emerges.

Resources and Reserves

In discussing mineral resources or evaluating the discussions of others, it is well to bear in mind the distinction between resources and reserves.

Minerals likely to be accessible to man are essentially limited to the outer 10 to 35 kilometers of the earth — its "crust," particularly to the silica-rich or continental parts of this crust. The deeper "mantle" rocks are likely to remain inaccessible to operations on a commercial scale. Moreover, unlike the continental crust, they have not undergone the successive episodes of melting, remobilization, and weathering that are responsible for the formation of most mineral deposits.

Thus we may think of the minerals in earth's crust as being the *total stock* from which our *resources* and *reserves* are drawn. We can estimate this total stock by multiplying the total mass of the crust by the average crustal abundance of each element as determined from chemical analyses of different rock types. Even though this outer crust represents only about 0.4 percent of the mass of the earth, it still amounts to about 2.4×10^{19} metric tons — 24 followed by 18 zeroes! Thus the total stocks are very large for abundant and widely distributed elements such as iron and aluminum. Even for a moderately common element like copper, the total stock is large. Copper ranges in abundance from about 6 parts per million in limestones to around 13 parts per million in granitic rocks and 90 parts per million in basaltic rocks. It has an average crustal abundance of about 55 parts per million. Multiplying the mass of the crust (2.4×10^{19} tons) by the average crustal abundance of copper (55×10^{-6}) gives a total stock for copper of about 1.3×10^{15} tons. What part of that is a resource?

Resources are that part of the total stock that man might find and recover practically under any likely combination of technological and economic conditions. Uncertainty about future developments in mineral exploration, materials science and technology, and economics makes any estimates of ultimate resources very uncertain. All we can be reasonably sure of is that they will be orders of magnitude less than total stocks — in the case of copper maybe about 5 or more orders of magnitude less (1/100,000), or perhaps an eventual 10 billion tons. We arrive at such a "guesstimate" on the basis of the following logic. Although total copper in earth's crust is believed to be about 1.3×10^{15} metric tons, it would be wildly optimistic to assume that more than 1 percent of that is accessible to mining by any means. That reduces the number to around 10^{13} tons. But this copper is very thinly disseminated in most rocks and the energy required for recovery rises to excessive levels at grades below about

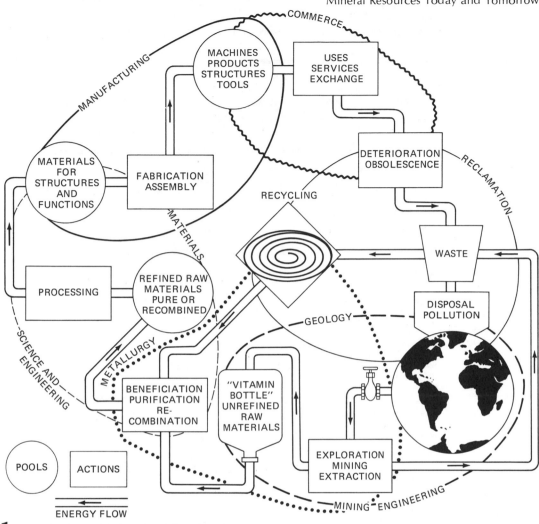

1 Cycle of Materials in Industrial Society.

Within the diagram the following labels appear:

COMMERCE

MANUFACTURING

MACHINES PRODUCTS STRUCTURES TOOLS

USES SERVICES EXCHANGE

DETERIORATION OBSOLESCENCE

RECLAMATION

MATERIALS FOR STRUCTURES AND FUNCTIONS

FABRICATION ASSEMBLY

RECYCLING

MATERIALS

PROCESSING

REFINED RAW MATERIALS PURE OR RECOMBINED

WASTE

DISPOSAL POLLUTION

GEOLOGY

METALLURGY

SCIENCE AND ENGINEERING

BENEFICIATION PURIFICATION RE-COMBINATION

"VITAMIN BOTTLE" UNREFINED RAW MATERIALS

EXPLORATION MINING EXTRACTION

MINING ENGINEERING

POOLS

ACTIONS

ENERGY FLOW

0.2 percent copper (we are now mining 0.4 percent copper). Thus it is unlikely that more than 1 percent of this 10^{13} tons can be recovered. That gives 10^{11} tons. Copper ores generally are caused by the enrichment in copper of upward-traveling hot vapors and fluids which concentrate it in the upper levels of the earth's crust. It is improbable that copper deep within the crust or mantle of the earth will be found to occur at concentrations much above crustal abundances. Thus the total extractable copper in sufficiently concentrated ore bodies is probably less than any of the preceding numbers. An optimistic outside limit might be about 10^{10} tons.

Reserves are that part of the resources and total stock known to be available under current technology at a given price. Known reserves of copper at a price near 60 cents per pound in 1970 were around 340 million tons. Estimates of reserves are based on actual physical exploration by drilling, trenching, and geophysical methods in geologically known ore bodies and districts. The degree of certainty is indicated by the terms measured, indicated, and inferred (or proved, probable, and possible). If, to known reserves of copper, we add marginal ores and inferred reserves, we get a total of about 1.4 billion tons of anticipated reserves, about four times known reserves. We have seen above that that quantity *might* eventually be increased to perhaps as much as 10 billion tons — about 30 times reserves now known, or

101

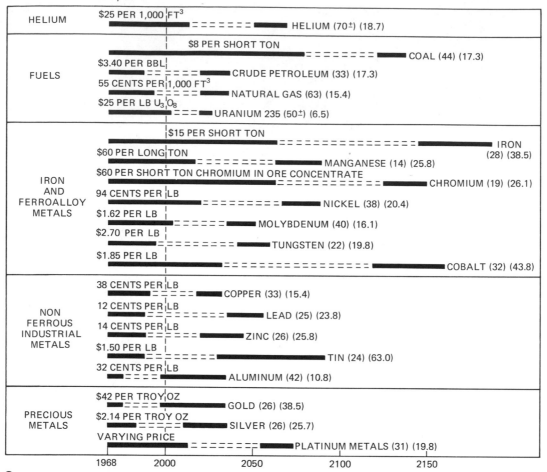

2 **Apparent Lifetimes of Estimated Reserves of Global Mineral Commodities.** Rates of demand are the anticipated averages to the year 2000. The largest reserve estimates warranted by the data are used for each mineral in combination with average projected rates of demand. The solid lines at the left show lifetimes of reserves at prices indicated in 1970 dollars; dashed lines show 5 times these reserves; the solid lines at the right extend lifetimes to the right at 10 times initially estimated reserves. The parenthetical numbers at the right are (on left) percentage of world consumption by U.S., and (on right) doubling times of consumption in years at average projected rates of demand. Data from References 1 (1970 data) and 2 (1968 data). Prices for mineral fuels are now in a fluid state, and their costs in turn will affect others shown here.

an extension of lifetime at projected rates of consumption (Figure 2) of about 80 years beyond exhaustion of currently known reserves. Costs in all terms will increase as grades decrease and the period of ample availability of copper fades to a close.

Everyone recognizes the need to consider undiscovered resources and those that may become reserves with technological advance or higher prices, as we have done above. The U.S. Geological Survey treats these potential

"future reserves" as (1) *conditional resources*, referring to deposits that could be exploited at higher prices or under improved technology, and (2) *undiscovered resources*, called *hypothetical* where expected as new deposits in known districts and *speculative* where hoped for on geological grounds as new mineral districts. All the possible additions are comprehended in the example of copper, used to illustrate this discussion.

We will not get into the detailed classification

of reserves and resources in this chapter, but that is the framework in which the following discussion is set.

Recoverable Mineral Resources

Consider now some aspects of the apparent lifetimes of estimated recoverable reserves of a selection of critical mineral resources and the position of the United States with regard to some of these. Figure 2 shows such apparent lifetimes for world reserves of different groups of metals and mineral fuels at minable grades and average projected rates of consumption as estimated by the U.S. Bureau of Mines (1970) from data available through 1968.[1] Many of the prices then prevailing are no longer realistic. Gold, for instance, was above $100 a troy ounce at the end of 1973 (and approaching $200 at the end of 1974), copper was fluctuating between 50 and 90 cents per pound, and oil was selling from $6 to $30 per barrel, if you could get it. Nevertheless, this graph suggests what is likely to be in short supply and what is relatively abundant. It takes both expected rates of consumption and probable additions to reserves into account. On geological grounds it seems reasonable to expect that as much as 10 times presently known reserves of many of these commodities will eventually reach the market place; much more than that for really abundant substances, such as iron, aluminum, magnesium; and the commercially important silicate minerals; perhaps somewhat more for copper and a few others. For these and the rest, substitution and intensive recycling will become necessary, in many instances with reduced performance levels.

The crucial questions are: (1) Can we increase and sustain production of industrial materials at a rate sufficient to meet the rising expectations of a world population that will reach 4 billion in 1975, growing at a doubling time of about 35 years? (2) Can we do so without irreparable harm to the environment? and (3) for how long?

As is readily apparent from recent history, the soundness of any national industrial potential over the long term can be gauged in terms of its indigenous resources as compared with those of other nations, plus its level of technological expertise. As for the United States, its level of technological expertise is high, but its current dependence on imports of materials is large. Despite actual increases in ore mined, the relative mineral production of the United States has decreased from about 40 percent of the world total in 1940 to under 20 percent in 1973 and is still trending downward. We are now almost completely dependent on foreign sources for 22 of the 74 nonenergy mineral commodities considered essential for modern industrial society. The Second Annual Report of the Secretary of the Interior under the Mining and Minerals Policy Act of 1970 (1973)[3] shows that, in terms of dollars across the board, in 1971, the United States imported about 20 percent of its total primary mineral requirements (including fuels), while using about 35 percent of global mineral production. Meanwhile, consumption is pushed upward both by growing populations and by increasing per capita demands. It escalates in such a way under these pressures that, even with expected levels of recycling (from around 60 percent for antimony and 35 percent for lead, nickel, and iron, to less than 1 percent for cobalt) and a substantially enlarged domestic production, it is expected that an end-of-century deficit of 54 percent of the U.S. demand for primary raw minerals will need to be made up from foreign sources. Thus our dependence on imports is increasing, and at the very time that general world demand has begun to rise rapidly, foreign sources grow increasingly precarious (e.g., Arabian oil), and Third World voices grow ever more insistent for a more equitable share in the earth's material bounty. The trends are explicit in the graphs, tables, and commodity summaries of the Interior Secretary's reports[2,3] as well as in the Final Report of the National Commission on Materials Policy, issued in June 1973[4] (Figures 3 and 4).

There are, of course, brighter aspects to the generally sobering picture. Ample low-grade sources of alumina other than bauxite are available with metallurgical advances and at a price. U.S. supplies of coal are ample for the immediate and intermediate future. Potassium and magnesium are easily extracted from seawater. And iron is in abundant supply the world over.

Helium, mercury, tungsten, tin, tantalum and the precious metals, however, appear to be in short supply throughout the world. And

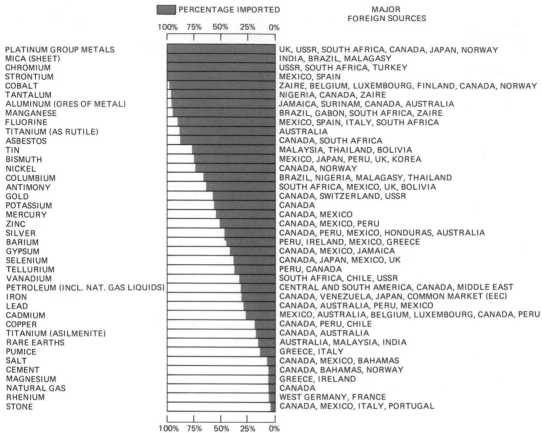

PERCENTAGE IMPORTED

MAJOR FOREIGN SOURCES

PLATINUM GROUP METALS	UK, USSR, SOUTH AFRICA, CANADA, JAPAN, NORWAY
MICA (SHEET)	INDIA, BRAZIL, MALAGASY
CHROMIUM	USSR, SOUTH AFRICA, TURKEY
STRONTIUM	MEXICO, SPAIN
COBALT	ZAIRE, BELGIUM, LUXEMBOURG, FINLAND, CANADA, NORWAY
TANTALUM	NIGERIA, CANADA, ZAIRE
ALUMINUM (ORES OF METAL)	JAMAICA, SURINAM, CANADA, AUSTRALIA
MANGANESE	BRAZIL, GABON, SOUTH AFRICA, ZAIRE
FLUORINE	MEXICO, SPAIN, ITALY, SOUTH AFRICA
TITANIUM (AS RUTILE)	AUSTRALIA
ASBESTOS	CANADA, SOUTH AFRICA
TIN	MALAYSIA, THAILAND, BOLIVIA
BISMUTH	MEXICO, JAPAN, PERU, UK, KOREA
NICKEL	CANADA, NORWAY
COLUMBIUM	BRAZIL, NIGERIA, MALAGASY, THAILAND
ANTIMONY	SOUTH AFRICA, MEXICO, UK, BOLIVIA
GOLD	CANADA, SWITZERLAND, USSR
POTASSIUM	CANADA
MERCURY	CANADA, MEXICO
ZINC	CANADA, MEXICO, PERU
SILVER	CANADA, PERU, MEXICO, HONDURAS, AUSTRALIA
BARIUM	PERU, IRELAND, MEXICO, GREECE
GYPSUM	CANADA, MEXICO, JAMAICA
SELENIUM	CANADA, JAPAN, MEXICO, UK
TELLURIUM	PERU, CANADA
VANADIUM	SOUTH AFRICA, CHILE, USSR
PETROLEUM (INCL. NAT. GAS LIQUIDS)	CENTRAL AND SOUTH AMERICA, CANADA, MIDDLE EAST
IRON	CANADA, VENEZUELA, JAPAN, COMMON MARKET (EEC)
LEAD	CANADA, AUSTRALIA, PERU, MEXICO
CADMIUM	MEXICO, AUSTRALIA, BELGIUM, LUXEMBOURG, CANADA, PERU
COPPER	CANADA, PERU, CHILE
TITANIUM (AS ILMENITE)	CANADA, AUSTRALIA
RARE EARTHS	AUSTRALIA, MALAYSIA, INDIA
PUMICE	GREECE, ITALY
SALT	CANADA, MEXICO, BAHAMAS
CEMENT	CANADA, BAHAMAS, NORWAY
MAGNESIUM	GREECE, IRELAND
NATURAL GAS	CANADA
RHENIUM	WEST GERMANY, FRANCE
STONE	CANADA, MEXICO, ITALY, PORTUGAL

3 **Percentages of U.S. Mineral Requirements Imported During 1972.** *Sources:* U.S. Bureau of Mines in Reference 4, Fig. 2.32; Reference 3, Fig. 2.

petroleum and natural gas will be exhausted or nearly so within the lifetimes of many of those alive today unless we decide to conserve them for petrochemicals and plastics (Chapter 5). Even the extraction of liquid fuels from oil shales and tar sands, or by hydrogenation of coal, will not meet energy requirements over the long term. If they were called upon to supply all the liquid fuels and other products now produced by the fractionation of petroleum, for instance, the suggested lifetime for coal, the ultimate reserves of which are probably the most accurately known of all mineral products, would be drastically reduced — and such a shift will be needed to a yet unknown degree before the end of the century.

The present situation with regard to U.S. and world reserves of selected salient mineral commodities is summarized in Figures 5 and 6.

Where the narrow bars project beyond the shaded boxes, sufficient known reserves exist at 1968 prices indicated to meet projected demands for new raw materials beyond the end of the century. Where the shaded boxes extend beyond the narrow reserve bars, projected demands exceed known reserves at indicated prices. It is obvious to knowledgeable geologists and mineral economists that reserves will be increased (as suggested in Figure 2) by new discoveries and by technological advances and price rises that will now bring noneconomic resources to the market place. The question is, at what costs in all terms and for how long?

It seems clear that a fresh and flexible materials policy must be formulated and continue to evolve if we are to cope with present challenges and expected changes. The concept of

4 **Developing Deficits Between U.S. Primary Mineral Demand and U.S. Primary Mineral Supplies.** In billions of 1971 dollars; figures for each year are at dollar values for that year. The dashed lines within patterns show demand and supply beyond 1971 at 1971 prices. (But note that the deficit for petroleum alone in 1985 is now expected to be about $64 billion.) *Source:* U.S. Bureau of Mines in Reference 3, Fig. 18.

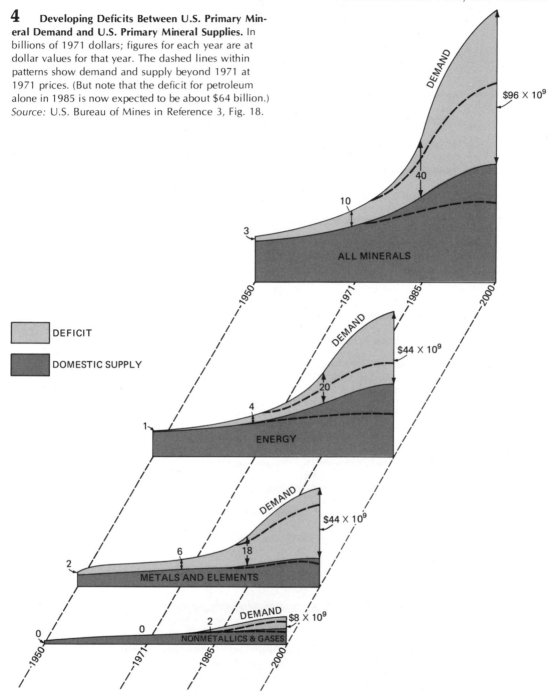

DEFICIT

DOMESTIC SUPPLY

continuing *material* growth as a keystone of economic policy needs to be reexamined, particularly where such growth does not demonstrably add to quality of life in terms of adding variety and flexibility of significant options for living generations and avoidance of their foreclosure for future ones. Environmental, social, and energy costs must be counted along with obvious fiscal costs as the total price we pay for continuing affluence. Growth must be seen

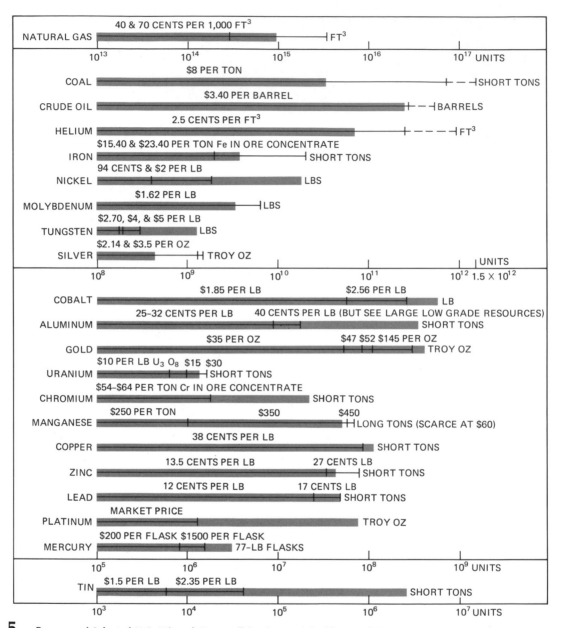

5 **Reserves of Selected U.S. Mineral Commodities Compared with Cumulative Demand 1968–2000.** The scale is logarithmic. Prices are in 1968 dollars as of 1970 estimates and are now in a state of flux. Note the variable nature of the units, with zero points for all lines far to the left of the chart. The median lines show optimistically estimated reserves as of 1968, and the shaded boxes indicate cumulative demand 1968–2000. Total reserves at different 1968 prices are indicated by the position of the vertical tick to the right of cost per unit. The dashed lines indicate speculative additions. Note that increased prices often do not bring commensurate additions in volume of commodity. *Source:* Data are from Reference 1, supplemented by data from References 5 and 6.

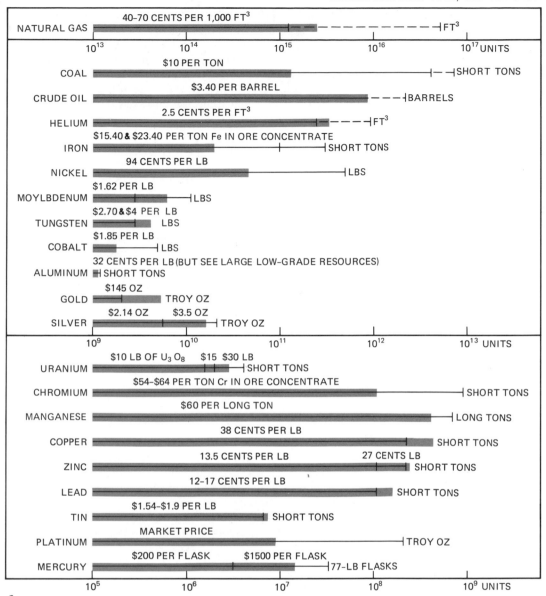

6 **Reserves of Selected Global Mineral Commodities Compared with Cumulative Demand 1968–2000.** The scale is logarithmic. Prices are in 1968 dollars as of 1970 estimates and are now in a state of flux. Note the variable nature of the units, with zero points for all lines far to the left of the chart. The median lines show optimistically estimated reserves as of 1968, and the shaded boxes indicate cumulative demand 1968–2000. Total reserves at different costs are indicated by the position of the vertical tick to the right of cost per unit. *Source:* Data are from Reference 1, supplemented by data from References 5 and 6.

not as a goal in itself but as one possible means by which real social goals may be achieved. Social goals need to be defined or redefined and the advisability of further material growth evaluated in terms of its contribution to their attainment.

The Cornucopian Premises

In view of the concerns outlined above, why do intelligent men of good faith seem to assure us that there is nothing in the mineral resource picture to be alarmed about? It can only be because they visualize a completely non-geological solution to the problem, or because they take a short-range view of it, or because they are compulsive optimists or are misinformed, or some combination of these things.

Let me first consider some of the basic concepts that might give rise to a cornucopian view of the earth's mineral resources and the difficulties that impede their unreserved acceptance. Then I will suggest some steps that might be taken to minimize the risks or slow the rates of mineral resource depletion, along with accompanying undesirable environmental and social consequences.

The central dilemma of all cornucopian premises is, of course, how to sustain an exponential increase of anything — people, mineral products, industrialization, or solid currency — on a finite resource base. This is, as everyone must realize, obviously impossible in the long run, no matter how large that base, and will become increasingly difficult in the short run. For great though the mass of the earth is, well under a tenth of a percent of that mass is accessible to us by any imaginable means (the earth's crust is about 0.4 percent of its total mass) and this relatively minute accessible fraction, as we have seen and shall see, is very unequally mineralized.

But the cornucopians are not naive or mischievous people. On what grounds do they deny the restraints and belittle the difficulties? The six main premises from which their conclusions follow are:

PREMISE I. The assumption of essentially inexhaustible cheap useful energy sources.

PREMISE II. The thesis that economics is the sole factor governing availability of useful minerals and metals.

PREMISE III. The hypothesis that metals occur in essentially continuous variation from those now constituting commercial ores down to average crustal abundance, which is inherent in Premise II, and from which emanates the misleading notion that quantity of a resource available increases as its grade decreases.

PREMISE IV. The crucial assumption of population control, without which there can be no future worth living for most of the world (or worse, the belief that quantity of people

is of itself the ultimate good, which, astounding as it may seem, is still held by a few people who ought to know better[7]).

PREMISE V. The notion of the technological fix.

PREMISE VI. The naive and unsupported faith that if all else fails, the sea will supply our needs.

Now these are appealing premises, several of which contain elements of both truth and hope. Why do I protest their unreserved acceptance? I protest because, in addition to elements of truth, they also contain assumptions that are gross oversimplifications, outright errors, or are not demonstrated. I warn because their uncritical acceptance contributes to a dangerous complacency toward problems that will not be solved by a few brilliant technological breakthroughs, a wider acceptance of deficit economy, or fallout of genius from unlimited expansion of population. They will be solved only by intensive, wide-ranging, and persistent scientific and engineering investigation, supported by new social patterns and wise legislation.

I will discuss these premises in the order cited.

PREMISE I

The concept of essentially inexhaustible cheap useful energy from nuclear sources offers the prospect of sweeping changes in the mineral resource picture.[8] It has been argued that such a development would banish many problems of environmental pollution and open up vast reserves of metals in common crustal rocks. There are, unhappily, some flaws in this roseate picture, of which it is important to be aware.

Uranium 235 is the only naturally occurring spontaneously fissionable source of nuclear power (Chapter 5). When a critical mass of uranium is brought together, the interchange of neutrons back and forth generates heat and continues to do so as long as the ^{235}U lasts. In a breeder reactor the "fast" neutrons kick common ^{238}U over to plutonium 239, which is fissionable and produces more neutrons, yielding heat and accelerating the breeder reaction. Even in conventional burner reactors some breeding takes place, and, if a complete breeding system could be put in place, the amount of energy available from uranium alone would be increased about 140-fold. If thorium also can be made to breed (to fissionable ^{233}U), energy generated could be increased about 400-fold over that now attainable. This would extend the lifetime of visible energy resources at demand levels anticipated by 1980 by thousands of years (assuming no further increase of demand) and gain time to work on contained nuclear fusion and the large-scale use of solar energy.

The main difficulties involve supplies of naturally fissionable ^{235}U for burner reactors, the hazards of the ^{239}Pu used in breeder reactors, and the problem of segregating radioactive by-products over the interval of 20 half-lives needed for them to reach near-background radiation levels (Chapter 12). The rate of radioactive decay is exponential and the half-life is the time it takes a radioactive substance to decay to half the initial mass. It ranges from a few seconds for some products to 24,360 years for ^{239}Pu. The worst foreseeable problem involves ^{239}Pu, of which large quantities could be in use and in transit by the end of the century if present plans for breeder reactors materialize. Plutonium 239 is carcinogenic in quantities as small as one-millionth of a gram. It is worth $10,000 per kilogram. It would take only 5 kilograms to make a crude atom bomb. The hazards of accidental dispersal or theft for terrorist activities are frightening. Any site of accidental dispersal would be contaminated for about half a million years. The use of uranium 233 (from thorium 232) instead of or in addition to ^{239}Pu may also become possible in breeder reactors of the future. Radioactive ^{233}U, however, has a half-life of 162,000 years, meaning contamination for over 3 million years in areas of accidental dispersal. Although I have previously argued differently,[6] I am now convinced that if man were wise he would abstain from further development of the breeder reactor and limit his use of nuclear fission in general. He would get by instead with the fossil fuels (including oil shale and tar sands), with all their lesser hazards, plus a limited use of burner reactors until a workable fusion reactor or solar-based economy (or some acceptable combination of sustainable energy sources) could be developed (Chapter 5).

Nevertheless, if and when a breeder reactor or contained fusion does become available as a practicable energy source, how might this help with mineral resources? It is clear immediately that it could take pressure off the fossil fuels so that it would become feasible to reserve them for petrochemicals, plastics, essential liquid propellants, and other special purposes not served by nuclear fuels. It is also clear that cheap massive transportation, or direct transmittal of large quantities of cheap electric power to, or its generation at, distant sources, would bring the mineral resources of remote sites to the marketplace — either as bulk ore for processing or as the refined or partially refined product.

What is not clear is how this very cheap energy might bring about the extraction of thinly dispersed metals in large quantity from "common rock." The task is very different from the possible recovery of liquid fuels or natural gas by nuclear fracturing. The procedure usually suggested is the breakup of rock in place at depth with a nuclear blast, followed by hydrometallurgical or chemical mining. The problems, however, are great. The prospect immediately arises that the health hazards may be so great as to call for a ban on all nuclear explosions (Chapter 12). Assuming no such ban, complexing solutions in large quantity, also from natural resources, would have to be brought into contact with the particles desired. This means that the enclosing rock must be fractured to that particle size. Then other substances, unsought, may use up and dissipate valuable reagents. Or the solvent reagents may escape to groundwaters and become contaminants. And the bacteria that catalyze reactions of metallurgical interest are all aerobic, so that, in addition to having access to the particles of interest, they must also be provided with a source of oxygen underground if they are to work there.

Indeed, the energy used in breaking rock for the removal of metals is not now a large fraction of mining cost in comparison with that of labor and capital. The big expense is in equipping and utilizing manpower, and, although cheap energy will certainly reduce manpower requirements, it will probably never adequately substitute for the intelligent man with the pick at the mining face in dealing with vein and many replacement deposits, where the sought-after materials are irregularly concentrated in limited spaces. There are also limits to the feasible depths of open-pit mining, which would be by all odds the best way to mine common rock. Few open-pit mines now reach much below about 1,500 feet. It is unlikely that such depths can be increased by as much as an order of magnitude. Moreover, the quantity of rock removable decreases exponentially with depth, because pit circumference must decrease downward to maintain stable walls.

It may also not be widely realized by nongeologists that many types of ore bodies have definite floors or pinch out downward, so that extending exploitative operations to depth gains no increase in ore produced. Even where mineralization does extend to depth, of course, exploitability is ultimately limited by temperature and rock failure.

Then there is the problem of reducing radioactivity so that ores can be handled and the refined product utilized without harm — not to mention the disposal of heat, waste rock, and spent reagents.

Altogether the problems are sufficiently formidable that it would be foolhardy to accept them as resolved in advance of a substantial nuclear capability plus a demonstration that either cheap electricity or nuclear explosions will significantly facilitate the removal of metals from any common rock, and that nuclear fracturing techniques do not present unacceptable hazards to public health.

Indeed, it is the uncommon features of a rock that make it a candidate for mining! Even with a complete nuclear technology, sensible people will seek, by geological criteria, to identify and work first those rocks or ores that show the highest relative recoverable enrichments in the desired minerals. In fact, recent studies of metal mining show that below certain grades (that vary with the mineral but are generally well above crustal abundance) the curves of increasing energy cost per unit extracted approach the vertical.

The reality is that even the achievement of ample cheap energy offers no guarantee of unlimited mineral resources in the face of geologic limitations, expanding populations with increased per capita demands, and growing environmental impact, even over the middle term. To assume such for the long term would be sheer folly.

PREMISE II

The thesis that economics is the sole, or at least the dominant, factor governing availability of useful minerals and metals is one of those vexing part truths that has led to much seemingly fruitless discussion between economists and geologists. This proposition bears examination.

It seems to have its roots in that interesting economic index known as the gross national product (GNP). Considering how small the current dollar value of the raw materials themselves is compared to the total GNP, it is logically deduced that the GNP could, if necessary, absorb a severalfold increase in cost of raw materials. The gap in logic comes when this is confused with the notion that all that is necessary to obtain inexhaustible quantities of any substance is either to raise the price or to increase the volume of rock mined. In support of such a notion, of course, one can point to diamond, which, in the richest deposit ever known, occurred in a concentration of only 1 to 25 million, but which, nevertheless, has continued to be available. The flaw is not only that we cannot afford to pay the price of diamond for many substances, but also that no matter how much rock we mine, we cannot get diamonds out of it if there were none there in the first place.

Daniel Bell[9] comments on the distorted sense of relations that emerges from the cumulative nature of GNP accounting. Thus, when a mine is developed, the costs of the new facilities and payroll become additions to the GNP, whether the ore is sold at a profit or not. Should the mine wastes at the same time pollute a stream, the costs of cleaning up the stream or diverting the wastes also become additions to the GNP. Similarly, if you hire someone to wash the dishes this adds to GNP, but if they are done by a member of the family it does not count. From this it results that mineral raw materials and housework are not very impressive fractions of the GNP. What seems to get lost sight of is what a mess we would be in without either!

Assuming continuous variation of ore grades from rich to poor, continuance of access to foreign markets, and stabilization of demand, mineral economists appear to be on reasonably sound grounds in postulating the relatively long-term availability of certain ores that are localized by sedimentary or weathering processes such as those of iron and aluminum. But the type of curve that can with some reason be applied to some deposits and metals is by no means universally applicable. This difficulty is aggravated by the fact that conventional economic indexes minimize the vitamin-like quality for the economy as a whole of the raw materials, whose enhancement in value through beneficiation, fabrication, and exchange accounts for such a large part of the material assets of society.

In a world that wants to hear only good news, some economists are perhaps working too hard to emancipate their calling from the epithet of "dismal science," but not all of them are. One dissenting voice is that of Kenneth Boulding,[10] who observes that "The essential measure of the success of the economy is not production and consumption at all, but the nature, extent, quality, and complexity of the total capital stock, including in this the state of the human bodies and minds included in the system." Until this concept penetrates deeply into the councils of government and the conscience of society, there will continue to be a wide gap between economic aspects of national and industrial policy and the common good, and the intrinsic significance of raw materials will continue to be inadequately appreciated.

Economic geology, which in its best sense brings all other fields of geology to bear on resource problems, is concerned particularly with questions of how certain elements locally attain geochemical concentrations that greatly exceed their crustal abundance and with how this knowledge can be applied to the discovery of new deposits and the delineation of reserves. Economics and technology play equally important parts with geology itself in determining what deposits and grades it is practicable to exploit. Neither economics nor technology nor geology can make an ore deposit where the desired substance is absent or exists in insufficient quantity.

The reality is that economics per se, powerful though it can be when it has material resources to work with, is not all powerful. Indeed, without material resources from some source to start with, no matter how small a fraction of the GNP they may represent, economics is of no consequence. The current

orthodoxy of economic well-being through obsolescence, overconsumption, and waste will prove, in the long term, to be a cruel and a preposterous illusion.

PREMISE III

Premise III, the postulate of essentially uninterrupted variation from ore to average crustal abundance, is seldom if ever stated in that way, but it is inherent in premise II. It could almost as well have been treated under premise II, but it is such an important and interesting idea that separate consideration is warranted.

If the postulated continuous variation of grade/tonnage ratio were true for mineral resources in general, volume of "ore" (not metal) would invariably increase with decrease of grade mined, the handling of lower grades would be compensated for by the availability of larger quantities of elements sought, and reserve estimates would depend only on the accuracy with which average crustal abundances were known. Problems in extractive metallurgy and environmental and social costs are, of course, not considered in such an outlook.

Were this simple picture true, it would supplant all other theories of ore deposits, invalidate the foundations of geochemistry, divest geology of much of its social relevance, and make the fate of the nonenergy mineral industry a simple function of economics and energy supply.

Unfortunately, this postulate is simply untrue in a practical sense for many critical minerals and is only crudely true, leaving out metallurgical problems, for particular metals, such as iron and aluminum, whose patterns approach the predicted form.[6] Sharp discontinuities exist in the abundances of mercury, tin, nickel, molybdenum, tungsten, manganese, cobalt, diamond, the precious metals, and even such staples as lead, zinc, and copper (from which the idea of continuously varying grade/tonnage ratios was originally derived).

Helium is a good example of a critical substance in short supply. Although a gas which has surely at some places diffused in a continuous spectrum of concentrations, particular concentrations of interest as a source of supply appear from published information to vary in a stepwise manner. Here I draw mainly on data summarized in the chapters on helium in Ref-

111

erences 1 and 11. Although an uncommon substance on earth, helium serves a variety of seemingly indispensable uses. Until recently a bit less than half of the helium now consumed in the United States was used in pressurizing liquid-fueled missiles as well as spaceships. Shielded-arc welding is the next largest use, followed closely by its use in producing controlled atmospheres for growing crystals for transistors, processing fuels for nuclear energy, and cooling vacuum pumps. Only about 5.5 percent of the helium consumed in the United States is now being used as a lifting gas. It plays an increasingly important role, however, as a coolant for nuclear reactors and a seemingly indispensable one in cryogenics and superconductivity. In the latter role, it could control the feasibility of massive long-distance transport of nuclear-generated electricity. High-helium low-oxygen breathing mixtures have important medical applications and may well be critical to man's long-range success in attempting to operate at great depths in the sea. Other uses are in research, purging, leak detection, chromatography, and so on.

Helium thus appears to be a very critical element, as the Department of the Interior once recognized in establishing its now-defunct helium-conservation program. What are the prospects that there will be enough helium in the twenty-first century?

The only presently utilized source of helium is in natural gas, where it occurs at a range of concentrations from as high as 8.2 percent by volume to zero. The range, however, in particular gas fields of significant volume, is apparently not continuous. Dropping below the one field (Pinta Dome) that shows an 8.2 percent concentration, we find a few small isolated fields (Mesa and Hogback, New Mexico) that contain about 5.5 percent helium, and then several large fields (e.g., Hugoton and Texas Panhandle) with a range of 0.3 to 1.0 percent helium. Other large natural gas fields contain either no helium or show it only in quantities of less than 5 parts per 10,000. From the latter there is a long jump down to the atmosphere, which has a concentration of only 1 part per 200,000.

Annual demand for helium in 1968 was 842 million cubic feet, with a projected increase in demand to between 1.4 and 3.6 billion cubic feet annually by the year 2000. It will be possible to meet such an accelerated demand for a limited time as a result of Interior's now-defunct purchase and storage program, which will augment recovery from natural gas. As now foreseen, if increases in use do not outrun estimates, existing sources will meet needs to somewhat beyond the turn of the century. When known and expected discoveries of reserves of helium-containing natural gas and nonburnable gases are exhausted, the potential sources of new supply will be from the atmosphere, from fusion reactor technology, or by synthesis from hydrogen, a process whose practical feasibility and adequacy remain to be established. The present cost of helium produced from the atmosphere is about 100 times higher than that from natural gas, although costs would certainly decrease to some extent on volume production. A fusion economy is unlikely before the year 2000, if ever, and is not likely to produce more than about one-fourth of the helium needed (oral communication from Charles Laverick, superconductivity specialist at Argonne National Laboratory).

A good hedge against future uncertainties would be to reinstitute and expand Interior's conservation program with better safeguards against excess profits. New sources of helium must be sought. Research into possible substitutions, recovery and reuse, synthesis, and extraction from the atmosphere should be accelerated. And we must curtail waste. Natural resources are the priceless heritage of all the people, including those yet to be born; their waste cannot be tolerated.

Problems of the adequacy of reserves exist for many other substances, especially under the escalating demands of rising populations and expectations, and along with them problems of environmental damage and international tension as we seek to meet growing demands from lean or foreign sources. It is the growing conviction of many geologists that time is running out. Dispersal of metals that could be recycled should be controlled. Imaginative incentive legislation should be generated to define permissible mixes of materials in manufactured products, to regulate disposal of "junk" metal, and otherwise to encourage wiser and more conserving practices. Above all, the wastefulness of war and preparation for it must be terminated if reasonable options for posterity are to be preserved.

The reality is that a healthy, nonpolluting, industrial economy can be maintained only on a firm base of geologic knowledge and geochemical and metallurgical understanding of the distribution, limits, processing, and production side effects of metals, mineral fuels, and chemicals in the earth's crust and hydrosphere.

PREMISE IV

The assumption that world populations will soon attain and remain in a state of balance is central to all other premises. Without this, the rising expectations of the poor are doomed to failure, and the affluent can remain affluent only by maintaining existing shameful discrepancies. Taking present age structures and life expectancies of world populations into account, it seems certain that, barring other forms of catastrophe, world population will reach 6 or 7 billion by about the turn of the century, regardless of how rapidly limitation of family size is accepted and practiced (Chapter 2).

On the most optimistic assumptions, this is probably close to the maximum number of people that the world can support on a reasonably sustained basis, even under strictly managed conditions, at a general level of living roughly comparable to that now enjoyed in the poorer parts of Western Europe. Far better to stabilize at a much smaller world population. In any case, much greater progress than is as yet visible must take place over much larger parts of the world before optimism on the prospects of voluntary global population control at any level can be justified. And even if world population did level off and remain balanced at about 7 billion (which is extremely unlikely short of global catastrophe), it would probably take close to 100 years of intensive, enlightened, peaceful effort to lift all mankind to anywhere near the level of material prosperity suggested or even much above the level of chronic malnutrition and deprivation. An eventual decline of numbers to 2 or 3 billion would offer much better prospects for the human race.

This is not to suggest that the predicament in which we find ourselves is one for panic or gloomy introversion. Rather it is a challenge to focus with energy and realism on seeking a truly better life for all people living and yet unborn and on keeping the latter to the minimum. On the other hand, an uncritical optimism is a luxury the world cannot, at this juncture, afford.

A variation of outlook on the population problem which, surprisingly enough, exists among a few nonbiological scholars is that quantity of people is of itself a good thing. The misconception here seems to be that frequency of effective genius will increase, even exponentially, with increasing numbers of people and that there is some risk, as a friend once put it, of "breeding out to a merely high level of mediocrity" in a stabilized population. The extremes of genius and idiocy, however, appear in about the same frequency at birth from truly heterogeneous gene pools regardless of size. What is unfortunate, among other things, about overly dense concentrations of people is that this leads not only to reduced likelihood of the identification of mature genius, but to drastic reductions in the development of potential genius, owing to malnutrition in the weaning years and early youth, accompanied by retardation of both physical and mental growth. If we are determined to turn our problems over to an elite corps of mental prodigies, a more sure-fire method of getting enough such prodigies may be just around the corner. Nuclear transplant from various adult tissue cells into fertilized ova whose own nuclei have been removed has already produced identical copies of amphibian nucleus donors and can probably do the same in man.[12] Thus, through a different kind of nuclear engineering, we may be on the verge of being able to make as many copies as we want of any particular genius as long as we can get a piece of his or her nucleated tissue and find eggs and incubators for the genome aliquots to develop in.

In reality, however, without real population control and limitation of material demand, mankind is headed for very serious trouble. If the bearing of children is a right, it is also a responsibility. The responsibility needs to be emphasized. More basic freedoms should be the right not to be born into a world of want and not to bear unwanted children. I am convinced that we must give up (or have taken away from us) the right to have more children

113

than are needed to replace a population of manageable size, or see all other freedoms lost for *them*. Nature, to be sure, will restore a dynamic balance between our species and the world ecosystem if we fail to do so ourselves — by famine, pestilence, plague, or war. It seems, but is not, unthinkable that this should happen. If it does, of course, mineral resources may then be or appear to be relatively unlimited in relation to demand for them.

PREMISE V

The notion of the "technological fix" expresses a view that is at once full of hope and full of risk. It is a gripping thought to contemplate a world set free by abundant cheap energy. Imagine soaring cities of aluminum, plastic, and thermopane where all live in peace and plenty at unvarying temperature and without effort, drink distilled water, feed on produce grown from more distilled water in coastal deserts, and flit from heliport to heliport in capsules of uncontaminated air. Imagine a world teeming with wanted children, who, of course, will grow up in large apartment complexes and under such germ-free conditions that they will need to wear breathing masks if they ever do set foot in a park or a forest. Imagine a world in which there is no balance-of-payments problem, no banks or money, and such mundane affairs as acquiring a shirt or a mate are handled for us by central computer systems. Imagine, if you like, a world in which the only problem is boredom, all others being solved by the state-maintained system of genius-technocrats produced by transfer of nuclei from the skin cells of certified gene donors to the previously fertilized ova of final contestants in the annual ideal-pelvis contest. Imagine the problem of getting out of this disease-free world gracefully at the age of 150!

Of course, such a view may have little appeal to people not conditioned to think in those terms. But the risk of slipping bit by bit into such a smothering condition as one of the better possible outcomes is inherent in any proposition that encourages or permits people or industries to believe that they can leave their problems to the invention of technological fixes by someone else. The problem, indeed, is not that future technology will not achieve great things but that too blind a dependence on it is likely to lead to unanticipated crises, to failure to organize for an appropriate use of its findings, or to misapplication of technological achievements.

Consider what would be needed in terms of conventional mineral raw materials merely to raise the level of all 4.0 billion people now living in the world to the average of the 209 million residents of the United States at the end of 1973. In terms of present mineral commodities, it can be estimated that this would require a "standing crop" of about 100 to 200 times the present annual production of major industrial materials, with a standing rate of world power production about 6 times the present, along with all the expected and unforeseen environmental consequences of such levels of production from ever-larger volumes of ore of ever-decreasing grade. To support the populations expected by the year 2000 at the same level would require, of course, a corresponding increase of all the above numbers or substitute measures. The iron needed could probably be produced over a long period of time, perhaps even by the year 2000, given a sufficiently large effort. But, the molybdenum needed to convert the iron to steel could become a limiting factor. The quantities of lead, zinc, and tin called for far exceed all measured, indicated, and inferred world reserves of these metals.

This exercise gives a crude measure of the pressures that mineral resources may come under. It seems likely, to be sure, that substitutions, metallurgical research, and other technological advances will come to our aid, and that not all peoples of the world will find a superfluity of obsolescing gadgets necessary for happiness. But this is balanced by the equal likelihood that world population will not really level off at 6.5 or 7 billion and that there will be growing unrest to share the material resources that might lead at least to an improved standard of living. The situation is also aggravated by the attendant problems of disposal of mine wastes and chemically and thermally polluted waters on a vast scale.

The technological fix, as its informed proponents well understand, is not a panacea but an anesthetic. It may keep the patient quiet long enough to decide what the best long-range course of treatment may be, or even to solve

114

some of his problems permanently, but it would be tragic to forget that a broader program of treatment and recuperation is necessary. The flow of science and technology has always been fitful, and population control is a central limiting factor in what can be achieved. It will require much creative insight, hard work, public enlightenment, and good fortune to bring about the advances in discovery and analysis, recovery and fabrication, wise use and conservation of materials, management and recovery of wastes, and substitution and synthesis that will be needed to keep the affluent affluent and bring the deprived to dignified levels of physical comfort. It will probably also take some revision of criteria for self-esteem, achievement, and pleasure if the gap between affluent and deprived is to be narrowed and demand for raw materials kept within bounds that will permit man to enjoy a future as long as his past, and under conditions that would be widely accepted as agreeable.

The reality is that the promise of the "technological fix" is a meretricious one, full of glittering appeal but often oblivious to environmental and social problems. Technology and "hard" science we must have, in sustained and increasing quality, and in quantities relevant to the needs of man. But in dealing with the problems of resources in relation to man, let us not lose sight of the fact that this is the province of the environmental and social sciences.

Having expressed these reservations, it is important to recognize that there is much technology can do and should be called upon to do, *particularly in the area of recycling, substitution, and environmental protection and restoration*. Indeed, recycling materials and substitution of common for uncommon metals are both important antipollution activities, and much more than is now being done can certainly be achieved in these fields.

Nevertheless, it is well to remember the social components and limitations of recycling and substitution. Although technological advances in waste management have reduced or eliminated the need for hand-sorting of wastes, a great deal of individual effort is still called upon to avert irrecoverable dispersal of materials that should be recycled. In an expo-

nentially growing system like ours, moreover, even total recycling (which is impracticable) would get only half of what was needed for each doubling of demand. The other half would still have to come from newly mined and processed (or stockpiled) materials. In thinking about recycling it is well to remember the great variability in recoverability of substances, depending on the nature of the material, price, energy requirements, and nature of use. Iron is highly recoverable because it is durable and used in large volume, lead and nickel because they are durable and relatively costly. Fossil fuels are not recyclable at all. When burned they are gone forever. The precious metals and copper tend to stay in use or be hoarded. Much silver is lost in dentistry and photography. The possibility thus ranges from perhaps 90 percent recyclability for some metals down to very low for others and zero for the fossil fuels.

As for substitution, one may wisely take a position somewhere between that of "the doctrine of infinite substitutability" and hand-wringing. With most of the world's naturally occurring elements and a good many of its "manufactured" ones already in use, it is clear that there are limitations. Still a great many substitutions will undoubtedly be made in using more-common for less-common things and in synthesizing new kinds of materials from recombination of old ones. In the end, one cannot disagree with those who say: "but surely some kind of industrial technology can be sustained on the universal and abundant elements alone — on iron, aluminum, magnesium, and the silicates with sunlight, hydrogen, oxygen, and nitrogen." The questions are: what kind of industrial economy, how to cope with population and pollution, what sort of performance would one get from the substituted materials as compared to those substituted for? What would take the place of gold in high-speed computers or helium in cryogenics, superconductivity, and special breathing mixtures? Nevertheless, the attempt to create such a technology is one of the great challenges down the road for materials science and engineering. A major effort should soon be begun to invent the future of an industrial economy based wholly or mainly on abundant, safe, materials whose extraction and

115

processing can be achieved with minimal environmental and social costs. But without control of populations and material consumption at manageable levels, even complete success in meeting this challenge would be to no avail.

PREMISE VI

What, finally, about marine mineral resources, often proposed as a veritable cornucopia, waiting only to be harvested?[13] This notion is so widely accepted that it is worth considering what we actually know about prospective mineral resources from the sea.

In 1964, mineral production from the sea represented about 10 percent of the total known value of the products recovered and about 5 percent of the entire world mineral output. Sizable quantities of oil and gas, sulfur, magnesium, bromine, salt, oyster shells, tin, and sand and gravel are now being produced from the sea.[6,14,15] What are the future prospects for these as well as for other substances not now being recovered in quantity?

Mineral and chemical resources from the sea may be found among the following (see also Figure 7):

1. Seawater.
2. Deposits concentrated as residual heavy substances (placer deposits) within or beneath now-submerged beach or stream deposits.
3. Sediments other than placers, and sedimentary rocks that overlie crystalline rocks (a) on the continental shelves and slopes (about 15 percent of the total sea floor) and (b) beyond the continental margins (about 85 percent of the total sea floor; the truly oceanic realm).
4. Crystalline rock exposed at the sea floor or lying beneath sediments (a) on the continental margins and (b) beyond the continental margins.

The sea contains about 1.3 billion cubic kilometers of seawater, an amount so vast that quantities of dissolved substances are large even where their concentrations are small. Yet recovery of such substances accounts for little more than 2 percent of current production of marine minerals and chemicals. Only magnesium, bromine, and common salt are now

being extracted in substantial quantities, and, for them, the seas do contain reserves that can be considered to be almost inexhaustible under any foreseeable pressures.

At the other extreme of accessibility is the 10 billion metric tons of gold in seawater, about 0.0013 troy ounce of gold in every million liters of water. Although capable people and corporations have worked intensively at the problem, the amount of gold so far reported to have been extracted from seawater is trivial.

Sixty-four of the 90 naturally occurring elements now known on earth have been detected in seawater. Of these only 15 occur in quantities of more than 1.8 kilograms per million liters. Only 9, all of them among the first 15 in abundance, represent 1965 values of more than $2.50 per million liters (chlorine, sodium, magnesium, sulfur, calcium, potassium, bromine, lithium, and rubidium). These 9 (or their salts), plus boron, fluorine, and iodine, offer the best promise for direct recovery from seawater through ion exchange, biological concentration, or more novel processes. Omitting these 12, and very few others, the metal elements we are likely to have greatest need to extract from seawater offer little promise for direct recovery.

Placer deposits now offshore were formed by gravitational segregation in and beneath former beach and stream deposits when the sea stood lower or the land higher. The outer limit at which such deposits can be expected is about 100 to 130 meters depth, the approximate position of the beach when Pleistocene glaciation was at its peak.

Diamonds, gold, and tin have been or are being recovered from submarine placers. Approximate 1964 global values, omitting cost of recovery, were as follows: diamonds, $4 million (a value exceeded by cost of recovery); tin, about $21 million; gold, unknown.

Submarine placers of tough heavy metals such as gold, tin, and platinum offer the best prospects for the practicable recovery of mineral resources from the sea other than magnesium (from seawater) and oil and gas (from continental shelf and slope sediments).

Continental-shelf sediments and salt-dome structures that penetrate them account for nearly 98 percent of the mineral and chemical wealth currently produced from within and beneath the sea — by far the greater part of this being oil and gas. About 20 percent of the world's petroleum and natural gas now comes from offshore, and this fraction will certainly increase. It has been estimated that about 700 to 1,000 billion barrels of liquid fuels are potentially recoverable from offshore areas worldwide, which is perhaps somewhat more than that remaining to be produced on land. This amount might be increased if the sediments of the continental slopes or mid-ocean rises were to be found rich in petroleum.

The oil-producing salt domes in the Gulf of Mexico also yield sulfur. More than 5 percent of the world's sulfur now comes from such sources. Although currently known reserves will have been depleted by about 1990, however, there is no shortage of sulfur. Indeed, its removal from various fossil fuels to comply with environmental standards may create a temporary sulfur surplus.

Sediments and sedimentary rocks beyond the continental shelves and slopes — the pelagic sediments — are strikingly different from those that comprise or blanket the continental margins. Here is where we find the curious and geologically interesting manganese nodules and crusts. Despite the extravagant claims that have been made, however, the prospects of these deposits for future mineral production remain to be established.

Finding and exploiting mineralized crystalline rocks on the sea floor, even on the continental shelves, involves the same problems as on land, and many more. Although it is clear from geological data that the substructure of the shelf and slope is of a continental rather than an oceanic type, and that mineral deposits comparable to those found on the adjacent land are to be expected, problems peculiar to the region hamper their discovery and exploitation. Nevertheless, it seems statistically certain that ore deposits exist in crystalline rocks somewhere on or beneath the continental shelves. How many of them will eventually be discovered and worked is another question.

Seaward beyond the continental margins the difficulties increase. The crystalline rocks of the ocean basin appear to be mainly volcanic. These rocks are of a type that is relatively rich in iron and magnesium (basalts) as compared to most continental rocks. These oceanic basalts are limited both in the variety of included minerals and in the degree of enrichment to be found. Nickel, chromium, copper, and platinum may be present, but, unless the manganese nodules prove to be such, the first submarine ore of any of these is yet to be found. Iron and magnesium are probably the most abundant metal elements in the oceanic rocks, but such ores, if any, could not compete with sources from dry land and from seawater.

Mineral and chemical resources of the sea that will be significant for man over the next half-century are those that can be extracted from seawater or recovered from the seabed of the continental shelf and slope (and perhaps the oceanic rises). About 5 percent of the world's known production of geological wealth came from the shelves and seawater in 1964 and the trend is upward. Oil and gas are by far the most important products, although their duration at expected rates of consumption will be limited. Seawater can supply ample magnesium and bromine, as well as common salt and some other substances. It will, however supply few important metal elements other than magnesium, sodium, potassium, iodine, and perhaps strontium and boron. Oyster shells, by no means an inexhaustible resource, are being taken from the shallower parts of the Gulf of Mexico in relatively large quantities for use in the recovery of magnesium from seawater and for road construction. The use of nearshore submarine sand and gravel, not yet large, will probably increase as coastal cities expand over and use up other local sources.

The sediments of the continental shelves and the crystalline rocks beneath them can be expected to produce mineral commodities similar to those of the immediately adjacent land. One way of guessing the magnitude of such resources is to reflect the hundred-fathom depth line across the shoreline. It then defines an onshore area about equivalent to the continental shelf and having roughly similar geology. For the equivalent onshore area in the United States (in 1966 dollars), the total cumulative value of minerals extracted since the beginning of the U.S. mineral industry, is about $160 billion (in 1968 dollars), exclusive of oil and gas, or $240 billion with oil and gas. Because of geological differences from the land, and other

117

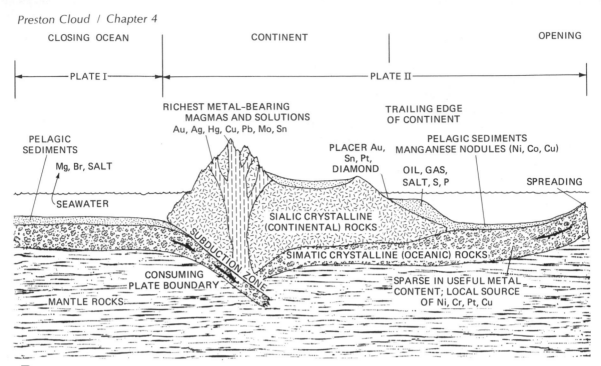

CLOSING OCEAN — CONTINENT — OPENING

PLATE I — PLATE II

7 **Schematic Vertical East-West Profile of Idealized Crustal Plates.** Sites of accumulation of different types of mineral resources are indicated. The patterns symbolize similar general rock types and sites of accumulation.

obstacles that complicate discovery and recovery in the subsea environment, however, this figure may be considerably larger than what we can expect from the continental shelves.

The ocean basins beyond the continental rises have little promise as sources of mineral resources.

A "mineral cornucopia" beneath the sea thus exists only in hyperbole. What is actually won from it will be the result of persistent imaginative research, inspired invention, bold and skillful experiment, and intelligent application and management — and such resources as are found will come mostly from the continental shelves, slopes, and rises. Whether they will be large or small remains to be seen. It is a fair guess that they will be substantial; but if present concepts of earth structure and of sea-floor composition and history are correct (Figure 7), minerals from the seabed (other than oil and gas) are not likely to compare with those yet to be recovered from the emerged lands. As for seawater itself, despite its great volume and the quantities of dissolved salts it contains, it can supply few of the substances considered essential to modern industry.

Rather than a cornucopia it is an arithmetic trap.

The Nub of the Matter

The reality of mineral distribution, in a nutshell, is that it is neither inconsiderable nor limitless, and that we just don't know yet, in the detail required for considered weighing of comprehensive long-range alternatives, where or how the critical elements are concentrated. Substances whose concentrations are controlled by sedimentary processes or structures, such as the fossil fuels and, to a degree, iron and alumina, we can comprehend and estimate within reasonable limits. Reserves, grades, locations, and recoverability of many critical metals, on the other hand, are affected by a larger number of less-well-understood variables.

A good beginning toward the assessment of the long-term capacity of the United States to convert potential mineral resources into reserves and eventually industrial materials has been made by the U.S. Geological Survey,[11]

118

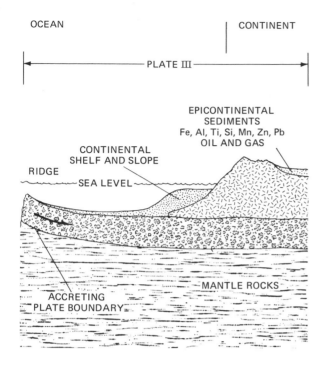

OCEAN | CONTINENT

PLATE III

EPICONTINENTAL
SEDIMENTS
Fe, Al, Ti, Si, Mn, Zn, Pb
OIL AND GAS

CONTINENTAL
SHELF AND SLOPE

RIDGE
SEA LEVEL

MANTLE ROCKS

ACCRETING
PLATE BOUNDARY

and that effort is continuing. A good grasp of metallogenic provinces and episodes is emerging from the new theory of plate tectonics[16,17] (in which the whole earth is divided up into eight or more moving plates that bump into or override one another along earthquake-prone boundaries). Downgoing plates undergo melting and remobilization that tends to generate fluids and vapors that move upward through the overriding continental blocks to become enriched in dissolved metals and compounds, which they deposit in the upper layers of the crust. From this we are beginning to understand better where to look for yet-undiscovered mineral resources (e.g., Figure 7). Yet we have a long way to go in gaining the kind of a global understanding about resource distribution, grades, and recoverability that is needed for the most effective global and national planning.

Advances in a number of areas must be achieved before we can face the future with the kind of calm and durable confidence that comes from secure knowledge and cooperative covenants rather than the aggressive confidence of temporary military or economic superiority or the bravado of graveyard whistling. A comprehensive global program of metallogenic mapping and geochemical census must be linked with advances in materials science and technology, social responsibility, environmental ethics, control of population and material consumption, legislative reform, and international collaboration toward the establishment of equity and harmony among peoples.

We in North America began to develop our rich natural endowment of mineral resources at an accelerated pace before the rest of the world. Thus it stands to reason that, to the extent that we are unable to meet needs by imports, we will feel the pinch sooner than countries, such as the USSR, which have a larger component of virgin mineral lands.

In some instances nuclear energy or other technological fixes may buy time to seek better solutions or will even solve a problem permanently. But sooner or later man must come to terms with his environment and its limitations. The sooner the better. The year 2052, by which time projected rates of consumption will have placed enormous strains on the world's most useful metals (Figure 2), is only as far from the present as the invention of the airplane and the discovery of radioactivity. In the absence of planned population control or catastrophe there could be 20 billion people on earth by then! Much that is difficult to anticipate can happen in the meanwhile to be sure, and to center faith in a profit-motivated technology and refuse to look beyond a brief "forseeable future" is a choice widely made. Against this we must weigh the consequences of error or thoughtless inaction and the prospects of identifying constructive alternatives for deliberate courses of long-term action (or inaction) that will affect favorably the long-range future. It is well to remember that to do nothing is equally to make a choice. It is well to remember also that little constructive is likely to happen without effective legislation and that effective legislation is not likely to be written until the people are ready for it.

Some of the things that such legislation should foster are:

1. Stabilization of populations at levels supportable at a high quality of life over the long term.

119

2. Slowdown of the rate of increase and eventual stabilization of material consumption.
3. Recycling at the maximal practicable level.
4. Conserving use of crucial materials in limited supply.
5. Regular monitoring of the entire supply–demand picture, with particular reference to adequacy of supply, control of demand, and environmental and sociopolitical side effects.
6. The creation of a nonpolluting industrial technology based as far as possible on common and abundant raw materials.

References

1. U.S. Bureau of Mines. 1970. *Mineral Facts and Problems*, 4th ed. (Bureau of Mines Bull. 650). Government Printing Office, Washington, D.C.
2. Secretary of the Interior. 1972. First Annual Report of the Secretary of the Interior Under the Mining and Minerals Policy Act of 1970. Government Printing Office, Washington, D.C.
3. Secretary of the Interior. 1973. Second Annual Report of the Secretary of the Interior Under the Mining and Minerals Policy Act of 1970. Government Printing Office, Washington, D.C.
4. National Commission on Materials Policy. 1973. Final Report. Government Printing Office (Stock No. 5203-00005), Washington, D.C.
5. U.S. Bureau of the Census. 1972. Statistical Abstract of the United States. 92d Congress, 2d Session, House Document No. 92-257. Government Printing Office, Washington, D.C.
6. National Academy of Sciences, Committee on Resources and Man. 1969. *Resources and Man*. W. H. Freeman and Company, San Francisco.
7. Clark, C. 1967, *Population Growth and Land Use*. Macmillan Publishing Co., Inc., New York.
8. Weinberg, A., and Young, G. 1966. The nuclear energy revolution. *Proc. Natl. Acad. Sci. (U.S.)* 57: 1–15.
9. Bell, D. 1967. Notes on the post-industrialist society II. *The Public Interest 7:* 102–118.
10. Boulding, K. E. 1966. The economics of the coming spaceship earth. In *Environmental Quality in a Growing Economy* (H. Jarrett, ed.). The Johns Hopkins Press (for Resources for the Future), Baltimore, Md.
11. U.S. Geological Survey. 1973. *United States Mineral Resources* (U.S. Geol. Survey Prof. Paper 820). Government Printing Office, Washington, D.C.
12. Lederberg, J. 1966. Experimental genetics and human evolution. *Bull. Atomic Scientists 22:* 1–11.
13. Mero, J. L. 1965. *The Mineral Resources of the Sea*. American Elsevier Publishing Co., Inc., New York.
14. Emery, K. O. 1966. Geological methods for locating mineral deposits on the ocean floor. *Trans. 2nd Marine Technol. Soc. Conf.:* 24–43.
15. Bascom, W. 1967. Mining the ocean depths. *Geosci. News I:* 10–11, 26–28.
16. Guild, P. W. 1972. Metallogeny and the new global tectonics. *Trans. 24th Internat. Geol. Congr., Sec. 4:* 17–24.
17. Rona, P. A. 1973. Plate tectonics and mineral resources. *Sci. Amer.* 229 (1): 86–95.

Further Reading

Brown, H. 1954. *The Challenge of Man's Future*. The Viking Press, Inc., New York. 290 p.

Landsberg, H. H. 1964. *Natural Resources and U.S. Growth*. The Johns Hopkins Press (for Resources for the Future), Baltimore, Md. 256 p.

Lovering, T. S. 1943. *Minerals in World Affairs*. Prentice-Hall, Inc., Englewood Cliffs, N.J. 394 p.

Park, C. 1968. *Affluence in Jeopardy*. Freeman, Cooper & Company, San Francisco. 368 p.

Skinner, B. J. 1969. *Earth Resources*. Prentice-Hall, Inc., Englewood Cliffs, N.J. 150 p.

Solar furnaces near Odeillo, France. Photo by Georg Gerster/Rapho-Guillumette.

5

Energy Resources

John P. Holdren   *is Assistant Professor in the interdisciplinary Energy and Resources Program at the University of California, Berkeley. He received his B.S. and M.S. degrees in engineering from M.I.T. in 1965 and 1966 and his Ph.D. in engineering and plasma physics from Stanford in 1970. He has worked in the controlled fusion program at the Lawrence Livermore Laboratory and in the Environmental Quality Laboratory and the Caltech Population Program at the California Institute of Technology. He is coauthor of two books (Energy and Human Ecology) and coeditor of three others. His research interests include the comparative environmental impact of advanced energy technologies, and the interactions among population, technology, and the quality of life.*

121

The flow of energy is the lifeblood of the biosphere and of the human civilizations that the biosphere supports. Energy from the sun warms the surface of the earth to life-sustaining temperatures, drives the massive machinery of wind and rain and ocean currents that shape global climate, and fuels the process of photosynthesis on which the pyramid of plant and animal life is built. The evolution of human societies over the millennia, in turn, has been shaped to an important degree by the evolution of ways to tap the biosphere's flows and stockpiles of energy to meet the perceived needs of people: first fire, for warmth and light and protection; then agriculture, permitting specialization, the birth of cities, and a vastly increased human population; then work from animals and wind and running water, to substitute for the push and pull of human hands and legs; and finally energy from fossil fuels, to build cities of concrete and steel and to heat them, light them, supply them, and connect them with each other.

Energy resources and energy technology are intimately woven into most of the great problems of contemporary existence: the global growth of population and consumption of materials, the prospects of the poor at home and the widening rich-poor gap among nations, the fragile balance of the international economic and political system, and the threats to human well-being associated with acute pollution and chronic ecological disruption. There is much reason to believe, therefore, that the ways in which mankind harnesses and manages energy will be as important in shaping the future of civilization as they have been in shaping its past.

Long ignored or taken for granted by most people, the complex role of energy in modern society has now been catapulted suddenly and forcefully into the public arena. It is likely to stay there. This chapter is intended to serve as an introduction to the subject: what energy is, what it does for us and to us, where we get it, and where we might get it in the future.

Energy Bookkeeping

Available energy is the capacity to do work: for example, to lift or pull a weight, to turn a wheel, to warm something up or cool it off. Energy takes many forms, some more available than others for doing work.

Kinetic energy is the energy of motion of an object: an automobile, a bullet, the earth spinning on its axis. When the kinetic energy is that associated with the random motion of the molecules in a substance, we call it *thermal energy* or *heat*. Temperature and heat are not the same thing. Temperature is a measure of how much energy there is in the random motion of an average molecule; the heat energy in a substance is the sum of the energy of random motion of all the molecules.

Whereas kinetic energy means something is moving, *potential energy* means something is waiting: the gravitational potential energy of a ball waiting to be dropped, the elastic potential energy of a stretched rubber band waiting to contract, the chemical potential energy of gasoline waiting to be burned. Some other examples are the electric potential energy of oppositely charged objects waiting to slam together and the nuclear potential energy of a uranium nucleus waiting to be split.

122

Electricity is a combination of kinetic and potential energy associated with the motion and position of electric charges. *Electromagnetic* or *radiant energy* is associated not with the motion of objects but with the motion of electric and magnetic forces. Light, radio waves, thermal radiation, and x-rays are all related forms of electromagnetic energy.

Albert Einstein theorized, and many experiments have subsequently verified, that any change in the energy associated with an object (regardless of the form of the energy) is accompanied by a corresponding change in mass. In this sense, mass and energy are equivalent and interchangeable. Because a small amount of mass is equivalent to a very large quantity of energy, however, the change in mass is only detectable when the change in energy is very large — in the case of nuclear explosions, for example.

Many processes in nature and in civilization involve the transformation of energy from one form to others. For example, combustion transforms chemical potential energy to light, thermal radiation, and heat energy of the combustion products. Photosynthesis transforms electromagnetic energy in the form of light to chemical potential energy stored in the bonds of carbohydrates. An electric drill transforms electric energy into kinetic energy of rotation of the bit. The imposing science of thermodynamics is really just the bookkeeping by means of which one keeps track of energy as it moves through such transformations. A rudimentary grasp of this bookkeeping is essential to an understanding of energy problems.

The essence of the accounting is embodied in two principles, known as the First and Second Laws of Thermodynamics. No exception to either one has ever been observed. The First Law, also known as the Law of Conservation of Energy, says that energy can neither be created nor destroyed. If energy in one form or one place disappears, the same amount must show up in another form or another place. In other words, although transformations can alter the *distribution* of amounts of energy among its many different forms, the *total* amount of energy when all forms are taken into account remains the same. When we burn gasoline, the amounts of energy that appear as light, thermal radiation, and heat energy of the combustion products are altogether exactly equal to the amount of chemical potential energy that disappears. The accounts must always balance. The immediate relevance of the First Law for human affairs is often stated succinctly as "You can't get something for nothing."

The Second Law of Thermodynamics conveys an equally significant message to the energy bookkeeper. It says that although energy itself is never destroyed, the fraction of energy that can be made to do useful work is always decreasing. What is consumed when we use energy, then, is not the energy itself but its usefulness or availability for doing work. Energy in forms such that a large fraction of it can be made to do useful work is often called "high-grade" energy. Correspondingly, energy of which only a small fraction can be made to do work is called "low-grade" energy, and energy that moves from the former category to the latter is said to have been degraded. (Electricity and the chemical energy stored in gasoline are examples of high-grade energy; the heat from a light bulb and the heat in an automobile's exhaust are corresponding examples of low-grade energy.) In these terms, the Second Law says that in all physical processes which involve a transformation of energy, some of the energy is degraded.

It is to be emphasized that this degradation is inevitable in technological processes as well as in natural ones. It cannot be evaded by technical tricks, it is going on all the time, and indeed its ultimate consequence is that *all* the energy we use is ultimately degraded to low-temperature, low-grade heat. Most of this heat is discharged to the environment at the point where the energy is used, and it cannot be recycled. In the same sense that the First Law says that we cannot get something for nothing, the Second Law tells us that we cannot break even or even get out of the game.

The scientists, engineers, and economists who do energy bookkeeping employ a bewildering array of units for the counting of energy and work, and they cannot seem to agree on a common approach. (There is no *need* to use different units for work and for energy in its various forms, because work and the various forms of energy are completely commensurable.) Engineers and the U.S. Department of Commerce usually use the British thermal unit (Btu), which is the amount of energy that warms 1 pound of water by 1 degree Fahrenheit. Biological scientists employ the calorie (which warms 1 gram of water 1 degree centigrade) or the kilocalorie (1,000 calories). Running the bodily machinery of an average, moderately active, adult human being uses about 2,500 kilocalories per day. Physicists employ the joule, a unit of energy in the metric system that corresponds to the kinetic energy of a mass of 2 kilograms moving at a speed of 1 meter per second. Universal adoption of the metric system is inevitable and desirable, and I will use units based on the joule throughout this chapter. (A kilojoule is 1,000 joules and will be the main energy unit relied upon here; elsewhere the reader may encounter the units megajoule and gigajoule, denoting 1 million and 1 billion joules.) Some amounts of energy associated with various natural and technological processes are shown in Table I.

It is useful in many applications to look not only at amounts of energy but at the rates with which energy flows or is used. (It is evident from perusal of Table I, for example, that the consequences associated with a transformation of energy depend very dramatically on how rapidly the transformation takes place — compare the TNT and the bread.) *The rate of energy flow or use is power.* A related definition, when one is talking about the rate at which use of available energy accomplishes work, is that *useful power is the rate at which work is done.* One can write these relations formally as follows:

$$\text{power} = \frac{\text{amount of energy that flows or is used}}{\text{time during which flow or use takes place}}$$

$$\text{useful power} = \frac{\text{amount of work done}}{\text{time during which work is done}}$$

Evidently, the units for power are units of energy divided by units of time — for example, Btus per hour, kilocalories per minute, and joules per second. A joule per second is called a watt. A kilowatt is 1,000 watts, and this is the unit I will use for power in this chapter. (Megawatt and gigawatt are units often encountered elsewhere, denoting 1 million and 1 billion watts.) These units are perfectly applicable to flows of nonelectric as well as electric energy, although you may be accustomed to them only in the electric context. Similarly, the kilowatt-hour, which denotes the amount of energy that flows in 1 hour if the rate (power) is 1 kilowatt, is an acceptable unit of energy outside the electric context. The power associated with some natural and technological processes is shown in Table II. Some conversion factors for transforming energy and power figures from one set of units to another are given in Table III.

A final important concept in energy bookkeeping is *efficiency.* Efficiency, a ratio, is the amount of energy delivered in useful form divided by the total amount of energy handled in a process. In the example of a Volkswagen moving at 60 mph (Table II), the rate of fuel consumption corresponds to a total power of 75 kilowatts or 100 horsepower. The rate at which work is being done by the engine, however, is only about 20 horsepower, corresponding to an efficiency of 20/100 = 0.20, or 20

I Energy of Various Processes

Process	Energy (kilojoules)
Bowling ball dropped 3 feet	0.067
Rifle bullet (20 grams, 1,000 meters per second)	10
Detonation of 1 pound of TNT	2,000
Metabolism of 1 pound of bread	5,000
Food energy for 1 adult for 1 day	10,000
Combustion of 1 gallon of gasoline	135,000
Average lightning stroke	1,600,000
Fuel consumption by Boeing 707, coast to coast	1,350,000,000
Average summer thunderstorm	160 billion
Hydrogen bomb (1 megaton)	4,000 billion
Strong earthquake	160,000,000 billion
Civilization's energy consumption in 1970	220,000,000 billion

II Power of Various Processes

Process	Power (kilowatts)
Electrical energy flow in 100-watt light bulb	0.100
Metabolic energy flow in average adult	0.120
Average rate of total energy consumption for U.S. citizen (1970)	12
Rate of fuel consumption of Volkswagen at 60 mph	75
Rate of fuel consumption of Boeing 747 at 600 mph	190,000
Rate of fuel consumption of large electric power plant	3,000,000
Rate of fuel consumption in Saturn V moon rocket at takeoff	260,000,000
Civilization's average rate of total energy consumption	7 billion
Average rate of global photosynthesis	80 billion
Winds, worldwide	1,000 billion
Solar input at top of atmosphere	174,000 billion

percent. The other 80 percent of the energy released by burning the fuel is discharged directly to the environment as heat — the same fate met by the useful fraction after it has done its job. Similarly, a large electric power plant consuming fuel at a rate corresponding to 3 million kilowatts delivers electric energy at a rate of 1 million kilowatts. (This efficiency, 33 percent, is about average for fossil-fueled power plants in the United States today.) Low-grade heat is discharged to the environment of the power plant at a rate of 2 million kilowatts, and the electricity itself is degraded to low-grade heat elsewhere after serving a useful function.

Patterns of Consumption and Supply

As noted in Table II, the average rate of energy use in all forms in the United States in 1970, per person, was roughly 100 times the average personal rate of intake of food energy. This total U.S. consumption per person is 2 to 4 times the corresponding figure for other "rich" countries (Europe, Japan, the Soviet Union, and a few others) and 25 times the average consumption per person for the 2.5 billion people living in "poor" countries (much of Asia, Africa, and Latin America).[1] In 1970, about one-third of all the nonfood energy used in the world was used in the United States.

Where did all this U.S. energy use go? Twenty-five percent of it was fuel burned in connection with transportation of people and goods. Of this 25 percent, more than half was used in automobiles, one-fifth in trucks, and the rest in buses, trains, aircraft, ships, and pipelines.

Twenty percent of the total U.S. energy use was in homes and apartments, more than half of this for heating. Other major residential uses were water heating, refrigeration, air conditioning, cooking, and lighting.

Thirteen percent of U.S. energy use in 1970 was in the commercial sector — office buildings, stores, hospitals, and schools. Major uses here were heating, lighting, and air conditioning.

The remaining 42 percent of U.S. energy use in 1970 took place in the industrial sector, including mining, manufacturing, and agriculture. Significant fractions of these energy expenditures were associated with the raw materials for and the manufacturing of automobiles, with residential and commercial construction, and with activities such as oil refining that are themselves a part of supplying energy. (This point and the energy expenditures associated with transporting fuels illustrate an important general phenomenon: It takes energy to get energy. In the special case

III Conversion Factors

1 Btu = 1,055 joules (J) = 252.5 calories (cal)
1 kilocalorie (kcal) = 4,184 J = 3.97 Btu
1 kilojoule (kJ) = 0.949 Btu = 0.239 kcal
1 watt (W) = 1 J/second = 14.3 cal/minute = 3.41 Btu/hour
1 kilowatt (kW) = 1.34 horsepower (hp)
1 kilowatt-hour (kWh) = 3413 Btu = 3,600,000 J = 860 kcal

125

of agriculture and the other activities associated with processing, storing, and delivering food, we spend about 6 kilojoules of fossil-fuel energy to get each kilojoule of food energy to the table.[2]) A somewhat more detailed breakdown of end uses of energy in the United States is given in Table IV.

Evidently, about one-third of U.S. energy use is directly in the hands of individual consumers, in the form of fuel burned in private automobiles and energy used in residences. The rest of the energy use, which occurs in the institutions and firms that produce and distribute the array of goods and services that Americans consume, is less subject to direct individual control. It is sometimes referred to, from the standpoint of personal consumption, as indirect energy use.

In Table IV and in the foregoing discussion, the use of energy in the form of electricity has been accounted for in terms of the amount of primary energy — fossil fuels, nuclear fuel, geothermal steam, or the energy of falling water — used to generate the electricity. About 25 percent of all primary energy use in the United States in the early 1970s was for generation of electricity, the energy for this enterprise amounting to 39 percent of the total use attributed to the residential sector, 50 percent of the total use attributed to the commercial sector, 26 percent of the total use attributed to the industrial sector, and a negligible fraction of the use attributed to fuel consumption in transportation.

What is the pattern of fuel supply that has been sustaining these patterns of U.S. energy use at a level 100 times the intake of food energy? The overwhelming characteristic of the pattern of supply is the extent of our continuing dependence on fossil fuels. As

IV   End Uses of Energy in the United States in 1970

	Percent of the general category	Percent of the whole
Transportation		25
Automobiles	55	
Trucks	21	
Commericial aircraft	8	
Other aircraft	4	
Railroads	3	
Pipelines	1	
Water freight	1	
Buses	½	
Other	4	
Space heating		20
Residences	60	
Commercial buildings	30	
Industrial buildings	10	
Industrial process steam		16
Industrial direct heat		11
Industrial electric drive		8
Nonenergy uses of fuels (lubricants, synthesis, asphalt, etc.)		6
Water heating		4
Residences	75	
Commercial buildings	25	
Air conditioning		3
Residences	33	
Commercial buildings	67	
Lighting		3
Residences	30	
Other	70	
Refrigeration		2
Residences	50	
Commercial	50	

Source: Reference 3.

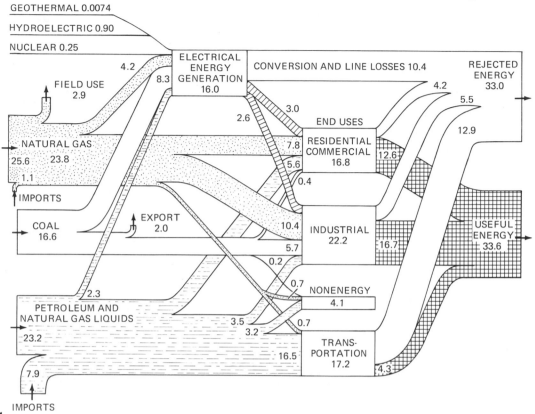

1   **Energy Flows in the United States, 1970.** Units are 10^{15} kilojoules. *Source:* Reference 5.

shown in Table V, natural gas and petroleum alone accounted in 1972 for about 75 percent of U.S. energy use, and petroleum, gas, and coal combined accounted for over 95 percent. Hydroelectric energy accounted for about 4 percent (measured as the amount of primary fossil-fuel or nuclear energy that would have been required to generate the same amount of electricity). Nuclear energy, belying the amount of public attention on the subject, accounted for only 1 percent of U.S. energy use in 1972. The pattern outside the United States was similar.[1,4]

Figure 1 is a flow diagram showing how the various primary energy sources were distributed among the major categories of end uses in the United States in 1970. The magnitudes of the flows are represented by the width of the pathways.

The efficiency of the various means of processing and delivery of energy (i.e., the fraction of energy delivered as useful heat or work) is roughly known. This information,

combined with the other data on which Figure 1 is based, leads to the conclusion shown there that the overall efficiency of the U.S. energy system is about 50 percent. It is important to note that not all the "waste" can be eliminated; a substantial amount of waste is the inevitable consequence of the Second Law of Thermodynamics. There are, nevertheless, significant opportunities for improvement in

V   **Patterns of Energy Supply**

	Percent used	
Energy supply	**U.S. (1972)**	**World (1970)**
Coal	17.2	32.1
Petroleum	41.9	42.8
Domestic	27.1	
Imported	14.8	
Natural gas	35.9	18.5
Hydroelectricity	4.2	6.2
Nuclear energy	0.8	0.4

Source: References 1 and 4.

127

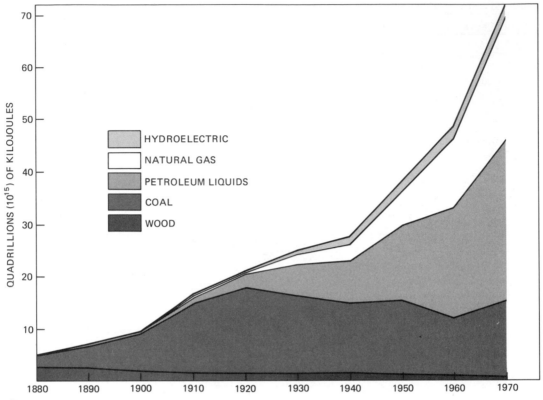

2   **U.S. Energy Use and Sources of Supply, 1880–1970.**

the present situation. For example, the efficiency of internal combustion engines in automobiles under average driving conditions is probably only about 15 percent, and we should be able to do much better than this. The 33 percent average efficiency of electricity generation can also be improved.

Growth and Change in Energy Flows

Many of the most interesting and problematical aspects of the energy situation involve not merely the size that civilization's energy flows have already attained, but the rate of growth of those flows and the rate of change of the patterns of supply behind them. The growth of energy use in the United States between 1880 and 1972, broken down into the main sources of supply, is shown in Figure 2. Total U.S. energy use in 1972 was 14.5 times that in 1880, and the U.S. population in 1972 was 4.2 times that in 1880. Hence energy use per person increased by a factor of 3.5 in this period,

and one can say as a crude first approximation that population growth and growth of energy use per person were about equally "responsible" for the increase in total energy use. The average annual rates of growth of U.S. energy use and population in different periods are given in Table VI.

Reliable data for worldwide energy use over so long a period are not available, but it appears that for the last two decades, at least, the rate of increase has been about 5 percent per year. In the poor countries it appears to have been 5.5 percent per year recently, corresponding to 2.5 percent per year population growth and 3 percent per year growth in energy use per capita.[1,7]

Projections of future levels of energy use are usually made by selecting an annual percentage growth rate based on recent experience or on a set of assumptions about how recent trends will be altered, and then carrying out the straightforward arithmetic of extending the curve of energy use into the future at this constant rate of growth. Such calculations are sometimes embellished and complicated by detailed analyses of economic trends and population growth, but the history of error in economic and demographic forecasting is such that one does well to admit the great uncertainty and simply examine the consequences of a range of possibilities. In no event should such projections or forecasts be regarded as reliable predictions. They merely give us an idea of what the future might be like under different sets of assumptions. If we do not like the picture of the future that some of the projections provide, we are at liberty to work to change the trends and assumptions that lead to these results.

With these disclaimers in mind, let us examine some projections of U.S. energy use. A rate of growth of energy use of 5 percent per year, such as prevailed in the United States between 1965 and 1970, leads (if continued) to a doubling every 14 years. (The general rule is, doubling time = 70 years ÷ average annual percentage growth rate; 70 ÷ 5 = 14.) This would correspond to a fourfold increase between 1972 and 2000 and a 16-fold increase between 1972 and 2028. This is a staggering increase, and I do not regard it as plausible. As will be discussed below, the energy crunch of 1973 and 1974 was in part an early manifestation of some of the sorts of difficulties likely to be encountered in attempting to maintain growth rates of this size.

If the average annual rate of growth is 3 percent rather than 5, the doubling time stretches out to 23 years. At this rate, the United States would face a doubling of energy use between 1972 and 1995 and a quadrupling between 1972 and 2018. The *difference* in energy use in the year 2000 between 5 percent and 3 percent annual growth after 1972 is equal to 1.7 times the 1972 use. The difference between these two cases in terms of cumulative energy use between 1972 and 2000 is equal to 17.3 times the 1972 use.

Extrapolation of the recent history of energy use in the poor countries is even more problematical. If somehow their recent rates of growth in population and energy use per person were sustained, it would take 107 years before the average energy use per person in the poor countries equaled that of the average U.S. citizen in 1970. The population in those countries would have increased in this period

from 2.5 billion to 36 billion, and their total annual energy use would be 50 times the world's annual use in 1970. This is not a plausible scenario.

It should be noted that *any* constant annual percentage growth rate, even 1 percent, leads if maintained for a long enough time to impossible consequences. Ten doubling times corresponds to an increase of a factor of over 1,000, twenty doubling times to a factor of over 1 million. Sooner or later, therefore, the necessity of stabilizing the level of energy use must be confronted. The question of whether it will, in fact, be sooner or later is a complicated one, and I will return to it.

The changes in the composition of energy supply in the last 100 years have been almost as dramatic as the changes in level of use. In 1880, more than half the U.S. annual energy use was being supplied by wood, and nearly all the rest by coal. By 1900, the share supplied by wood had dropped to 20 percent; petroleum, natural gas, and hydroelectric energy had appeared on the scene to divide 10 percent of the market among them; and coal accounted for the remaining 70 percent. By 1920, the fossil fuels had increased their combined share to 90 percent, a degree of dominance that has persisted up to the present. However, whereas the bulk of the growth in energy use between 1880 and 1920 was fueled by coal, the dramatic increases since that time have been based on rapid expansion of the exploitation of petroleum and natural gas. As is apparent from Figure 2, the amount of coal consumed was about the same in 1970 as in 1920, although its share of the greatly expanded total energy use had been much reduced.

It has been widely assumed until recently

VI Rates of Growth of U.S. Energy Use and Population

Period	Average annual rate of growth of:	
	Energy use (%)	Population (%)
1880–1910	3.8	2.0
1910–1929	2.6	1.5
1929–1932	−8.6	0.8
1932–1945	4.7	0.9
1945–1965	2.5	1.7
1965–1972	4.5	1.0

Source: References 4 and 6.

that nuclear energy would provide the next major shift in the composition of energy supply. Substitution of nuclear for fossil fuels is possible with today's technology only in the electricity-generation component of the energy economy, however, inasmuch as no commercial techniques exist as yet for producing portable fuels in nuclear reactors, or for using nuclear heat directly in industrial processes. Some projections hold that the fraction of energy used to generate electricity will increase from 25 to 50 percent by the year 2000, and that nuclear reactors will then be providing half of this electrical half of the energy budget (or one-fourth of the total). A variety of difficulties that have been encountered with nuclear fission (Chapter 12) now make so large a role for it in the year 2000 seem questionable to many observers. Recent events suggest that the composition of energy supply in the next 20 to 30 years may be altered more by expanded use of coal, exploitation of oil shales, and introduction of solar technologies than by the expansion of fission.

Energy Resources

Among the major questions that arise in connection with present and projected patterns of energy use are those concerning adequacy of energy resources in terms of size. Which resources are scarce and which abundant? What does it mean to "run out" and how accurately can such eventualities be predicted? What is the potential for harnessing inexhaustible sources?

It is useful at the outset to distinguish between stock-limited or nonrenewable resources and flow-limited or renewable ones. The first category refers to fuels of which the earth is endowed with fixed stocks; once these stocks are depleted, no more will be available on any time scale of practical interest. The main examples are the fossil fuels (principally coal, petroleum, natural gas, tar sands, and oil shales) and the nuclear fuels (principally uranium, thorium, deuterium, and lithium). The fossil fuels were accumulated over hundreds of millions of years, when plant material incorporating stored solar energy was separated from the energy cycle of the biosphere by biological and geophysical happenstance.

The nuclear fuels are thought to have originated when the constituents that eventually became our sun and its planets were fused together from elemental hydrogen in more distant stars. The duration of the fixed stocks of these materials depends on how much is ever found in exploitable form (it is not obvious for some of them that all that exists will be found) and on the rate at which civilization chooses to exploit what is found.

Flow-limited resources, by contrast, are virtually inexhaustible in duration but limited in the amount of energy that is available per unit of time. The most important example is solar energy, including not only the incoming radiation from the sun but also the various other harnessable forms of energy into which sunlight is converted by natural processes: falling water, wind, waves, ocean currents, temperature differences in the oceans, and plant materials. A much smaller flow-limited source is the energy of the tides, which is derived from the kinetic and potential energy of the earth–moon–sun system.

Geothermal energy is difficult to classify as stock-limited or flow-limited; it has some characteristics of both. At present, the exploitable form of geothermal energy consists of isolated pockets of hot water or steam. These localized stocks are depleted by use, but on a time scale of decades or perhaps centuries they may be replenishable from the much larger stock of geothermal energy in the earth's molten core. Future methods to tap the geothermal energy stored in relatively deep dry rock will also be stock-limited in a short-term sense but flow-limited (by the rate at which heat flows from surrounding regions into depleted rock) in the long term.

Analysis of the depletion of conventional stock-limited energy resources is itself a complicated enterprise. Much confusion is sometimes engendered in this connection by failure to distinguish clearly between reserves and resources. The term *reserves* generally refers to material whose location is known (proved reserves) or inferred from strong geologic evidence (probable reserves), and which can be extracted with known technology under present economic conditions (i.e., at costs such that the material could be sold at or near prevailing prices). The resources of a substance include the reserves and, in addition, material whose location and quantity are less well established, or which is not extractable under prevailing technological and economic conditions. The term *ultimately recoverable resources* describes an estimate of how much material will ever be found and extracted, implicitly including an assessment of how good technology will ever get and how much civilization will ever be willing to pay for the material. Such estimates are necessarily very crude. The relation between reserves and resources is illustrated in Figure 3.

Probably the least sophisticated approach to the analysis of depletion is to estimate the lifetime of the supply by dividing present proved reserves by the present rate of consumption. This approach is the origin of the often-heard statements to the effect that "We have X years' worth of petroleum left." The method errs because consumption is not likely to stay constant and because proved reserves often bear little relation to ultimately recoverable resources. Sometimes it simply does not pay to invest money to locate and evaluate deposits that will not be extracted and turn a profit until 15 to 20 years hence.

A somewhat more instructive way to assess the lifetime of a fuel is to use available estimates of the ultimately recoverable resources, together with a level of consumption several times the present one (on the assumption that consumption will level off before too long) or a level of consumption continuing to grow as it has in the recent past. A shortcoming in the first case is the sensitivity of the result to the highly uncertain estimates of the resource and of the equilibrium level of consumption; in the second case the result is not so sensitive to errors in the resource estimate but is very sensitive to the growth rate chosen.

The most realistic approach seems to be the one devised by geologist M. King Hubbert.[8] He noted that the production cycle for any stock-limited resource is likely to be characterized by several phases: first, increasingly rapid growth in the rate of exploitation as demand rises, the industry becomes more efficient, and costs per unit of material fall; then a leveling off of production as the resource becomes scarcer and starts to rise in price; and finally a continuous decline in the rate of exploitation, as increasing scarcity and declining quality proceed more rapidly than can be

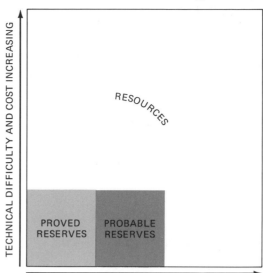

3   The Concepts of Reserves and Resources.

compensated for by improving technology, and as substitutes are brought fully to bear. Hubbert's approach incorporates explicitly the fact that we never really run out of anything suddenly or completely; we simply use the most concentrated and accessible supplies first, gradually working our way into material of declining quality until what finally remains is so dilute or so deep or so hard to find that it no longer pays to look for it and extract it.

A generalized illustration of Hubbert's production cycle is shown in Figure 4. Note that the area under the production curve at any time is the cumulative production up to that time, and the amount under the completed curve is the magnitude of the ultimately recovered resources. The important measures of the lifetime of the fuel in this approach are the year in which the level of consumption reaches its peak (here denoted Y_p) and the year at which 90 percent of the ultimately recoverable resources has been extracted (here denoted Y_{90}).

Conventional Fossil Fuels

Results of the application of the various approaches to depletion analysis discussed above are given for the conventional fossil fuels in

131

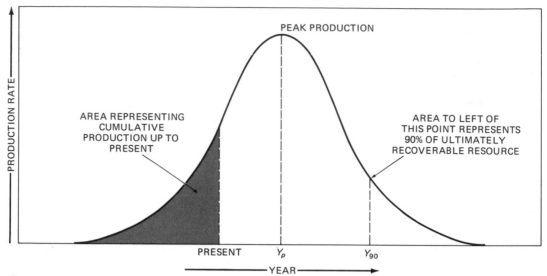

4 **Hubbert Production Cycle for Stock-Limited Resources.**

Tables VII and VIII. Table VII gives recent consumption rates, reserves, some estimates of ultimately recoverable resources, and lifetimes of the reserves and resources at constant and growing consumption rates. Table VIII gives Hubbert's estimates of ultimately recoverable resources and the corresponding years, Y_p and Y_{90}. Reserves and resources are given in all cases in units of 10^{18} kilojoules (10^{18} = 1 billion billion). The conversions from the units usually found in the geologic literature (tons of coal, barrels of oil, cubic feet of natural gas) are given in Table IX. Differences

in the high and low estimates of ultimately recoverable resources reflect varying interpretations of the geologic evidence and different degrees of optimism about future extractive technologies.

Several clear messages emerge from Tables VII and VIII. By any method of analysis, U.S. supplies of petroleum and natural gas are severely limited, and world supplies of these fuels are almost as poor in relation to projected demands as the U.S. ones. Both for the U.S. and for the world, any significant increase in consumption of oil and gas will lead to the

VII Reserves, Resources, and Lifetimes for Conventional Fossil Fuels

	Proved reserves (10^{18} kJ)	Remaining ultimately recoverable resources (10^{18} kJ)	Recent annual consumption (10^{18} kJ)	Reserve lifetime[a] (years)	Resource lifetime, increased constant consumption[b] (years)	Resource lifetime, exponential growth[c] (years)
U.S. petroleum liquids	0.31	1.1–3.5	0.035	9	7–20	19–36
U.S. natural gas	0.33	0.9–1.7	0.026	13	7–13	20–29
U.S. coal	10.2	42.8	0.015	660	570	99
World petroleum liquids	3.7	10.9	0.10	37	22	37
World natural gas	2.1	10.6	0.04	53	53	53
World coal	124	220	0.09	1,380	490	96

[a]Proved reserves divided by recent annual consumption.
[b]Remaining resources divided by 5 times recent annual consumption.
[c]Duration of remaining resources if annual consumption grows 5 percent per year.
Source: References 8, 9.

132

substantial depletion of the recoverable resources of these materials by early in the next century. U.S. domestic production of petroleum was probably near or even past its peak in 1974, and domestic natural gas will not be far behind.

U.S. and world supplies of coal are much larger, roughly 10 times the combined supplies of petroleum and natural gas. These coal resources could sustain consumption rates several times larger than today's for hundreds of years. As large as these amounts are, however, they could sustain a level of consumption growing continuously at 5 percent per year for only about a century — another example of the power (and danger) of high growth rates. The more realistic analysis by Hubbert suggests that coal consumption might peak around the year 2200, and that coal would be near the end of its importance as a major energy source some 200 years later.

A greatly expanded role for coal, which seems to be suggested by the data on fuel supply, poses a number of problems. For coal to relieve the burden on petroleum and natural gas, it must be interchangeable with these fuels in a variety of applications. This interchangeability can be achieved by means of coal gasification and liquefaction, processes that are known to be technically feasible and capable of delivering two-thirds to three-fourths of the energy of the coal in the final gaseous or liquid synthetic fuel. The facilities needed to accomplish these transformations are large, expensive industrial installations, comparable in complexity to oil refineries; the problem is not whether it can be done at all, but how quickly and at what cost in capital investment a significant transition from natural gas and petroleum to synthetic fuels from coal can be accomplished. Another problem, of course, is the environmental liabilities of greatly increased reliance on coal.

Tar Sands and Oil Shale

Tar sands consist essentially of a mixture of sand and asphalt. The material is mined like rock and processed to convert the hydrocarbons into a synthetic crude petroleum. The principal known deposits are in Alberta, Canada, and the energy recoverable from them amounts to about 1.8×10^{18} kilojoules.[8] This is of the same order of magnitude as the remaining recoverable resources of petroleum in the United States. A small commercial plant has been in operation at the Athabasca tar sands since 1966, and several other commercial operations were scheduled for the mid-1970s.

Oil shale consists of fine-grained sedimentary rock permeated by a rubbery, solid hydrocarbon called kerogen. Oil shale is much more abundant than tar sands, although technologically more difficult to convert into synthetic crude oil. The Green River shales of Colorado, Utah, and Wyoming alone contain the equivalent of 2,000 billion barrels of oil, or about 12×10^{18} kilojoules, at concentrations between 10 and 100 gallons of oil per ton of shale. Total U.S. resources at greater than 10 gallons per ton have been estimated at about 160×10^{18} kilojoules and world resources of similar quality outside the United States at $1,900 \times 10^{18}$ kilojoules.[8,9] How much of this vast amount of potential fuel will ever be extractable economically is completely uncertain at this time. Rising costs of conventional pe-

VIII	Hubbert's Analysis of Conventional Fossil-Fuel Supplies		
	Estimate of original ultimately recoverable resources (10^{18} kJ)	Year of peak production, Y_p	Year when 90% is gone, Y_{90}
U.S. petroleum liquids	1.4	1970	2000
U.S. natural gas	1.6	1980	2015
U.S. coal	42.8	2220	2450
World petroleum liquids	12.4	2000	2030
World coal	218	2150	2400

Dates are approximate; where Hubbert gave a range of resource estimates, the higher figure was used here. *Source:* Reference 8.

troleum in the early 1970s have sent major corporations scurrying to develop the richer parts of the Green River shales, however, and substantial commercial operations there seem likely.

Nuclear Fuels: Fission

Nuclear fission is the splitting of certain heavy elements into lighter ones, accompanied by the conversion of a small part of the mass originally present into energy. Fission is induced when a nucleus of a suitable isotope is struck by a free neutron. (An *element* is characterized by the number of protons in the nucleus; different *isotopes* of the same element have the same number of protons but different numbers of neutrons in the nucleus.) Among the fragments from each fission event are some new free neutrons — an average of about 2.5 per fission — making possible a chain reaction: one or more of the new neutrons induces a new fission, which produces new neutrons, which induce more fissions, and so on.

If the chain reaction grows very rapidly, one has a fission bomb. In a 20-kiloton weapon, 1 kilogram of nuclear fuel is fissioned in a fraction of a millionth of a second, converting about 1 gram of the original mass into 79 billion kilojoules of heat, light, x-rays, and blast waves. In a fission reactor, by contrast, the chain reaction is nurtured carefully to the desired level and then maintained there. In a large reactor, it takes about 7 hours to fission 1 kilogram of nuclear fuel, releasing the 79 billion kilojoules as heat at a rate of 3,000,000 kilowatts. To obtain the same 79 billion kilojoules released by the fissioning of 1 kilogram of nuclear fuel would require the combustion of almost 3 million kilograms (about 3,000 tons) of coal.

Isotopes that can sustain a fission chain reaction are called *fissile*, and there are three important ones: uranium 235 ($_{92}U^{235}$), plutonium 239 ($_{94}Pu^{239}$), and uranium 233 ($_{92}U^{233}$). (The subscript gives the number of protons in the nucleus, which characterizes the element; the superscript gives the number of protons plus neutrons, which characterizes the particular isotope.) Of the fissile isotopes, only ^{235}U occurs in appreciable quantities naturally. It comprises 0.7 percent of the element uranium in nature. Plutonium 239 is produced by a series of nuclear transformations that may ensue when ^{238}U (comprising the other 99.3 percent of natural uranium) is struck by a fast-moving neutron. Isotopes that can be transformed into fissile ones this way are called *fertile*. The other important fertile isotope is thorium 232 ($_{90}Th^{232}$), which upon being struck by a slow-moving neutron may undergo a series of transformations, ending with fissile uranium 233. If struck by a particularly energetic neutron, the fertile isotopes ^{238}U and ^{232}Th may also fission, but they are incapable of sustaining a fission chain reaction on their own.

Most present-day nuclear fission reactors use ^{235}U as the primary nuclear fuel. A mixture of ^{235}U and ^{238}U is present in the reactor, and some of the neutrons released by the fission of ^{235}U strike ^{238}U nuclei and initiate the transformation to ^{239}Pu. Some of this new plutonium is subsequently fissioned itself while still in the reactor, thus contributing to the chain reaction. (This process can be thought of as indirect fission of ^{238}U.) Some plutonium remains when, after a year or so, the fuel is removed from the reactor for reprocessing (see Chapter 12). This plutonium, recovered at the reprocessing plant, may be recycled as fuel for the reactor that produced it, it may be saved for use in future reactors, or it may be used to manufacture nuclear bombs. This nuclear fuel cycle and the similar one involving thorium are diagrammed in Figure 5.

When all the details are taken into account,

IX	Fossil-Fuel Conversion Factors		
Coal	1 metric ton = 1,000 kilograms = 1.1 short tons		
	= 28,800,000 kilojoules = 2.88×10^{10} joules		
Petroleum	1 barrel = 42 gallons = 0.137 metric ton = 5,900,000 kilojoules		
	= 5.90×10^9 joules		
Natural gas	1,000 cubic feet = 28.4 cubic meters = 1,120,000 kilojoules		
	= 1.12×10^9 joules		

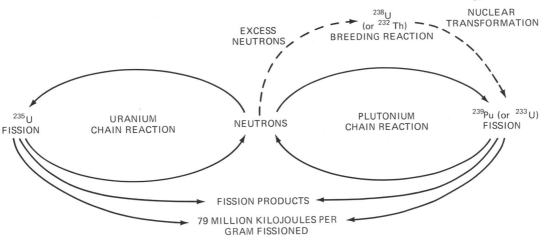

5 Nuclear-Fission Fuel Cycles.

one finds that a present-day commercial light-water-cooled reactor, when recycling its plutonium, manages to fission about 10 grams of nuclear fuel (uranium or plutonium made from uranium) for every kilogram of natural uranium that is mined. This is a fuel-utilization efficiency of 1 percent, compared to the unattainable ideal in which every uranium nucleus mined is fissioned, directly or indirectly. More advanced, gas-cooled reactors relying on the ^{232}Th/^{233}U conversion process can achieve a fuel-utilization efficiency of about 2 percent, fissioning 20 grams of nuclear fuel for every kilogram of uranium mined.[10] (For this type of reactor, uranium is the limiting resource even though some thorium is used; the fuel cycle requires the mining of only 1 kilogram of thorium for every 5 kilograms of uranium, and thorium is about three times as abundant in nature as uranium is.) One commercial-size demonstration plant of this type was in operation in 1974.

Water-cooled and gas-cooled reactors of the kind just described are called *converters*, in reference to their ability to convert a modest amount of fertile material to fissile fuel. By means of drastic changes in design, it is possible to build a reactor that converts fertile material to fissile fuel faster than the reactor consumes fissile fuel in its own chain reaction. Such a reactor is called a *breeder*. In terms of the fission-fuel cycles depicted in Figure 5, breeder reactors permit elimination of the loop that depends on scarce U-235. The fuel utiliza-

tion of possible future breeder reactors could be perhaps 40 percent for fuel cycles based on the ^{232}Th/^{233}U conversion and 60 percent for those fuel cycles based on the ^{238}U/^{239}Pu conversion.[10]

With this background, the effective energy content of nuclear fuel resources can be evaluated. Figures compiled by the U.S. Atomic Energy Commission for recoverable supplies of uranium and thorium, at various costs of extraction, are summarized in Tables X and XI. The figures given are the sum of what the Commission calls "reasonably assured" and "probable additional" reserves. The 1972 price of uranium oxide (U_3O_8, the form in which raw uranium is sold) was about $16 per kilogram, providing little incentive for exploration for deposits in the more expensive categories. Most authorities believe that an extensive exploration program would yield significant addition to the figures listed in Table X. The cost of electricity generated in nuclear reactors is so insensitive to the cost of raw uranium that the use of U_3O_8 costing $110 per kilogram is not out of the question even with the inefficient 1970 reactor technology. Breeder reactors could use the U_3O_8 listed at $220 per kilogram of U_3O_8 easily; probably they could also use the 4 trillion kilograms of uranium in the oceans and the even larger quantities of uranium and thorium dispersed in such common rocks as granite. (Preliminary studies suggest that uranium could be extracted from seawater at $220 per kilogram or less, and that uranium and thorium could be extracted from granite at a few times this figures.[11,12]) The situation for thorium is similar; even the most expensive supplies listed in Table XI could be utilized without serious impact on the price of electricity.

More detailed analysis has shown that the most dramatic projections of the growth of U.S. nuclear electricity generation could be met through the year 2020 with 1970 reactor technology (no breeder reactors, or even gas-cooled advanced converters), without new discoveries of uranium, and with only a 10 percent increase in the delivered price of electricity due to use of the more expensive known uranium resources.[13] Breeder-reactor technology could support a level of electricity generation much larger than recent levels for thousands of years, even without resorting to the abundant but dilute uranium in the oceans and in common rock. (Energy use for electricity generation in the United States in 1974 was about 0.02×10^{18} kilojoules.)

The rest of the world has been even less thoroughly explored for uranium and thorium than has the United States, and no meaningful figures are available above the $33 per kilogram range. The types of geologic formations that contain significant concentrations of these fuels are widely distributed around the globe, however, and one might expect to find resources, on the average, roughly in proportion to land area. If this is even approximately true, then it is apparent for the world as well as for the United States that the real difficulties with fission involve, not the supply of fuel, but the environmental questions discussed later in this chapter and in Chapter 12.

Nuclear Fuels: Fusion

The energy of the sun and of the hydrogen bomb comes from the fusing together of light nuclei to form somewhat heavier ones rather than from the fissioning of heavy nuclei. Harnessing fusion reactions in a nonexplosive way on the surface of the earth is far more difficult than harnessing fission, but controlled fusion — when it is finally achieved — should offer impressive advantages over fission with respect to safety, magnitude of long-lived radioactive wastes, and accessibility of the enormous fuel supply.

The difficulty of fusion is associated with three conditions that must be met: (1) the fusion fuel must be heated to a temperature of tens of million to billions of degrees; (2) the density of the fuel must be high enough to yield a significant reaction rate; and (3) the fuel must be confined under these conditions for a time long enough that the energy output from fusion reactions exceeds the energy input for heating and confinement. A quarter century of research on fusion has led to many ingenious approaches to the problem, but none has yet attained the combination of conditions that would be needed for a fusion reactor to produce more energy than it consumes. If this "breakeven" set of conditions is achieved in the early 1980s, as many scientists now think likely, the first commercial fusion reactors

X U.S. Uranium Supplies and Energy Content

Cost per kilogram of U_3O_8	Millions of kilograms of U at this cost	Energy at 1% utilization[a] (10^{18}kJ)	Energy at 2% utilization[b] (10^{18}kJ)	Energy at 60% utilization[c] (10^{18} kJ)
$18 or less	460	0.36	0.73	21.8
$18–33	670	0.53	1.06	31.8
$33–55	610	0.48	0.96	28.8
$55–110	6,000	4.74	9.48	284
$110–220	12,000	9.48	18.96	568

[a]Contemporary commercial light-water-cooled reactor (1970 technology).

[b]Advanced gas-cooled reactor (1980 technology).

[c]Liquid-metal-cooled breeder reactor (1990? technology).

Source: Data from Reference 10.

FUSION

D + T ——————FUSION—————→ HELIUM + n + 1,700 MILLION KILOJOULES PER MOLE OF T CONSUMED

6D ——VARIOUS FUSION REACTIONS——→ 2 HELIUM + 2 ORDINARY HYDROGEN + $2n$ + 690 MILLION KILOJOULES PER MOLE OF D CONSUMED

TRITIUM BREEDING

n + ^{6}Li ——————FISSION—————→ HELIUM + T + 460 MILLION KILOJOULES PER MOLE OF ^{6}Li CONSUMED

n + ^{7}Li ——————FISSION—————→ HELIUM + T + n - 240 MILLION KILOJOULES PER MOLE OF ^{7}Li CONSUMED

6 **Principal Fusion and Tritium-Breeding Reactions.** n denotes a neutron; 1 mole = 6.02 × 10^{23} atoms and is 2 grams of D, or 3 grams of T, or 6 grams of ^{6}Li, or 7 grams of ^{7}Li.

might be ready for service around the year 2000.

The most suitable fuels for earthbound fusion are the two heavy isotopes of hydrogen, deuterium and tritium. Deuterium ($_1^2$H, usually abbreviated D) is nonradioactive and occurs in seawater in the ratio of 1 deuterium atom for every 6,700 atoms of ordinary hydrogen. This does not sound like much, but it is a great deal; the deuterium in a gallon of seawater is equivalent in energy content to more than 300 gallons of gasoline. Tritium ($_1^3$H, usually abbreviated T) is radioactive with a half-life of 12.3 years (see Chapter 12) and is almost nonexistent in nature. It can be produced readily by bombarding lithium with neutrons, however. Lithium is thus a *fertile* material for fusion in much the same way that ^{232}Th and ^{238}U are fertile materials for fission.

The principal fusion reactions are summarized in Figure 6. The D–T reaction is significantly easier to achieve than reactions involving D alone, and early fusion reactors will almost unquestionably rely on it; hence the tritium breeding reactions and the supply of lithium are of great importance. The element lithium in its natural form consists of 92.6 percent lithium 7 and 7.4 percent lithium 6. Some calculations of the effective energy content of lithium as a fusion fuel have erroneously assumed that only the rarer ^{6}Li could be used and thus have underestimated the energy available. The ^{7}Li reaction consumes some energy but far less than is obtained when the resulting tritium undergoes fusion. It can be shown, taking into account the use of both ^{7}Li and ^{6}Li, that the net energy derived from converting 1 gram of natural lithium to tritium and then fusing the tritium with deuterium is about 90 million kilojoules.[14] This is roughly three times the energy one expects from 1 gram of natural lithium if only the contribution of the ^{6}Li is considered.

Lithium occurs in mineral ores called pegmatites, in heavy brines, and in seawater at a concentration of 0.17 gram per cubic meter (versus 33 grams per cubic meter for deuterium). Known and inferred U.S. reserves in the brines and pegmatites amounted in 1973 to about 4 billion kilograms, or 2,000 times the annual (nonfusion) U.S. consumption of lithium at that time. Additional estimated re-

XI **U.S. Thorium Supplies and Energy Content**

Cost per kilogram of ThO$_2$	Millions of kilograms of Th at this price	Energy at 40% utilization (10^{18} kJ)
$22 or less	500	15.8
$22–$33	170	5.4
$33–$55	8,300	263
$55–$110	20,800	656

Source: Data from Reference 10.

sources amounted to another 3 billion kilograms. Reserves outside the United States in 1973 were about 380 million kilograms (350 times annual non-U.S. consumption), with additional estimated resources of about 2.7 billion kilograms.[15] These high ratios of reserves to consumption mean there has been little incentive to search for more lithium, and it is very likely that significant additional amounts would be found if the effort were expended. If it were ever necessary, lithium could probably be extracted from the oceans; even if this cost 100 times as much as we now pay for lithium, it would still be cheap fuel.[16]

Table XII summarizes the availability of the basic fusion fuels. Deuterium is so abundant as to be almost inexhaustible — it could supply a world energy demand much larger than today's for billions of years — and the technology for extracting it cheaply is already well established. Although lithium is less abundant, today's reserves on land could support a greatly increased energy demand for centuries.[16] That should be long enough to make D–D fusion work, or to learn to extract the lithium from the oceans. As with fission, limits on the use of fusion are much more likely to be based on environmental or social restraints than on availability of fuel.

Geothermal Energy

The size of geothermal energy resources is extraordinarily difficult to estimate, because the amount that can be recovered depends completely on the details of technology that has only begun to be developed. A 1972 review by the U.S. Geological Survey gives identified and inferred reserves of geothermal heat in the United States, recoverable with existing technology, as about 0.01×10^{18} kilojoules.[9]

At an efficiency of conversion to electricity of 10 percent (very low, owing to the low temperature associated with geothermal heat), this modest amount of energy could deliver electric power at a rate of 1 million kilowatts for only 33 years. This is equivalent to the production of two or three moderate-sized fossil-fuel power plants in their operating lifetime of 30 years. The only operating geothermal power-plant in the United States, at The Geysers in northern California, had an electrical capacity of a few hundreds of thousands of kilowatts in the mid-1970s.

The United States has not been carefully explored for geothermal potential, however, and some experts believe that the potential of a single field in California's Imperial Valley exceeds the Geological Survey's figure for recoverable reserves in the entire United States. The Survey's 1972 estimate for *undiscovered* U.S. geothermal resources in the ground was 40×10^{18} kilojoules.[9] If these resources materialized, if 50 percent of the energy could be brought to the surface (versus 15 percent with existing technology), and if this energy were converted to electricity at 20 percent efficiency, it would amount to 670 times the 1970 U.S. electricity consumption. Obviously, the practical potential of geothermal energy depends very much on highly uncertain estimates about the size of the exploitable stock and the efficiency with which it can be harnessed. The *theoretical* potential is big enough to make the problem well worth pursuing.

Viewed as a widespread flow resource rather than a stock to be tapped in particularly favorable locations, geothermal energy is not encouraging. The rate of heat flow to the surface of the earth, corresponding very roughly to the rate of heat generation by the nuclear disintegration of radioactive elements in the earth's crust, is only about 0.063 watt per

	Millions of kilograms	Energy (10^{18} kJ)	**XII**	Fusion Fuels and Energy Content
U.S. lithium resources	6,900	620		
Lithium resources outside the United States	3,100	280		
Lithium in oceans	250,000,000	22,000,000		
Deuterium in oceans	50,000,000,000	17,300,000,000		

Source: Data from References 15, 16.

square meter of surface area (63 kilowatts per square kilometer).[8] At 20 percent conversion efficiency, a 1 million electrical kilowatt geothermal power plant relying on this flow would have to find a way to harness it over 80,000 square kilometers.

Solar Energy

Energy from the sun is a flow-limited resource: the energy income will continue for the remaining lifetime of the sun — some billions of years — but the rate of energy flow and hence the amount received in any given span of time is fixed. This fixed flow amounts to 1,400 watts per square meter facing the sun at the top of the earth's atmosphere. The average flow reaching the surface of the earth (averaged over day and night, summer and winter, good and bad weather, and all latitudes) is 180 watts per square meter, or 180,000 kilowatts per square kilometer.[17] At this rate, the amount of solar energy falling on the surface of the 48 coterminous United States in a year is 44.5×10^{18} kilojoules, or 650 times the total 1970 U.S. energy use.

The important questions are for what uses and with what technologies solar energy can be harnessed economically. The principal difficulties that arise with direct harnessing of solar energy are intermittency, requiring a means of storing the energy for nighttime and periods of bad weather, and the energy's low concentration, requiring large collector areas. These problems are not insurmountable.

Direct use of solar energy for space heating and water heating in individual residences was already technically feasible and potentially competitive economically with electricity in many U.S. locations in the early 1970s.[18] The technology consists of simple flat-plate collectors amounting to about half the roof area, and an insulated water tank of a few thousand gallons capacity for energy storage. Some auxiliary heating from fuel or electricity is required for prolonged cloudy weather, but it is easy to meet about three-quarters of the total year-round heating burden with solar energy. Such systems were not in widespread use in the early 1970s because the first costs were higher than for conventional electric systems and the total costs higher than for heating with fuel.

Rising fuel prices, threats of unavailability, and recognition that total costs for solar heating are already competitive with electric systems should lead to heavy residential use of solar heating in the late 1970s. Solar air-conditioning systems for individual buildings should be available and economically attractive by the early 1980s.

Large-scale harnessing of solar energy for the generation of electricity or the production of a portable chemical fuel such as hydrogen is more difficult. If a 20 percent efficiency of conversion to electricity can be achieved (various published estimates fall mostly in the range from 10 to 25 percent), a plant with an average capacity of 1 million kilowatts would occupy 28 square kilometers (about 11 square miles). The problem here is not land use — the 1970 U.S. electricity consumption could be met from an area 72 kilometers on a side — but rather is the cost of covering so large an area with the sophisticated collectors needed to achieve this efficiency. Storage requirements to maintain generating capacity at night and in bad weather are formidable. Estimates based on existing technology suggest that the cost of generating electricity with such a system would be 4 to 6 times generation costs in contemporary fossil-fuel and nuclear plants.[18] New technology might bring the cost down enough to be attractive, but it is most unlikely that large-scale, commercial, solar-electric plants will be in use before the year 2000. Conceivably, large solar-to-hydrogen conversion plants could come a bit sooner, but this is highly speculative.

Solar energy makes itself available in a variety of indirect forms as well as in the form of light striking the earth's surface. These forms include the chemical energy stored via photosynthesis in plants; the energy of the hydrological cycle; the energy of wind, waves, and ocean currents; and the energy represented by the temperature difference between deep and shallow ocean waters (ocean thermal gradients).

Some approaches to the indirect harnessing of solar energy via photosynthesis are: using wood or other forms of cellulose, including municipal trash, as combustible fuels; fermenting plant material to make alcohol; and processing agricultural and human wastes to produce methane and culturing algae for direct com-

139

bustion or conversion to methane. Because the net efficiency of plants in converting to chemical energy the sunlight incident on a given area of land is rarely more than 1 percent, the land requirements for such schemes as wood-burning power plants or alcohol-producing plantations are quite large. Careful study is also required of the nonsolar energy that must be supplied to such systems to maintain suitable growing conditions. Extraction of energy from municipal and other wastes is probably worthwhile, although there are some collection problems. The total extractable energy content of U.S. municipal trash in 1970 amounted to about 3 percent of our total energy consumption; the extractable energy content in our agricultural wastes in 1970 amounted to 15 percent of our total energy consumption.

Of the nonbiological processes that provide opportunities for indirect harnessing of solar energy, by far the most heavily used is the hydrological cycle. Developed hydroelectric capacity worldwide in 1967 was 243 million electrical kilowatts (kWe), compared to an estimated potential of about 2,900 million kWe.[8] U.S. hydro capacity in 1970 was 55 million kWe compared to a potential of 160 million kWe and a total electrical capacity in all forms of 360 million kWe.[4] Most of the world's undeveloped hydro capacity is in Asia, Africa, and South America. Although the flow of rivers can be expected to be roughly constant over many millenia, it is often noted that much of the potential of specific hydroelectric sites is typically destroyed on a time scale of one to three centuries when the reservoirs fill with silt. No real solution for this problem is yet known.

The practical energy potential of wind, waves, and ocean thermal gradients is largely unknown because no technology is yet in hand for harnessing these indirect forms of solar energy on a large scale. The energy flow in wind, waves, and ocean currents has been estimated variously as between 250 billion and 2,500 billion kilowatts.[17] It seems unlikely that as much as 1 percent of the total amount could ever be harnessed, but even a tenth of 1 percent of the higher figure is more than one-third of civilization's 1970 rate of energy use.

Despite the formidable obstacles, efforts to harness the direct and indirect flows of solar energy merit the increasing intensity now being devoted to them, both because of the long duration of these sources and because of the apparent environmental advantages of many of them over the stock-limited alternatives.

Tidal Energy

Tidal energy is the least significant of the flow-limited energy sources. The total rate of tidal energy flows in shallow seas, where this energy is most accessible, is about 1.1 billion kilowatts, and only about 13 million kilowatts is harnessable with existing technology.[8] Prospects for significant improvement in this figure are poor, on fundamental physical grounds.

MHD, Gas Turbines, and Hydrogen

Magnetohydrodynamics (MHD), gas turbines, and hydrogen are not primary energy sources; their usefulness is for making existing and future primary sources more efficient or more versatile.

An MHD generator is a device that converts the thermal and kinetic energy of a flow of a hot gas or liquid metal to electricity, without going through the intermediate steps of rotating machinery, such as a conventional turbine–generator combination. A primary energy source — generally fossil fuels at the present time but perhaps nuclear energy in future systems — is needed to heat the working fluid. The most attractive application of MHD seems to be in combination with a conventional steam power plant, such that the hot exhaust from the MHD generator is used to produce the steam. Here the role of MHD is as a topping cycle—it harnesses combustion-product energy at higher temperatures than would otherwise be possible — and the overall efficiency of the plant in converting fuel energy into electricity may reach 60 percent. A variety of technical problems with MHD remain to be worked out, and it will probably not be in large-scale commercial service before 1990.

A less exotic combined cycle employs gas turbines, much like those that power jet aircraft, at the topping stage. Like MHD systems, modern gas turbines can harness energy at higher temperatures than is possible for steam turbines. Combined-cycle power plants using gas turbines exhausting into a steam generator can achieve overall efficiencies of 45 to 55 percent. A number of large commercial plants of this kind were on order in the mid-1970s.

Hydrogen is a portable chemical fuel that can be manufactured by dissociating water using any of several processes, given a primary energy source. It can also be manufactured from fossil fuels. Hydrogen is relatively clean burning, producing only water as a combustion product and oxides of nitrogen as a side product if it is burned in air. It offers a means of storing and transporting energy obtained from intermittent sources such as sunlight, and it could permit electricity from nuclear fission and nuclear fusion to serve as the energy source for aircraft and other applications requiring portability, long after the fossil fuels are gone. An important drawback of hydrogen produced using electricity is the low overall efficiency of the process. Several approaches to ameliorating this problem are being intensively investigated. The safety of hydrogen as an everyday fuel for vehicles also needs further investigation.

Energy Use and Environmental Impact

The growth of civilization's energy consumption is likely to be limited sooner by the capacity of the environment to absorb abuse than by the availability of energy resources. Indeed, the heart of the energy dilemma in the long term is energy's dual role in the economy and the environment: use of energy is both a fundamental ingredient of well-being, as the prime mover of industrial society, and a fundamental threat to well-being, through the degradation of environmental services that industrial society cannot replace. No means of supplying energy is completely free of environmental liabilities, but *degree* of freedom from them is certain to become an increasingly important criterion in society's choices among the alternative energy sources.

The cycle of energy use in society usually involves many steps, the main ones being the following: exploration for and discovery of the resource, harvesting, processing (e.g., upgrad-

ing and sorting), conversion (e.g., to electricity), transportation (usually more than once), final consumption, and management of wastes. Significant environmental problems may occur at several or all of these stages. These problems, in turn, can usefully be classified according to the character of the impact on human well-being: direct assaults on life and health, damage to property, social disruption, other impact on perceived quality of life, and indirect effects on human well-being through disruption of natural environmental services. Table XIII arranges some of the major environmental impacts of energy technology according to the stage of the fuel cycle and the character of the impact on well-being.

Many of these impacts are discussed in detail in Chapters 8 through 15, but a few general observations are in order here. First, although direct impact of energy technology on health and quality of life has rightly received much attention, the threats that are ultimately most serious may well be found in the fifth category, the disruption of environmental services. These services include the fertility-maintenance and waste-disposal functions of the biosphere's nutrient cycles, natural controls on crop pests and disease organisms, and the regulation of climate.[19] For example, the considerable direct health effects of sulfur dioxide and nitrogen oxides from combustion of fossil fuels (Chapter 8) may be less important than the ecological effects of the acid rain that these pollutants create, which disrupts the nutrient-cycling and waste-handling functions of entire ecosystems (Chapter 1).

Second, the distinction between acute and chronic phenomena is an important one. The policy maker trying to balance the day-in, day-out (chronic) health impact of burning fossil fuels against the small probability of a catastrophic (acute) accident at a nuclear reactor or related facility faces a genuine dilemma.

Third, the degree of irreversibility associated with different environmental problems should certainly be taken into account in making choices among energy technologies. We cannot completely avoid making environmental mistakes, but we should strive to minimize the chance of making irreversible ones.

Fourth, we should be wary of the notion that technical "fixes" for polluting energy technologies can render the environmental

141

impact of energy use innocuous (Chapter 19). It is true that in many respects energy's impact can be dramatically reduced, but for each proposed "fix" it is important to ask several questions: How fast can it be implemented (compared, for example, to the rate of expansion of the offending technology)? What degree of control does the fix provide, and what will we do for an encore after the gains have been erased by further growth? How much will the fix cost, and who will pay for it? What *new* environmental problems will the fix create?

Finally, there is an "ultimate" pollutant associated with all energy use, in the form of the heat discharged to the environment not only at conversion steps but at the final point of consumption. This consequence of the laws of thermodynamics cannot be circumvented, and its impact on global climate will eventually put a stop to growth of energy consumption if nothing else does first (Chapter 15).

XIII Some Environmental Impacts of Energy

	Health	Property	Social	Quality of life	Environmental services
Exploration					
Oil/gas	—	—	—	Invasion of wilderness	—
Harvesting					
Coal mining	Accidents, black lung	Loss of farmland, subsidence	Use of Indian lands	Defaced landscape	Acid drainage
Offshore oil	Accidents	—	—	Oil on beaches	Oil as a biocide
Hydroelectric dam	Dam collapse	Loss of farmland	Displacement of residents	Loss of wild rivers	Fish passage, wildlife breeding grounds
Processing					
Oil refining	Air/disease	Air/crops	—	Smells, visibility	Pollution of estuaries
Shale processing	Air/disease	Water consumption	—	Waste piles	Water pollution
Conversion					
Coal power plant	Air/disease	Air/crops-buildings	—	Noise, visibility	Acid rain, CO_2 particles/climate
Fission reactor	Reactor accident that breached containment would produce all classes of impact				
Transportation					
Oil tanker	Fire	Fire, collision	—	Oil on beaches	Oil as biocide
Electrical transmission	Electrocution	Restriction on land use	—	Unsightly towers	—
Plutonium	Leak/cancer	Land contamination/quarantine	Terrorism, nuclear bombs	—	—
Consumption					
Automobile	Air/disease	Air/crops	Suburbanization	Noise, visibility	Paved environment, heat/climate
Waste management					
Radioactive wastes	Leak/mutations	Land use	Terrorism, sabotage	—	Groundwater contamination

The Energy Crunch

The "energy problem" can usefully be analyzed in terms of three sets of factors: (1) the ingredients of the *short-term predicament* of the United States and many other industrialized nations; (2) a set of *underlying phenomena* that played a role in setting the stage for the present predicament, and whose existence guarantees that the energy problem will not go away soon or easily; and (3) the *long-term dilemma* that energy poses for civilization as a whole.

The short-term current predicament of the mid-1970s consists of a supply squeeze and associated economic problems, an overreliance on environmentally disruptive technologies, and a system of energy regulation and responsibility that has been aptly described with the words, "no one is in charge."[20] The supply squeeze — the "energy crisis" of the headlines — has been a crisis of flow, not of stock. That is, the problems have been caused not because of absolute shortage of energy resources (stocks), but because the rate of harvesting and processing and delivering energy supplies (all flows) did not grow as rapidly as did demand. The economic and social fallout of these bottlenecks of flow have included rapidly increasing prices for energy, the specter of windfall profits in the energy industries, unemployment in some other industries, and, almost certainly, a disproportionate burden on the poor.

On the surface, there were many causes that converged to produce the energy squeeze of the mid-1970s at this particular time: lack of government action to prepare for a transition from increasingly less accessible domestic petroleum and natural gas to more abundant energy sources; misdirected government action in fostering the rapid growth of demand by subsidizing or regulating the price of energy at artificially low levels; the maneuvering of oil companies in pursuit of maximum profit; the actions of oil-producing Arab nations; and the efforts of environmentalists to bring energy-related environmental costs and risks into the balance sheets. There is no doubt that these factors in combination governed the exact timing and dimensions of the energy squeeze in the United States, although one can argue at great length about the specific most important factor and how the various parties *should* have behaved.

It is probably more fruitful, though, to look at the underlying phenomena, which suggest that if the 1973–1974 energy squeeze had not materialized as it did, a superficially different combination of factors would have led to a similar result not long thereafter. The first of these underlying phenomena is that, beyond a certain point, growth gets harder as the level already achieved gets higher. Doubling a daily consumption of 20 million barrels of oil in the space of a decade is much more difficult than doubling a daily consumption of 5 million barrels in a decade, even though the percentage increase is the same in both cases. One begins to encounter limitations in terms of how quickly new facilities for harvesting and processing energy resources can be constructed (e.g., 3 to 5 years for refineries and power plants, not including site selection, acquisition, and licensing); stresses are felt in other sectors of the economy, as capital and labor are diverted to the energy sector; and hasty attempts to maintain a high rate of growth foster more frequent and more serious economic and environmental errors. The second phenomenon underlying the energy situation is that no energy technology is completely free of environmental liabilities. Some are better than others, but there is no technical panacea that could support the long-continued growth of energy consumption without encountering environmental barriers. The third phenomenon is the slowness of both technology and large-scale social behavior to change in the face of altered conditions. The combination of rapid growth, inadequate information about the future, and inability to respond quickly to new problems as they materialize is a sure prescription for an increasingly disruptive sequence of energy crunches, even after most of the apparent causes of that of the mid-1970s have been removed.

The major dilemma associated with energy in the long term arises from the foregoing basic phenomena and from energy's conflicting roles in affecting human well-being: on the one hand, energy is a major ingredient of economic prosperity; on the other hand, it is a major contributor to the disruption of environmental processes that provide services on which all prosperity ultimately rests — maintenance of soil fertility, controls on pests and disease, disposal of wastes, and regulation of climate. The threat of major interference with these services is likely to slow and then halt the growth of energy consumption well before any absolute shortage of resources does. The real contribution of the *positive* economic role of energy to the dilemma is this: leveling off global consumption before closing the gap in well-being between the rich and the poor in the world seems ethically untenable and in practice most unlikely.

What To Do About It: Some Proposals

Both in the short term and thereafter, the mainstay of a rational energy policy must be energy conservation. Comparison of different countries and different sectors of the U.S. economy shows that the relation between energy consumption and prosperity is not an inflexible one. Large savings in energy are possible in transportation, in space heating and air conditioning, in the materials industries and other industrial processes, and indeed throughout the economy. A 1972 report by the Office of Emergency Preparedness indicated that measures of relatively modest social and economic impact could cut the projected increase in U.S. energy consumption between 1970 and 1985 in half.[3] Such measures include car pooling, more small cars, shifting of freight from truck and aircraft to rail, better insulation, more efficient heating and cooling systems for homes and commercial businesses, more recycling of materials such as metal and paper, and leak plugging in industrial processes. Shifting economic activity from energy-intensive to labor-intensive enterprises has great potential for reducing energy consumption without increasing unemployment.

Environmentally, the first step is to clean up the mainstays of the present energy budget, the fossil fuels. Special attention should be given to finding environmentally tolerable ways to exploit the abundant resources of coal, and perhaps oil shale. The environmental risks of fission (Chapter 12), including the threat of terrorism and sabotage at the facilities or using stolen nuclear materials, deserve searching reexamination before a commitment is made to massive expansion of this approach to energy supply. A transition to the most environmentally benign energy sources available — for example, various forms of solar energy now, perhaps supplemented by fusion later — should be begun and pushed.

The economic costs of cleanup and transition will in some cases be considerable, adding to the increase in energy costs as inappropriate subsidies from the past are removed. The rising price of energy will stimulate some of the needed conservation measures, and the least desirable effects (windfall profits, a disproportionate burden on the poor) can be alleviated by legislative measures.

In the longer term, energy consumption will certainly have to be stabilized. This should take place first in the rich countries, as extravagant uses there are curtailed and necessity-oriented uses in the poor countries expanded. Stabilized consumption will mean both zero population growth and zero growth in energy consumption per person, but it need not mean no growth in well-being. What is required is to learn to squeeze more and more prosperity out of each bit of energy, a task that has barely been begun but has great potential.

The foregoing is a great deal to ask. Many things must be done at once — some expensive, some technically difficult, some socially or politically unpalatable. Unfortunately, there is no reason to believe that anything less will be enough.

References

1. United Nations, Statistical Office. 1971. *Statistical Yearbook of the United Nations.* U.N., New York.
2. Hirst, E. 1974. Food-related energy requirements. *Science 184:* 134.
3. Office of Emergency Preparedness. 1972. *The Potential for Energy Conservation.* Government Printing Office, Washington, D.C.
4. U.S. Department of Commerce. 1973. *Statistical Abstract of the United States.* Government Printing Office, Washington, D.C.
5. Cook, E. 1971. The flow of energy in an

industrial society. *Sci. Amer.*, September: 135.

6. U.S. Department of Commerce. 1960. *Historical Statistics of the United States: Colonial Times to 1956.* Government Printing Office, Washington, D.C.

7. Darmstadter, J. 1968. *Energy in the World Economy.* The Johns Hopkins Press, Baltimore, Md.

8. Hubbert, M. 1969. Energy resources. In *Resources and Man.* W. H. Freeman and Company, San Francisco.

9. Theobald, P. K., Schweinfurth, S. P., and Duncan, D. C. 1972. *Energy Resources of the United States* (U.S. Geol. Survey Circ. 650). Government Printing Office, Washington, D.C.

10. U.S. Atomic Energy Commission. 1970. *Potential Nuclear Power Growth Patterns* (WASH-1098). Government Printing Office, Washington, D.C.

11. U.S. Department of Interior. 1971. *Availability of Uranium at Various Prices from Resources in the United States* (U.S. Bureau of Mines Inform. Circ. 8501). Government Printing Office, Washington, D.C.

12. Brown, H., and Silver, L. T. 1955. The possibilities of obtaining long-range supplies of uranium, thorium and other substances from igneous rocks. In *United Nations International Conference on Peaceful Uses of Atomic Energy.* U.N., New York.

13. Holdren, J. P. 1974. Uranium availability and the breeder decision (EQL Memorandum 8). Environmental Quality Laboratory, California Institute of Technology, Pasadena, Calif.

14. Lee, J. D. 1969. Tritium breeding and energy generation in liquid lithium blankets. *Proceedings of BNES Conference on Nuclear Fusion Reactors.* British Nuclear Engineering Society, Culham, England.

15. Norton, J. J. 1973. Lithium, cesium, and rubidium — the rare alkali metals. In *United States Mineral Resources* (U.S. Geol. Survey Prof. Paper 820). Government Printing Office, Washington, D.C.

16. Holdren, J. P. 1971. Adequacy of lithium supplies as a fusion energy source. In *Controlled Thermonuclear Research.* Hearings before the Subcommittee on Research, Development, and Radiation of the Joint Committee on Atomic Energy, Pt. 2. Government Printing Office, Washington, D.C.

17. Sellers, W. D. 1965. *Physical Climatology.* University of Chicago Press, Chicago.

18. Morrow, W. F., Jr. 1973. Solar energy: its time is near. *Technol. Rev.*, December: 30.

19. Holdren, J. P., and Ehrlich, P. R. 1974. Human population and the global environment. *Amer. Scientist*, May–June.

20. Freeman, S. D. 1972. Issues associated with the use of energy. In *Issues Associated with the Use of Energy: Toward a National Energy Policy* (ORNL-CF-72-8-4). Oak Ridge National Laboratory, Oak Ridge, Tenn.

Further Reading

Hammond, A. L., Metz, W. D., and Maugh, T. H., II. 1973. *Energy and the Future.* American Association for the Advancement of Science, Washington, D.C.

Healy, T. J. 1974. *Energy, Electric Power, and Man.* Boyd & Fraser Publishing Company, San Francisco.

Holdren, J., and Herrera, P. 1971. *Energy.* Sierra Club Books, San Francisco.

Hottel, H. C., and Howard, J. B. 1971. *New Energy Technology. Some Facts and Assessments.* The MIT Press, Cambridge, Mass.

6
Water Resources

Tinco E. A. Van Hylckama *is a hydrologist with the U.S. Geological Survey and part-time Professor of Hydrology at Texas Tech University, Lubbock, Texas. Born in The Netherlands, he studied forestry there at the Agricultural University. He spent nine years in Sumatra, Indonesia, doing research for a private company and later as a prisoner of war during World War II. Forced to leave Indonesia during the Sukarno regime, he came to the U. S. and eventually joined the Johns Hopkins Laboratory of Climatology. He took his present post with the U. S. Geological Survey in 1958. He is the author of* The Water Balance of the World.

In 1973 large areas in mid-Africa suffered severe drought, causing millions of dollars damage to farmers, including the death of thousands of cattle and the loss of human lives.

In the same year the Mid-continent of North America suffered extremely heavy and prolonged flooding. Many lives were lost, many homes evacuated and destroyed, and crop production was severely curtailed —again at the cost of millions of dollars.

Properties of Water

Water is the most important fluid on earth. Without it, life as we know it on this planet would be impossible. Not only would all living things dry up, the earth itself would be subject to such extreme temperature fluctuations as to make it uninhabitable. Let us, therefore, examine a few of the important characteristics of this remarkable fluid.

In the first place, water is abundant. If we take the crust of the earth to be about 5 kilometers deep, we have a shell of 2,540 million cubic kilometers, of which about 1,500 million, or more than half, are water.

Second, water has an unusual molecular structure. A water molecule consists of one atom of oxygen and two atoms of hydrogen. There is a strong bond between these three atoms, which results in a molecule that, unlike most other molecules, is asymmetric and electrically charged. Because in most minerals that occur in nature the atoms are held together by electrical attraction, the water molecule, with its positive and negative charges, can easily squeeze in between atoms in other molecules. This ability is the cause of the enormous dissolving power of water; no other fluid can match it. It also explains, in part, the action of water upon the earth's surface, known as erosion. Mountains are continuously dissolved and, over the eons, washed into the sea.

The irregular shape of the water molecule has other consequences of great importance to our lives. When water becomes a solid (freezes), the molecules become arranged in open, crystal-like structures. Ice is therefore less dense than water and floats on the water's surface. If ice melts, that is, goes from the solid to the liquid phase, the molecules move around more freely, filling up the holes, and the water gets denser. The densest point is reached at 4°C. With rising temperatures, the molecules begin to move faster and faster, occupying more and more space until at 100°C the water boils, that is, goes from the liquid to the gaseous phase. However, because of the strong molecular attraction, it takes enormous energy to bring water into the gaseous stage.

The opposite is also true. When water condenses, enormous quantities of energy are released. This happens in the summer in our cumulus clouds, which in only an hour or so may build up into a thunderstorm. One average thunderstorm has the same energy as is created by the burning of 6,000 metric tons of coal.

Biological Importance of Water

It is fairly certain that the first living things on earth were formed in watery surroundings, and the inheritance of this origin is still with us. There are only a few living things that contain less than 10 percent of their weight in water, for example plant seeds and spores of bacteria and fungi. Most of the vegetable matter that we use for food, such as tomatoes, potatoes, lettuce, and carrots, contains at the time of harvest 85 to 90 percent water by weight. Even such derived foods as bread contain more than 30 percent water.

Plants use large quantities of water, but only a small part is used for building plant material. The rest is transpired into the air via small openings in the leaves called stomata (mouths). One of the first to have an inkling of this process was Jean Baptiste van Helmont, who in the early 1600s planted a willow tree in a large tub after carefully having weighed the amount of soil in the container and the plant itself. After 5 years he weighed both objects again. The willow had increased by 164 Dutch pounds, but the container has lost only 2 Dutch ounces.[1] Assuming a rainfall of about 75 centimeters per year, it is possible to figure that for each kilogram of plant material formed, about 70 kilograms of water were transpired into the air. This proportion of 70 to 1 is called the transpiration ratio. Figures one encounters in modern literature are much higher because investigators take into account only the dry matter of a crop or only those parts that are edible. For instance, the mean ratio for alfalfa is 844, but that is water use versus total *dry* matter; for corn it is 1,405 but that is water use versus crop product, thus only the ears are included.[2]

Man himself is extremely watery. This author weighs 65 kilograms; about 70 percent of this, or more than 45 kilograms, is water and he must continuously strive to keep this ratio as is or he will die of dehydration long before the water is completely evaporated out of his body. The amount of water man needs varies with the circumstances in which he lives and works. The very minimum probably lies between 2 and 4 liters per day, including the water that is used for cooking his food.

Distribution and Movement of Water

We mentioned already that there is on and in the crust of the earth the staggering amount of 1,500 million cubic kilometers of water. Such a volume is impossible to visualize, but the following may be useful. Roughly seven-tenths of the earth's surface is occupied by the oceans. As the top part of Table I shows, these oceans contain 1,457 million cubic kilometers of water. This water is spread over 361 million square kilometers of ocean surface. So the *average* depth of the oceans is about 4,000 meters.

The other numbers of Table I can be visualized as in Figure 1. The large cube represents all the water in and on the earth's crust, about 1,500 million cubic kilometers. The next smaller cube represents the water that exists in the form of ice caps, (such as on the Antarctic Continent and on Greenland) and in the form

of glaciers, a total of slightly more than 33.5 million cubic kilometers, or 2.24 percent. If this would all melt, it would raise the ocean surface by more than 70 meters. The next smaller cube represents all the water in the land, a mere 0.61 percent. The water in sweet-water lakes is a considerably smaller quantity: only 0.009 percent. The quantity of water present at any one time in the atmosphere is estimated to be about 15,000 cubic kilometers and is represented by the next smaller cube. Finally, the water present in rivers at any one time is just barely visible, a mere drop compared to the total quantity. Yet 1,500 cubic kilometers (see Table I) is a lot of river. Just think: If we take the average river to be 500 m wide and 10 m deep, there are 300,000 km of river, which is actually about the total length of all the rivers of the world.

You will have noticed the term "at any one time" when the quantities of water in rivers and in the air are mentioned. Why? Because water is continuously moving and changing from one phase into another. We all know that water does not have to boil to produce vapor. We can see vapor rising from warm water long before it boils, and even outdoors this is observable; for example, on a cool morning, warm irrigation water produces a layer of vapor in the air above it. The energy of such processes comes almost entirely from the sun, as symbolized in Figure 2. Because of the sun's energy, water rises from oceans, lakes, and other surfaces, condenses in the air to form clouds, and the water returns to the earth in the form of rain, snow, or hail.

Most of the evaporated water is derived from the oceans, but about 14 percent comes from the land surfaces, lakes, streams, moist soils, and the leaves of transpiring vegetation. (The yearly totals are given in the bottom part of Table I.) A large part of the water that is evaporated eventually falls back on the ocean, but about 24 percent falls on the land surfaces. Notice that this is a larger percentage than what evaporates from the land. However, 10 percent does not remain long on the land surfaces. It soon reaches streams and rivers and in a comparatively short time, varying between a few hours and a few weeks, it again arrives in oceans. This is the runoff given at the bottom of Table I: 40,000 cubic kilometers per year, or about 3 million cubic meters per

149

second. At this rate Lake Erie would empty in less than 1 week or Lake Tahoe in 1 day.

Still 14 percent remains on the land in the form of ice and snow or penetrates into the soil and becomes available to the plants. But soil, like a sponge, can hold just so much water. When the water-holding capacity is reached, water can enter only at the rate that it percolates down under the influence of gravity and capillary attraction. The part that has entered the soil and is not picked up by plant roots can continue to travel down and reach what is called the *water table*. If you take a glass bowl, fill it with some sand and then pour water in it, you can see a distinct line marking the position between the wet and dry sand. This line is the water table. Below the line, the sand is saturated; above it, unsaturated or dry. In moist climates such a table can be found at depths of only a few meters or even less; in dry climates there may be no water table at all. In both places, however, water may have collected in prehistoric times and, owing to processes of erosion, sedimentation, and other geological activities, be buried. This is the groundwater, which plays an immensely important role in large areas of the western United States, of North Africa, and many other places. Such water-carrying layers are called *aquifers*. Sometimes this water is under enormous pressures, owing to the overlying rock and soil formations, and when wells are dug, water may spout several meters into the air, forming a fountain that will last until the pressure in the underlying layers is released. Such layers are called *artesian aquifers*.

I Water Data

	World (km³)	Percent	United States (km³)
Where is the water?			
Water on the land			
Freshwater lakes	135,000	0.009	20,000
Rivers (at any one time)	1,500	—	70
Glaciers	230,000	0.015	40,000
Ice caps	33,380,000	2.225	—
Saline lakes	100,000	0.007	10,000
Water in the land			
Soil moisture	38,500	0.003	4,000
Groundwater			
Within 1,000 meters	4,550,000	0.303	300,000
In the next 1,000 meters	4,550,000	0.303	300,000
Atmosphere (at any one time)	15,000	0.001	300
World oceans	1,457,000,000	97.134	—
	1,500,000,000	100.000	674,370
Where does the water go?			
Annual evaporation			
From land	56,000	0.004	4,000
From oceans	344,000	0.025	—
	400,000	0.029	
Annual precipitation			
On land	96,000	0.007	6,000
On oceans	304,000	0.022	—
	400,000	0.029	
Runoff of all rivers per year (water yield)	40,000	0.003	2,000

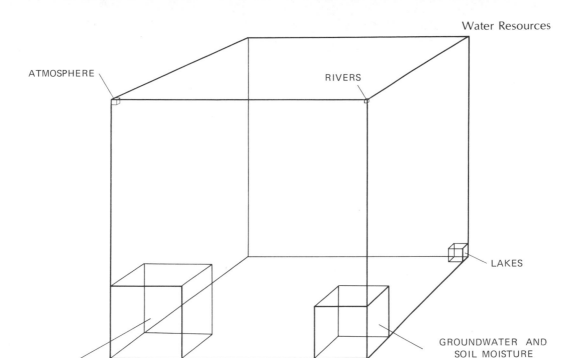

ATMOSPHERE

RIVERS

LAKES

GROUNDWATER AND
SOIL MOISTURE

ICE AND GLACIERS

1 **Waters of the World.** The large cube represents all the waters of the world. The smaller ones represent the quantities occupied by different kinds of water. The rest of the large cube represents ocean water and saline lakes.

It is clear from the above that a molecule of water can spend hundreds, even millions of years in one place, for example when it falls on the Greenland ice cap. On the other hand, a drop of rain falling on a hot pavement evaporates and becomes part of the moisture in the air in less than 1 second. Although the time that a water molecule remains suspended in the atmosphere before it again becomes part of a raindrop or snowflake can vary widely, on the average this *residence time* is about 12 days.

SALT WATER

Salt water is discussed in detail in Chapters 11 and 15, so we shall only mention it here and point out that ocean water contains on the average 35 grams of dissolved solids for each liter. By far the largest part of these solids, more than 70 percent, is sodium chloride, which we know as kitchen salt. Because of this salt, the properties of oceanic water differ from those of sweet water. Oceanic water does not have its greatest density at 4°C and it does not freeze at 0°C. It freezes at about −2°C and has at that temperature also its greatest density. In consequence, oceanic water, when it cools,

sinks. The resulting mixing is very important and necessary for plant and animal life, but considered as a direct resource of water, the oceans are of little value because the water is unsuitable for agricultural, domestic, and most industrial uses.

SWEET WATER

Since man cannot use salt water directly for his own use or for agriculture, his life depends on the availability of sweet water. If we add all the water, mentioned in Table I, that is on the land, in the land, and in the atmosphere, we find that there are more than 42 million cubic kilometers of water in these places. However, it is immediately evident that at least 33.5 million (nearly 80 percent) is not readily available, because it is locked up in ice caps and glaciers. Furthermore, of the water in the land, nearly half occurs below 1,000 meters. Much of this water is under heavy pressure from the earth above it and has in consequence a high temperature and a high dissolving power. Hence water that is pumped from great depths is often too hot and too briny for use on an economic scale.

151

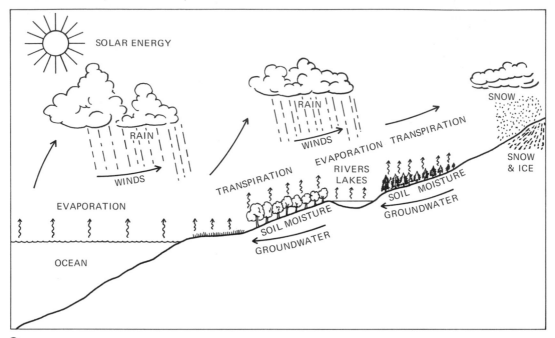

2   **The Hydrologic Cycle.** *Source:* Reference 4.

Water in the atmosphere, at any one time, is only available when it actually comes down as precipitation. In some areas it is carefully captured from the rooftops and stored in barrels. Elsewhere smaller or larger ponds are available or dug so that cattle can use the captured water. By far the larger part, however, runs off over or through the ground to become river water or groundwater. Rivers, in turn, keep the lakes full and to a lesser degree also contribute water to aquifers. It is for these reasons that the "living water" of the rivers and streams always was and still is the main source of supply for our agricultural, industrial, and domestic needs.

A little arithmetic will show that, on the face of it, the flowing amount of water is an enormous quantity. Suppose that the world carries 6 billion people, a number that has been estimated to be the world population by the year 2000. This number divided into 1,500 cubic kilometers would give 250,000 liters per person. A city dweller in the United States uses about 600 liters per day for his personal use. Thus it seems that there is plenty of sweet water on hand at any time. For several reasons that is not so. Some of the biggest rivers flow in areas that are practically uninhabitable, such as the Amazon; the three big Siberian rivers, Ob, Yenesey, and Lena; and the MacKenzie in northern Canada. Moreover, a large part of the world population, especially in the United States, lives or wants to live in areas where there are no rivers or only small rivers but lots of sunshine. In other places the rivers have been so intensely used as sewage canals that the water is unfit for human consumption unless it were to be thoroughly and expensively treated. Such is the case with the Mississippi in the United States, the Rhine in Europe, the Ganges in India, and the Volga in the USSR.[3] Since all the water that is out of reach or unusable is part of the 1,500 cubic kilometers, the actual available amount is much less than 250,000 liters per person, maybe only 50,000 liters per person.

Furthermore, we considered only the use of water by city dwellers, but this is just the water for personal use. Far more water is used in industry and agriculture. It has been estimated that in the United States 6,000 liters of water per day per person are used to provide each citizen with food, clothing, and the amenities of life. A large part of this water is being used over and over again, such as water used for cooling in the steel industry. On the other hand, water used in strip mining and in the paper and pulp industries is unfit for further consumption unless it is purified. Even though water appears to be in oversupply, this is only true as an *average* condition, and to properly evaluate the abundance of water we have to examine its distribution, as we do below after we discuss groundwater.

The seemingly large source of water in the ground must be discussed separately because it is a source with very different characteristics. Until recent times only shallow groundwater could be used, but modern pumping techniques have made it possible to open up huge sources of groundwater such as found on the High Plains of Texas, Kansas, and Nebraska; in the San Joaquin Valley in California; and in the deserts around Lake Chad in North Africa and those in Pakistan. As a look at the hydrologic cycle (Figure 2) shows, it is possible that the water withdrawn from underground can be replaced by rainfall seeping in through the layers above the aquifer or laterally from hills or mountains in the neighborhood. Such was expected to be the case in California, but it should be realized that water moves through the ground only very, very slowly, a few meters a day at the most. It has been calculated that at the present rate of water seeping to the aquifer, a process that is called *natural recharge*, it would take 150 years to build up the estimated U.S. supply of 300,000 cubic kilometers mentioned in Table I.

So, unless the withdrawal is equal to the recharge, groundwater is being mined and becomes a nonrenewable resource. This happened north of Phoenix, Arizona, where the desert was converted into cotton fields, which had to be abandoned when the water levels dropped so far that pumping became uneconomical. A similar situation may soon exist to the south of Phoenix in the fertile valleys around Casa Grande.

The High Plains of Texas owe their prosperity to waters accumulated there several millions of years ago in an aquifer called the Ogallala Formation. This water is now being pumped out faster than it is renewed by infiltration of snow- or rainwater. The water table is dropping, and sooner or later all the water will

be used up. Unless surface water is imported one way or another, this area of the country will have to abandon its present type of agriculture.

A moment ago we talked about average conditions of the water availability, but at the beginning of this chapter we mentioned two extremes occurring in the same year. Clearly, it is the *local* water supply that is crucial, not the average supply. The Dutch have too much water and have developed an intricate drainage system; at the same time they have to prevent seawater from seeping in under their land and contaminating the groundwater. The South Africans, in a much drier country, have diverted rivers to bring the water where they think it is most needed; the same has been done in Australia (see below) and on a very much larger scale in the United States. A sparse population, living mostly on hunting and food gathering, like the Australian aboriginals, can survive on very little water, while a city like Los Angeles is vigorously trying to provide itself with more and more water.

Cassandras who predict that in certain parts of this or other countries people within 20 or 60 years will fall gasping of thirst on their grimy pavements or dusty fields[5] overlook the fact that man has developed the capacity to lead water to where he wants it. It is sad but true, though, that undesirable side effects may occur and have occurred, some of which will be discussed later, while pollution problems are dealt with in Chapters 10 and 11.

The most severe potential shortages occur in dry areas, where the population is mining groundwater, that is, using it at a rate faster than the rate of natural or artificial recharge. But it is in such areas also that people are looking for means to import river water from areas where water is abundant. If such importation were not possible, then, of course, certain areas would run out of water and would not be able to sustain the dense population they have now or the type of agriculture that is now practiced there. Arizona and environs are already considered to be low on water supply (hence the urgency of the Central Arizona Project, mentioned below). Areas to the northwest of there, Nevada and surroundings, may be running low in the 1980s, whereas Texas and Oklahoma may run short by the year 2000. But, to repeat, such predictions are mere speculation and do not take into account the resourcefulness of man or his capacity to live in equilibrium with nature.

Water Use

The way man utilizes water resources can be separated into three types. In the first place, there is *withdrawal*, the amount of water that is taken out of a stream or pumped out of an underground or surface reservoir. If this water is used in industry, say for the purpose of cooling, it can, after use, be stored in a reservoir and be used again as soon as it has cooled off sufficiently, or it can be released into the stream from which it was taken. If the quality and the temperature have not changed drastically, the released water can be reused and the only loss is that due to evaporation.

Another type of use also starts with withdrawal, but only a small part of the water withdrawn is returned because most of it is used by the plants for transpiration and evaporates from the soils and ditches during the irrigation of crops. The part that is so lost is called *consumptive use*. Many farmers, just to be on the safe side, use more water than is actually needed. This tail water, as it is called, runs off at the low side of an irrigated field, sometimes flooding roads and running to waste, although a small part may eventually reach the groundwater again. A considerable amount of this tail water is, at least theoretically, reusable. However, too often during the irrigation process water has taken up such quantities of salt and silt that it is no longer fit for further use. The Colorado River provides a

most revealing example and will be discussed in some detail later. Another disturbing example is found in Pakistan. Thirty million people live on the Indus plain, where an enormous irrigation network delivers water to 90 million hectares of land, but more than 2 million of these are already lost because of salinity; current annual losses are about 40,000 hectares.[6] If this holds true elsewhere, such fertile areas as the Imperial Valley in California may be in danger.

Then there is *nonwithdrawal use*. This is the use that is made of water for navigation on rivers; for swimming, boating, and fishing in lakes and reservoirs; and for other recreational purposes. If these were the only nonwithdrawal uses, they would affect the quality of the water only slightly, but the greatest nonwithdrawal use of water is made by industry, cities, and agriculture when they dump sewage, waste, and everything they want to get rid of into the rivers, lakes, and streams.

Water use is also classified by type of supply. By *municipal* supply we mean a publicly or privately owned water system that mainly serves a city or suburb and may supply some water to local commerce or industry. *Self-supplied industrial* water is derived from a system established by an industry for its own use, such as wells on its own or leased property or a diversion dam in a river. Water withdrawn for hydroelectric power is usually not included in this category. Water supplied to a crop by a system of sprinklers or ditches is called *irrigation* supply, while water on the farm (stock and poultry watering, lawn sprinkling, etc.), not obtained from a municipal system, is called *rural* supply. The last two are often combined for statistical purposes.

II U.S. Water Use (1970)

Supply	For Public Supplies			Rural Use		Irrigation		
	Withdrawn	C.U.[a]	Per Person	Withdrawn	C.U.[a]	Total	Lost[b]	C.U.[a]
Surface water	68.1	—	655	3.4	—	307	—	—
Groundwater	35.6	—	587	13.6	—	170	—	—
Reclaimed sewage	—	—	—	—	—	1.4	—	—
	103.7	22.3	629	17.0	12.9	478	83	276

All figures are units of 10^9 liters/day, except use per person, which is given in liters per day.
[a]Consumptive use (evaporation and transpiration).
[b]Lost in transit (evaporation and seepage).
Source: Reference 7.

Water for domestic use is only a fraction of that used by industry. The withdrawal for domestic use is on the order of 7 percent of the total, but this is an exceedingly important 7 percent because of the requirements of purity, or better, drinkability of this water. The standard requirements vary from country to country, and the deviations from standard probably even more.

After such water has been used for food and drink supply, for other uses in the household including that in the garden and bathroom, it enters a sewage system (if such is available), undergoes a biological and chemical treatment, and is again usable or could be released into streams as clean water. The Ohio River's water, for instance, is used on an average of four times before it gets to the Mississippi. It is the sad truth that returning the water to the river clean, cool, and drinkable is far from the general practice (Chapter 10).

WATER USE IN THE UNITED STATES

Water use over the world is difficult to estimate, but data for the United States are regularly published. Table II presents some data on water use for various purposes according to the latest available source.[7] The table shows that about two-thirds of water used for public supplies (i.e., mostly municipal water) comes from surface-water sources. Of 103.7 billion liters total use per day, 22.3 billion, or slightly over 21 percent, are consumptively used. Of the 17 billion liters withdrawn for rural use (i.e., both domestic and livestock) about 80 percent is derived from groundwater and an equal amount is consumed. It is an interesting and disturbing fact that 83 billion liters per day, or more than 20 percent of all water used for irrigation, is lost while the water is in transit from the source to the fields (that is, by evaporation from open ditches and by seepage from unlined canals). An additional 56.5 percent is used for evaporation and transpiration in the field itself.

Electric utilities can and do use a considerable amount of saline water, and Table III shows that nearly 80 percent of all the water is used in the production of thermoelectric power. The consumptive use is just 3 percent, very small compared to the consumptive uses mentioned in Table II.

Table IV is an attempt to look into the future.[7-9] Noteworthy is the nearly *fivefold* increase in water use by industry expected by the year 2000. However, a very large part of this can be supplied by multiple reuse if the water is treated properly. In other words, a fivefold predicted increase is not necessarily indicative of disaster.

By the year 2000 water withdrawn for rural and irrigation use is expected to have increased to 256 percent of the 1960 value. The less water that a plant uses for producing the same amount of useful material, the more efficient is the use of water. Plant scientists are trying to breed crops that produce the same amount of useful material, with a lower consumption of water. Also, attempts are being made to cover the soil, and even the plants, with plastic coating or other materials to reduce transpiration and evaporation, thereby lowering the transpiration ratio and the consumptive use percentage. We see from Table IV that no significant results are being expected, because the amount of water for consumptive use remains at 56 percent of the withdrawal for rural and irrigation purposes.

It is now useful to compare the data of Table IV with some from Table I. The total withdrawal for the year 2000 is estimated to be 3,537 billion liters per day or about 1,300 cubic kilometers per year. Table I shows that the water yield (runoff) for the United States is 2,000 cubic kilometers per year, a great deal more than the estimated demands. Moreover, the 1,300 cubic kilometers is withdrawal; therefore, a good part of this water can be used two or more times if it is handled wisely. Also, the total water consumed, that is, the water that is

III Self-Supported Industrial Water Use

Supply	Electric utilities	Other	Total	C.U.[a]
Surface water (sweet)	454.2	117.3	571.5	—
Surface water (saline)	174.1	26.9	201.0	—
Groundwater (sweet)	5.3	30.3	35.6	—
Groundwater (saline)	—	3.8	3.8	—
Reclaimed sewage	—	0.6	0.6	—
	633.6	178.9	812.5	24

Data for water use in the United States in 1970 by self-supported industry, given in 10^9 liters/day.
[a]Consumptive use.
Source: Reference 7.

lost for further use until it returns as precipitation, would diminish from 24 to 17 percent by more efficient use.

All these are generalized figures that summarize data valid for a large part of a whole continent. On a smaller scale, the situation can be much more complicated and, at times, less reassuring as far as the future is concerned. As an illustration, two examples follow, one dealing with a predominantly rural area in a dry climate and the other with a very densely populated district but in a moist climate.

THE LOWER COLORADO

The head waters of the Colorado River and its tributaries are in the high mountains of Wyoming, Colorado, and Utah, parts of which are shown in Figure 3. The river enters Arizona just north of Lees Ferry, runs through the spectacular Grand Canyon, and from there winds its way south toward the Gulf of California, a distance of more than 2,200 kilometers. Before 1935 the Colorado was an unpredictable stream because the quantities of snow and rain, and the rates of snow melt, varied so much that the river was a roaring current at some times and a mere trickle at others. From 1905 to 1923 the average yearly runoff at Lees Ferry was 19.5 cubic kilometers per year, and when in 1922 the seven states that claimed a right to Colorado River water got together and drew up the Colorado River Compact, it was decided that the basin area below Lees Ferry would be entitled to about half that amount and the area above would get the other half. But after 1923 and up until 1966, the river never again reached this average. In only 6 years was the runoff larger than 19.5 cubic kilometers per year, while 24 years

yielded less than 15 cubic kilometers. In other words, the states claimed water that simply was not there.

In 1935 Hoover Dam was built with the idea of creating a reservoir from which during dry times water could be withdrawn so that the area below the dam could be ensured of a dependable supply. Because the water from upstream carries a heavy load of silt, Lake Mead, as the reservoir was called, began to collect mud at a rate faster than was expected. In 1963 Glen Canyon Dam was built, not only to provide power and water for irrigation to areas above Lees Ferry, but also to diminish the rate of sedimentation in Lake Mead. Lake Powell, as the reservoir above Glen Canyon Dam is called, began to fill up in 1963 and it was hoped that there would be enough water to start generating electricity by June 1964. But 1961 and 1962 had been dry years and, with Glen Canyon Dam withholding water, the water level in Lake Mead got so low that, unless something was done, it would be impossible to keep the generators at Hoover Dam operating. So the Bureau of Reclamation, which is in charge of these works, opened the gates at Lake Powell in order to fill Lake Mead. This made people in the upper states unhappy because they had to wait so much longer before they got their electricity and irrigation water. In the meantime Davis and Parker Dams downstream from Hoover Dam had been built, again with the idea of providing dependable water supplies not only to farmland but in particular to Los Angeles from behind Parker Dam. Thus two more lakes had been created: Lake Mohave and Lake Havasu. All these reservoirs are in hot country and evaporate great quantities of water, estimated

IV Future U.S. Water Use

	Withdrawn					Consumed					
	1960	1980	% increase over 1960	2000	% increase over 1960	1960	%	1980	%	2000	%
Public supplies	78	125	160	176	226	13	17	26	21	28	16
Industry	530	1,375	260	2,510	473	12	2.3	42	3.1	91	3.6
Rural and irrigation	332	632	190	851	256	197	59	355	56	477	56
Total or average	940	2,132	227	3,537	376	222	24	423	20	596	17

In billions of liters per day. Extrapolations into the future partly made by the author. Percentage consumed indicates amount of withdrawal in same year.
Source: References 7, 8, and 9.

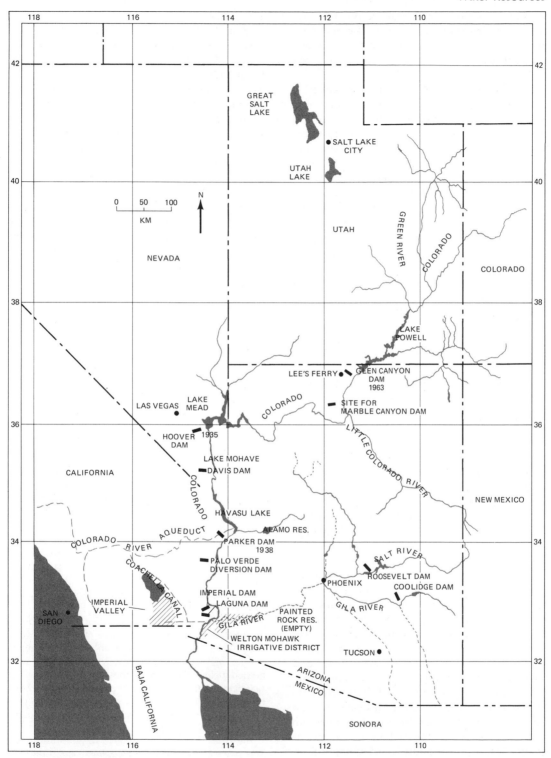

3   **Location of Dams and Reservoir Along the Colorado River.** Only a few of the tributaries of
the river are shown.

157

at slightly more than 2 cubic kilometers per year or 10 percent of the high average flow from before 1923. As a result of the evaporation, salt concentrations in the river water are high.

These salts consist mostly of sulfates, chlorides, and bicarbonates of calcium, magnesium, sodium, and potassium. The quantities are usually expressed in milligrams per liter total dissolved solids. For drinking water the maximum recommended by the U.S. Public Health Service is 500, and most public water supplies in the United States have much less than that. By contrast, seawater, as we saw before, contains about 35,000 milligrams per liter. With these figures as a guide, we can now look at the quality of the Colorado River water and how it deteriorates downstream.

Below Hoover Dam the water contains 700 milligrams per liter dissolved solids, at Lake Havasu 800, and at Imperial diversion Dam just north of Yuma, 900. From behind Imperial Dam this water is diverted via the All-American Canal to the Imperial Valley in California. On the Arizona side of the river is the Wellton Mohawk irrigation district, which also uses Colorado River water, but because of poor drainage conditions, the fields there were in danger of becoming saturated with saline water. To combat this salinity, groundwater is pumped in this district to flush out the soils. The water is then pumped out into the Colorado River behind the Morelos diversion dam, from where it can be delivered to Mexico. However, the tail water that is diverted back into the Colorado River has salinity figures at times as high as 4,000, and the water that reaches the farmers in Mexico was reported to have had (in 1966) more than 3,000 milligrams per liter total dissolved solids. The Mexican farmers protested that a salt level of well over 1,200 milligrams per liter[10] ruined their fields, and now the water from the Wellton Mohawk district is often carried around Morelos Dam into the Colorado River and hence directly to the Gulf of California. Farmers in the Imperial Valley and in the Wellton Mohawk district say that the Mexicans use bad irrigation practices, but the Mexicans maintain that they have contractual rights to good-quality water. The dispute is still going on.

We might say that the Colorado River is "overworked." The demand for water is simply greater than the river can deliver, even with multiple reuse. To make matters more complicated, work has started on the Central Arizona Project, which, when it is finished by about 1985, will divert 1,500 million cubic meters of water per year from behind Parker Dam and bring it to the Phoenix and Tucson areas. There are plans (at least temporarily canceled) to build two more dams, Bridge Canyon Dam above Lake Mead and Marble Canyon Dam a few miles north of the up-river boundary of Grand Canyon National Park below Glen Canyon Dam. They might be useful for generating power but would not create more water. It has been suggested that water be diverted from the Columbia River to the Colorado area, but this belongs in the realm of the "grandiose water plans" and is a different story.

NEW YORK CITY

New York City, located on the Hudson River, should have plenty of water both for municipal and industrial use, yet this is not the case. The story, splendidly told by Schneider and Spieker,[11] is quoted here in full (see Figure 4).

In its early years, New York's water was supplied by shallow wells and small reservoirs, all privately owned. None of these sources was satisfactory, and epidemics were frequent. By the 1820's it was clear that a public supply was needed, but there were no adequate reservoir sites nearby. New York's population was then approaching 300,000. A proposal to build a 60 kilometer long aqueduct to a reservoir site near Croton was first considered preposterous but gradually became accepted as a necessity. A cholera outbreak in 1832 killed 3500 people and dramatized the necessity of a new supply, which was authorized in 1834. A disastrous fire in 1835 further demonstrated the desperate need and construction was accelerated. The system was completed in 1842.

At that time the Croton Reservoir was no doubt regarded as the ultimate answer to New York's water needs. Within 20 years, however, it had to be enlarged. Several new reservoirs and a large aqueduct were needed before the turn of the century. By then all satisfactory sites in the Croton watershed had been exhausted, and the demand was fast catching up with the available supply. Clearly, new sources of supply would have to be sought.

The Catskill Mountains, about 120 miles from New York, were chosen for the new reservoir sites. Construction began in 1907, and the system was completed in two stages: the Ashokan Reservoir was completed in

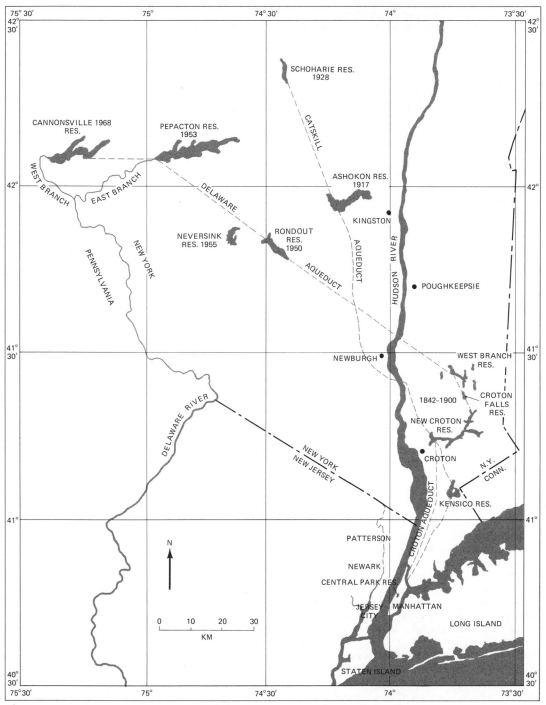

4 **Location of Reservoirs and Aqueducts that Provide Water to the New York Area.**
Tributaries to reservoirs and rivers are not shown.

1917 and the Schoharie Reservoir, in 1928. Although addition of the Catskill system more than doubled the previous supply, the new supply was barely able to keep up with the rapidly increasing demand. By the late 1920's another water crisis was in sight.

This time alternatives were considered. The Hudson

159

River was ruled out because of its allegedly inferior quality. New Yorkers insist on drinking pure mountain water. The Adirondacks were eliminated because of the excessive distance. In 1928, then, it was decided to expand the Catskill system and to develop new reservoirs in the headwaters of the Delaware River basin. The Delaware River is an interstate stream, so the consent of New Jersey and Pennsylvania was needed to divert water from this basin. The issue was resolved after considerable litigation in 1931 by a decree of the U.S. Supreme Court that allowed New York to divert no more than 1665.4 million liters per day from the Delaware River Basin. First the depression, then World War II delayed construction. The first operational phase of the expansion consisted of an emergency diversion from Rondout Creek to the new Delaware aqueduct from 1944 to 1951.

In the meantime, yet another crisis occurred. The postwar urbanization explosion strained the Croton and Catskill systems almost to their limit. Average pumpage exceeded 4 billion liters/day. Abundant rainfall deferred the day of reckoning until 1949, when reservoir levels dropped to the danger point. Stringent water conservation measures were enforced, and for the first time the Hudson River was tapped at Chelsea as an emergency source of supply.

Rondout Reservoir, an expansion of the Catskill system, became operational late in 1950 and the diversion from the Hudson was discontinued. Neversink and Pepacton Reservoirs, with their diversion appurtenances, began being used in 1953, but full use of the Delaware system was not achieved until 1955.

History repeats itself. The crisis of 1949 and the forecasts of the even greater population explosion to come made the water planners all too painfully aware that even the Delaware River basin supply system under construction would only temporarily satisfy the city's needs. An additional source would be needed. Thus, planning began for a new reservoir in the Delaware River basin. In 1954 the Supreme Court authorized New York to increase its diversion and in 1955 construction started on the Cannonsville Reservoir. Planners estimated that, with this new addition, the system would have total capacity of 6.8 billion liters/day, sufficient to meet demands through 1980.

The record breaking drought of 1961–66 occurred, however, before the Cannonsville Reservoir was completed. At one time the existing reservoirs were drawn down to 26 percent of capacity (near the minimum safe drawdown for which the system was designed). The most stringent water-use controls in the city's history were put into effect, and the Chelsea pumping station of the Hudson River was rebuilt. By 1967 abundant rainfall eased the crisis, and the situation returned to "normal." But if history can be taken as any guide, it will not be long before New York is again faced with a water crisis. Indeed, planning has already started on alternative means of meeting expanded needs.

The history of the New York water system has been one of continuing crisis in order to satisfy the demands of the population explosion. Yet part of the water demand might be regarded as unnecessary or artificial. Wasteful and inefficient use of water is encouraged by the absence of metering and unrealistic pricing. While the elaborate network of reservoirs and aqueducts has been built at great cost, the Hudson River, which might supply New York's needs many times over, has, like many other rivers, been allowed to degenerate in quality. The state's Pure Water Program improvements show promise of effecting some regeneration. Planning decisions must be sensitive to economics, politics, and public attitudes. The citizens of New York have become conditioned to drinking "mountain water," and any change in established practices of water supply would require a massive campaign of public information and education.

The Search for Water

The old civilizations that developed along the rivers did not have to search for water. They saw it, and there was usually plenty of it. By ingenious systems of ditches and dams, they could divert this river water to their land and grow crops on irrigated fields, although climatic changes have forced people for longer or shorter time, to leave such fertile valleys. An interesting case in point is the irrigation system near Phoenix, Arizona, where at least from A.D. 100, but possibly earlier, and up to about 1400, a tribe of Indians flourished who diverted river water from the Salt and Gila rivers and used it to irrigate their fields. Traces of the canals are still visible, and many modern ditches of the Roosevelt irrigation district follow the courses used by the Indians. Later Indians who saw the remnants of irrigation canals and dwellings called the tribe the Hohokam, meaning the vanished one.

When man began to understand where groundwater came from and how it was formed (which comprises the science of hydrogeology or geohydrology), scientific exploration for water could begin, but large-scale exploration still had to wait the invention of engines capable of drilling fast enough and deep enough. An early drilling occurred between 1829 and 1841 near Paris, France. The drill was driven by man or horses walking on a tread mill. It took 12 years to reach a depth of slightly over 500 meters, but the results were spectacular because this was an artesian well and water spouted 30 meters high in the air. We know now that such wells do not continue to produce their initial rate of water unless they are continuously fed, and this well was no exception. By 1910 it was no longer artesian and water had to be pumped, because by that time numerous wells were tapping the same aquifer.

In arid lands, of course, the situation might well be much worse, and in many cases is. We have cited cases where groundwater is being mined, and when it is used up, the population will have to move or different means of water importation sought.

There are other possible sources of augmenting a water supply. Some of them at the present time seem fanciful, but lots of things that have happened now seemed fanciful only 15 years ago.

There is, for instance, the possible use of the frozen resources, which constitute as we have seen two-thirds of all the sweet water in the world. It has been proposed to tow icebergs from Greenland or Antarctica to thirsty cities. Greenland icebergs are irregularly shaped and have a habit of going topsy-turvy at times which makes towing rather risky, but the Antarctic icebergs are flat-topped, very much more stable, and are larger. A small one may measure 3 × 3 kilometers and be 250 meters thick. It has been figured that one good-sized tugboat making 1 knot could bring such a berg from Antarctica to Australia in 6 to 7 months at the cost of 1.5 million dollars. If we assume that about half of the ice would melt during the trip, there still would be 1 billion cubic meters of sweet water, or enough for 4 million people for a year at an average cost per family of about $1.50. Such water would be very much cheaper than that of even the most efficient desalinization plant now available. This of course does not include the cost of purifying, pumping, and piping. By comparison, the U.S. Public Health Service estimates that a person who depends on municipal water supply systems pays between $5 and $15 per year, depending on the locality. Fantastic as these plans may be, they have the attractiveness of being "natural." Icebergs float around anyway while melting, and to have them melt where we want them would not affect climate or ecology to a noticeable degree.

We have seen that in many places water is available but the quality is such that it cannot be used as is. It is sometimes economically possible to remove the undesirable chemicals from the water by one of the several processes of desalinization. This is especially so when there is a large demand for domestic and industrial use. Big desalting plants can be found in Aruba and Curaçao in the Netherlands Antilles, and in Kuwait on the Persian Gulf. In these three cases it is mainly the oil industry that makes this expensive process feasible.

All in all, there are at the moment at least 700 desalting plants in operation around the world. The land-based ones use saline or brackish groundwater and are comparatively small. Those near the oceans or saline lakes can be more than 100 times as large because of the unlimited supply of water. The one near Tijuana, Mexico, for instance, is supposed to produce 35 million liters per day at a cost of less than 15 cents per 1,000 liters, compared to about 80 cents for plants built in the 1950s.[12] Yet 15 cents per 1,000 liters is still much too high for agriculture, for which 2 cents per 1,000 liters is considered extravagant. (Also, 15 cents per 1,000 liters does not include transportation.)

The cost of lifting water is generally estimated to be 1 cent per 100 meters of lift per 1,000 liters. At this rate a city of 150,000 inhabitants at an elevation of 1,000 meters, such as Lubbock, Texas, which used 1,000 liters per person per day would have to pay 10 cents per person per day or more than $5 million per year for lifting alone, and at least another 5 million for horizontal delivery (if desalting costs were to go down to 10 cents per 1,000 liters).

In another type of search, man looks to the heavens for more water. When the knowledge of cloud formations and cloud physics developed and the necessity for the presence of condensation nuclei became known, many scientists, among them Vincent Schaefer and Irving Langmuir, introduced the idea of providing clouds with extra nuclei on which the cloud vapor could condense, thus forming drops that would fall near the area where the cloud was "seeded." Some apparently spectacular successes between 1947 and 1950 resulted in the formation of a large number of private enterprises which claimed, rightfully or not, that they could make it rain on demand. The enthusiasm has cooled considerably; the summary of the Third National Conference on Weather Modification, held in Rapid City, South Dakota, in June 1972, complains about a lack of good data and an overemphasis on conjectures.[13]

Finally, there is the possibility of "importing" water. When groundwater reservoirs become exhausted and when more withdrawal from nearby surface water, because of increase in population, industry, and agriculture, becomes impossible, the populace of such an area fixes its eyes on the big rivers that apparently run uselessly into the sea in other areas. Attempts to transfer water from one basin to another may be costly, but it still may be economically feasible if authorities or owners of the losing basin do not object to the gaining basin. For example, from the Snowy Mountains about 100 kilometers south of Canberra, Australia, the Snowy River runs south directly into the ocean. By a system of gigantic tunnels this water is diverted to the northern side of the mountains to help irrigate the fertile valleys of the Murrumbidgee and the Murray. On their way the waters drive gigantic underground turbines to provide electricity for Melbourne, Sydney, and the area beyond.

In California, waters from the north are successfully diverted to the fertile valleys and densely populated areas of the south, and plans are being made to divert water from the Columbia River to the south, but the adjacent states seem to object to such a drastic change.

No such objection would occur if the plans to divert the Pechora and Vychegda become a reality. These two rivers, west of the Urals in northern Russia, at present meander northward on their way to the Arctic Ocean. The plans are to lead the waters into the Volga, which flows south into the Caspian Sea. This not only would provide more irrigation water for the warmer valleys in the south, it would also stop further decline in the level of the Caspian Sea. Probably neither of these two plans would affect climate, but this is not necessarily so with larger endeavors. Worthy of mention among such proposals are the North Atlantic Water and Power Alliance (NAWAPA), which hopes to divert huge amounts of water from northern Canada and Alaska to the United States and Mexico; and

the South America Lakes Plan, in which dams in the Amazon would create huge lakes that would connect numerous points in South America. The execution of such plans may create ecological problems of unprecedented scale.

A different approach to adding to the water supply is to increase the reservoir system. Storing water in reservoirs is attractive because such artificial lakes quite often can be stocked with fish and provide popular recreation areas for boating, water skiing, and diving, but reservoirs eventually silt up, and large quantities of the water are lost through evaporation. Attempts have been made to cover lakes and reservoirs with thin films (one molecule thick) which would prevent water molecules from leaving the surface but would not prevent oxygen and carbon dioxide from entering or leaving the water. Hexadecanol (which, incidentally, is the base for lipstick) and other fatty alcohols have been tried, but wind and wave action make the method far from effective.

If it were possible to store water underground either in artificially created caverns or in the existing aquifer formations that are now being mined, at least some water that is now consumptively lost could be saved, and less land would be drowned by reservoirs. This method of storing is called artificial recharge. It is practiced successfully in many locations in the United States (e.g., Los Angeles, California) and abroad (e.g., Israel; Dortmund, Germany). However, there are technical and legal problems with this method as with the others.

Problems

We must now consider the present and future problems related to water resources. As we have seen, there is on the average enough sweet water available even if the world population doubled and even if the water use per person doubled, but this is *on the average*. Arid and semiarid lands, such as the western part of the United States, do not have enough water; elsewhere there may be too much water. In many humid tropical lands drainage is the main problem. In general, we want water when we want it, where we want it, and of an agreeable quality.

There are only a few places in the world

where rainfall is dependable enough and falls at the time when needed. Water is therefore captured directly (stock tanks) or indirectly (a dam in a river forms a reservoir). Building dams creates three kinds of problems. More water is lost by evaporation than would have occurred without the reservoir; the storage capacity will gradually diminish because of silting; and people downstream, especially when they are of different states or nations, may feel deprived of water which they claim is rightfully theirs. If river and dam are big enough, there may also be serious ecological and sociological problems.

The evaporation problem is minor compared to all the water that is lost from natural surfaces and vegetation, including agriculture. All man-made reservoirs over the whole world are a mere puddle compared to reservoirs in natural lakes, streams, and groundwater reservoirs. At the moment, the knowledge and techniques of evaporation suppression are inadequate, but significant progress can be expected in the next 10 to 20 years.

Silting is sometimes combated by dredging, but this is just a matter of a stay of execution. Sooner or later, a reservoir will fill with silt just as natural lakes will; man only hastens the process. The real problem may be in the future, even if only four generations from now, but it is a serious one. Dam-building organizations in this and other countries are beginning to run out of dam sites.

The complaints filed by the downstream people are often met by promising them return water, but the difficulty is to provide them with water of the desired quality. We already discussed the problem on the Colorado River; another example is that of the Pecos River, where New Mexico assures a dependable supply to Texas. This river gets some of its water from groundwater that seeps into the river bed. Pumping groundwater near the river deprived the Pecos of part of its supply and created difficulties in regulating the river flow to restore the promised water supply to Texas. In August 1974, Texas was in the process of suing New Mexico in the U.S. Supreme Court over the issue of Pecos River water.

Ecological problems have rarely been investigated completely, but we are more and more aware of the fact that if we manipulate nature, we may cause irreversible damage. For in-

stance, before the Aswan Dam was built, sediments and nutrients, which fed the coastal sardines, used to be carried to the delta and the Mediterranean Sea. This flow is now stopped and the coastal fishing industry has been destroyed. The delta itself is dwindling as a result of erosion, and land-reclamation projects farther inland are endangered.[14]

Another real danger caused by dams and reservoirs is that they may trigger earthquakes. One example is the earthquake in India near Koyna Dam (about 130 kilometers southeast of Bombay) in 1967. It caused great damage, including 200 lives; but there was no previous history of tremors in the area.[15]

The second means of obtaining water when we want it is by developing groundwater resources, but we must remember that groundwater is a reservoir and the only part that is replenishable is what comes from natural or artificial recharge. If the draft on the reservoir is larger, the area will eventually run out of water. If a reservoir supplies water to a river, as it often does, overpumping will affect the stream flow. This has happened many times, the Pecos River being one example. Here, again, there are problems of law and tradition.

In the western parts of the United States, the owner of the land is usually considered also to be owner of the water underneath the land. Under the appropriation doctrine, water is presumed to be a replenishable resource, and the appropriative right has been defined as the right to use a specified rate of flow annually and forever. The quantity of water taken from a stream cannot exceed the quantity that is in it, except for multiple reuse, but for groundwater it is possible to overdraw, and the term "annually and forever" becomes meaningless.[16] What makes things worse in certain states is that the landowner has the right to the amount of water he "always" used to use. So the farmer is advised by his lawyer to pump much more than he actually needs so that he can maintain the right to more water when he expands his operation.

Another interesting problem occurs when artificial recharge is successfully practiced. Whose water is this? It is impossible to solve such a problem except by changing the existing laws and traditions. Water sources and reservoirs on and under the land surface will have to be managed by natural units to the benefit of all involved.

Problems connected with the desire to have water where we want it are sometimes different in nature but are often the same. Along rivers and streams the "where" can often be met by multiple use. The problem here is one of degradation of the quality of the water, as happened along the Colorado River. There is also an increase in consumptive use, as can be expected. Communities in many arid areas in the United States and elsewhere want a transfer of water to their dry lands. This may become extremely expensive and create enormous ecological problems.

If the USSR ever succeeds in reversing all the big Siberian rivers, and Canada and the United States work together to turn around the Canadian rivers now flowing into the Arctic Ocean, we may expect changes in oceanic currents. This, in turn, may change climates drastically, and thus affect the flora and fauna in the oceans, and even the food chain, including the one from fish to man.

Desalting plants, also a matter of "where we want the water," create a different kind of problem. It is proposed that nuclear power plants be used for desalting water at the rate of 400 million liters per day.[17] This means that per day 14,000 tons of salt will have to be disposed of. Even if this is done far offshore, a large area of the oceans will be uninhabitable for plant or fish.

The last problem to be mentioned is in connection with rainmaking. Present experiments have mostly been made in remote areas and on a comparatively small scale. Suppose that an effective method (there is none yet) were to be applied to a large area, with the result of disastrous flooding in one place and/or serious drought in another. The legal complications alone are staggering, to say nothing of ecological changes that might occur.

Finally, it should be noted that in the United States we could endure in equilibrium with our water supply by using the available "water yield" (amount of runoff) wisely. By the year 2000 we may be approaching an equilibrium condition when daily withdrawals reach 5,300 billion liters, not counting reuse. All of the additional means of obtaining water described in this section involve greater or smaller environmental disturbances, including possible climatic changes and, without proper costly precautions, perhaps dangerous increases in radiation (Chapter 12).

Summing up, then, we might conclude that there are water problems, but they are not those of quantity and not even necessarily of quality. It is the attitude of people and their concern for each other that become the problem. Man will have to learn to live with man, not on a competitive but on a cooperative basis, and this seems to be the biggest problem for man and his environment.

References

1. van Helmont, J. B. 1648. *Ortus Medicinae* (The Rise of Medicine). Edited in Amsterdam by Franciscus van Helmont.

2. Todd, D. K. (ed.). 1970. *The Water Encyclopedia*. Water Information Center, Port Washington, N.Y. pp. 88–89.

3. Abelson, P. H. 1970. Shortage of caviar. *Science 168:* 199.

4. van Hylckama, T. E. A. 1956. The water balance of the earth. *Publ. Climatol. 9:* 58–177.

5. White, G. F. 1969. *Strategies of American Water Management*. University of Michigan Press, Ann Arbor, Mich. p. vii.

6. Nace, R. L. 1969. *Water and Man: A World View*. UNESCO, New York. 46 pp.

7. Murray, C. R., and Reeves, E. B. 1972. *Estimated Use of Water in the United States in 1970*. (U.S. Geol. Survey Circ. 676). Government Printing Office, Washington, D.C.

8. MacKichan, K. A., and Kammerer, J. C. 1961. *Estimated Use of Water in the United States, 1960* (U.S. Geol. Survey Circ. 456). Government Printing Office, Washington, D.C.

9. Piper, A. M. 1965. *Has the United States Enough Water?* (U.S. Geol. Survey Water-Supply Paper 1797). Government Printing Office, Washington, D.C.

10. Irelan, B. 1971. *Salinity of Surface Water in the Lower Colorado River-Salt Sea Area* (U.S. Geol. Survey Prof. Paper 486-E). Government Printing Office, Washington, D.C.

11. Schneider, W. J., and Spicker, A. M. 1969. *Water for the Cities — the Outlook* (U.S.

Geol. Survey Circ. 601-A). Government Printing Office, Washington, D.C.

12. Grua, C. 1970. New water through desalting. In (R. B. Mattox, ed.) *Saline Water* (Commission on Desert and Arid Zone Research, S.W. and Rocky Mts. Div.). *Amer. Assoc. Advan. Sci. Contrib. 13:* 48–105.

13. Dennis, A. S. 1972. Conference summary. *Bull. Amer. Meterol. Soc. 53:* 878–879.

14. McCaull, J. 1969. Conference on the ecological aspects of international development. *Nature and Resources (UNESCO) 2:* 5–12.

15. Rothé, J. P. 1968. Fill a lake, start an earthquake. *New Scientist 39* (605): 75–78.

16. Thomas, H. E. 1961. *Groundwater and the Law* (U.S. Geol. Survey Circ. 446). Government Printing Office, Washington, D.C.

17. Clawson, M., Landsberg, H. H., and Alexander, L. T. 1969. Desalted sea water for agriculture: is it economic? *Science 165:* 1141–1148.

Further Reading

Batton, L. J. 1962. *Cloud Physics and Cloud Seeding.* Doubleday & Company, Inc., Garden City, N.Y. 144 pp.

Davis, K. S., and Day, J. A. 1961. *Water, the Mirror of Science.* Doubleday & Company, Inc., Garden City, N.Y. 195 pp.

Furon, R. 1963. *Le Problème de l'eau dans le monde.* Payot, Paris, 251 pp. Translated in 1967 by P. Barnes as *The Problem of Water.* American Elsevier Publishing Co., Inc., New York. 208 pp.

King, T. 1963. *Water, Miracle of Nature.* Macmillan Publishing Co., Inc., New York, 238 pp.

Kuenen, P. H. 1955. *De kringloop van het water.* Leopold, N.V., The Hague, 350 pp. Revised translation in 1963 by M. Hollander as *Realms of Water.* John Wiley & Sons, Inc., New York. 327 pp.

McGinnies, W. G., and Goldman, B. (eds.). 1969. *Arid Lands in Perspective.* The University of Arizona Press, Tucson, Ariz. 421 pp.

Editor's Commentary— Chapter 7

WE HAVE CHOSEN to treat world fisheries separately because they illustrate several problems very nicely. First, unlike most other resources, biological resources are renewable in the short term, and the rate at which they are renewed depends to a great extent on how carefully they are harvested. Much of agriculture is highly artificial, and if we ruin a crop one year, we can start over again the next. But there is little we can do to help ocean fisheries except to avoid making mistakes in the first place. So fisheries present scientific problems of great interest, as John Gulland shows in this chapter.

Secondly, fisheries are a very nice example of what Garrett Hardin has called a "commons"; that is, most fisheries are not "owned" by any one person or group, and many different people have access to them. They therefore pose especially difficult problems of management because it is in the short-term interest of each fisherman to maximize his harvest, but if all fishermen do this it may result in overfishing, which is against everyone's long-term interest.

Because fisheries are a commons, I find Dr. Gulland's chapter one of the most encouraging in this book, for he shows that, with sound scientific advice and cooperation, it is quite possible to manage these fisheries successfully. Two examples illustrate this. The Peruvian anchoveta, which is the major resource of a single nation, seems to have been saved by sensible harvesting, following a period in which the untrammelled development of private fisheries and some unfortunate oceanic changes caused severe declines in the stocks. The answer here was nationalization of the industry. The second example concerns whaling, where the problems of the commons are magnified because the stocks are international. Yet, even here, recent management has been successful in allowing a slow recovery of the whale stocks, including blue whale, fin whale and sperm whale, and these populations are not, under current regulations, in danger of extinction. Indeed, Dr. Gulland points out that our recent performance at international cooperation and management of whaling is much superior to Man's earlier efforts, which resulted in the demise of the gray whale in the Atlantic and severe depression of the population of right whales.

If this chapter is encouraging because of its evidence of at least limited international cooperation and sound management, it is also sobering because it contains one of the clearest examples of limits to growth: The fisheries of the oceans (at least those taking presently-eaten sorts of fish), even with careful management, are not going to yield more than about double their current harvest. At present growth rates, we will see the limits to ocean fisheries around the turn of this century.

Fish meal factory near Paracas, Peru. Photo by Georg Gerster/ Rapho-Guillumette.

7

The Harvest of the Sea

John Gulland   *works in Rome for the Department of Fisheries of the Food and Agriculture Organization of the United Nations. He received a B.A. in mathematics from Jesus College, Cambridge. Since 1951 his work has centered on fisheries, first with the British Government, and since 1966 with FAO. He has studied the population dynamics of fish stocks, particularly the effects of fishing. Recently he has been particularly concerned with the management and rational utilization of fish stocks, both nationally and through international councils and commissions.*

The Growth of World Fisheries

Over an area equal to two-thirds of the world's surface, man has so far had little visible impact. Although man has caused enormous changes to the land and its natural features, often with damage to his own long-term interests, the oceans still appear almost as they did a thousand or a million years ago. The oceans offer great opportunities for profitable use, and also a great challenge to ensure that this use is wise and not carried out only in pursuit of immediate advantage. The possible uses are several. To some the seas offer the promise of enormous increases in food production to match the enormous human population threatened for the end of the century; to others the deep oceans are seen as providing a convenient dump for the masses of waste that the land cannot absorb. Great hopes are also expressed for the extraction of minerals on and under the seabed, and the production of oil and natural gas from several areas of the continental shelf is already considerable.

In this chapter I will try to show that, although some increase in yield of fish from the sea is possible (perhaps a doubling), we are not far from the limit of readily usable animals that the oceans can provide, and that careful management of the resources will be needed to produce that harvest.

For the present the main use of the sea, other than for transport, is in the harvest taken from the natural populations of fish and other animals. How well the fisheries of the world have performed in the last quarter-century, in relation to the rising world population, is shown in Figure 1. This shows the trend in total world catch of marine fish. Because much of the recent fluctuations about a steady trend have, in recent years, been due to changes in the catches from one stock, the Peruvian anchovy; the total catch less the anchovy is also shown. For comparison the trends in world population are also given. Fish production (even excluding anchovy) has been increasing approximately 5 percent per year, and this has, in this period, consistently outstripped population growth.

This steady growth has been the result of harvesting a steadily increasing number of different fish stocks. Catches from individual stocks do not grow continuously over a long period, but rather in the typical pattern of growth within certain definite restrictions: a slow beginning, a period of rapid exponential growth, and then a flattening out at a level determined by the potential of the particular stock concerned. No other fishery has reached such a high annual yield as that of the Peruvian anchovy, and very few have achieved as high a rate of growth during the exponential phase (nearly doubling each year between 1954 and 1960), but many fisheries have maintained striking rates of growth. For example, the trawl fisheries of Thailand have maintained an annual growth of 20 percent, or better, since about 1960. With an annual catch of 1.5 million tons (about equal to the combined catches of West Germany and the United Kingdom, and more than half that of the United States), the value of the Thai catch is greater than that of Peru, because most of it is sold for direct human consumption, at higher prices than those obtained for reduction to fish meal.

The growth of any particular fishery has usually been triggered off by one of three factors: (1) the introduction of existing techniques into countries or areas where they were unfamiliar, for example purse seining plus fish meal production into Peru or trawling into Thailand; (2) the identification of new stocks available to local fisheries, for example the growth of mackerel catches by Norway from 50,000 tons in 1964 to 850,000 tons in 1967; and (3) the aggregation of mobile long-range fishing fleets at newly discovered stocks, for example the sudden jump in catches of Patagonian hake (*Merluccius hubbsi*) from 100,000 tons in 1965, solely by South American countries, to 600,000 tons in 1967, 80 percent of which was caught by the USSR.

Rarely has new technology in itself been the immediate cause of the growth of a new fishery, although an important exception might be claimed for the fishery for Alaska pollack (*Theragra chalcogramma*) in the North Pacific. This fishery, principally by Japan and now one of the biggest in the world, with annual catches well in excess of 3 million tons, owes its recent expansion to the development of minced fish as a product by which an intrinsically unattractive fish can be used in large quantities in such products as fish sausages.

The basic techniques of catching fish have not altered much since the development of the purse seine and otter trawl (the latter over a century ago) enabled machine power to take over from sails or manpower. The more significant development on the catching side has been the hydraulic power block, which has enabled bigger purse seines to be handled, and handled more easily and quickly, and which has greatly improved the efficiency of purse seining, particularly for large and active fish such as tuna.

Although the modern 5,000-ton stern trawler factory ship, which is capable of catching and processing 50 tons of fish or more every day anywhere in the world, is in some ways a triumph of modern technology (like the SST), it is not a very efficient tool for catching fish. This is most clearly shown by the dozens of factory ships that cross the North Sea each week on their way to the distant grounds off Greenland, Newfoundland, or northwestern Africa. These cannot afford to stop in the North Sea, although smaller trawlers — not so different from those in operation 50 years ago — are making a good living.

The more significant developments have been in the processing and use of fish, the

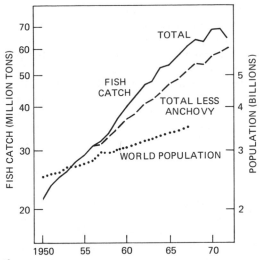

1 **Increases in World Population and in the Total Fish Catch (With and Without Peruvian Anchovy).** Note that the scales for both fish catch and world population are logarithmic. Since the data for both population and fish catches were (until recently) almost straight lines, this shows that both have been expanding exponentially, but the slope of the catch line is greater, showing that fish catches have increased more rapidly than population.

general supporting facilities, and especially in the general growth in personal standards of living and in national infrastructure (transport, marketing, etc.) that allow, for example, sea fish to be delivered to inland markets of Thailand at a price that the average man there can now afford to pay. The most striking of these developments has been the freezing and processing of the catch at sea.

Until the introduction of ice in the nineteenth century the effective duration of fishing trips was, with a few exceptions, only a few hours. A notable exception was the salt cod fishery, which, very shortly after Columbus, was largely instrumental in opening up northeastern America to European influence. The use of ice allowed ships to stay at sea for 2 to 3 weeks before the fish got too stale to sell. This opened up to intense fishing great areas of the North Atlantic and North Pacific adjacent to the markets of Europe, North America, and Japan. Processing at sea removed all restrictions of distance, and the large fleets of great factory trawlers of Japan, the USSR, and other countries, supported in the case of the USSR by numerous supply ships and other vessels, can operate anywhere in the world where the catches are good enough. Modern products such as fish sticks and minced fish remove much of the need to worry what particular species is being caught.

The other important development has been the growth of the broiler chicken industry and the use of fish meal in animal feeds. This has spurred on the growth of several big fisheries — Peru being the most outstanding — wholly or mainly for the production of fish meal. Much of the rapid rise in total world catch in the 1960s was due to the growth of these fisheries. Although their product was used for human nutrition only indirectly, via chickens or pigs, the catches were largely taken from stocks previously fished lightly, if at all, so that these fish meal fisheries did add significantly to the protein supply of mankind. The more regrettable fact is that this supply has largely gone to the richer countries of Europe and North America rather than the protein-short Third World.

This rapid and sustained growth of world catches raises two vital questions: First, how long can new resources be discovered and brought under exploitation to maintain the growth? Second, how well can we manage the stocks already being harvested so as to keep the yield from them at a high sustained level? The final section will attempt to answer these questions, but before doing so, the problems of managing and conserving fishery resources will be examined in more detail, with particular reference to two of the world's biggest fisheries, for Antarctic whales and for Peruvian anchovy.

Antarctic Whaling

PRINCIPLES OF GOOD MANAGEMENT

The whaling industry in the Antarctic is one of the classic examples of the conservation and management (or failure to achieve good management) of a major natural resource. The ultimate tests of good management are that the stocks concerned are maintained at their optimum level and that no more is taken each year than will be replaced by the natural production. What is considered the optimum level will differ according to the objectives of the

conservation and management and will be different for whales and, say, wolves. To satisfy these ultimate tests a number of intermediate conditions must be achieved. These include:

1. Sufficient data for scientists to work on.
2. Clear and comprehensive advice from the scientists on the magnitude of the optimum stock abundance, and the allowable removals, including evaluation of the effects of departing from these values.
3. A forum to discuss and decide on the action required, including representation of all interests (different countries, long-term and short-term considerations, etc.).
4. Identification of balanced measures that will limit the removals but do not have unacceptable side effects (e.g., undue economic losses to some or all participants).
5. Timely action at each stage.

At various periods the conservation of whales ran into great difficulties through failure of four of these five conditions (inadequate advice, an unrepresentative forum, unbalanced measures, and the lack of timely action).

Whaling provides a striking example of conservation problems, for two reasons: first, the natural drama of hunting the largest animals that ever lived in the world's stormiest and most remote ocean; and second, that the biology of whales makes the penalties for failure in managing whale stocks much more severe than in the case of most fish stocks. Although in terms of introducing controls on the amount and types of animals caught, the International Whaling Commission has done more than most of the numerous other international commissions charged with the conservation of fish stocks in many parts of the world, the lack of resilience of whale stocks has caused it to be much less successful in maintaining the stocks and the yields from them at a high level.

Like those of any natural population, certain biological characteristics of a stock of whales will change if the abundance of the stock changes. Thus since whaling started in the Antarctic and the total numbers have decreased, whales have matured earlier[1] and breed more frequently[2]. These changes will allow whale stocks to maintain themselves in the face of moderate rates of harvesting. The harvest that can be taken *without changing the stocks* (the sustainable yield) will be a function of the abundance of the stock, being small at both small stocks and large stocks, with a maximum sustainable harvest being available at an intermediate level of abundance. Figure 2 illustrates the relation between the sustainable yield and the stock abundance of whales, as well as that between sustainable yield and the rate at which they are harvested. The maximum sustainable yield is some 6,000 to 7,000 whales, taken from a stock of a little over 100,000 whales.[3] Since whales produce no more than one young every other year, the numbers of females born is no more than 25 percent of the mature female stock. Allowing for losses between birth and maturity, and natural deaths of mature animals, the sustainable harvesting rate of whales can be no more than 10 percent. Rates in excess of this, if maintained over a period, will result in the extinction of the stock. In contrast, most fish have enormous fecundities (a large female cod may produce 1 million or more eggs) and may be able to withstand very high harvesting rates without long-term damage.

In the short run, high harvesting rates of whale stocks are possible. When harvesting starts, catches in excess of the current sustainable yield are inevitable while the stock is being thinned down to the optimum level (around that giving the maximum sustained yield). At this level, catches slightly in excess of the sustainable yield will cause a decrease in stock that is at first slow, but as the stock and thus also the sustainable yield decrease, the gap widens between the actual catch and what the stock can sustain, leading to an accelerating decline in the stock.

The intense exploitation of the Antarctic whales started soon after World War I. The subsequent history of the industry is shown in Figure 3. The check in the 1931–1932 season was purely economic, owing to the world depression and the drop in oil prices, but a sign of more serious trouble was soon becoming apparent in the statistics. This was the drop in the number of the preferred species, the blue whale. From about 1933 onward, increased catches of whales were attained only by increased attention being paid to the less-favored fin whales.

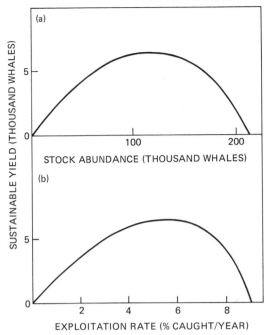

2 **Sustainable Yield of Blue Whales.** The yield is plotted as a function of the stock abundance (a) and the proportion caught each year (fishing rate) (b).

INTERNATIONAL COOPERATION

The drop in the availability of blue whales, which was most clearly perceived by the gunners and expedition managers in the Antarctic, caused considerable concern, particularly to the Norwegians, who then dominated the industry and who were familiar with the disappearance of whales from their more accessible haunts in the North Atlantic. Like any other sensible businessmen, while looking mainly at the current conditions and the year's profit-and-loss account, they also worried about the long-term future and the foundations on which their business was based. As a result, there were extensive discussions before World War II on measures that should be taken to conserve the whale stocks, and hence ensure the long-term prosperity of the whaling industry. These resulted in some useful measures, such as the complete prohibition of the killing of gray and right whales (already reduced to very low levels), setting of minimum size limits for blue whales (70 feet), fin whales (55 feet), and humpback whales (35 feet), as well as regulations on the operations of the whale

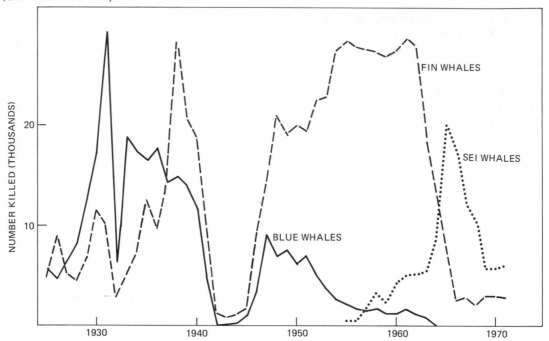

3 **Catches (Numbers) of Whales in the Antarctic.** *Source:* Data from International Whaling Commission statistics.

factory ships aimed at making the full use of all whales caught. However, the central problem of controlling the numbers caught was not at that time resolved.

A major step toward solving this problem was taken when the International Whaling Commission was established in 1946. Despite its later failures to keep the catches within the limit that the stocks could withstand, the IWC deserves credit and recognition for the action taken, a quarter of a century before conservation and the proper use of the environment became a fashionable matter of concern, to control the harvest taken by a major international industry. The nature of the decision taken by the IWC is clearly expressed by Meadows and his colleagues.[4] "The basic choice that faces the whaling industry is the same one that faces any society trying to overcome a natural limit with a new technology. Is it better to try to live within that limit by accepting a self-imposed restriction on growth? Or is it preferable to go on growing until some other natural limit arises. . . . For the last several hundred years human society has followed the second course so consistently and successfully that the first choice has been

all but forgotten." Meadows and his collaborators apparently were not aware that the IWC chose the first alternative as early as 1947, when they limited the total catch (omitting the rather small catches taken by land stations), to 15,000 Blue Whale Units (1 blue was counted as the same as 2 fin whales, 2.5 humpback whales, or 6 sei whales). That this was a real restriction is shown by the fact that this was barely half the catch taken in the 1937–1938 season, and that by 1952 the quota was reached halfway through the season (i.e., without restrictions the catch taken during the season could have been doubled).

Acceptance of a limit to growth is unfortunately not sufficient in itself to ensure success in the rational utilization of natural resources. There were a number of shortcomings in the arrangements for Antarctic whaling which were to prove nearly fatal for the IWC, the whaling industry, and the whales themselves. The chief of these were: first, that the machinery for adjusting the quota was complex; second, that there were no simple provisions for the separate management of different stocks, and third, that with no provision for allocation of the quota among nations, the economic

172

success of the whaling industry was not assured.

The first shortcoming to be tackled was the lack of allocation. Since all catching of whales had to stop once the quota was reached, there was enormous incentive for all participants to do all they could to get the greatest possible share for themselves while the open season lasted. During these "Whaling Olympics" the number of expeditions, and the number of catching vessels attached to each expedition increased steadily, with the natural result that the quota was reached earlier each season.

When the expensive capital equipment of a whale factory ship and associated vessels could only be used for a steadily decreasing period each year — only 58 days in the 1955–1956 season — operations were clearly becoming uneconomic. The obvious solution was to allocate shares to each of the limited number of participants in the fishery and allow each to take his share in the most efficient manner. Given the obvious differences in economic interests between the USSR and Japan, and between the growing whaling industries of these two countries on the one hand and the stagnant or declining industries of Norway, The Netherlands, and the United Kingdom on the other, agreement was difficult and took some time to reach. Not until agreements on national allocation were reached, in time for the 1959–1960 season, did the delegates to the IWC turn their full attention to the status of the stocks. By then blue and humpback whales were very scarce. Attempts had been made to give them special protection, while allowing some catches, by reducing the length of the open season, but these had not been effective. By the mid-1960s the whaling industries had been forced to accept a complete ban on catching these species.

The great crisis that faced the IWC as it emerged from its quota-allocation wrangles in 1960 concerned the allowable catch of fin whales. Unfortunately, this crisis found the scientific advisers to IWC in a poor position to assist. The advice given to the Commission at this critical period had two important failings: (1) there was not complete agreement among scientists, which provided an excuse for not acting on the advice of the majority, and (2) the advice did not include descriptions of the effects of alternative actions. Although the ma-

jority of scientists called clearly for a reduction in catches, this tended to be phrased as an instruction or moral duty: the quota *should* be reduced to, say, 14,000 Blue Whale Units. This form of advice had little effect on the whaling industries, who felt that their interests, and the interests of their shareholders (or directors of national plans) and employees, were that catches should if possible be increased, nor did it persuade many government officials or administrators. The more persuasive advice was providing the Commission and the whaling industries with a clear alternative: should they take big catches now, and have only small whale stocks and low catches in the future, or should they reduce present catches so as to have bigger stocks and moderately large sustained yields in the future?

The Commission tackled the problem of disagreement between its scientific advisers by setting up in 1960 a Special Committee of Three. This Committee presented in 1963 its full report,[3] which set out in detail the status of the different stocks and the effects of different action that might be taken by the Commission. In particular, the report made clear to the Commission members the implications of taking for any period more than the sustainable yield. It then became accepted as a minimum target that the catch taken from any but the most lightly exploited stock should not exceed the sustainable yield and that, for heavily exploited stocks, the catch should be as far as possible below the sustainable yield to rebuild the stock as fast as possible to its optimum level (i.e., around the level giving the maximum sustained yield).

Agreement to cut catches the necessary amount, however, was not reached at the 1963 Commission meeting. Any action to conserve resources involves some conflict between short-term and long-term interests and the foregoing of some immediate advantages to obtain some long-term benefits. In relation to whales, this had long been accepted in principle by the setting up of the IWC. It had also been followed in practice by the industry accepting short-term sacrifices (catching only half the quantity that could be taken each season) for longer-term benefits.

In the early 1950s it would not have been difficult to reduce the actual catches of fin whales (around 25,000 whales) to the level of the sustainable yield from the stocks at that time (around 20,000 whales). But by the time the Commission had resolved the problem of quota allocation, had failed to get fully agreed advice from the whale biologists, and had received the report of the Committee of Three, the gap between current catch and sustainable yield had widened very seriously as the stock had declined. The estimate of the sustainable yield of fin whales in 1963 was only 4,800 animals. The number of fin whales killed (18,600) was four times the estimated sustainable yield, even though in the 1962–1963 season the whalers had for the first time failed by a clear margin to catch their quotas.

With a difference of this size it was not surprising that none of the active Antarctic whaling countries could agree, at the 1963 Commission meeting, to a reduction of catch to the level of the estimated sustainable yield. No one, and no business, can find it easy to accept a cut of three-fourths of their income. It was particularly difficult for Norway because, although whaling was only of relatively minor national interest, it was concentrated in one district, and for Sandfjord, the whaling capital of the world, whaling in 1960 provided nearly all the economic life of the town.

Despite these real practical problems, the need to reduce the quota to close to the level of the sustainable yield was obvious, and after a period of intense discussions in the Commission and equally intense, if less public, negotiations in each country, the quotas for the 1965–1966 Antarctic season were brought close to the value of the combined yield from the fin and sei whale stocks.

Since 1966 the central problem of the IWC has been to resolve this conflict between, on the one hand, the long-term interests (of the whaling industry itself as well as of the world as a whole) to bring the catches down below the level of the sustainable yield, so as to rebuild the stocks, and, on the other hand, the short-term interests of the whaling industry in maintaining the current catches at a high level.

The basic argument concerned the rate at which the fin whale stock should be rebuilt, but has been confused by arguments on the precise value of the sustainable yield, the extent to which current stock levels were, if at all, below the optimum level, the need to set limits for each species separately, and the effectiveness with which the recommendations of the IWC were enforced. The real difference lies between the short-term interest of catching many whales now with only a slow rate of recovery, and catching few, and allowing the stocks to rebuild quickly. The extreme version of the latter position is the call for a complete moratorium on all whaling for a period of 10 years. This blanket moratorium has little scientific justification, since it ignores the great differences between the condition of different species: blue whales, which are in any case already protected, require much more than 10 years to recover, whereas sei whales (probably) and minke whales (certainly) are near or above their optimum levels of abundance.

The argument between slow or fast recovery rates remains largely unresolved and seems likely to remain so as long as those currently engaged in whaling see little guarantee of continuing in whaling, and therefore see no advantage in accepting short-term sacrifices for the long-term benefit of someone else. The IWC has therefore no clear policy on the rate of rebuilding of the stocks; nevertheless, the annual catch limits have been brought down to levels that are (except for minke whales, which have been so far very lightly exploited) below the values of the sustainable yield for each species, and the stocks are being slowly rebuilt.

LESSONS FROM THE HISTORY OF WHALING

Whales and whaling have been cited as the test of whether man can come to terms with his environment. How well are we measuring up to this task? The answer is moderately encouraging. Of the whales that have been the subject of intensive whaling, the sei whales (the last to become heavily exploited — the minke whales are still barely touched, at least in the Antarctic) are at about the optimum level and are providing close to the maximum sustainable yield. The fin whales have been depleted well below the optimum but can still sustain small catches — probably rather more than the present catches, so that the stocks are slowly rebuilding. The status of sperm whales is less clear but is probably intermediate between sei and fin whales (i.e., depleted somewhat below the optimum but in reasonably healthy shape). Blue whales had been reduced to a very low level, but information accumulating since they received complete protection suggests that they are not as scarce as was once thought, are certainly not (so long as the IWC regulations are effective) in danger of extinction, and are probably increasing.[5]

On this evidence twentieth-century man need not be ashamed of himself. Despite the assistance that modern technology can give to thoughtless destruction, his record is better than that of his ancestors. Prehistoric man probably exterminated the gray whale in the Atlantic; open boat whalers in the nineteenth century and earlier reduced the right whales to such a low level that even now they are only beginning to recover their numbers in some areas. The sperm whales were saved in the nineteenth century chiefly by the development of the petroleum industry and the collapse of the market for sperm oil.

What general lessons can be learned from the whale story? First, that whalers are, like any other group of people, not entirely occupied with short-term profit and will, as in 1946 and again in 1966, accept immediate sacrifices for some clear long-term objective. On the other hand, they are less willing to accept proposals that do not offer some definite benefit, even in the long run. The failure to agree to the recommendations of the majority of the scientific committee for a reduction in the quota between 1955 and 1963 were due less to the lack of unanimity among the scientists (although this provided a convenient excuse) than to the manner in which the advice was presented. This often appeared as a direct instruction — "the quota should be reduced." The reaction of the industry to such a statement tended to be that, from their point of view, the quotas should be increased. The Committee of Three's proposals were much more persuasive because they set out the immediate and long-term results of alternative actions. Knowing these results, the commission could make a rational choice among the alternatives. Proper utilization of the whole stocks required close contact and understanding of each other's problems and responsibilities among the scientist, the administrator,

and the industry. The same is required for any natural resource.

Peruvian Anchoveta

The Peruvian anchoveta fishery presents some interesting contrasts with Antarctic whaling. Instead of dealing with the largest animals that ever existed, the mind can be staggered by the numbers involved; in the 1969–1970 season more than 500,000,000,000 anchovies, at an average weight of less than 20 grams each, were caught. Antarctic whaling has been carried out on the high seas by rich, developed countries, and even to Norway is of only local or minor importance. The anchoveta is fished only by Peru (and to a lesser extent by Chile), in waters under the direct control of the Peruvian government, and is of great importance to the national economy, providing, in good years, a third or more of Peru's foreign-exchange earnings. These differences have affected the ways in which the same objective of rational exploitation has been approached through scientific studies and advice and the implementation of management measures. Despite the differences, the same tests of good management and the same intermediate conditions apply equally to anchovy as to whales.

The riches of the waters off Peru are due to the Peru, or Humboldt, Current (see Figure 4). As this moves north, Coriolis force, aided by the shape of the coastline, makes it swing away from the coast, and cool, nutrient-rich water wells up from the depths. As this water reaches the sunlit surface zones, there is a rich blooming of plant life, which in turn supports in abundance the animals that eat the plants, the animals that eat them, and so throughout the food web.[6] The anchoveta comes at an early stage in this web, eating both phytoplankton and zooplankton. Its most conspicuous predators are the guano birds, but anchoveta is also a major food item of many fish, and of squids. Similar upwelling systems occur off the western subtropical coasts of all continents (with on-land corresponding deserts or near deserts in each case), but the Peruvian system is about the biggest, probably equaled only by that off southwestern Africa.[6]

To the visitor to Peru the clearest evidence

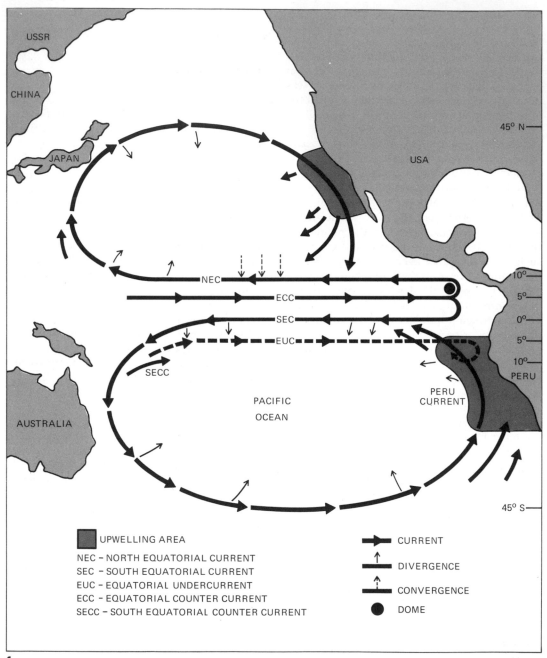

4   **Pacific Ocean Circulation.** *Source:* Reference 6.

of the richness of the ocean is the masses of seabirds, and it was the guano produced by these birds at their nesting colonies that supported the first industry to harvest Peru's ocean wealth. During the nineteenth century the guano industry was engaged in mining the accumulated droppings of centuries to supply fertilizer to European farmers. Only as it became clear that the resources were being exhausted was the government's attitude to the industry changed from considering it as mining (i.e., using a nonrenewable resource) to treating the guano on a sustained yield basis, taking no more each year than the annual production. Controls were set on the exploitation, preference was given to use in local agriculture rather than to exports, and positive steps were taken to increase the bird stocks, through reducing human disturbance and predation by foxes, rats, and so on.

GROWTH OF THE FISHERY

Although the ancient coastal civilization of Peru did catch fish, mostly from the shore, modern commercial fishing was slow to start. During World War II bonito and other fish were canned for export. The waste from these fish was turned into fish meal, but it was not until 1950 that the first plant was set up purely to produce fish meal. After a quiet start the fish meal industry underwent, in the decade from 1956, the most rapid and striking expansion of any fishery (Tables I and II). Catches doubled each year, and in 1962 Peru caught a greater weight of fish than any other country in the world. In 1972 over 12 million tons of anchovy were caught!

Without the natural riches of the Peru Current, this expansion could not have occurred, but several other factors helped the industry expand at the speed it did. The concurrent growth of mass rearing of animals in the United States and Europe, especially the development of the broiler chicken industry, provided a market that could absorb even the enormous Peruvian fish meal production. At the same time a number of technical developments — particularly synthetic fibers which allowed the construction of much larger purse seines, hydraulic power blocks to ease the handling of these big nets, and fish pumps to speed up the transfer of the catches of a hundred tons or more to the fish hold — all made the fishery more efficient and profitable. Once growth started it was largely financed by the reinvestment of profits. As a result, although some American and British companies have had important interests in the Peruvian fish meal industry, it is not dominated by foreign companies to nearly the same extent as many primary industries in developing countries. By far the largest company, before the whole industry was nationalized in 1973, was Peruvian-owned.

THE NEED FOR MANAGEMENT

By 1962 the classic signs of heavy fishing were becoming apparent — falling catch rates (i.e., catch per unit effort) and a slowing down of the increase in total catch. Within a few years the Peruvian government was faced with the need to introduce appropriate regulations to manage the fishery and to conserve the resource. Wisely, it had already laid the foundation for the preparation of the necessary scientific studies and advice required for the best choice of resources. With the assistance of the United Nations Development Programme and the Food and Agriculture Organization of the United Nations, a research institute (the Instituto del Mar del Peru) was set up to study the marine resources of Peru. This has since 1960 carried out a variety of research programs, including as a particularly important

I Catches of Anchovy and Production of Fish Meal in Peru*

Year	Catches	Meal production	Year	Catches	Meal production
1951	a	7.2	1962	6,275	1,120.8
1952	a	9.2	1963	6,423	1,159.2
1953	a	12.1	1964	8,863	1,152.2
1954	a	16.5	1965	7,242	1,282.0
1955	a	20.0	1966	8,530	1,470.5
1956	a	30.9	1967	9,825	1,816.0
1957	a	64.5	1968	10,263	1,922.0
1958	a	126.9	1969	8,961	1,610.8
1959	a	332.4	1970	12,277	2,257.1
1960	a	558.3	1971	10,277	1,934.5
1961	4,579	863.8	1972	—	894.9

*Figures given in thousands of metric tons.
[a]Detailed statistics of anchovy catches not available.

element the collection of the basic statistical data of the anchovy fishery.

From these studies a good working knowledge of the life history of the anchovy has been obtained. Spawning can occur throughout the year but usually reaches a peak in the Peruvian winter (around September). The offspring from this spawning become available to the fishery around the end of the year, when they are 8 to 9 cm long. For the next couple of years they are fished, but few survive for more than 2 years. Figure 5 shows the growth in length and weight of the individual fish, as well as the changes in total numbers and weight of a typical brood of fish. For about the first 12 months of life (i.e., for some time after they enter the fishery), the reduction in numbers of a brood is more than made up by growth of the survivors, so the total weight of fish in the brood is increasing even though it is being fished.

The greatest weight from a given brood of fish would be obtained if they were not touched until the total weight had reached maximum, and then all the brood were harvested at once. This is obviously impracticable for anchovy, although this is the normal way of cropping domestic animals. Something quite close to the maximum can be obtained by starting fishing moderately hard shortly before the brood reaches its greatest total weight. There will be a loss of catch if fishing starts too early, or is too intense, because most fish will be caught before they have had a chance to grow to a decent size. Catching the very small anchovy (peladilla) is particularly wasteful because their yield of meal and oil per ton of fish is less than that from larger fish. It is also economically wasteful to fish harder than a certain level, because costs are increased without any commensurate increase in the value of the catch.

The management of the anchovy fishery has been largely concerned with making the best use of each brood of fish. This requires, first, the determination of the optimum rate of capture and the optimum size at which to start fishing. Second, these optimum rates have to be implemented by appropriate actions. Management of the anchovy (and of most other fisheries) has in the past been less concerned with ensuring the strength of future broods. Most fish lay enormous numbers of eggs (a large cod over 1 million, and even the anchovy some tens of thousands). Even a very small number of females could, if conditions are favorable, produce a brood of offspring of average or better abundance. Most fish stocks can therefore be reduced quite substantially without any reduction in future generations. But there must be a limit below which the size of future broods will be affected; very likely even at this limit all will be well if other conditions are favorable, but if the natural conditions for the young fish are poor, a very weak brood could result.

The studies on the Peru anchovy showed that by about 1965 the rate of fishing had reached about the optimum; that is, any further increase would not give any greater yield on weight from a brood of fish, and that a bigger weight would be harvested by avoiding the capture of the smallest fish. There was not, at least until 1972, any evidence that the stock had been reduced so far that the numbers of young produced had been affected.

MANAGING THE FISHERY

Since 1965 the Peruvian Government has introduced a range of measures to attain the two biological objectives: (1) to limit the total amount of fishing to that recommended by the scientists, and (2) to protect the very small fish. A third objective has been economic: to maintain the operating efficiency of the industry. Many measures that reduce the amount of fishing cause loss of efficiency and do not

II Exports of Fish Meal from Peru, by Country of Destination*

	1971	1972
West Germany	421	292
The Netherlands	124	158
Italy	74	73
Spain	83	91
East Germany	194	46
Czechoslovakia	111	77
Other Europe	266	176
United States	171	317
Latin America	173	132
China	66	83
Others	67	79
	1,751	1,524

*Figures given in thousands of metric tons.
Source: Data from Sociedad Nacional de Pesqueria

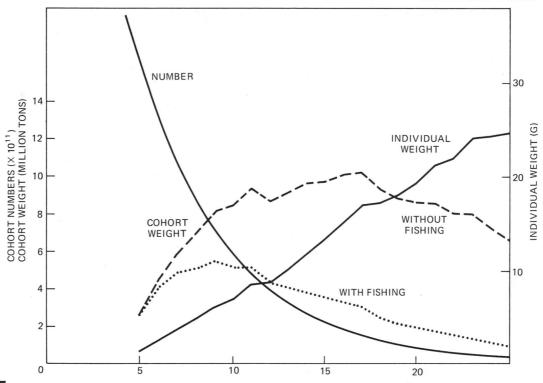

5 **Anchovy Growth and Survival.** Changes in weight of an individual anchovy and in numbers and weight of a typical cohort (i.e., a group of fish of the same age). Initially there were nearly 2×10^{12} (2,000 billion) 5-month-old individuals.

allow the costs to be reduced in proportion to the reduction in the amount of fishing.

Early measures were a closed season for a couple of months in the middle of the year, when the shoals tended to be dispersed and catch rates low. This gave an opportunity for overhaul, maintenance, and modernization of the fish meal plants and the fleet, and was generally welcomed by the industry. Equally welcomed by the fishermen was a rule prohibiting fishing on Saturday and Sunday. Another closed season, at the beginning of the year, was introduced specifically to avoid the capture of the small fish, the peladilla, which are particularly abundant then, although this was less immediately acceptable because catch rates then are usually high. Finally, as these proved insufficient, limits were set on the total season's catch. Fishing was stopped in May or June when the season's limit was reached, and reopened at the beginning of the new fishing season in September.

Under this system, which developed gradually from 1965 onward, with the precise set of regulations being different each year, it appeared that the fishery was being well managed. Two of the three objectives were being met. The amount of fishing was apparently close to the recommended figure, and the small fish were being protected. However, there were two reasons for concern.

The first related to the economics of the fishery and the failure to meet the third objective. Although the amount of fishing was controlled, there was no control on the capacity of the fleet or of the processing plants. The fleet in particular continued to grow in real capacity continually during this period. New large boats, carefully designed as highly efficient anchovy killers, replaced the old small and rather crude vessels which were thrown together in the early days of the fish-meal boom. As has been noted above, the regulations had to become steadily more extensive and restrictive to keep the amount of fishing at the desired level. The economic wastage caused by the excess capacity of fleet and processing plant—which could turn the whole annual U.S.

catch into fish meal in about 1 month — is enormous. It has been estimated[7] that the resulting imposed inefficiency has added at least $50 million to the annual costs of the industry. Until these excess costs have been eliminated, or greatly reduced, the anchovy fishery is not making its full potential contribution to the economy of Peru.

The other doubt was whether the stock itself was in as healthy a state as the simple analysis suggested. Spectacular collapses of other fisheries, such as those for sardine off California, and for herring off Norway and northern Japan, showed that these species of shoaling pelagic fish, closely related to anchovy, could be highly unstable. Also there were suspicions that the efficiency of a unit of fishing was not being very well measured and that the amount of fishing in the most recent years was higher, and the stock abundance lower, than estimated.

THE RECENT CRISIS

These two factors — excess capacity of the fishery and possible instability of a heavily exploited stock — could, some scientists began to suspect, act together to produce a serious crisis in the fishery. These fears were very clearly expressed by Paulik[8] as follows:

Thinking about unthinkable catastrophes is one way of preventing them. Destruction of the world's greatest single species fishery is unthinkable. Could it really occur?

Consider the following scenario: Heavy fishing pressure has so reduced the abundance of 1- and 2-year-old anchoveta that the bulk of the catch is taken from recruits before their first birthday. Most of the catch is made during January through May. A moratorium on vessel construction has held the size of the fleet to 300,000 tons of hold capacity. In the year 197(?) adverse oceanographic conditions cause a near failure of the entering year class and low catches in the southern spring and early summer (September through December) cause fish meal prices to climb to $300 per ton. Weak upwelling currents in January, February, and March concentrate the residual population in a narrow band extending about 10 kilometers from the coast. The total catch during the first 4 months of the season is 1 mmt (1 million metric tons) and the total biomass of the entire residual population plus the entering recruits is about 5 mmt. Although many danger signals are flashing and scientists warn the industry that fishing effort must be curtailed immediately, the industry's creditors are clamoring for payment of short-term loans, and many factories and fleets are unable to meet payrolls and pay operating expenses while fish meal prices are at an all time high.

The January catches total 0.5 mmt instead of the 2.00 mmt expected. The industry rationalizes that recruitment is later than usual and will occur during the February closure. When the season reopens in March, the entire fleet is poised and the residual population is vulnerable. Within 2 months the entire population of 4.5 mmt is caught and sold at record prices. The government imposes an emergency closure when the catch-per-unit drops to zero at the end of April. The fishery is closed during May but it is too late. The Peruvian anchoveta stocks have been fished into oblivion. Only small scattered schools of anchoveta remain; dead birds, rusting bolicheros, and idle fish meal factories will soon litter the beaches from Chicama to Arica. Just how unlikely is this apocalyptic vision?

The actual events in 1972 and 1973 have followed this scenario with astonishing precision. The spawning in mid-1971 was a failure, but this was not immediately obvious at the beginning of 1972, because catches of older fish were good. The actual numbers were reasonably high and were concentrated near the coast, so very good catches were taken in March and for much of April (the fishery was closed in January and February to protect the peladilla, which were in fact very rare that year). Thereafter catches dropped off quickly as the old fish were caught and were not replaced by recruits. By May the seriousness of the situation was clear to all, and the fishery was closed in June. At that time the scientists warned that the chances of a rapid and complete recovery of the stocks would be greatest if catches could be kept at a very low level (if not stopped altogether). The status of the stock should be monitored in as many ways as possible, including surveys by chartered commercial vessels using echo sounders (these Eureka surveys had been carried out, on varying scales, for a number of years), and fishing should not start again until there was good evidence that the stock had increased (or that the low estimates of stock size made in June–July 1972 had been too pessimistic). The subsequent Eureka cruises in the second half of 1972 confirmed the low abundance, except in the south, and the fishery remained closed for the rest of 1972 except for a short period in December and for some fishing in the south of Peru.

In 1973 the fishery remained closed in January and February (when it might be hoped that new recruits might be entering the fishery), but was opened in March for 3 weeks, after which it was closed again to allow the scientists to analyze the results. These analyses, by the Peruvian scientists and by the FAO panel of experts which met at the end of March, were again pessimistic. Although catches had been fairly good, this had again been due more to the concentration of the fish in a few areas close to the coast than to a high overall abundance. Also, the fish were mainly old, survivors of the good 1970 spawning, and the weak 1971 spawning, with few new recruits. Again the scientists warned of the possibility of permanent collapse of the fishery and the need for drastic action to reduce this possibility, particularly to maintain the spawning stock at the highest possible level. Accordingly, the fishery was closed in April, and at the time of writing (February 1974) has not yet been reopened.

The fit of these events to Paulik's scenario is remarkable, with a couple of heartening exceptions — the government has been quicker than predicted to react to the crisis by imposing controls, and as a result the stock has not been reduced so far to quite such low levels as Paulik feared. The agreement does raise a number of important questions, such as to what degree has the fishery induced a dangerous instability into the stock?

PROBLEMS AND PROSPECTS
OF THE FISHERY

What have been the cause or causes of this present disaster? A change in oceanic conditions, or overfishing? What action should have been taken to prevent it happening, and how might future disasters be avoided? What are the immediate prospects of the fishery? How has the Peruvian government coped with the immense social and economic problems of the disappearance (if, hopefully, only temporary) of one of the country's major industries?

Most of these questions are easier to put than to answer. The immediate cause of the crisis was the failures of the 1971 and 1972 spawning, coupled with the rapid fishing away of most of the survivors of the 1970 spawning in March –April 1972, but this only raises the question of why the recruitment failed. One simple explanation is that oceanographic conditions were unusual at that time. Periodically the normal pattern of ocean circulation off Peru is disturbed, and it seems that this may be part of a general disturbance over much of the eastern tropical Pacific. Instead of cool, nutrient rich water, the area off Peru is occupied by warmer, less productive water typical of most tropical areas. This phenomenon has such widespread effects that it has its own local name, El Niño. Such a disturbance occurred in 1972 and the first half of 1973. However, this seems to be too late to account for the poor survival of the brood spawned in the middle of 1971. Also, El Niño occurred in 1965 and had no obviously harmful effect on recruitment.

There are similar weaknesses in an explanation based wholly on overfishing and reduction of the adult stock to too low a level. The spawning stock in 1972 was certainly very low, but during the 1971 spawning season anchovy were very abundant (owing to the contribution of the good 1970 year class), although it was possible that an unusually high proportion of these were still just too small to participate in the spawning. A more likely explanation is a mixture of both overfishing and oceanic change, for example that heavy fishing has reduced the stock to such a level that while recruitment could be maintained under average or better environmental conditions, the unfavorable conditions of El Niño would cause a recruitment failure that would not have occurred if the adult stock had been larger.

If this explanation is broadly true we can with hindsight see what went wrong with the management of the anchovy. Strategically, fishing was allowed to increase to too high a level, so the stock was reduced to an unstable position. Tactically, the failure of the 1971 year class was recognized too late to prevent the serious depletion of the survivors of the 1970 year class. In taking advantage of the knowledge of events after they occurred to review the Peruvian performance, it must be recognized that this performance should be measured not against the ideal, as determined later, but against knowledge at that time and against the performance in other fisheries. On this basis the performance is quite good. Although catches not infrequently exceeded those recommended by the scientists, the excess, in relative terms, was not large (10 to 15 percent). It was not until very recently, at the Third Session of the FAO Stock Assessment Panel in 1972, that warnings issued were at all explicit concerning the dangers and possible collapse of the stock. The delay in action over the recruitment failure was due more to lack of early warning that failure had occurred rather than of action once the warning was made. In the future it may be possible to forecast the occurrence of El Niño, and also the impact this will have on recruitment, although both are at present little more than hopes. If both the problems are solved, this could give warnings of poor recruitment a year or two in advance. Shorter warning periods can be almost equally helpful. Techniques of surveying the small prerecruit fish (e.g., using acoustic techniques) are now being developed in Peru and elsewhere. If these had been available and used in the period November 1971–February 1972, catches in March–April 1972 could have been reduced so as to maintain a much higher breeding stock in 1972 than actually occurred.

The future remains uncertain at the time of writing (February 1974). Oceanographic conditions have returned to normal, and the stock, although low compared with past numbers, is still remarkably high by most non-Peruvian standards — several million tons. There is no strong reason to suppose that this stock, under average or better conditions, could not produce near-average recruitment. If fishing is kept low in the immediate future, the stock could return to its former level within a couple of years.

NATIONALIZATION AND RATIONALIZATION

The anchovy crisis has hit Peru hard. The general economy has obviously been harmed to some extent by the loss of a major export, but the effect is much more serious in places like Chimbote (already hit by the earthquake of May 31, 1970), which is the fish meal capital of the world and has virtually no industry except fish meal and supporting activities. Between June 1972 and February 1974 the Chimbote plants have operated for barely 30 days. Extreme distress among the factory workers and fishermen has been avoided by government requirements that employees should be kept on and paid a certain minimum wage.

This shifted the problem onto the fish meal companies. At first they could keep going on their financial reserves and the sales of meal produced earlier, but by mid-1973 all the companies were, to varying extents, in financial difficulties. How these might have been resolved, in the absence of further income from fish meal sales for perhaps another 12 months, will not now be known, because in the middle of 1973 the Peruvian government nationalized the whole industry (for some time previously all exports had to be made through a government sales organization).

The reasons for this nationalization are probably a common mixture of political, economic, and social motives, but as far as the anchovy and its conservation is concerned, the Government's action has provided an excellent opportunity for rationalization of the industry and reducing the capacity of the fleet and plants closer to a balance with the potential yield of the resource. Thus rationalization is being achieved by the complete removal from the industry of the older and least efficient vessels and plants, and the transfer of some vessels to fishing for other species for direct human consumption. The development of this food fish fishery (for hake, mackerel, etc.) is receiving high government priority. Operating a more limited number of fishing vessels and processing plants at something approaching full capacity will obviously mean a much more profitable fishery. Also with a larger stock and with an industry with rather less than the capacity of that in 1973 — which could handle rather more than 100,000 tons per day and remove perhaps 15 percent or more of the stock in 1 week — the risks of collapse and of failure to recognize the incipient collapse and to take appropriate action to reduce catches before the stock has declined to a serious extent are all reduced. All this suggests that although the management by the Peruvian government cannot be claimed as a resounding success (although it should not be forgotten that the two successive year-class failures could have been due wholly to natural factors and might have occurred whatever the management policy), some measure of success had been achieved before the present crisis, to keep the amount of fishing within proper limits. If, as is to be hoped, the stocks soon recover, there is a reasonable hope that future management policies will ensure both a healthy stock and an economically successful fishery.

The Situation of World Fisheries

The histories of the whale and anchovy fisheries have a number of features in common, including a short period of rapid growth and a period of apparent stabilization within limits, which turned out to be set too generously, giving rise to a period of crisis. Both now seem to be in a period of convalescence after crisis. Most major fisheries have also experienced the periods of growth and stabilization but have avoided severe crises. This is chiefly because unlike mammals, and possibly certain shoaling pelagic fish such as herring and anchovy, most fish can withstand large reductions in adult stock without much impairment of recruitment.

All those concerned with fisheries are thus used to accepting a limit in catches from any particular stock and do not expect unlimited growth in catches from one stock. This is, however, not true of fisheries as a whole. Almost without exception all countries that make explicit national plans for 5 to 10 years ahead, and include fisheries as a separate element in these plans, expect their national catches to increase. Taken together this means an expectation that the world fish catch will continue to expand at about the same rate as in the past.

This is not necessarily unreasonable, for a finite period, despite the limits already experienced in many individual stocks. Past expansion has resulted mainly from the development of new fisheries, or fisheries on new stocks, and all will be well if a sufficient number of new stocks can be discovered. How reasonable is this? In other words, what is the potential yield of fish from the sea.

THE POTENTIAL HARVEST FROM THE SEA

A considerable number of estimates have been made of this, in varying degrees of detail. One of the most extensive and recent has been that made by FAO in collaboration with fishery scientists and institutions throughout the world.[9] Table III shows some of the results by main sea areas together with some esti-

mates of the present and projected future degree of utilization as estimated by the Indicative World Plan of the FAO.[10]

A total potential annual catch of 100 million tons estimated in that study has become generally accepted as a reasonable figure for the more familiar types of fish. This includes all fish that currently appear on the fish markets of the world, also species that may not appear at present but are quite similar in size and appearance to those that do. Roughly, they include all fish down to the size of a small anchovy.

The estimates of the potential from individual areas or individual stocks used in compiling the figure of 100 million tons were obtained by a variety of methods and are subject to error, which may be considerable for the less-well-known areas. However, there was a good agreement between the methods used (surveying by fishing gear or sonar, examination of the food web and the magnitude of primary production,[11,12] etc.). Also, developments in the fisheries since the estimates were made have shown them to be at least without any systematic error. Thus it can be fairly confidently stated that the figure of 100 million tons for the potential annual catch of the familiar types of fish is accurate within a possible error of ±50 percent.

Excluded from this total are the smaller animals that are not readily harvested at reasonable cost or do not fetch a good price. To a large extent these are the fish that the larger animals —cod, tuna, whales, and so on—feed on, and being closer to the basic primary production, the total production is likely to be appreciably larger. The best-known example is the Antarctic krill (*Euphausia superba*) on which the baleen whales feed. The lowest estimate of the potential yield from this stock is 50 million tons, and both the Russians and Japanese are actively engaged in finding practicable means to catch and use it.[13] If such a krill fishery does develop, it will be confined to the more technically advanced countries, will be in most ways quite separate from the fisheries known now, and will serve little to solve their problems.

These problems are set quite simply by (1) the arithmetic of a growth rate of 5 percent per year (and none of the pressures of more people and the need to feed people better that have caused this growth in the past is likely to be less in the near future); (2) a present marine catch of 60 million tons; and (3) a limit of 100 million tons (whether it is actually 90, 120 million or even 150 million makes little difference). A 5 percent growth rate means a doubling roughly every 14 years, so in 14 years we will have reached the limit at current growth rates. The growth phase of fisheries is coming to an end, taking the world as a whole, even if many important fisheries can still expand. The world fishery community must very soon come to terms with a limited supply in the same way (and hopefully more successfully) as those concerned with Antarctic whaling did in 1946 or with the Peruvian anchovy around 1965.

Fortunately, the penalties for failure are seldom as severe as they have been in whaling and may be in the case of anchovy. Many fish stocks can withstand a high and uncontrolled level of fishing without much damage in terms of any significant fall in total catch. For example, the stocks of bottom-living fish (cod, plaice, etc.) in the North Sea have been heavily fished for the best part of a century, apart from pauses during the two world wars, with the only controls being on the sizes of fish that can be landed and the size of mesh used in the nets. Despite this, catches of some major species in the last few years have been at record levels.

MANAGING THE WORLD'S FISHERIES

With this background of world catches approaching the productive capacity of the stocks of the traditional kinds of fish, the prospects for the management and conservation of world fisheries may be judged against the tests listed in the whale section. These were basic data, scientific advice, a representative forum, identification of suitable measures, and timely action.

Marine area	Potential (million tons)	Utilization (%)			
		1962	1965	1975 (proj.)	1985 (proj.)
Atlantic Ocean					
Northeast	15.7	46	58	60	68
Northwest	6.2	47	50	65	78
Western Central	7.3	14	16	20	40
Eastern Central	4.2	23	30	57	81
Southwestern	11.2	3	4	27	47
Southeastern	9.1	16	23	51	75
Total Atlantic	53.7	26	32	46	63
Pacific Ocean					
Northeast	8.4 }	48	52	60	76
Northwest	7.5 }				
Western Central	17.2	31	34	67	98
Eastern Central	2.3	24	27	33	44
Southwestern	2.4	6	6	50	79
Southeastern	17.7	39	46	65	72
Total Pacific	55.5	37	41	63	80
Indian Ocean					
Eastern	2.1	31	38	62	83
Western	5.2	19	21	48	70
Total Indian	7.3	23	26	52	74
Mediterranean Sea	1.7	50	54	61	76
World total	118.2	31	36	54	72

III Utilization of Fish Potential by Marine Area 1962, 1965, 1975, and 1985

Source: Reference 10.

The first point to be acted on must be the institution of some *forum for discussion*. Where a fish stock is only harvested by one country, or spends all its life in waters under the jurisdiction of one country, the forum can be provided by a suitable organ of the government of the country concerned. For the anchovy, the relatively small catches taken by Chile, probably from the same group of fish as are harvested in southern Peru, have been ignored, and the management of the anchovy has been treated as a Peruvian problem. Fortunately, Peru has a strong Ministry of Fisheries capable of considering all the wide range of factors involved. This is not necessarily so in all countries, and development of suitable national machinery may be needed.

For whales, it is clearly essential to bring together at least all the countries actively engaged in Antarctic whaling. To a large extent, and for most of the time, the International Whaling Commission (IWC) has provided a forum in which all relevant interests can be discussed. However, the Commission, especially its decision-taking procedures, are weighted toward the immediate short-term interests of those currently engaged in whaling. Long-term interests are less well represented. A result is that the fin whale stocks are being rebuilt more slowly than is probably desirable in the interests of the world as a whole.

Most fish stocks demand a similar multinational forum. Many of the world's most important fisheries take place on the high seas. This pattern may change in the near future with the current tendency toward wider limits of national jurisdiction, which were discussed during the recent U.N. Conference on the Law of the Sea. Even with extended limits, many fish stocks will move between regions of different jurisdiction, and this will still require some coordinated international action.

A large number of international bodies similar to the IWC have been established to tackle the problems of international fisheries on a regional or subject basis. The oldest of these, the International Council for the Exploration of the Sea, was set up at the beginning of the century purely to coordinate marine research, although of late much of this research has been directed specifically to answering problems put to it by a younger body, the North

East Atlantic Fishery Commission (NEAFC), which has the immediate responsibility for recommending appropriate management action. Bodies similar to NEAFC have been set up in many parts of the world. Most of these, particularly those with direct management responsibility, are independent bodies, set up by special international conventions, but others are subsidiary bodies of the Food and Agriculture Organization of the United Nations (FAO). As forums for discussion these bodies perform an adequate job. In most regions, all or most of the main participants in the fisheries now meet regularly to discuss their problems. Only exceptionally do severe political difficulties prevent such meetings. The most obvious example is in the northwestern Pacific, for the waters in the arc from the Korean peninsula to the island of Taiwan. Where the discussions are not productive of proper conservation actions, the reason is usually among those discussed below.

Supply of *adequate data* on which a scientific study of the stocks can be based is clearly essential. The basic requirements are statistics on the catch and the amount of fishing, with details of where fishing was done and what species are caught. Data on the sizes and ages of fish caught are also invaluable, and as studies become more advanced, so are a range of other data on other species of fish, the physical and chemical environment, and similar considerations.

The data for the whale and anchovy fisheries were very good, and at no time have these fisheries been mismanaged because of the lack of data. These fisheries are exceptional, although it is because they are exceptional in having good data available that they make such good textbook examples. Far too often the data are inaccurate or incomplete, even with regard to total catch. Provision of data is clearly particularly difficult when several countries fish the same stock, because data must be collected in a compatible form from all. Nevertheless, the actual data available from international fisheries are often better than those from purely national fisheries. This is because the existence of the various international commissions has underlined the responsibility of all those concerned to provide the necessary data and has provided on their staff people with the time, interest, and authority to ensure that the necessary data are collected. For example, some of the best basic statistical data for U.S. fisheries are those for the Pacific tuna fisheries collected by the Inter-American Tropical Tuna Commission. Still, for most fisheries, both national and international, the basic data are still far from adequate, and much has to be done to improve matters. This improvement is receiving high priority from all the regional fishery commissions, as well as from the FAO (which has responsibility for statistics and other data on a worldwide scale) and from most governments.

Shortcomings in *scientific advice* led to trouble at different times for both Antarctic whales and Peruvian anchovy. For whales, the advice in the period around 1960 did not take account of the concepts of population dynamics that were then available in fisheries, nor were the recommendations framed in a way that made clear the choice that needed to be made by the Commission. In Peru the advice was probably as good as could be provided, and valuable under average conditions. The scientists failed, however, to understand the degree of instability of the anchovy population and to advise accordingly.

What about the rest of the world? The first mistake, failing to take advantage of ideas and methods already in existence, is the less excusable. It only occurred in whaling because whale biology had become cut off from the quantitative ideas common in fishery research. Communications among various parts of the fishery field are better. The various regional fishery commissions and councils provide a good means of bringing the best talent to bear on problems of individual fisheries. A few bodies (e.g., the Inter-American Tropical Commission) have their own research staffs, but most depend for advice on scientists in national institutions. A cheering feature of fishery research since its earliest days has been the close cooperation that has always existed between scientists who were studying the same stocks. One cannot be quite so optimistic about the development and use of new ideas. Fishery scientists as a group are too often engaged so closely and for so much of their time with immediate problems of, for example, advising on the catch that should be taken

next year that they have little time for new ideas. The ideas used and generally used very successfully in the day-to-day problems are usually of a single species, steady-state situation. The growing need is to be able to give advice on fisheries or groups of fisheries in the same area that exploit a whole range of species. It would also be most valuable for fishery science to use the whole range of techniques, such as computer simulation, that are being developed in related fields of science. The need is for better contact between the scientists engaged in fishery problems and those working in related terrestrial subjects.

Both the whale and anchoveta fisheries got into trouble by the failure to find *suitable regulations* that allowed the industries concerned to be economically successful. The discussions to modify the single quota system in the Antarctic and to introduce national quotas distracted attention at the crucial time when reduction in the fin whale catches could have been effective and fairly painless. Failure to reduce the overcapacity in Peru prevented the fish meal industry from getting into a strong position from which to withstand the troubles that it met in 1972–1973.

In principle, these difficulties did not concern the quantity of the catch or the effect on the stock, only how the catch was taken and the economic performance of the fishery. In practice they made it less easy for more than minor reductions in the total catch to be accepted.

The same has occurred in many other fisheries. Far too often conservation and management have been conceived of as purely scientific (or, more narrowly, biological) problems. As a result, regulations have been formulated which fail to take account of the economic interests. At best, technical or other improvements may be discouraged, and often they are actually prohibited. An outstanding example of legislated inefficiency is the Pacific salmon fishery, where, among other rules, it was forbidden until comparatively recently to use motorized boats to fish for salmon in Bristol Bay. It is hardly surprising that the fishing industry has often greeted management proposals with a marked lack of enthusiasm!

Matters are improving. It is becoming generally accepted that management is complex and that a wide range of interests have to be accommodated. It is, however, a further and more difficult step to determine how the interests can be met; however, the experience of Antarctic whaling, in which a balance was found between very diverse interests in the matter of country allocation, gives a reasonable hope for other fisheries.

Of the conditions listed, *timeliness* is probably the most critical. Data can be collected and analyzed and a regional commission set up with all interests represented and carefully balanced regulatory measures formulated, but all this is of little value if it comes too late. The right kind of scientific advice came too late to save the blue whale from commercial extinction, and was acted on too slowly to prevent a catastropic fall in the fin whale stocks. The failure of the recruitment to the Peruvian anchovy fishery was only recognized after the stock of older fish had been seriously reduced.

The performance in other fisheries has been no better. The story of most heavily fished stocks is of measures being applied only after the fishery concerned had run into difficulties. Partly this has been due to an absence of data, of scientific analysis, of a suitable forum to discuss the problem, or of suitable measures, but mostly the failure has been an unwillingness to take action.

The world fisheries community has been no more willing than most groups to make unpleasant decisions until forced to do so. Most international fishery bodies, especially those like the North Atlantic bodies with their complex mixtures of species of fish, gears used, and countries concerned, have started their management program quietly. The first regulations introduced have only interfered slightly with fishermen's activities, through controls on the sizes of fish that can be landed or of the meshes that can be used in trawls. Only recently have more than a few bodies (usually those with restricted membership and interests) moved into the more difficult field of preventing some (or all) fishermen from fishing when they want to (e.g., by setting catch limits). Both the fisheries we have examined in detail have been somewhat unusual as regards the relatively early stage at which restrictions were accepted. More typical of most regional commissions and indeed of many govern-

ments has been the slowness with which any control of the amount of fishing has been introduced into North Atlantic fisheries. Only in 1972 were catch limits set for any of the major stocks (herring and cod) in these waters, and then only on the western side.

PROGNOSIS FOR THE FUTURE

Progress is being made. After the breakthrough in 1972, the International Commission for the Northwest Atlantic Fisheries (ICNAF) has been moving fast, introducing catch limits not only for those stocks for which they are already demonstrably needed and the levels at which they should be set have been clearly established, but also for some stocks (e.g., mackerel) for which the evidence is still scanty. This Commission has thus advanced past the stage of dealing with crises only after they have occurred and is implementing precautionary measures to prevent crises. Undoubtedly, these advances by ICNAF have in large measure been achieved, when they were, through external pressure. Most of the stocks concerned lie off the U.S. coasts but have been to a large and increasing extent harvested by foreign vessels. U.S. fishermen, seeing their share of the catch steadily decrease, have been pressing for wide extensions of territorial limits, which might well have been introduced (with the consequent exclusion of foreign fishing) if the ICNAF had not acted. At the same time, these pressures were effective because there was general agreement that precautionary measures were desirable in principle, even if it took pressure to make them accepted in practice.

This achievement in the northwestern Atlantic, by which limits have been set for most of the major stocks, represents the most striking progress in the last few years, but significant advances in the approach to managing fish stocks and in the measures actually taken have occurred elsewhere. Actions to limit the amount of fishing under international, bilateral, or national measures now cover such diverse stocks as yellowfin tuna in the eastern tropical Pacific, king crab in parts of the northern Pacific, and capelin in the Barents Sea, as well as the anchovy and whales already described.

Unfortunately, the urgency and magnitude

187

of the management problems are increasing as fast, if not faster, than the ability and willingness to deal with them. In the past many of the difficulties that arose from failures to manage one stock were mitigated by the diversion of much of the excess effort to other, underutilized stocks (the switch from blue whales to fin whales, and later to sei whales, is a good example of this process). These opportunities to avoid the penalties of mismanagement are becoming steadily fewer as the number of underexploited stocks of fish declines.

The penalties of mismanagement are usually mainly economic — involving excessive costs and some drop in total production — and rarely involve the complete disappearance of the fishery. However, a few fisheries have died, and their ghosts haunt the fishery manager and his scientific advisors. These include many of the fisheries for marine mammals — whales, seals, and sea otters — in some of which the species themselves have only just avoided extinction along with the industries that exploited them. A number of fisheries on small pelagic fishes, related to the anchovy, have also disappeared. The Californian sardine fishery, which 30 to 40 years ago was among the biggest of the world, has gone, and catches of the Atlantic–Scandian herring (which inhabits the area between Norway and Iceland) dropped from 1.7 million tons in 1966 to barely 20,000 four years later. In these vanished pelagic fisheries, the cause of death for each particular case is not completely established, but the common correlation with intense fishing is at least highly suggestive. Fortunately, the Norwegian fishermen could turn to other species (mackerel and capelin), and California is not a poverty-stricken area. Elsewhere the disappearance of a fishery can be a disaster, and even the moderate drop in catch and the economic loss more typical of unmanaged fisheries should be avoided if possible.

World fisheries are thus engaged in a race. On the one side are the fishermen and fishing industries who collectively are pressing toward the limits of catch that the seas can sustain and the possible disaster if these limits are exceeded. On the other side are those concerned with the collection of data, its analysis, and the resulting preparation of scientific advice, especially those concerned with using this advice to set limits on the fisheries and ensuring that these limits are accepted and observed. It is still unclear who will win. Although better data are being compiled and analyzed in better ways, and, most important, there is increasing acceptance that management measures are necessary, at the same time the intensity of fishing pressure on the stock is increasing rapidly and the gap between current yields and total world potential is decreasing. The two most critical factors are probably the degree to which scientists can obtain a clear and rapid understanding of what governs the changes in abundance of fish stock, and the willingness of fishermen, fishing industries, and governments to accept the measures, often including short-term sacrifice that may be severe, necessary to conserve the stocks and hence the fisheries based on them.

References

1. Lockyer, C. 1972. The age at sexual maturity of the southern fin whale (*Balaenoptera physalus*) using annual layer counts in the ear plug. *J. Cons. Internat. Explor. Mer* 34(2): 276–294.
2. Laws, R. M. 1962. Some effects of whaling on the southern stocks of baleen whales. In *The Exploitation of Natural Animal Populations* (E. D. le Cren and M. W. Holdgate, eds.). Blackwell Scientific Publications, Ltd., Oxford, England. pp. 137–158.
3. Chapman, D. G. 1964. Report of the Committee of Three Scientists on the Special Scientific Investigations of the Antarctic Whale Stocks. *Rept. Internat. Comm. Whaling 14*: 32–106.
4. Meadows, D. H., et al. 1972. *The Limits to Growth.* Universe Books, New York.
5. Gulland, J. A. 1972. Future of the blue whale. *New Scientist 54*(793): 198–199.
6. Cushing, D. H. 1969. Upwelling and fish production. *Advan. Marine Biol. 9:* 255–334.
7. Anon. 1970. Report of expert panel on the economic effects of alternative regulatory measures in the Peruvian anchoveta fishery. *Informe. Inst. del Mar del Peru 34.*
8. Paulik, G. J. 1971. Anchovies, birds and fishermen in the Peru Current. In *Envi-*

ronment: Resources, Pollution and Society (W. W. Murdoch, ed.). Sinauer Associates, Inc., Sunderland, Mass.

9. Gulland, J. A. 1971. *The Fish Resources of the Ocean.* Fishing News (Books) Ltd., London.
10. United Nations, Food and Agriculture Organization. 1969. *Provisional Indicative World Plan.* FAO, Rome.
11. Schaefer, M. B. 1965. The potential harvest of the sea. *Trans. Amer. Fish Soc. 94*(2): 123–128.
12. Ryther, J. H. 1969. Photosynthesis and fish production in the sea. *Science 166*(3901): 72–77.
13. Lyubimova, I. G., Naumov, A. G., and Lagunov, L. L. 1973. Prospects of the utilization of krill and other nonconventional resources of the world ocean. *J. Fish. Res. Bd. Canada 30*(12): pt. 2.

Further Reading

Anon. 1973. Papers presented at the FAO Technical Conference on Fisheries Management and Development. *J. Fish. Res. Bd. Canada 30*(12): pt. 2.

Christy, F. J., and Scott, A. 1965. *The Common Wealth in Ocean Fisheries.* The Johns Hopkins Press, Baltimore, Md.

Gulland, J. A. 1974. *The Management of Marine Fisheries.* Scientechnica (Publishers) Ltd., Bristol, England.

———. 1972. *Population Dynamics of World Fisheries.* University of Washington, Seattle, Wash.

Roemer, M. 1970. *Fishing for Growth: Export-led Development in Peru, 1950–1967.* Harvard University Press, Cambridge, Mass.

Environmental Degradation

EPA DOCUMERICA PHOTO BY DAVID HISER

Editor's Commentary—
Chapters 8-11

CHAPTERS 9 THROUGH 13 describe the variety of pollutants that industrial Man produces, their major sources and where they go. Each of these chapters deals briefly with the broadly environmental effects of pollution and also mentions human health effects, but the latter are drawn together and treated in detail in Chapter 8 by Drs. Epstein and Hattis. The health effects of radiation are discussed in Chapter 12 by Dr. Cook.

The problem of pollution and human health is very complex, and we have by no means developed a satisfactory and complete understanding of how pollutants affect our health and life expectancy. It is easy to understand why: There are almost 60,000 chemicals for sale in the United States, most of them complex synthetic organic chemicals, and most of them have *not* been analyzed for their human toxicity. We often do not have a good accounting of where these chemicals move in the environment, or what are the properties of the products they break down into. Furthermore, it is extremely difficult to separate out the effects on our health of any particular chemical in the environment from the total effects of all of the substances we are exposed to, as well as the things we eat, the contribution of our genes and our way of life. Unfortunately (although fortunately in some respects) we cannot perform controlled experiments with people and must therefore extrapolate from other animals. In addition, some of the health effects, including some cancers, that might be caused by pollutants develop only after a long exposure or a long time after exposure. As an illustration of these difficulties, consider how long it has taken to build a good case that cigarette smoking is a causal agent in lung cancer, and yet cigarette smoking is much easier to study than most pollutants because we can specify who is exposed and who is not.

In spite of all of these difficulties, Chapter 8 reaches an extremely important conclusion: a large fraction, and perhaps the great majority, of cancers are caused by environmental factors. And, of course, environmental pollutants also cause, or worsen, a sizable but unmeasured fraction of the incidence of other diseases, from mental retardation to emphysema. Two conclusions flow from these facts. First, too high a proportion of medical research money is spent on curing, rather than preventing, environmentally induced disease. Second, as Drs. Epstein and Hattis stress, we need much more rigorous government regulation to protect us from involuntary exposure to hazardous pollutants.

The material in Chapters 9 through 11 is perhaps more straightforward and less in need of an introduction. In these three chapters on air, fresh water, and ocean pollution, each chapter deals with the major types of pollutants, their sources, and their effects. The most serious of these, with respect to human health, is air pollution, while the least serious from any point of view is ocean pollution, at least at current rates.

David Lynn, in Chapter 9 on air pollution, makes the significant point that social control of air pollution is even more important than engineering solutions, and he discusses some of the possibilities. He is somewhat optimistic that, given the needed changes in attitude, we

193

can achieve greatly improved air quality. This task may be much easier for stationary sources, where the major problems are sulfur oxides and particulate matter, than for mobile sources, especially the automobile. In contrast to the car, several options are available for stationary sources, including changing the industrial process and installing efficient control devices; furthermore, stationary sources are easier to police, and there is the additional incentive of reclaiming the materials removed from the effluent. The automobile and its internal combustion engine pose more serious difficulties for several reasons: the engineering solutions are more restricted; three different types of pollutants are involved (carbon monoxide, hydrocarbons and nitrogen oxides) and it is not possible to get low emissions of all three simultaneously; each polluting unit is small, thus relatively little money can be spent on its control device; and the existence of millions of mobile sources makes policing very difficult. In fact, as David Lynn points out, the best solution to automobile pollution is to substitute better-designed mass transit systems and better-designed cities, while at the same time reducing the use of cars with internal combustion engines. If these changes can be brought about, then the outlook for air pollution in the intermediate future (10 to 20 years) can be greatly improved, though the recent postponement of clean air compliance requirements and pressure from the energy crisis make the short term outlook gloomy.

Although some serious health hazards do occur from water pollution, its greatest effect is on freshwater ecosystems and human amenities (Chapter 10). Even more than in the case of air pollution, this reflects a wastage of resources. Dr. Edmondson notes that, not only could it be economic to recycle nutrients from sewage and feedlot runoff, but that freshwater ecosystems have a remarkable ability to recover once pollution has been reduced. Dr. Edmondson demonstrates this for Lake Washington, and the Thames river in England is perhaps an even more dramatic example. Fifteen years ago the Thames in and near London stank and was so heavily polluted that for several months each year it lacked oxygen. Only one fish species (the eel) was present in the metropolitan Thames. In the early 1960s a program was begun to clean up the effluent going into the river. This has been so successful that by 1970 more than 50 species of fish had been found! Unfortunately, the cost of adequate waste water treatment in the U.S. would run into many billions over the next decade, and one of the first actions of President Ford was to cut several billions out of the water treatment program.

Perhaps the most astonishing pollution, though it does not seem very hazardous at present, is ocean pollution (Chapter 11). Dr. Goldberg shows that man is causing colossal amounts of materials to be moved about the earth over vast distances. Via rivers, the atmosphere, ships, pipes and outfalls, billions of tons of materials move into the world's oceans, especially in the Northern hemisphere. A few materials, such as the heavy metals, can cause at least local concern, while oil is essentially everywhere in the oceans and its effects on ecosystems, if any, are unmeasured.

I should like to add a comment on one area that deserves special attention, namely estuaries. These are tremendously productive areas (Chapter 1) and support great populations of organisms, including hundreds of species of birds. They are also especially important as nursery areas for many ocean fish and shellfish (especially in the United States). Unfortunately they also suffer greatly from Man's activities, for several reasons. Because of their position and water currents, they are natural "sinks" for pollutants. They are also very attractive areas for people and industry, and throughout the world there is not only a high concentration of population and industry along the sea, but there is a movement towards the sea, and estuaries in particular. So estuaries are not only heavily polluted, some are also obliterated as cities grow over them. In short, estuaries are perhaps the single most endangered ecosystem in the United States.

Discarded cans of agricultural poison litter on Irish beach near Dublin. Photo by Paolo Koch/ Rapho-Guillumette.

8

Pollution and Human Health

Samuel S. Epstein   *is Swetland Professor of Environmental Health and Human Ecology at Case Western Reserve School of Medicine. After qualifying in medicine (1950) from London University, he served in the Royal Army Medical Corps, then worked as a tumor pathologist and cancer researcher in London. From 1960 to 1972 he was chief of the Laboratories of Environmental Toxicology and Carcinogenesis at the Children's Cancer Research Foundation and Harvard Medical School. An authority on toxic and carcinogenic effects of chemical pollutants, Epstein has edited two volumes on* Consumer Health and Product Hazards, *and has served as a consultant to a wide range of public interest groups, congressional committees and federal agencies. He is President of the Society of Occupational and Environmental Health, Washington, D.C., Chairperson of the Commission for the Advancement of Public Interest Organizations, Washington, D.C., and President of the Rachel Carson Trust.*

Dale Hattis   *is a Sears Public Interest Fellow in the Complex Systems Institute at Case Western Reserve University. He received his B.A. in biochemistry at the University of California at Berkeley in 1967 and earned a Ph.D. in genetics at Stanford University in 1974. In 1970, while working as a student intern at Ralph Nader's Center for Study of Responsive Law, he initiated a request to the Food and Drug Administration for access to information supplied by manufacturers relating to the safety of food additives and pesticide residues. This request (and a subsequent lawsuit) eventually led to the opening of the files. At present his work is primarily in the field of occupational health and safety.*

Health and Chemical Pollutants

To what extent do environmental chemical pollutants contribute to human health problems? Despite considerable recent research, that question deserves a better answer than can be given at the present time. What we will do in this chapter is:

1. Define what is meant by "pollutants" of the "environment" and outline the routes and modes of human exposure.
2. Classify chemical pollutants.
3. As background, survey the overall incidence of the major categories of disease in the United States.
4. Discuss the reasons for growing concerns that pollutants are contributing to our overall burden of disease.
5. Describe available laboratory methods for investigating possible adverse health effects of particular environmental pollutants.
6. Describe available epidemiological methods for investigating possible adverse health effects of particular environmental pollutants in human populations.
7. Discuss the nature of the "social gambles" implicit in technological change and the need for reform of current controls and decision-making processes.

Throughout the chapter it will be apparent that what is not known about this area is far more important than what is known. In addition to a sense of the facts, we hope to convey a sense of the struggles to ascertain the facts and secure appropriate social action in the light of the facts and our ignorance.

DIRECT AND INDIRECT POLLUTANTS

Pollution is a very pervasive phenomenon. Chemical or physical agents capable of adversely affecting man or other living organisms (*pollutants*) may be released directly or indirectly into any part of the *environment* — air, land, water, or biota — by a wide variety of possible mechanisms and routes. Factory smokestacks and municipal sewage outfalls come readily to mind when one thinks of direct pollution, but it is important to be aware of the existence of many other less obvious and less easily detectable mechanisms, such as the evaporation of spilled mercury from the floors of dental offices, the leaching of plasticizers into blood stored in plastic bags, and the release of toxic substances into food by contaminating bacteria and molds in the ordinary process of food spoilage.

Even more difficult to detect, in general, are indirect forms of pollution, where potentially harmful substances may be chemically transformed or concentrated at locations distant from the source of original environmental release. For example, DDT, polychlorinated biphenyls, heavy metals, and other chemicals that are poorly metabolized and excreted by animals tend to be concentrated in biological food chains. Plankton and other lower organisms take up these pollutants originally from water, and with each successive phase of predation (phytoplankton, eaten by small fish, eaten by larger fish, eaten by birds) a substantial portion of the pollutant becomes concentrated in a much smaller total weight of living organisms. For such bioconcentrating chemicals, it is not unusual to find concentrations in fish-eating birds or larger fish many thousand times in excess of the concentrations found in the water where the plankton grow. Humans can be affected by this process to the extent that they rely on fish grown in contaminated waters for food. An illustrative and serious example was the recent epidemic of paralysis due to eating fish contaminated by industrial mercury discharges in Minimata Bay in Japan.

ROUTES OF HUMAN EXPOSURE

There are three major routes of human exposure to chemicals: skin or mucous-membrane contact, ingestion, and inhalation. For drugs, there is a fourth route: the parenteral, such as intravenous or intramuscular injection. The effects of chemicals may be restricted to the initial site of contact and exposure, such as the skin or lung, or may become generalized by subsequent absorption.

For any particular chemical, of course, it is possible that exposure will occur through more than one route. For example, with domestic formulations of insecticide aerosols, exposures occur by direct skin contact, by eating food contaminated with the aerosol, and most importantly, by inhalation. Similar multiple routes of exposure also occur for hair sprays, deodorants, and other products dispensed in aerosol form. Children living in dilapidated inner-city housing are often exposed to serious risk of lead poisoning by a combination of lead exposures from eating peeling paint and inhalation of lead particulates from automobile exhaust.[1] Exposure to asbestos, an established human carcinogen (cancer-causing agent), may occur from inhalation in insulation workers, by way of drinking water, or by eating Japanese rice dusted with talc contaminated with asbestos. A water pollution case recently in litigation involved the contamination of Lake Superior (and hence the water supply of cities including Duluth) with asbestos from taconite processing at Silver Bay, Minnesota. Lesser quantities of asbestos are present in the water supplies of other cities, probably from asbestos-cement water pipes.

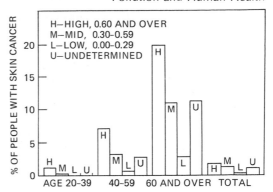

1 **Incidence of Skin Cancer.** Age-specific prevalence rate (percent) as a function of arsenic concentration (ppm) in well water. *Source:* Reference 3.

Potentially Harmful Chemicals

More of the difficulty in dealing sensibly with health hazards from environmental chemicals is apparent when one attempts to categorize our chemical world. The spectrum of chemicals naturally present in the biosphere is notable in itself for its complexity, and the actions of modern civilization have added greatly to the confusion with the introduction of thousands of new synthetic chemicals into the environment.

NATURAL SIMPLE INORGANICS

Some of the simplest and most ubiquitous materials are by no means necessarily innocuous, particularly when present in unusually large amounts. Inorganic arsenic, for example, is present normally in soil in the parts-per-million range (1 part per million of arsenic is 1 gram of arsenic in 1 million grams of soil) and in seawater in the parts-per-billion range. Some sea animals tend to concentrate the arsenic, prawns up to as much as 170 parts per million.[2] Freshwater supplies occasionally become contaminated with unusually large amounts of arsenic, either naturally or as a result of human activity. Figure 1 shows the results of a survey of the incidence of skin cancer among 40,000 people in southwestern Taiwan, some of whom drink water from artesian wells heavily contaminated with arsenic.[3] Skin cancer was observed in nearly 20 percent of the people over age 60 in the group with the highest exposure.

Other studies of people exposed by inhalation and/or ingestion to arsenical insecticides and by occupational inhalation of inorganic arsenic in copper smelter operations[4] also show greatly increased incidences of skin and internal tumors. There has, as yet, been no published study of people eating large numbers of contaminated prawns.

Nitrate and nitrite present another example of a pair of simple inorganic ions, widely distributed in nature, which pose recognized health hazards. Babies' stomachs sometimes contain, by virtue of being less acid than adult stomachs, bacteria capable of reducing nitrate to nitrite. Nitrite combines with and temporarily inactivates hemoglobin, the oxygen-carrying pigment in blood, leading to hypoxia and, in severe cases, death. Drinking water that is sufficiently contaminated to cause this is usually the result of the overuse of nitrate fertilizers in agriculture or inadequate protection of wells from septic tank seepage. Substantial levels of nitrate can also be found in some vegetables, such as spinach, particularly when nitrate levels in vegetation are increased by the use of phenoxy herbicides such as 2,4-D. Additionally, considerable quantities of nitrate and, more importantly, nitrite are deliberately added to cured meats to preserve the red color during storage and cooking. Examples are ham, bacon, and sausage. Some other uses of nitrite in foods are ostensibly for the purpose of preventing botulism, but the need and usefulness of nitrite for all these purposes is questionable and inadequately documented. The unnecessary introduction of nitrate/nitrite additives in foods have aroused special concern because of the tendency of nitrite to combine with secondary and tertiary amines, other chemicals naturally present in various foods, to form nitrosamines — a class of substances with high carcinogenic, teratogenic, and mutagenic potential.[5,6] Recent experiments suggest that substantial nitrosamine formation may occur catalyzed by bacteria in the intestine and in sewage, as well as by acid catalysis in the stomach.

NATURAL ORGANICS

The rise in ecological concerns in recent years has been accompanied by a tendency to think of the natural world as a kind of Garden of Eden, completely wholesome for people if it were not for regrettable interferences in the natural world by people themselves. Particularly in the area of food, consumers have come to associate safety and wholesomeness with the minimum possible cooking, processing, and artificial ingredients. This association may be, in general, correct, but it is a potentially dangerous myth that natural or "organic" foods are hazard-free.

Evolution did not, in fact, produce a living world free of organic chemicals with toxic effects on *Homo sapiens*. Over the years, it has evidently been of considerable selective advantage to plants to discourage excessive animal predation by synthesizing, and incorporating into various structures, chemicals poisonous to animals (Chapter 1). An elaborate array of such substances is known[7]:

1. Peach pits, bitter almonds, lima beans, and cassava contain particular sugars (cyanogenic glycosides) which can be hydrolyzed to release hydrogen cyanide by enzymes liberated by crushing. Despite cooking and other procedures to remove most of the cyanide, peripheral neuropathy from long-term cyanide poisoning is a major problem among poor people in Nigeria, for whom cassava makes up a large portion of the staple diet.

2. A number of foods contain substances that destroy or make unavailable certain vitamins. Raw soybeans have an enzyme (lipoxidase) that destroys carotene. Raw egg white contains a protein, avidin, which binds biotin so strongly as to make it unavailable for use.

3. Some fungi, particularly *Aspergilus* species, produce potent toxins. Aflatoxin, produced by *Aspergilus flavus* mainly on groundnuts, is one of the most powerfully carcinogenic materials known. Another famous composite of fungal toxins is produced in rye infected with *Claviceps purpurea*. Human consumption of contaminated rye gives rise to ergotism, characerized by hallucinations and convulsions or by a gangrenous necrosis of the limbs known in previous centuries as Saint Anthony's fire. However, most antibiotics are fungal toxins whose major activity is against bacteria rather than animals.

4. A number of plants contain substances with

198

various actions on the central nervous system: from the stimulants in coffee (caffeine) and tobacco (nicotine) to the more profound mind-altering substances in poppies (morphine), peyote cactus (mescaline), and Mexican mushroom (psilocybin).

This list could be greatly extended, but the point is clear that "natural" is not necessarily synonymous with "harmless." What can be said in general about naturally occurring toxic organic chemicals is that, at least in the amounts and locations where they occurred in the millennia before civilization, they have coexisted on earth with surviving species of plants and animals without totally exterminating all members of any individual life-form. In addition, to the extent that human beings and their forebears were exposed to selective pressure from particular natural toxicants during evolution, it can be presumed that behavioral and metabolic mechanisms were developed, if biologically feasible, to reduce the toxic effects. Weak as these assurances are, they cannot, of course, apply to synthetic chemical pollutants whose large-scale introduction into the biosphere has resulted from recent human activity.

WELL-DEFINED MAN-MADE CHEMICALS

The well-defined man-made chemicals can be grouped into a wide variety of functional categories. *Agricultural chemicals*, notably fertilizers and pesticides, have received a great deal of attention (pesticides are discussed in detail in Chapter 13). *Food additives* may be incorporated in food either intentionally — primarily to improve flavor, color, texture, or aid in processing of mass-produced items[8] — or unintentionally as a consequence of use in packaging materials or by accidental contamination with pesticides or industrial chemicals. Much less recognized, however, are many other categories of synthetic chemicals, including fuel additives, optical brighteners, plasticizers, and general industrial chemicals.

A recent chemical industry directory lists a total of 58,000 different chemicals for sale by one company or another, mostly synthetic organic chemicals.[9] The era of cheap, plentiful petroleum and rapid industrial expansion since World War II has seen an unprecedented expansion both in the number of available synthetic organic chemicals and in the quantities manufactured. Shown in Table I are the quantities of some major classes of synthetic organic chemicals produced in 1970 and the percent increase in production since 1967 and since 1949.[8,10]

It is impossible at this time to define the proportion of these novel chemicals that pose potential hazards to man. Except for some special-purpose regulation in the areas of pesticides, food additives, and drugs, the massive expansion of applied chemical technology that has occurred since World War II has unfortunately taken place largely unrestricted by governmental controls. There has been no general requirement for screening of chemicals prior to commercial introduction for possible adverse effects on workers handling the materials in

I   Synthetic Organic Chemicals, 1970 Production and Percent Increase in Production from 1967 and from 1949

Class of chemical	1970 production (billions of pounds)	Increase 1967–1970 (%)	Increase 1949–1970 (%)
Cyclic intermediates	28.3	38	N.A.[a]
Plastics, resin material, and plasticizers	20.6	36	1,130
Synthetic rubber and rubber processing	4.7	16	350
Surface-active agents	3.9	12	810
Pesticides	1.0	−2	710
Others	79.7	33	550
Total or weighted average	138.2	32	N.A.

[a]N.A., not available.
Source: Data from References 8 and 10.

the course of manufacture and use or on the general population. As a consequence, not even the most basic information is available on the possible toxicity of the great majority of these compounds. In this situation, much of the scanty information now available, particularly with respect to chronic hazards, has only been obtained at great human cost, in many cases decades after much damage has occurred.

A recently appreciated example of this is vinyl chloride monomer, the basic chemical that is polymerized to make polyvinyl chloride plastics. First introduced into large-scale production in the 1950s, synthesis grew at about 15 percent per year until about 4 billion pounds were manufactured in 1970 (included in the category "Others" in Table I). Recently, however, several cases of a previously very rare form of liver cancer (angiosarcoma) have been detected among workers engaged in polymerizing vinyl chloride. This same chemical was shown to produce cancer in animals a few years ago. The latency period in the plastic workers (the time between the onset of exposure and detection of the tumor) has in all cases been longer than 12 years. Minimum latency periods longer than this are common in human carcinogenesis. Average latency periods for all tumors that may ultimately result from a given carcinogenic exposure can be as much as 30 or 40 years. The practical result of this delay in recognizing the human health hazard from vinyl chloride is that while the first tumors were developing, many other people have been exposed — both as a result of the massive expansion in the manufacture of plastics derived from vinyl chloride and from miscellaneous uses of vinyl chloride, such as a cheap propellant in some domestic aerosol products. It cannot be foreseen how great this particular pollution problem will prove to be or how many similar problems may remain to be discovered.

POORLY DEFINED MAN-MADE
CHEMICALS

If there are substantial deficiencies in available information on the hazardous properties of the thousands of well-defined chemical entities, chemicals that occur mainly in ill-characterized mixtures, of course, pose even greater difficulties. Air pollution is made up of hundreds of chemical entities, whose proportions change with time and from place to place even within the same city. Usually measured are certain major components, such as carbon monoxide, ozone, nitrogen oxides, sulfur dioxide, and total particulate weight. There has been some progress in relating the levels of some of these measured components to the incidences of specific health effects, particularly bronchitis, emphysema, and lung cancer, but the complexity of the wide range of more obscure components in the pollutant mixture makes even chemical analysis difficult. In addition to community air and water pollution, some industrial processes produce very complex mixtures of hazardous substances. Notable in this regard are volatile chemicals driven off in the process of converting coal into coke for the steel industry (coal-tar-pitch volatiles). Men exposed to these volatiles have very much larger incidences of cancer of the lung and kidney than steel workers as a whole.[11]

Aspects of Disease in the United States

Now that we have examined chemical pollutants of the environment, we can turn to adverse health effects. In this section we shall not attempt to relate any of the adverse health effects specifically to pollutants. (An examination of the basis for concerns about connections between adverse health effects and pollutants will be presented in the following section.) Rather, we shall present here a general survey of (1) the numbers of people who suffer different degrees of impairment of health due to various categories of diseases, and (2) the limitations of the available data base.

It is unsatisfactory to treat human "health" in negative terms, as the absence of demonstrable disease ("adverse effects"). Ideally, one could wish that the focus of social thought and action in this field could be on promoting the *best* physical and mental functioning rather than treating or avoiding the gross malfunctioning represented by disease. Unfortunately, our knowledge of what would constitute truly excellent human functioning, and what environmental influences impair us in achieving it, is fragmentary indeed.

MORTALITY

The most severe and final expression of disease is, of course, death. Shown in Table II are the major categories of illnesses identified on death certificates as the major underlying causes of death for people in different age groups in the United States in 1969. These figures should be viewed with some reservations, outlined below, but they can serve as a reasonably valid approximate indicator of the frequencies with which different broad categories of diseases are the major identifiable immediate cause of death in people of different ages. Cardiovascular diseases make the largest contribution to overall mortality, being preeminent causes of death in elderly people. As one looks at younger age groups (45–65 years and 5–44 years), the contribution of cardiovascular diseases declines relative to the contributions of cancer and accidents. For children under 5, certain causes of death in early infancy (including birth injury and certain anoxic and hypoxic conditions), congenital anomalies, and infectious diseases are of primary importance.

In considering the data in Table II, however, the reader must bear in mind some possible distortions implicit in such data. For one thing, not all underlying causes of death are diagnosed with adequate skill, care, and information. It is likely that considerable numbers of deaths from comparatively obscure or poorly recognizable causes are misreported. Even for major subdivisions within cardiovascular diseases, a review of evidence supporting diagnoses on a sample of 1960 death certificates revealed that only about 30 to 74 percent (depending on the cardiovascular category) of the classifications were based on information adequate to lead a panel of cardiovascular experts to believe the diagnosis in question to be "a reasonable inference or better."[13]

An even more important difficulty with the official mortality data is the failure to list contributing causes of death, although more than half of all death certificates specify more than one cause. The true impact of chronic respiratory diseases on mortality seems to be particularly understated for this reason. Respiratory ailments frequently result in increased blood

II   Thousands of People Dying from Various Causes in 1969

	All ages	5 yr.	5–44 yr.	45–65 yr.	65 yr. and over
Major cardiovascular diseases	1,014	1	28	218	767
Heart 739					
All other 274					
Cancer	323	1	24	118	180
Accidents	116	7	57	24	28
Motor vehicle 56					
Accidental poisoning 5[a]					
All other 55					
Pneumonia, influenza, and other infectious and parasitic diseases	85	11	7	16	51
Certain causes of mortality in early infancy	43	43	—	—	—
Diabetes	39	—	2	10	27
Chronic lung diseases (bronchitis, emphysema, and asthma)	31	—	1	9	21
Cirrhosis of the liver	30	—	6	17	7
Suicide	22	—	10	8	4
Congenital anomalies	18	13	3	1	1
Homicide	15	—	11	3	1
All other causes	187	10	27	51	100
	1,923	86	176	475	1,187

[a]1968.

Source: Data from Reference 12.

pressure in the pulmonary artery, in partial compensation for lowered oxygen uptake by blood passing through the lung. The need for increased pressure in the pulmonary artery forces the right ventricle to increase its work, with consequent thickening, dilation, and ultimate failure of the right ventricle.[14]

PERMANENT DISABILITY

Not all illnesses, of course, are fatal. Table III shows the major causes of permanent total disability among preretirement age men (data for women, unfortunately, are not available) for which the Social Security Administration awarded support in the years 1959–1962.

Because they exclude women and those men not covered by Social Security, these figures must be regarded as serious underestimates of the numbers of people under 65 years becoming permanently unable to work due to physical or mental disease. It is useful, however, to note changes in the relative importance of different conditions in these as compared with mortality data. In particular, important categories of illness are apparent in the disability data which are not major direct causes of mortality, such as mental disorders, nervous system diseases, and arthritis.

LIMITATION OF ACTIVITY

Very much larger numbers of people, although not permanently incapacitated, are lim-

ited, by chronic conditions, in the amount and kinds of activity they can perform. Table IV contains data from a 1969–1970 survey[17] of U.S. households on the prevalence and severity of limitation of activity from different causes. Because this was a household survey, it should be borne in mind that institutionalized individuals were excluded.

RECURRENT SUBJECTIVE DISCOMFORT

Even these figures, however, do not reflect the full impact of chronic disease. Shown in Table V, for illustration, are the results of a similar survey[18] to the one measuring limitation of activity, only this time the people were asked how frequently they were "bothered" (not further defined in interviews) by different chronic respiratory conditions. Note the differences between these data and those concerning the effects of respiratory conditions on limitation of activity.

ACUTE, SHORT-TERM ILLNESS

Nearly all people from time to time are restricted in their activities by various acute conditions (Table VI). On the average, people of all ages suffer 3.8 days of bed disability per year from acute conditions, chiefly upper respiratory ailments of short duration.

CANCER

Cancer will be discussed in depth later, so it is useful here to summarize its effects on soci-

III Numbers of Men Under 65 Granted Disability Awards for Various Conditions, 1959–1962

Condition		Thousands of awards
Cardiovascular disease		261
Chronic lung diseases		65
Emphysema	52	
Occupational pneumoconioses	10	
Bronchitis	3	
Mental disorders (except mental deficiency)		61
Cancer		60
Nervous system diseases[a]		51
Pulmonary tuberculosis and other infectious and parasitic diseases		43
Arthritis		39
Diabetes		16
Other		95
		691

[a]Encephalopathy, Parkinson's disease, multiple sclerosis, epilepsy, etc.
Source: Data from Reference 15.

IV Millions of People with Limitation of Activity from Selected Classes of Chronic Conditions, According to Degree of Limitation and Age, 1969–1970

Condition	Degree of limitation				Age distribution[a]		
	All degrees	Severe[b]	Considerable[c]	Moderate[d]	Under 45	45–65	Over 65
Cardiovascular diseases	6.2	2.1	3.2	0.8	0.7	2.5	2.9
Physical or sensory impairments[e]	6.0	1.4	2.8	1.7	2.4	1.9	1.7
Musculoskeletal disorders	4.2	1.0	2.5	0.7	0.7	1.7	1.8
Arthritis and rheumatism 3.3							
All other 0.9							
Chronic respiratory diseases	2.5	0.7	1.2	0.6	1.0	0.8	0.7
Asthma 1.0							
Emphysema 0.6							
Bronchitis 0.2							
Upper respiratory 0.3							
Other 0.5							
Mental and nervous conditions	1.0	0.2	0.5	0.4	0.4	0.4	0.2
Diabetes	0.9	0.3	0.4	0.1	0.1	0.4	0.4
Cancer	0.4	n.a.[f]	0.2	0.1	0.1	0.2	0.1
Other	2.3	n.a.	n.a.	n.a.	n.a.	n.a.	n.a.
Total	23.5	5.7	12.3	5.3	7.3	8.0	8.0

[a]Of people with all degrees of activity limitation.
[b]Unable to carry on major activity (ability to work, keep house, or engage in school or preschool activities).
[c]With limitation in amount or kind of major activity.
[d]With limitation, but not in major activity.
[e]Chronic or permanent defects, usually static in nature, representing decrease or loss of ability to perform various functions of the sense organs and musculoskeletal system (visual, 1.1; hearing, 0.4; back or spine, 1.6; other, 2.8).
[f]n.a. Data not available.
Source: Data from References 16 and 17.

ety. Assuming that the current trend of cancer rates continues, more than 25 percent (53 million) of the 210 million people now living in the United States will develop some form of malignant neoplasm. With available methods of treatment, 35 million of these can be expected to die of cancer.[19] In 1972, approximately 610,000 new cases of cancer were diagnosed and 1,000,000 were under treatment.[20]

Even looked at in a narrow economic sense, the overall impact of cancer is very substantial. The direct and indirect costs of cancer, including loss of earnings during illness and during the balance of normal life expectancy, have

V Millions of People Affected with Different Frequencies by Chronic Respiratory Conditions, 1970

Chronic condition	Frequency of "bother"					
	All the time	Often	Once in a while	Not bothered	Unknown or not specified	Total
Chronic sinusitis	2.1	3.4	13.8	0.2	1.1	20.6
Asthma with or without hay fever	0.9	1.3	3.1	0.1	0.7	6.0
Hay fever without asthma	0.7	1.6	6.7	0.1	1.7	10.8
Emphysema	0.6	0.1	0.3	0.1	0.2	1.3
Chronic bronchitis	0.4	0.7	4.3	0.4	0.7	6.5
Other	—	—	—	—	—	4.1

Source: Reference 18.

been estimated at a total of $15 billion for 1971.[20] On the basis of analysis of selected cases, it appears that the total direct costs for a particular patient may range from $5,000 to over $20,000 at least.

MUTATIONS AND GENETIC CONTRIBUTIONS

There are three main categories of genetic contribution to disease:

1. Gross chromosome abnormalities, leading, in most cases to multiple congenital anomalies and early death — present in one-fourth of all unintentional abortions and affecting about 0.5 percent of all live births.
2. The numerous, but individually rare, disorders caused directly by simply heritable single genes — affecting approximately 1 percent of all births.[21] Nearly 2,000 single-gene diseases have been described,[22] including, for example, muscular dystrophy, sickle-cell anemia, many types of cataracts, and hemophilia.
3. Disorders caused by interactions among many genes or by combinations of genetic and environmental influences — affecting a larger but unknown percentage of the population.

In aggregate, Joshua Lederberg[23] has estimated that:

If we give proper weight to the genetic component of many common diseases which have a more complex etiology than the textbook examples of Mendelian defects, we can calculate that at least 25 percent of our health burden is of genetic origin. This figure is a very conservative estimate in view of the genetic component of such griefs as schizophrenia, diabetes, atherosclerosis, mental retardation, early senility, and many congenital malformations. In fact, the genetic factor in disease is bound to increase to an even larger proportion, for as we deal with infectious disease and other environmental insults, the genetic legacy of the species will compete only with traumatic accidents as the major factor in health.

Because individuals with genetic diseases generally have fewer offspring, such genes tend to be eliminated by natural selection. These deleterious genes may be maintained in the population either by compensating fertility advantages to some bearers of the genes (as in sickle-cell anemia, which confers resistance to malaria) or by the process of random mutation of normal genes to the defective types. All chromosome abnormalities and most dominant genes that cause simply heritable diseases are probably the result of continuous production by new mutations.

A mutation is defined as any inherited change in the genetic material. This may be a chemical transformation at a single DNA base, designated a *point mutation*, which may cause the affected protein to have altered function, or alternatively the change may involve a rearrangement, or a gain or loss, of large parts of chromosomes. These may be microscopically visible and are referred to as *chromosome aberrations*. Mutations may occur in somatic (body) cells or in germ cells responsible for reproduction of the organism.

It seems likely that some subclass of mutations that occurs in somatic cells may be involved in the generation of cancer, since to a

VI Millions of Days of Bed Disability Associated with Acute Conditions, United States, 1970

Acute conditions	Under 16 yr.	17–44 yr.	45 yr.	All ages
Respiratory	156	127	127	410
Upper respiratory 150				
Influenza 200				
Other 59				
Injuries	16	43	40	99
Infective and parasitic diseases	58	23	14	95
Digestive-system conditions	10	15	20	45
All other acute conditions	24	50	36	110

Source: Reference 18.

large degree the same types of chemical agents can cause both mutations and cancer and since it is increasingly being found that carcinogenic materials, like mutagens, in general can be shown to interact with the genetic material *in vivo*.

Mutations that occur in germ cells, of course, have the greatest potential for ultimate harm, as those changes may be transmitted to future generations. The severity of the effects of any particular mutation and the number of descendants ultimately affected depend on the particular change in genetic information that has occurred. Major change, such as most gross chromosome abnormalities, manifest themselves as early fetal loss (abortions) or serious dominant genetic disorders with a lethal or sterilizing effect, which thus persist only one generation. On the other hand, mutations that cause only slight impairment of health or are recessive may be transmitted through numerous generations and affect many different individuals. If, as suggested by experiments in *Drosophila*, mildly harmful mutations occur with much greater frequency than more severe mutants, the major impact of a mutational increase might be a subtle and multigeneration addition to the prevalence of adverse health effects.

BIRTH DEFECTS

Teratology is the study of congenital malformations (birth defects). Birth defects are generally defined as structural abnormalities which can be recognized at or shortly after birth and which can cause disability or death. Less restrictedly, teratology also includes microscopical, biochemical, and functional abnormalities that arise during gestation. Such defects, occurring long after conception, do not involve genetic change in reproductive cells and thus are not genetically heritable, although genes can influence the likelihood of birth defects.

The incidence of human congenital malformations is unknown in the absence of a comprehensive national surveillance system and registry; it has been variously estimated as ranging from 3 to 7 percent of total live births. Congenital malformations pose incalculable personal, familial and social stresses. The financial cost to society of one seriously retarded child approximates $250,000, computed on the basis of remedial and custodial care alone, excluding deprivation of earnings.[24]

CONCLUSION

No inference should be drawn from this discussion that environmental chemical pollutants are known to make substantial contributions to all, or even most of the disease conditions listed above. These data do, however, illustrate:

1. The massive importance and costs of adverse health effects for society.
2. The many-faceted nature of ill health, and the impossibility of adequately characterizing it by any single measure, such as death rates.
3. The inherent defects and uncertainties in available statistics.

Reasons for Concern About Pollutants

It has already been pointed out that in recent years we have seen massive increases in the numbers and quantities of synthetic chemicals produced in the United States, and that in general there is little or no information about possible hazards from these novel substances. Now we shall explore some of the reasons for concern that toxic effects of various categories are likely to be occurring.

ACUTE TOXICITY

The most obvious way in which people are adversely affected by chemicals is acute poisoning. About 5,000 deaths per year in the U.S.A. are officially attributed to this cause,[25] and there is reason to believe this may be too conservative a figure. The numbers of nonfatal poisonings and other toxic reactions from chemical consumer products in the home each year are listed in Table VII.

The majority of these poisonings occur in children under the age of 5. Prevention of accidental acute poisonings of children from consumer products requires repackaging and the use of less hazardous materials.

Acute toxic effects from chemical pollutants also occur with considerable frequency as a result of dermal (skin) and inhalation exposures in the course of work. Statistics in this area, however, are inadequate to allow de-

205

velopment of reasonably quantitative estimates of prevalence and severity.

More subtle than direct acute poisoning are the indirect acute (and chronic) effects of community air pollutants, chiefly sulfur dioxide and particulates, on the incidence of common respiratory diseases. Although it is well recognized that these are primarily infectious conditions, increased frequencies of acute respiratory illnesses have been clearly associated with increased pollutant levels in both chidren and adults, after corrections for weather and other variables.[27] Results of a recent study[28] indicating the magnitude of the excess of acute respiratory illnesses with higher pollutant levels for different family-member groups in Chicago and New York are shown in Table

VIII. In several of the adult groups, apparent increases are more than 30 percent.

CHRONIC TOXICITY

Chronic toxicity can be any manifestation of an adverse health effect from a chemical or physical pollutant which develops slowly over time. This can occur either when the pollutant itself (or a metabolite) is poorly excreted and builds up to high concentrations in sensitive organs with repeated exposure, or when single or repeated exposures to the pollutant produce irreversible damage in small incremental steps.

The first type of toxicant is typified by heavy metals such as lead and mercury. Because of the slow accumulation over months of exposure, and the gradual onset of nonspecific

VII **Poisonings and Other Toxic Reactions from Chemical Consumer Products per Year**

	Number injured
Laundering, cleaning, and polishing products	
Poisonings from soaps, detergents, cleaners	40,000
Poisonings from bleaches	35,000
Poisonings from disinfectants and deodorizers	20,000
Poisonings from furniture polish	20,000
Poisonings from lye corrosives	15,000
Other poisonings from laundering, cleaning, and polishing products	20,000
Other injuries from laundering, cleaning, and polishing products	100,000
	250,000
Pesticides	
Poisonings from insecticides	35,000
Poisonings from rodenticides	20,000
Other poisonings from pesticides	20,000
	75,000
Cosmetics[a]	
Injuries from perfume and toilet water	20,000
Injuries from lotions and creams	10,000
Other injuries from cosmetics	30,000
	60,000
Miscellaneous	
Poisonings from airplane glue	25,000
Poisonings from lighter fluid	15,000
Poisonings from kerosene	15,000
Poisonings from turpentine	25,000
Poisonings from other flammable liquids	10,000
	90,000
Total all products	475,000

[a]Injuries from cosmetics are chiefly skin eruptions, loss of hair, severe allergic reactions, etc., sufficiently serious to restrict activity for 1 day or require medical attention.
Source: Data from Reference 26.

symptoms (gastric distress, irritability), lead and mercury toxicity is often unrecognized for long periods. Ultimate damage is generally to the nervous system and the kidneys. Mental retardation is a well-recognized consequence in many cases of excessive lead absorption in children. Additionally, recent evidence has associated the childhood hyperactivity syndrome with increased body burdens of lead, in the absence of a previous history of frank poisoning.[29]

Many kinds of dust present in the work environment cause long-term debilitating lung diseases, called *pneumoconioses*.[30] About 125,000 coal miners suffer from coal workers' pneumoconiosis ("black lung"), and the disease is estimated to contribute to about 3,000 to 4,000 deaths each year.[8] Similar conditions are asbestosis from asbestos, byssinosis from cotton, bagassosis from sugar cane, and silicosis from various silica-containing dusts. Overall, 9,000 deaths per year have been attributed to these occupational dust diseases.

The chronic lung diseases that are widely prevalent in the population as a whole — bronchitis, emphysema, and asthma — are also exacerbated by occupational and community

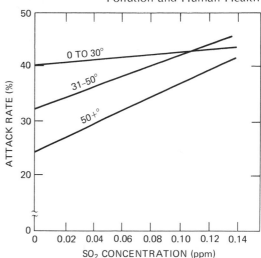

2 **Relationship Between Asthma and Pollution.**
Shown is the asthma attack rate versus daily sulfur dioxide concentration within three temperature ranges. *Source:* Reference 31; by permission.

air pollutants. Such effects often interact with the effects of weather conditions, which also affect the severity of chronic respiratory illnesses. Shown in Figure 2, for example, is an observed interaction between temperature and

VIII   Acute Respiratory Disease Attack Rates

| | Community air pollution exposure[a] | | |
	Low	Intermediate	Highest
Chicago			
Fathers	—	1.00[b] (2.80)[c]	1.33
Mothers	—	1.00 (4.76)	1.25
Older siblings	—	1.00 (7.04)	1.18
Nursery school students	—	1.00 (0.35)	1.02
Younger siblings	—	1.00 (9.41)	1.37
New York			
Fathers	1.00 (1.58)	1.41	—
Mothers	1.00 (1.72)	1.55	—
School children	1.00 (3.97)	1.09	—
Preschool children	1.00 (6.12)	1.10	—

After corrections for the effects of smoking.
[a]Families lived at least 3 years in these communities.
[b]For example, Chicago fathers living in areas with the highest air pollution suffered 1.33 times as many acute respiratory illnesses as their counterparts in areas with intermediate pollution; that is, the illness rate increased by one-third.
[c]Figures in parentheses indicate the base rate of respiratory illness per 100 person-weeks of observation; for example, Chicago fathers in intermediate air pollution areas experienced an average of 2.8 attacks per 100 weeks.
Source: Data from Reference 28; copyright 1973, American Medical Association.

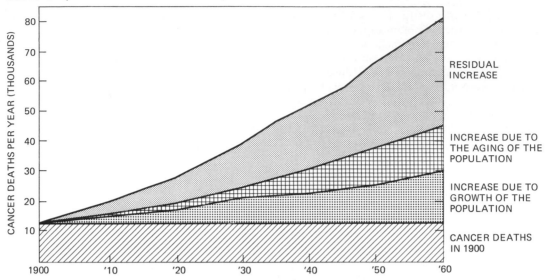

3   **Deaths from Cancer.** The number of people dying from cancer each year has been increasing. Some of the increase appears to be caused by environmental factors. *Source:* Reference 32.

sulfur dioxide in influencing the incidence of asthma attacks. At lower temperatures, SO_2 had no effect; at higher temperatures, increased SO_2 levels were strongly associated with increased numbers of attacks.

CANCER

Overall rates of cancer have been increasing in recent decades. Figure 3 indicates that this increase in new cancer cases is real and is over and above that due to the increase in average age of the population. That many cancers are environmental in origin, and hence ultimately preventable, is further suggested by the wide variations in the rates of particular kinds of cancer in different areas of the world. In a number of studies, the role of specific environmental carcinogens contributing to this variation has been implicated or identified.[33] For example, the high incidence of cancer of the oral cavity in Asia, representing some 35 percent of all Asiatic cancers, in contrast to 1 percent of all European cancers, is clearly related to the chewing of betel nuts and tobacco leaves. The high incidence of liver cancer in the Bantu and in Guam may well be due, respectively, to dietary contamination with aflatoxin, a potent fungal carcinogen, and to eating cycad plants, which contain azoxyglucoside carcinogens. The high incidence of gastric cancer in Japan, Iceland, and Chile has

been associated with high dietary intake of fish; suggestions have been made implicating nitrosamines, formed by reactions between secondary amines in fish and nitrite preservatives.[5] The high incidence of cancer of the esophagus, in Zambians drinking Kachasu spirits and in the Calvados area of France and in other clearly defined geographic areas, may be related to contamination of alcoholic drinks or food with nitrosamines, some of which produce esophageal cancer in experimental animals. Tables IX and X show the range of variation between cancer death rates for common cancer sites for different countries.

Such large differences are unlikely to be due to discrepancies in diagnostic procedures. Furthermore, studies of people who have migrated from one nation to another show a tendency for the migrants to assume, although not always completely, the pattern of cancer deaths present in the host countries,[35,36] indicating that genetic differences between countries are unlikely to be a major source of the variation. Diagnostic and genetic differences not being clearly significant, there is now a general consensus, partly from such data, that *the great majority of cancers must be due to environmental factors*, which vary appreciably for populations in different parts of the world.[33] This is a profoundly important conclusion, perhaps the most significant in this chapter.

208

The best-documented and most significant data on carcinogenesis due to environmental chemicals are those on tobacco. It has been suspected for several decades that heavy tobacco smoking is directly and causally related to chronic lung disease, especially cancer. Over 29 retrospective epidemiologic studies on lung cancer, generally substantiating the causative role of smoking, have been published from 1939 to 1964.[37] These studies demonstrated unambiguously that the incidence of lung cancer was positively correlated with cigarette smoking. Mortality rates from lung cancer have increased exponentially over the last few decades, in both men and women, particularly in the United States, and have now reached epidemic proportions. There are also many studies which demonstrate the contributory role of urban air pollution in lung cancer, in addition to the identification of numerous classes of chemical carcinogens in polluted urban air. These studies have shown that there is an excess of lung cancer deaths in smokers living in polluted urban areas, when contrasted with those living in nonpolluted rural areas.[38]

More restricted data on chemical carcinogenesis in humans derive from followup studies on patients treated with carcinogenic drugs.[33] These include immunosuppressive agents, used in transplant therapy; alkylating agents, used in treatment of cancer and to a lesser extent autoimmune disease; radioactive phosphorus, used in treatment of polycythemia vera; estrogens, used for treatment of prostatic cancer and for hormone-replacement therapy in women; and, diethylstilbestrol, used for treatment of threatened abortions and as a "morning-after" pill.

Data on occupational carcinogenesis extend back to the eighteenth century, with the discovery by Sir Percival Pott of a high incidence of scrotal cancers in young British chimney sweeps exposed to soot. A very wide range of occupational cancers has since been identified and studied in detail.[33,39] These include bladder cancer in the aniline dye and rubber industry, caused by such chemicals as 2-naphthylamine,

IX Mortality Rates for High- and Low-Risk Populations and for U.S. White Males by Cancer Site, 1964–1965

Cancer site	High-risk population	U.S. white population	Low-risk population
Lung	75.6 (Scotland)	38.9	3.4 (nonsmoker, U.S. sample)
Colon	16.1 (Scotland)	13.7	3.3 (Japan)
Prostate	22.0 (U.S. nonwhite)	12.7	1.9 (Japan)
Stomach	68.6 (Japan)	9.4	9.4 (U.S. white)
Pancreas	9.6 (U.S. nonwhite)	8.1	3.8 (Italy)
Leukemia	8.6 (Denmark)	7.6	3.7 (Japan)
Bladder and kidney	8.3 (South Africa)	5.0	2.2 (Japan)
Oral cavity	9.2 (France)	4.4	1.4 (Japan)
Esophagus	13.3 (France)	3.3	2.3 (Norway)
Larynx	10.2 (France)	2.0	0.6 (Norway)
	241.5	105.1	32.0

Deaths per 100,000 population per year.

Source: Reference 34; by permission of Academic Press, Inc.

benzidine, 2-aminodiphenyl, and 2-nitrodiphenyl; lung cancer in uranium miners of Colorado (Chapter 12), in coke oven workers, in nitrogen mustard factories in Japan, and in U.S. workers even briefly exposed to bis(chloromethyl) ether; skin cancer in cutting and shale oil workers; nasal sinus cancer in woodworkers; lung cancer and pleural mesotheliomas in insulation workers and in others, such as construction workers, exposed to asbestos; cancer of the pancreas and lymphomas in organic chemists; and cancer of the cervix in prostitutes.

MUTATIONS

Ethyleneimines are good examples of highly mutagenic chemicals which are used for many purposes, including therapy of neoplastic and non-neoplastic diseases, chemosterilant insecticides, pigment dyeing and printing, fireproofing and crease proofing of fabrics and textiles, ingredients in solid rocket fuels, intermediates in many industrial syntheses, and as cross-linking agents in starches and shampoos.

Ethyleneimines have also been proposed for rodent and plant control.

Another example of a mutagen in common use until recently is trimethylphosphate (TMP), which was added to gasoline for controlling surface ignition and spark plug fouling due to the use of alkyl lead additives. It also is used as a methylating agent, chemical intermediate in the production of polymethyl polyphosphates, flame-retardant solvent for paints and polymers, and catalyst in preparation of polymers and resins. TMP is mutagenic in the dominant lethal assay in mice (see the section "Animal Testing") and produces chromosomal damage in bone marrow cells of rats following oral or parenteral administration of subtoxic doses. It is difficult, however, to estimate mutagenic hazards to which the general population has been exposed in the absence of information on the concentration of unreacted TMP and of biologically active pyrolytic products in automobile exhaust. Such problems hitherto have not received adequate consideration. According to a recent

X Mortality Rates for High- and Low-Risk Populations and for U.S. White Females by Cancer Site, 1964–1965

Cancer site	High-risk population	U.S. white population	Low-risk population
Breast	25.6 (Netherlands)	21.8	3.8 (Japan)
Colon	15.3 (Canada)	12.8	3.2 (Japan)
Ovary	11.1 (Denmark)	7.3	1.7 (Japan)
Cervix	12.9 (Venezuela)	6.1	1.0 (Israel)
Lung	11.4 (Scotland)	5.8	2.2 (Portugal)
Leukemia	5.7 (Israel)	5.0	2.7 (Chile)
Pancreas	5.7 (U.S. nonwhite)	4.8	2.3 (Italy)
Stomach	39.0 (Chile)	4.7	4.7 (U.S. white)
Bladder and kidney	2.6 (U.S. nonwhite)	1.7	1.0 (Japan)
Esophagus	5.3 (Chile)	0.8	0.8 (Italy)
	134.6	70.8	23.4

Deaths per 100,000 population per year.
Source: Reference 34; by permission of Academic Press, Inc.

statement by a major manufacturer: "We have not attempted to detect trimethylphosphate in exhaust, but it is reasonable to assume that a portion (maybe 1 percent) may appear unreacted in the exhaust." Similar considerations apply to fuel-additive formulations, based on triaryl phosphates, some of which have been used to replace TMP.

An additional source of prolonged exposure of large population groups to TMP results from the extensive use of the insecticide Dichlorvos, or Vapona (Shell No-Pest Strip). TMP is one of the usual ingredients of Vapona, being present as an impurity together with other unlabeled "related compounds" in commercial formulations. Additionally, Vapona itself is mutagenic in microbial systems. Most serious is the fact that Vapona itself alkylates DNA both *in vitro* and *in vivo*; that is, it reacts chemically with the genetic material. Most carcinogens and mutagens, or their proximate activated forms, are alkylating agents; hence the property of alkylation strongly suggests carcinogenic and mutagenic hazards in the absence of specific evidence to the contrary.

James Crow, an eminent population geneticist at the University of Wisconsin, has concluded that social priorities on mutagenic hazards have not been uniformly set between chemicals and radiation[40]:

There is reason to fear that some chemicals may constitute as important a risk as radiation, possibly a more serious one. Although knowledge of chemical mutagenesis in man is much less certain than that of radiation, a number of chemicals — some with widespread use — are known to induce genetic damage in some organisms. To consider only radiation hazards may be to ignore the submerged part of the iceberg.

BIRTH DEFECTS

Three major categories of human teratogens (agents that produce birth defects) have so far been identified: viral infections, x-irradiation, and chemicals, such as mercurials, thalidomide, and diethylstilbestrol.

Thalidomide is a drug that was widely marketed in Europe in the late 1950s and early 1960s for use by women for "morning sickness." An alert medical practitioner associated the drug with the birth of thousands of babies with malformed limbs of a peculiar type which had begun to appear. It should be noted that thalidomide was detected as a human teratogen only because of the dramatic and unusual nature of the birth defects it produced. If it had been responsible for a similar incidence of more common defects, such as congenital heart anomalies, in all likelihood it would still be in use as a "safe" drug to this date.

The phenoxy herbicides 2,4,5-T and 2,4-D have aroused considerable concern because of findings of teratogenicity.

Animal Testing

The most desirable way to investigate the potential effects of environmental chemicals is prior to human exposure, by appropriate animal testing. Before testing can begin, there are questions to be answered. Is it enough to test only the actual chemical in question? Do other substances need to be studied together with the material of prime interest? Experience has shown that toxicity testing cannot always be arbitrarily confined to a single parent chemical, and it is often important to collect information on chemical and metabolic derivatives of the parent compound, its pyrolytic and other degradation products, and contaminating chemicals. Examples are the occurrence of dioxin contaminants in phenoxy herbicides and cyclohexylamine as a contaminant and human metabolic product of the sweetening agent cyclamate. In both of these cases, the contaminants/metabolites are more toxic and of greater general concern than the parent materials.

ROUTES OF ADMINISTRATION

The route of administration of the chemical should, of course, reflect the routes of likely human exposure, although other routes can also be used to gather collateral information. The most frequent mode of administration of agents in current toxicological testing is by mouth, as is appropriate for testing of food contaminants, which are of major regulatory concern at present. For many environmental pollutants of the home and workplace, however, the most logical route for testing would appear to be inhalation. Respiratory exposure is of particular human significance for pesticide and other aerosols, and many exposures of workers to industrial chemicals are through the lungs. Surprisingly, there are as yet scanty toxicological data that utilize respiratory routes of exposure.

ACUTE AND SUBACUTE TOXICITY TESTS

In acute tests, the agent is administered on a single occasion. In subacute tests, the chemical is given repeatedly for brief periods, extending from 1 week to 3 months. Acute oral toxicity is an index of the immediate lethality of an agent, administered by mouth, and frequently constitutes the only toxicological data available on a particular material. The LD_{50} is that dose of chemical which kills 50 percent of the animals tested, by any practical route of administration (usually oral). Doses are generally expressed on a body weight basis; for example, Y mg/kg would be a dose of Y milligrams of chemical *per* kilogram of body weight of the animal. Note that with this way of expressing toxicity, the *lower* the LD_{50}, the *more acutely lethal* is the agent in question. Table XI gives approximate quantities necessary to produce death in people by materials of different toxicity, expressed in LD_{50}.

EYE IRRITATION

Short-term eye-irritation tests in animals are standard for agents, such as detergents, which are likely to be splashed into the eye. In the standard Federal Hazardous Substances Act (FHSA) Tests, a small volume, 0.1 ml, of the detergent is instilled into the conjunctival sac of six rabbits whose eyelids are held closed for 1 second. Eye damage is scored for up to 2 weeks. Variations of the eye-irritation test include washing out the test agent from the eye 1 second after its instillation, as rabbits do not have a protective tear response, as do primates and humans.

PRIMARY SKIN IRRITATION AND SENSITIZATION

Procedures for primary skin-irritation tests are based on application of pastes of the test agent, such as detergents, to the shaved skin of rabbits for periods up to 1 hour, followed by rinsing and observation for up to 3 days.

More modern procedures are based on patch tests in albino rabbits and human volunteers. Small volumes of solutions are applied to discrete areas on the intact and abraded shaved skin on the backs of albino rabbits. The solutions are allowed to dry and then covered with small patch bandages for 24 hours and the skin is then inspected for primary irritation. The reactions are graded from slight redness to swelling and blistering. Human volunteers can be similarly studied, sometimes with the repeated-insult method, in which serial applications of test solutions are made at the same site.

TOXICITY AND CARCINOGENICITY TESTS

The object of chronic toxicity tests is to determine the effects on experimental animals of life-long or long-term exposure to a particular chemical or product. It is standard practice to use at least two species, generally mice and rats, and four groups of animals; an untreated or solvent-treated control group, and high-, intermediate-, and low-dose test groups. Effects are measured by death and weight loss and also by more subtle changes, such as alterations in liver enzyme function. At the end of experiments, animals are carefully autopsied and various organs are histologically examined, particularly for the numbers and distribution of tumors and other lesions.

For this kind of testing, chemicals are usually administered at several concentrations, ranging up to much higher levels than those to which people are likely to be exposed. This has led to considerable confusion and controversy. High dosage levels are generally necessary in carcinogenicity tests to overcome statistical insensitivity inherent in attempting to detect effects in small groups of animals, as opposed to the very large populations of humans at presumptive risk. Generally, if a material is to be detected as carcinogenic in groups of approximately 50 animals, a dose must be found which induces tumors in 10 to 20 percent of those in the treated group.

XI   Quantities Required to Produce Lethal Effects

Acute oral toxicity, LD_{50} (mg/kg)	Quantity
Less than 5	A few drops
5–50	A pinch to one teaspoonful
50–500	1 teaspoonful to 2 teaspoonfuls
500–5,000	1 ounce to 1 pint (1 pound)
5,000–15,000	1 pint to 1 quart (2 pounds)

Source: Reference 41.

For illustration, Table XII shows data on the incidence of liver tumors from a 2-year feeding experiment in which the diets of rats were supplemented with various levels of dimethylnitrosamine. Note that although there is no obvious cutoff point below which tumors are not found, at lower doses the decreasing incidence of induced tumors makes the difference between experimental and control groups much more difficult to establish with certainty, especially with small numbers of animals. The one animal developing a liver tumor at the 2-part-per-million level in this experiment is not sufficient to prove that that was not a spontaneous event. A much larger group of animals, possibly in the thousands, would be necessary to measure the frequency of tumors at the 2-ppm level with any accuracy. The task of detecting infrequent carcinogenic effects becomes practically impossible as, with even lower doses than this, the incidence drops even further. If at some dose in the parts-per-billion range the incidence were 0.05 percent, a group of 10,000 rats would only contain five rats with induced tumors. Even an incidence as difficult to detect experimentally as this, however, would be an effect of very great significance to the human population of the United States. Among 200,000,000 Americans, 0.05 percent translates into 100,000 people.

It should be emphasized that testing at high dosages does *not* produce false positive carcinogenic results. There is no basis whatsoever for the allegation that all chemicals are car-cinogenic at high doses.[33] For example, in a study by the Bionetics Laboratory,[43] 140 different pesticides were tested orally in mice of both sexes and two separate strains at maximally tolerated doses from the first week of life until sacrifice at 18 months. In this extensive series of investigations, less than 10 percent of the tested compounds were found to be carcinogenic.

TERATOGENICITY TESTS

The ability of chemicals or other agents to produce birth defects is experimentally determined by their administration to pregnant animals, generally rodents, during the phase of active organ development of the growing embryo, organogenesis. Agents are administered to animals by oral, or less commonly by other routes, either on a single occasion or repeatedly during this period. Shortly before anticipated birth, embryos are harvested by caesarean section and examined. Parameters considered in test and concurrent control groups include the incidence of abnormal fetuses per litter, the incidence of specific congenital abnormalities, the incidence of fetal mortality, maternal weight gains in pregnancy, and maternal and fetal organ/body weight ratios. Additionally, some pregnant animals may be allowed to give birth in order to identify abnormalities that may otherwise be manifest only in the perinatal period or in subsequent adult life.

Teratogenic responses may be widely vari-

XII Incidence of Liver Tumors from Various Levels of Dimethylnitrosamine in the Diet

Concentration of dimethylnitrosamine[a] in diet (parts per million)	No. animals in group	No. rats with liver tumors	Rats with liver tumors (%)
0 (control)	41	0	0
2	37	1	3
5	68	5	7
10	5	2	40
20	23	15	65
50	12	10	80

Groups of male and female animals have been combined.
[a]Dimethylnitrosamine is one of the carcinogenic substances mentioned earlier which can result from the interaction of nitrite and secondary amines, in this case dimethylamine.
Source: Data from Reference 42; by permission of the editor and publishers of the *British Journal of Cancer.*

able between different species, and in any particular instance, humans may be more or less sensitive than the usually tested rodents. Meclizine, for example, a drug used in treatment of morning sickness, is teratogenic in the rat but not apparently in the small number of women studied.[44] By contrast, for thalidomide the lowest effective human teratogenic dose is 0.5 mg/kg body weight/day; but the corresponding values for the mouse, rat, dog, and hamster are 30, 50, 100, and 350 mg/kg/day, respectively.[45] Thus humans are 60 to 700 times more sensitive to thalidomide than these other species.

MUTAGENICITY TESTING[46]

The most generally applicable test for mutagenic activity of chemicals in mammals, the dominant lethal test, detects largely or solely gross structural or numerical chromosome aberrations. The agent to be tested is given to male mice or rats either once or several times within a few days or weeks. Following the chemical dosing, the treated males are mated for several weeks with untreated females. The pregnancies are interrupted at midterm, and the number of fetuses that have died early in pregnancy is counted. Any increase in early fetal deaths is taken as an indication that the sperm producing those offspring carried an increased frequency of dominant lethal mutations. The time period following chemical administrations where the increase occurs can be taken as an indication of the stage(s) of spermatogenesis which are susceptible to the mutagenic action of the chemical under consideration. Other mammalian mutagenesis tests include the host-mediated assay (in which increased mutation frequency is measured in indicator bacteria or yeasts injected into an animal dosed with the test agent) and *in vivo* cytogenetics (in which dividing cells from lymphocytes, bone marrow, or testes are microscopically examined for chromosome abnormalities).

LIMITATIONS IN TESTING PROCEDURES

If these methods are used to screen new chemical additions to commerce in the future, much human suffering may be avoided. However, as can be seen, they have many limitations, including the previously stressed statistical insensitivity attendant on the use of small groups of animals.

An additional difficulty, of particular concern, is a tendency for the effects of different environmental chemicals to interact in poorly foreseeable ways. Such interactions can be either additive, antagonistic (effects tending to cancel one another), or *synergistic* (combined effects greater than the simple sum of the individual effects). Several examples of synergism can be cited. Malathion is a pesticide which ordinarily has very low acute toxicity (LD_{50} greater than 2,000 mg/kg) because it is normally rapidly converted to less harmful metabolites by mammalian liver enzyme systems. When the metabolizing enzymes are inhibited by some other pesticides of the same general class, though, the toxicity of malathion greatly increases, and human fatalities have been recorded, on occasion, in farm workers. Owing to less well understood interactions, the carcinogenicity of benzo(*a*)pyrene or benzo(*a*)anthracene when applied to mouse skin is increased 1,000-fold when they are dissolved in noncarcinogenic *n*-dodecane.[47] Intratracheal instillation of benzo(*a*)pyrene and ferric oxide in adult hamsters elicits a high incidence of lower respiratory tract tumors only in animals pretreated at birth with a single low dose of diethylnitrosamine. Nutrition can also play a substantial modifying role in the results of animal testing. The acute toxicity of the fungicide captan is increased 26-fold in rats on protein-deficient diets and more than 2,000-fold in rats completely deprived of protein for 4 weeks.[48] With the wide variability in nutrition among the human population, and the simultaneous exposures of people to large numbers of potentially interacting chemical agents, the results of single-chemical testing in otherwise healthy, well-nourished animals must clearly be regarded as tentative and incomplete indicators of possible hazards associated with actual human exposure.

Epidemiological Methods

Even with well-planned and well-executed toxicological testing, it is likely that unexpected adverse effects from toxic chemicals will occur in humans, reflecting the insensitivity or inappropriateness of test systems. Occasionally, epidemiological studies of human populations may provide *post hoc* information on hazards from chemicals already present in the environment.

Epidemiological methods are basically statistical procedures designed to:

1. Detect clusters of cases of particular diseases which would be unlikely to occur by chance (such clusters may be either in time or space).
2. Associate the clusters with unusual exposures to suspect environmental agents.

Several examples of successful epidemiological linkages between chemical pollutants and particular diseases have been given earlier in this chapter (e.g., smoking and lung cancer). Such linkages can be made in either

1. Retrospective studies — in which one "looks backward" to compare groups of people who have already manifested a particular disease with control populations to pick out those environmental factors which occur more frequently in the histories of those with the disease, or
2. Prospective studies — in which one begins with a general population without disease, and monitors the characteristics of those who develop the disease in the course of the study.

It is important, however, to emphasize the difficulties and limitations in epidemiological approaches, as the absence of data definitely establishing possible human health effects for particular chemicals is often inappropriately used to support assertions that health effects do not occur. For example, industrial proponents of the continued use of the chlorinated hydrocarbon insecticides, DDT and dieldrin, have frequently attempted to minimize the significance of data showing that DDT and dieldrin are carcinogenic in rodents by making statements to the effect that, as yet, no cases of human cancer have been clearly linked to these agents. Such statements ignore the fact that the following very special set of circumstances must prevail before epidemiological

identification of adverse health effects (such as carcinogenic) from any particular environmental pollutant is possible:

1. There must be clearly definable differences in exposure to the agent among different groups of people large enough to study. For example, it was relatively easy to define population groups with graded exposures to cigarettes by simple questionnaires. For such a substance as a widely used food additive, pesticide, a minor but relatively ubiquitous air pollutant, or a contaminant from car upholstery, defining groups of people with substantially different degrees of exposure may often be difficult or impossible. In the absence of such definable exposure differentials, even if the agent is responsible for substantial portions of particular adverse health effects, there is no way to establish this if populations differing in their risk cannot be defined or do not exist.
2. Sufficient time must have elapsed for the effects of the agent to be manifested in the exposed populations. For diseases with long latent periods, such as cancer and pneumoconioses, the latent period may extend to two or three decades. For mutations, the latent periods are measured in generations.
3. The amount of some specific disease attributable to the environmental agent must be substantial when compared to other causes of the same disease. Generally if an agent is responsible for only a few percent of the cases within any specific disease category in the population at presumptive risk, random statistical variations, measurement errors, and variability in the presence of the major causes of the disease will completely obscure the contribution of the environmental agent, even though in absolute terms that contribution may be far from negligible. Just how great an increased risk must occur to be detectable will, of course, depend on the variability of the other factors mentioned, but the most easily recognized effects will always be large relative increases in the incidence of what are otherwise very rare conditions, such as mesotheliomas among asbestos workers, the thalidomide babies, or liver

215

angiosarcoma among vinyl chloride workers. Smaller relative increases in the incidence of common diseases are most likely to be completely undetected, even though many more people may be adversely affected by such increases.

4. A competent epidemiologist must, in fact, uncover the association. With the large number of important health parameters outlined in previous sections of this chapter, and the even larger number of possible suspect chemical pollutants, it is clear that only a tiny fraction of possible associations can be investigated. In all likelihood, many associations between health effects and environmental chemicals are missed simply for lack of suitable study.

Certainly, it cannot be expected that this combination of factors will be present in every case where environmental agents cause disease in the human population. The fact of long, apparently safe, use of a particular chemical must therefore be interpreted with caution, particularly in the presence of experimental data from other species which indicates the possibility of some insidious hazard.

The Social Gamble

There is no way for society to reap the benefits of technological innovation without incurring some risk of unanticipated, undesirable side effects. Still the magnitude of what is at stake in the post-World War II mass introductions of new chemical agents is enough to give pause to the hardiest of gamblers. What criteria should be used in determining which gambles should be accepted?

RISKS

Clearly there should be different standards for risks that are assumed voluntarily and with full knowledge by individuals, and those over which people have little or no individual control. In cases where people can be expected to be able to make choices based on reasonably complete information and their own subjective values, the rights of individuals to choose the risks they wish to accept must be respected. For example, to the extent that smokers, knowing the hazards of cigarettes, neverthe-

less wish to continue smoking, it would be as inappropriate for society to attempt to force other behavior on them as it would be politically and technically impossible. On the other hand, it is not inappropriate for society to

1. Protect the rights of nonsmokers not to be exposed involuntarily to cigarette smoke.
2. Vigorously educate the community on the suicidal nature of smoking.
3. Prevent the use of public media for manipulative conditioning of people, particularly young people, to smoke.
4. Assist in the development of cigarettes with less risk of adverse health effects.

The great majority of health risks from environmental pollutants are, however, for practical purposes, involuntary. By and large, individual people do not know, and cannot readily (if at all) choose to avoid, health hazards such as those from pesticides in foods, community air pollutants, or chemical contaminants in the places where they work. In such cases it is imperative that society ensure that precautions are taken to minimize such risks through governmental regulations.

EFFICACY AND SOCIAL UTILITY

One important concept that has as yet found only limited regulatory application is, if society is going to accept some risk from a new chemical product, there should be a showing of clear social benefit to be derived from that product. There are two classes of questions of relevance with regard to efficacy and social utility. First, does the innovation or chemical product achieve its narrowly stated objective, as dictated by the Federal Trade Commission's requirements for truthful advertising? Is there evidence that the product does, in fact, do whatever it explicitly claims to do? Unfortunately, even in regulatory areas such as drugs, where strong legal requirements for proof of this kind of efficacy marketing were imposed by the Kefauver amendments of 1962, enforcement has been less than optimal. A report of the National Academy of Sciences in 1968 concluded that for about 60 percent of prescription drugs and combination antibiotics in common use, there was no evidence to support a conclusion of efficacy. Additionally, 80 percent of all label claims for prescription

216

drugs were found not to be substantiated. There is also similar evidence on the lack of efficacy of many animal feed additives.[49]

A second more demanding test of societal efficacy has recently been proposed by Senator Gaylord Nelson with respect to food additives. That is, does the additive accomplish a technical effect which is of broad social utility and which cannot be accomplished as well, more safely, or less expensively by existing alternatives? Excluded by such criteria, for example, are certain food additives with solely cosmetic effects, such as the food dye Citrus Red 2, used to redden Florida oranges, and nitrate–nitrite additives used to improve the appearance of cured meats. These improve marketability of products by catering to consumer preferences which are probably mainly artificially engendered by promotional practices. Artificial sweetening agents such as saccharin are also open to some question with regard to general usefulness. Saccharin does substitute for a very few calories, but the actual amount of weight reduction achievable in this way is quite small.

OCCUPATIONAL EXPOSURES TO POLLUTANTS

A very unfortunate ethical dichotomy currently exists between the regulatory practices for protection of the general public against potential adverse health effects from environmental pollutants, and occupational regulatory practices for protection of workers who are in general exposed for prolonged periods to very much larger amounts of specific pollutants.[50,51] There is no general requirement for safety testing and standard setting for industrial chemicals before worker exposure. Out of the tens of thousands of toxic substances to be found in the workplace, official standards for worker exposure exist for less than 500. Even these few standards are set, in general, only with a view toward preventing the most obvious and immediate toxic effects, and little or no margin of safety is allowed between the minimal levels of pollutants known to produce such effects and permissible levels. This often results in a wide disparity between the levels considered acceptable for the general community. Table XIII shows some examples.

SAFETY TESTING PRIOR TO HUMAN EXPOSURE[49]

It is currently required that food additives, pesticide residues, and drugs be tested for safety by the manufacturers of these products prior to governmental registration and marketing, using some of the animal tests outlined above. (Mutagenicity testing is not yet a reg-

XIII   Environmental Versus Occupational Standards

Pollutant	Environmental standard[a]		Occupational standard[b]
Sulfur dioxide (in parts per million)	Annual arithmetic mean	0.03	5
	Maximum 24 hr once/year	0.14	
	Maximum 3 hr once/year	0.50	
Carbon monoxide (in parts per million)	Maximum 8 hr once/year	9	50
	Maximum 1 hr once/year	35	
Nitrogen dioxide (in parts per million)	Annual arithmetic mean	0.5	5
Particulates (in milligrams per cubic meter)	Annual geometric mean	0.075	Respirable fraction 5
	Maximum 24 hr once/year	0.26	Total dust 15
Lead (in micrograms per cubic meter)	30-day mean	1.5	200

[a]These are Environmental Protection Agency standards for community air pollution exposure, except for the lead standard, which is from the California Air Resources Board.
[b]These are Occupational Safety and Health Administration Standards, based on 8 hours per day of exposure.
Source: After Reference 50; Copyright Scientists' Institute for Public Information, 1973.

ulatory requirement, nor are carcinogenicity tests uniformly required for all categories of these products.) Such registration requirements should be extended to cosmetics, detergents, miscellaneous household products, and, as previously stated, generally to all industrial chemicals with which people come into contact. However, the current system of having manufacturers themselves test their products either in in-house laboratories or by contract has lent itself to some abuse in the past. One possible remedy would be the introduction of a disinterested advisory group or agency as an intermediary between manufacturers who pay for tests and the commercial testing laboratories. Manufacturers could notify the intermediary group when safety evaluation was required for a particular product, and the intermediary group would then solicit contract bids on the open market. Bids would be awarded on the basis of economics, quality of protocols, and technical competence. At the conclusion of the studies, the advisory group would comment on the quality of the data, make appropriate recommendations, and forward these to the concerned regulatory agency for routine testing. This approach minimizes constraints on objectivity in generation of the data and would serve to upgrade the quality of testing. Additionally, claims of efficacy could be validated prior to marketing by the same mechanism.

Any positive findings of carcinogenicity, mutagenicity, or teratogenicity in the animal testing should, in the absence of extraordinary circumstances, be grounds for requiring the most rigorous exclusion of the agent from human contact. As described above, these tests are all quite insensitive and are only capable of detecting activity of materials which induce very large incidences of cancer, mutations, or birth defects in the test animals. Furthermore, there is every reason to believe that the initiating event in carcinogenesis and mutagenesis may be a single molecular interaction of a chemical agent with the genetic material (DNA). This being the case, effects may be expected to occur, with decreasing frequency, however low the dosage to people is made. At present there is no scientific basis for specifying a safe level for any carcinogen, mutagen, or teratogen. Particularly in view of the variation in susceptibility between differ-

ent species and the long time that would elapse before even severe effects in humans could be detected epidemiologically, the risk to society from large-scale exposure to such materials would almost invariably be greater than any likely benefit from a new chemical product.

PUBLIC ACCESS TO INFORMATION[49]

If there is to be any hope of linking particular current chemical exposures with current health effects, there must be much more publicly available information on the composition and properties of household, industrial, and other chemical products. Data on the composition of particular products are now generally regarded as confidential trade information. The lack of information on exposures of people precludes many types of epidemiological investigations, deprives experimental toxicologists of knowledge of what chemicals are most worthy of immediate investigation in animal studies, and precludes public choice of safer materials in the marketplace and citizen action to stimulate the elimination of unjustifiably hazardous materials by regulatory agencies. All safety and efficacy data gathered in the course of the scientific investigations leading to registration of products should also be available for public review. This is currently true of data supporting registrations of pesticides and food additives but not drugs.

PUBLIC INTEREST ADVOCACY[49]

In addition to open access to data on all issues of public safety, it is important that the consumer and public interest be adequately represented at all informal and formal stages of the decision-making process and in agency–industry discussions. Decisions made by regulatory agencies solely under the lobbying influence of affected industries have a poor history. It is clear that the system of checks and balances, essential to the democratic process, is largely absent from current regulatory practice. Apart from limited *post hoc* recourse, the citizen and consumer, and those who represent their interests, scientifically and legally, are virtually excluded from anticipatory involvement in decisions vitally affecting them. (The establishment of a Consumer Protection Agency, as proposed by Senator Abraham Ribicoff in S707 in February 1973, would be a great step forward in this regard. This agency would represent general citizen interests in the bureaucracy and support the efforts of independent citizens' groups.) The concept of matching benefits against risk has been generally applied to maximize short-term benefits to industry, even though this may entail minimal benefits and maximal risks to the consumer. Although such an approach is of course detrimental to the consumer, it is also often detrimental to the long-term interests of industry, which may suffer major economic dislocation when hazardous products, to which it has improperly developed major commitments, are belatedly banned from commerce. Such problems are in large measure attributable to crippling constraints which have developed and which still dominate the decision-making process within regulatory agencies. Responsibility for these constraints must be shared with regulatory agencies, by the legislature, by the scientific community, and by consumers and citizens, who have not yet developed adequate mechanisms for protecting their own vital rights and interests.

References

1. Control of lead additives in gasoline — fuel regulations, *Federal Register 38:* 33734 (Dec. 6, 1973); EPA's position on the health implications of airborne lead, *Federal Register 38* (Nov. 28, 1973). Available from the Publications Section, Environmental Protection Agency, 401 M Street SW, Room 238W, Washington, D.C.

2. Hueper, C. 1966. *Occupational and Environmental Cancers of the Respiratory System.* Springer-Verlag New York, Inc., New York.

3. Tseng, W. P., et al. 1968. Prevalence of skin cancer in an endemic area of chronic arsenicism in Taiwan. *J. Natl. Cancer Inst. 40:* 453.

4. Lee, K. M., and Fraumeni, J. F. 1969. Arsenic and respiratory cancer in man: an occupational study. *J. Natl. Cancer Inst. 42:* 1045.

5. Lijinsky, W., and Epstein, S. S. 1970. Nitrosamines as environmental carcinogens. *Nature 225:* 21.

6. Magee, P. N. 1971. Toxicity of ni

trosamines: their possible human health hazards. *Food Cosmetics Toxicol. 9:* 207–218.

7. Food Protection Committee, National Academy of Sciences. 1973. *Toxicants Occurring Naturally in Foods,* 2nd ed. The Academy, Washington, D.C.

8. Panel on Chemicals and Health of the President's Science Advisory Committee. 1973. *Chemicals and Health.* Science and Technology Policy Office, National Science Foundation, Washington, D.C.

9. *Chem Sources — USA,* 1974 ed., Directory Publishing Company, Inc., Flemington, N.J.

10. U.S. Tariff Commission. 1973. *Synthetic Organic Chemicals — United States Production and Sales, 1970* (TC Publ. 479). Government Printing Office, Washington, D.C.

11. Redmond, C. K., et al. 1972. Long-term mortality study of steelworkers. VI. Mortality from malignant neoplasms among coke oven workers. *J. Occup. Med. 14:* 621.

12. National Center for Health Statistics, U.S. Public Health Service, 1969. *Vital Statistics.* Government Printing Office, Washington, D.C.

13. Moriyama, I. N., et al. 1966. Evaluation of diagnostic information supporting medical certification of deaths from cardiovascular disease. In *Epidemiological Study of Cancer and Other Chronic Diseases* (Monograph 19), National Cancer Institute, Bethesda, Md.

14. Filley, G. F. 1972. The effects of chronic respiratory disease on the function of the lungs and heart. In *Environmental Factors in Respiratory Disease* (D. H. K. Lee, ed.). Academic Press, Inc., New York.

15. U.S. Department of Health, Education, and Welfare. 1967. *Occupational Characteristics of Disabled Workers, by Disabling Condition* (Public Health Service Publ. 1531). Government Printing Office, Washington, D.C.

16. U.S. Public Health Service. 1973. *Limitation of Activity Due to Chronic Conditions, United States, 1969 and 1970* (Ser. 10, No. 80; DHEW Publ. HSM-73-1506). Government Printing Office, Washington, D.C.

17. U.S. Public Health Service. 1973. *Prevalence of Selected Chronic Respiratory Conditions, United States — 1970* (Ser. 10, No. 84, DHEW Publ. HRA-74-1511). Government Printing Office, Washington, D.C.

18. U.S. Public Health Service. 1972. *Current Estimates, from the Health Interview Survey, United States — 1970* (Ser. 10, No. 72, DHEW Publ. HSM-72-1054). Government Printing Office, Washington, D.C.

19. American Cancer Society. 1973. *Cancer Facts and Figures — 1973.* The Society, New York.

20. National Cancer Program. 1973. *The Strategic Plan* (DHEW Publ. NIH-74-569). Government Printing Office, Washington, D.C.

21. *Report of the United Nations Scientific Committee on the Effects of Atomic Radiation.* 1958. General Assembly, Official Records, 13th Session, Suppl. 17 (A/3838). U.N., New York.

22. McKusick, V. A. 1971. *Mendelian Inheritance in Man.* The Johns Hopkins Press, Baltimore, Md.

23. Lederberg, J. 1971. Forward to *Mutagenicity of Pesticides: Concepts and Evaluation* (S. S. Epstein and M. Legator, eds.). The MIT Press, Cambridge, Mass.

24. Oberle, M. W. Lead poisoning: a preventable childhood disease of the slums. *Science 165:* 991–992.

25. National Safety Council. 1972. *Accident Facts.* Chicago.

26. National Commission on Product Safety. 1970. *Final Report.* Government Printing Office, Washington, D.C.

27. Higgins, I. T. T., and Ferris, B. G., Jr. 1973. Epidemiology of sulphur oxides and particulates. In *Proceedings of the Conference on Health Effects of Air Pollutants.* Assembly of Life Sciences, National Academy of Sciences–National Research Council (prepared for the Committee on Public Works, U.S. Senate), Washington, D.C.

28. French, J. G., et al. 1973. The effect of sulfur dioxide and suspended sulfates on acute respiratory disease. *Arch. Environ. Health 27.*

29. David, O., et al. 1972. Lead and hyperactivity. *Lancet 2:* 900.

30. Selikoff, I. L. 1972. Occupational lung diseases. In *Environmental Factors in Respiratory Disease.* Academic Press, Inc., New York.

31. Cohen, A. A., et al. 1972. Asthma and air pollution from a coal-fueled power plant. *Amer. J. Public Health 62:* 1181.

32. U.S. Public Health Service. 1964. *Cancer Rates and Risks* (PHS Publ. 1148). Government Printing Office, Washington, D.C.

33. Epstein, S. S. 1974. Environmental determinants of human cancer. *Cancer Res. 34:* 2425.

34. Wynder, E. L., and Mabuchi, K. 1972. Etiological and preventive aspects of human cancer. *Preventive Med. 1:* 300.

35. Buell, P. 1973. Changing incidence of breast cancer in Japanese-American women. *J. Natl. Cancer Inst. 51:* 1479.

36. Reid, D. C., et al. 1966. Studies of disease among migrants and native population in Great Britain, Norway and the United States: III. Prevalence of cardiorespiratory symptoms among migrants and native born in the United States. *Natl. Cancer Inst. Monog. 19:* 321.

37. U.S. Department of Health, Education, and Welfare. 1964. *Smoking and Health* (Report of the Advisory Committee to the Surgeon General of the PHS; PHS Publ. 1103). Government Printing Office, Washington, D.C.

38. National Academy of Sciences. 1972. *Particulate Polycyclic Organic Matter.* The Academy, Washington, D.C. p. 213.

39. Hueper, W. C. 1972. Environmental cancer hazards. *J. Occup. Med. 14:* 150.

40. Crow, J. F. 1968. Chemical risk to future generations. *Scientist and Citizen 10:* 113.

41. Hayes, W. J. 1963. *Clinical Handbook on Economic Poisons* (PHS Publ. 476). Government Printing Office, Washington, D.C.

42. Terracini, B., et al. 1967. Hepatic pathology in rats in low dietary levels of dimethylnitrosamine. *Brit. J. Cancer 21:* 559.

43. Innes, R., et al. 1969. Bioassay of pesticides and industrial chemicals for tumorigenicity in mice: a preliminary note. *J. Natl. Cancer Inst. 42:* 1101.

44. Yerushalamy, J., and Milkovich, L. 1965. Evaluation of the teratogenic effects of meclizine. *Amer. J. Obstet. Gynecol. 93:* 553.

45. Kalter, H. 1968. *Teratology of the Central Nervous System.* University of Chicago Press, Chicago.

46. Hollaender, A. 1971. *Chemical Mutagens — Principles and Methods for Their Detection,* Vols. I and II. Plenum Publishing Corporation, New York.

47. Bingham, E., and Falk, H. L. 1969. Environmental carcinogens. Modifying effect of carcinogens on the threshold response. *Arch. Environ. Health 19: 779.*

48. Boyd, E. M., et al. 1970. Endosulfan toxicity and dietary protein. *Arch. Environ. Health 21:* 15.

49. Epstein, S. S. and Grundy, R. D., eds. 1974. *The Legislation of Product Safety: Consumer Health and Product Hazards.* Vol. 1. *Chemicals, Electronic Products, Radiation.* Vol. 2. *Cosmetics and Drugs, Pesticides, Food Additives.* MIT Press, Cambridge, Mass.

50. Commoner, B. 1973. Workplace burden. *Environment 19*(6): 15.

51. Epstein, S. S. Current problems with occupational health standards. Testimony before the U.S. House of Representatives Committee on Education and Labor, Select Committee on Labor, April 25, 1974.

Dense black smoke rises
from burning methane
(natural gas) in a Libyan
oilfield. Photo by Georg
Gerster/Rapho-Guillumette.

9

Air Pollution

David A. Lynn *is a Senior Research Engineer and Statistician with the GCA Corporation in Bedford, Massachusetts. He received undergraduate degrees in physics and mathematics from Heidelberg College (1960) and civil engineering from Case Institute of Technology (1962). After a brief stint with a sanitary engineering design firm, he joined the air pollution division of the U. S. Public Health Service, the forerunner of the current federal air pollution program in the Environmental Protection Agency. As head of a group working with ambient air pollution data, he became concerned with the use of such information in formulating public policy, and returned for graduate work in statistics at Harvard University. In the last three years, he has resumed his work in federal air pollution research, doing contract research in the Technology Division of GCA.*

Like all problems, the problem of air pollution can be looked at from a variety of viewpoints, and the results can be quite different. Yet all views may have equal validity, for such is the nature of a complicated society. In the broad environmental viewpoint that directs the overall contents of this volume, the pollution of our air is one of several ways in which our urban, industrialized society wastefully disposes of the by-products derived from our resources. From a narrow, technical viewpoint, one held predominately for many decades and still fairly prevalent today, the air pollution problem in an engineering problem, and a very challenging one of late — the problem of designing or modifying the processes and equipment that we use for industrial production and energy conversion in such a way that pollution is reduced. This chapter is written from another viewpoint, one intermediate to these. Air pollution can be viewed as a single societal problem in and of itself — with particular reasons for controlling pollution; groups in the society who hold varying pro and con positions; a sizable body of technical, especially engineering, knowledge; and a body of legal, governmental, and other social arrangements within which the problem and its solution are considered and decisions made.

These viewpoints are of course simply differences in the degree of detail involved. In the pursuit of broad environmental solutions, we must turn to air pollution, water pollution, noise, solid waste disposal, each in some degree of detailed isolation; similarly, in solving the air pollution problem, we must turn at some point to the requisite body of detailed technical knowledge. The broad environmental overview, if the reader is to find it, will emerge from one's own synthesis of the material in these chapters. So here I view air pollution as a problem unto itself to some extent; and although I consider the social rather than the technical problems as the key to progress, I shall emphasize the concrete physical facts about air pollution — the sources, chemistry, and technical control of emissions — because the philosophical, social, and technological roots of the problem are considered elsewhere.

DEFINITION

A typical definition says that air pollution is the presence in the outdoor atmosphere of one or more substances, put there directly or indirectly by an act of man, in such an amount as to interfere with his health or welfare, or the full use and enjoyment of his property. Each of the phrases in this definition has something to say about the scope of our subject, and each is subject to change as the scope changes. In limiting our topic to the outdoor atmosphere, we have excluded the workplace and the indoor domestic air environments. A worker's air is studied and regulated by those in the field of occupational health; a homemaker's is not considered very thoroughly by anyone at present but is probably gradually coming under the aegis of the air pollution control field. The word "substances" excludes such phenomena as noise, an environmental concern that may yet come to be part of air pollution, and radio and television signals, about which silence is the wisest position. Not long ago, the phrase "foreign substance" would probably have been used, but the case of carbon dioxide, a perfectly normal atmospheric constituent, has made that point obsolete. Similarly becoming obsolete is the qualification "in such an amount as to interfere"; since we are becoming increasingly aware of the very subtle involvement of our species and our wastes in the overall ecological context, many people are becoming much less willing to draw an arbitrary line between harmless and harmful amounts of chemicals in the air. The phrase "by an act of man" is also one that poses problems of fuzziness. Included primarily to distinguish between smokestacks and volcanoes, the phrase still leaves us uncertain when we consider such phenomena as forest fires and the Oklahoma dust bowls.

BRIEF HISTORY

Early historical references to air pollution are scattered from Roman times through medieval times to the beginning of the Industrial Revolution. It is apparent that whenever enough people came to live closely enough together, their activities, including their cooking and heating fires, managed to bother each other, and at least sometimes the problem was bad enough to find its way into the limited writings of the day. With the passing of medieval society, the potential for air pollution problems as we now know them began to slowly accelerate with the coming of the Industrial Revolution. As agriculture began to give way to crafts and later to industrialized manufacture, the cities began to grow, and the artisans and merchants, and later the industrialists and their workers, not only crowded together in a newer, denser type of habitat, but in the process of his daily living and working, each used fuel for cooking and heating far beyond his former consumption. A parallel development was the replacement of wood by coal as the dominant fuel of European civilization. As the forests of Europe were depleted, increasing population and industrialization spurred the development of coal resources, which in turn led to continued growth of industrial society.

By 1661, the situation was bad enough in London to prompt a classic essay[1] by John Evelyn, one of the prominent scientists of the day. Through various reigns, there were occasional attempts to regulate coal use, often with harsh penalties, but the rewards of industrialization were great, and the problem persisted. In England, "modern" laws concerning industrial furnaces were promulgated in 1820, and as the United States joined in industrialization, smoke control laws began in American cities in 1881. Until the mid-twentieth century, the air pollution problem was primarily one of the black smoke and soot and the sulfurous odors that resulted from the burning of coal. The word "smog" was coined in London sometime around the turn of the century, as a contraction of smoke and fog, to define the effect of the famous London fogs combined with the effluents from the myriad of coal-burning fireplaces that Londoners used in the heating of their homes.

The visible thread of history through these times is a series of unusually severe episodes of smog, much as political or military history is so often seen as a series of distinct events and prominent, named, battles. In the early part of the twentieth century, the focus was on the London smog, seemingly worse with each winter. Then in December 1930, a five-day smog episode in the industrial Meuse River valley of Belgium killed 63 people and reminded us that London was not unique. Heightened citizen concern through the Depression years culminated in this country in major smoke control efforts, in St. Louis in the late 1930s and after World War II in Pittsburgh and other areas. Known for decades as the Smoky City, Pittsburgh's problem had been so bad that on occasion street lights and auto headlights were needed at midday.

The late 1940s saw the recognition, if not the beginning, of another of the major air pollution problems that we have at the present time. The Los Angeles Basin of southern California had a major share of the growth that was stimulated by World War II; and after the war, the inhabitants began to notice a haze in the air that obscured the nearby mountains and which made one's eyes water and feel irritated. After a new state law in 1947, and some serious control of traditional sources of pollution, it became obvious that the Los Angeles smog was not the same as the traditional air pollution in other cities. Although not clearly understood then, it has since been discovered to be the by-product of a totally different set of pollutants, which originate largely from automobiles and react photochemically in the air to create the smog.

Not to be upstaged by a newcomer, however, the London smog reasserted its presence. In 1950, the worst documented episode in London killed 4,000 people and led to the adoption of the present air pollution control laws of the United Kingdom. In 1949, the first clear killer smog in the United States struck the tiny (12,000) industrial valley town of Donora, Pennsylvania, not far south of Pittsburgh. Within 2 days 40 percent of the population was sick, and by the third day, 20 had died. The Donora episode did, however, have the positive effect of prompting the federal government to begin the air pollution program currently conducted by the Environmental Protection Agency (EPA).

Similar episodes have occurred since but none of the magnitude or clarity of the earlier ones. And at some point in the 1950s and early 1960s, the general concept of air pollution as a subject jelled into a certain form, which it still largely maintains, or has maintained until the early 1970s, when the awakening of a broader environmental awareness began to revise this into part of a more general perspective.

Several general areas of knowledge and expertise make up the structure of the air pollution problem: (1) identification of the pollutants and their sources, (2) the meteorology of the atmosphere (which affects how the pollutants are transported from the source to the receptor), (3) the effects of the pollution on the health and welfare of the receptor (human or otherwise), and (4) control of the pollution, which I would subdivide into (a) the actual engineering control of the pollutants at their source and (b) the social control or legal–governmental process which ensures that engineering control takes place (Chapter 17). Except for the health effects of the pollutants, these areas lie ahead of us, one by one, through the rest of this chapter. The health effects of air pollutants are in some cases hard to separate from other chemical and biological insults that we have chosen to put into our environment, and in any case ought not be treated in isolation, so these topics have been largely set aside for treatment within the overall context of environmental health presented in Chapter 8.

Pollution and Its Sources

The myriad of chemicals that we put into our air can be classified in many ways for many purposes. Some are solid particles, ranging from simple carbonaceous black soot to complex organic compounds and heavy metals. There is a wide variety of gases and some liquids. Most pollutants are emitted into the air directly from human activities, but some very important ones are formed in the air from other chemicals, often with sunlight providing the energy. Some are seasonal concerns, others year-round problems, and some are greater problems in one part of the country than another. Probably the simplest overview classifies the overall air pollution problem into three fairly distinct problems: "London smog," "Los Angeles" or photochemical smog, and the problem of specifically toxic substances.

London smog (which occurs in cities all over the world) is of concern because it has a wide variety of adverse effects. As evidenced in several air pollution episodes, it can have severe acute effects, including fatalities, especially among the aged, very young infants, and persons with respiratory handicaps. At lower, less acute, exposures over longer periods of time, it can also cause, or significantly contribute to, a variety of chronic lung and respiratory ailments, including emphysema, chronic bronchitis, and lung cancer. It also has adverse effects on materials, damaging metal, stone, paint, and other surfaces, and on a variety of plants; and it involves haze and smoke and other visibility and aesthetic detriments.[2,3] This type of smog is primarily a problem in dense industrial urban areas that use large quantities of fuel, especially coal and heavy oil. It is more of a problem in the midwestern and northeastern portions of the country, where such fuels are widely used, and during the winter, when space-heating requirements increase the total amount of fuel used.

Photochemical smog, on the other hand, is a problem only in the summer, generally May through October, and more in the Southwest and in southern California than elsewhere; this is because formation of the smog depends on sunlight to provide energy, as its name implies. Because the principal ingredients come primarily from motor vehicles, it is also primarily an urban and suburban problem. Photochemical smog also has a wide variety of health and other effects. At midafternoon on a bad day, acute smog levels cause severe eye irritation, a very noticeable haze, and the aggravation of a number of the symptoms and difficulties of respiratory patients. It has fairly severe effects on certain plants, especially in the economically important citrus groves, and it has adverse effects on some materials, especially organic fabrics and dyes.[4] In all, the effects seem less drastic than those of London smog in the Northeast, but they occur in parts of the country that were relatively pristine and clear within the memory of people still alive, and so seem to them much worse.

The third problem is actually a collection of specific problems that are more or less similar in structure if not in detail. Many chemical substances are specifically toxic in the human system, participating in the body's biochemistry, in contrast to the components of the two types of smog, which are primarily respiratory irritants. The variety of such biochemical poisons is great, as is the nature of their sources, distribution, and effects. The two most common, carbon monoxide and lead, both come from automobiles and so are problems in most sizable urban areas. Carbon monoxide has only one effect, its biochemical toxicity. It reacts with the blood's hemoglobin, reducing its ability to carry oxygen to the body tissues. Similarly, the other compounds in this third problem category typically have only their biochemical toxicity to give them a place as pollutants of concern. Lead, widespread because of its use as an antiknock additive in gasoline, accumulates in the body and involves itself in enzyme biochemistry; other heavy metals — mercury, beryllium, cadmium, nickel — have similar effects. Asbestos, the mineral used for heat insulation, can cause lung cancer, as can polycyclic organic matter, complex organic chemicals formed during coal combustion.

Thus dividing the air pollution problem into three smaller problems helps in understanding the total problem, but action to reduce the pollution levels and the adverse effects they cause requires us to consider the specific emission of particular substances from specific sources. To do this, we turn first to a more detailed classification of airborne substances, focusing on their physical and chemical properties and the sources from which they come. The concept of a pollutant as it is used at this level of detail is somewhat ambiguous. Although it may imply a single chemical compound, the word "pollutant" is also used to embrace combinations and classes of chemicals

of various sizes, ranging from a very few to the very large category that encompasses all nongaseous matter, regardless of chemical composition.

The federal EPA has singled out nine such pollutants for primary control and abatement efforts at the present — six as major ubiquitous problems, the other three as less widespread but far more hazardous materials. Two of the major pollutants are the principal constituents of London smog: *sulfur oxides* and *particulate matter*. Three others are associated with photochemical smog: *hydrocarbons* and *nitrogen oxides* are the principal ingredients of the smog milieu, and *photochemical oxidant* is the principal secondary pollutant formed in the air by the photochemical reactions. The sixth major pollutant is *carbon monoxide;* it is one of the third group of pollution problems, a material with specific biochemical toxicity.

The other three substances that EPA regulates are asbestos, beryllium, and mercury; these are officially designated as Hazardous Pollutants, meaning that they are very toxic or harmful and must be regulated closely, although they are not necessarily very prevalent. Asbestos is of concern because it can cause a particular form of lung cancer (Chapter 8). Beryllium and mercury are highly toxic metals, able to interfere with body enzymes; they are not by any means common air pollutants, although mercury is a serious water pollutant.

With the exception of carbon monoxide and possibly lead, however, none of these biochemically toxic pollutants are anywhere nearly as widespread as the more major pollutants, and they are generally considered of much less concern. Some view them as at best marginal in terms of being prevalent enough in the air to actually cause effects and justify use of the term "pollutant." Others are more cognizant of the possibility of very long term cumulative health effects, food-chain concentration and ecological effects, especially on a global scale, and tend to view these, especially the heavy metals, much more seriously. The careful focus of these concerns on solutions is handicapped by the fact that they are often simultaneously air, water, and food contaminants, with the resultant total body burden being the factor that governs the degree of effect noted. Resolution of this type of fragmentation and the problems it causes is one of the primary benefits to be gained from a total-environmental viewpoint, so we may anticipate growing efforts to deal with such problems.

Nonetheless, it is still true that the vast majority of the air pollution research and development efforts, and essentially all of the actual pollution control effort, is directed at the six major pollutants, or more accurately at the five of them that are primary pollutants, emitted by human activity. Table I lists estimated national emissions of these major pollutants from various categories of human activity, and the subsequent discussion considers them one by one in terms of their chemical and physical nature and the circumstances of their emission.

PARTICULATE MATTER

Beginning with particulate matter, we note first that although the name comes from the word "particle," the commonly accepted definition includes not only solid particles but also what little liquid there is in the air (not counting rain or fog). Note also that nothing is said about the chemical identity of the particles or droplets; particulate matter is an aggregate of a great number of things, from simple carbon smoke to road dust and soil, from pollens and insect wings to toxic metals and asbestos fibers.

Aside from natural sources, of which only windblown dust is occasionally a problem, most of the particulate matter comes from industrial processes and fuel combustion, as shown in Table I. The particulate matter emitted from fuel combustion sources is either black smoke and soot from the incomplete burning of the carbon in the fuel, or suspended ash resulting from the incombustible material in the fuel; both are more of a problem with coal than with other fuels.[6]

Industrial process emissions, as distinct from the industrial use of fuels, are quite varied. As seen in Table II, the most significant sources are those industries that handle dry, granular materials, such as stone- and rock-crushing operations, portland cement mills, grain-handling operations, and so on, while some seemingly major industries, such as oil refining and almost all manufacturing industries, are not major sources of process-loss particulate emissions. Some of the latter, of course, do cause problems from fuel consumption or, as in automobile manufacturing, from other indirect problems.

The estimated national total of particulate emissions is about 25 million tons per year, or about 250 pounds per capita. In the case of particulates, it turns out, however, that the amount directly emitted by particulate sources does not provide a good indication of the particulate problem and all the ramifications involved. Because of size factors, the particulates that we breathe include only a part of those emitted; more significantly, what we breathe does include some particulates that are not directly emitted but are formed in the air — such as nitrates formed in photochemical smog and sulfates formed from emitted SO_2.

The most important property of particulate pollution is the size of the particles (more precisely, their equivalent aerodynamic size, which also involves the particle's shape and density), because that determines how long they will remain suspended in the air, and consequently how far away from their source they will disperse. The largest, heaviest particles settle out relatively close to the source, causing problems of laundry and windowsill

I National Emission Estimates, 1970*

	Sulfur Oxides	Particulate Matter	Nitrogen Oxides	Hydro-carbons	Carbon Monoxide	Total
Fuel combustion in stationary sources	26.5	6.8	10.0	0.6	0.8	44.7
Transportation	1.0	0.7	11.7	19.4	110.8	143.6
Solid waste disposal	0.1	1.4	0.4	2.0	7.2	11.1
Industrial process losses	6.0	13.3	0.2	9.5	11.4	40.4
Agricultural burning	0.0	2.4	0.3	2.8	13.8	19.3
Miscellaneous man-made	0.3	1.0	0.1	0.4	3.0	4.8
	33.9	25.6	22.7	34.7	147.0	263.9

*Figures given in millions of tons per year. *Source:* Reference 5.

II Particulate Emissions from Industrial Process Losses*

	Emissions (thousands of tons per year)	
Industry type	1950	1970
Iron and steel industry		
Blast furnace	1,560	74
Open-hearth furnace	437	20
Basic oxygen process	Negligible	15
Electric arc furnace	21	20
Coke production		
Beehive ovens	580	140
By-product ovens	123	166
Mining, miscellaneous	805	1,650
	3,526	2,085
Primary metal smelting		
Aluminum	30	125
Copper	107	280
Lead	15	16
Zinc	200	20
	352	441
Secondary metal processing		
Gray iron foundries	—	233
Aluminum	1	51
Lead	11	4
Zinc	4	5
Copper	13	51
Brass and bronze	11	—
	40	344
Mineral products and industries		
Cement	961	1,120
Sand and gravel	2	47
Stone and rock crushing	2,330	4,190
Lime	559	1,060
Carbon black	486	72
Coal cleaning	19	218
Other	115	794
	4,472	7,501
Miscellaneous		
Oil refining	24	42
Other chemical industry	Negligible	14
Cotton ginning	77	21
Pulp and paper	903	549
Grain and feed handling	1,410	2,103
Ferroalloys	—	47
Silicon metals	—	17
Fertilizers	—	166
	2,414	2,959
Grand total (rounded)	10,800	13,300

*As distinguished from the use of fuel by industry. The increases from 1950 to 1970 generally reflect growth in an industry; the decreases are usually the result of process improvements. *Source:* Reference 5.

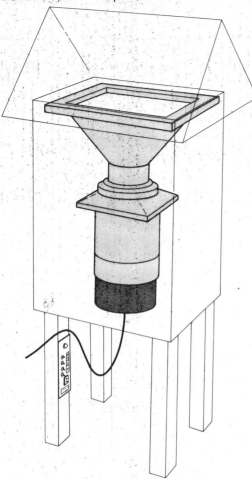

1 **High-Volume Air Sampler.** Uses a powerful motor to draw air through a glass-fiber filter, which collects the suspended particulate matter.

soiling, deterioration of materials, and the like. These aspects of particulate pollution are particularly apparent and annoying to people, and since the very large particles are relatively easy to prevent or to collect with pollution control devices, this dustfall pollution, as it is called, is less of a problem than formerly. Since it is concentrated in the older, more industrialized areas of cities, however, it is a sizable contributor to the image of center cities as dirty, squalid, slummy places to live, and hence to the social problems that such neighborhoods support. Dustfall levels are measured simply by setting out a wide-mouthed jar or "dustfall bucket" and weighing the collected material. This relatively unglamorous method perhaps contributes to the general neglect and lack of concern over dustfall, which I feel is unfortunate.

Somewhat smaller particles, ranging very roughly from 5 to 10 micrometers down to less than 1 micrometer in diameter, can be bounced about by turbulence in the air and remain suspended for long periods, the smaller ones essentially indefinitely. These particles are called suspended particulate matter, and they are measured by filtering them out of the air with a motor-filter arrangement called a high-volume sampler (Figure 1). A typical level in an urban setting would be about 100 micrograms of particulate matter per cubic meter of air ($\mu g/m^3$); this ranges from 5 to 10 $\mu g/m^3$ in very remote nonurban areas to 300 to 400 $\mu g/m^3$ in a dirty, coal-burning city on a bad day.[7]

The size of the particles affects not only their ability to stay suspended in the air, but also their ability to be deposited in the human lung. The largest particles, 5 to 10 micrometers and more, are deposited in the upper portion of the respiratory tract, where they are most easily removed by natural clearance mechanisms; the smaller particles can penetrate much more deeply, are less readily cleared, and consequently cause disproportionate damage. I have always found it interesting that the lungs can best handle the size range of particles that includes those naturally produced and have the most trouble with the smaller, man-made particles.

In particular, among the very smallest particles, in the 0.1- to 0.3-micrometer range, are particles that are produced in the air from gaseous pollutants rather than being emitted as particles directly from sources. Called secondary particles, these include most prominently organic nitrate particles, which are produced in photochemical smog and are believed responsible for its visible haze, and sulfate-sulfuric acid particles that are formed from the SO_2 and other pollutants emitted in London smog and believed responsible for most of the associated lung irritation.

SULFUR OXIDES

The other major pollutant involved in London smog, sulfur oxides, is also a combination of several chemicals but much more closely related ones than in the case of particulates. Sulfur oxides include sulfur dioxide (SO_2) and sulfur trioxide (SO_3), both gases, and various sulfates (SO_4^{2-} ion), including particulate sulfate salts and sulfuric acid, which is present either as an extremely fine mist or adsorbed onto particles. The various oxides of sulfur are considered together because, as we just noted, they can be changed about by atmospheric oxidation and reaction with other substances.

As seen in Table I, the vast majority of sulfur oxides are emitted from fuel combustion activities, the direct result of burning sulfur-bearing fuels. The sulfur in the fuel is oxidized, mostly to SO_2, with small amounts of SO_3 and even less sulfate, as the exhaust gases are emitted up the smokestack. As the plume rises and disperses, however, and apparently for some time afterward, the sulfur dioxide continues to oxidize, and in other, only partially understood processes, reacts to form various sulfates. The sulfate particles and acid mist are very small and remain suspended very well; in polar regions of the earth, a large share of the low particulate levels there is sulfate. Even when removed from the atmosphere, the sulfurous pollutants can have effects; being quite soluble, sulfates and SO_2 are easily removed by the rain, which then becomes acid. With enough sulfurous fuel combustion and enough acid rain, the acidity of surface waters goes up. This has actually been observed in the rivers and lakes of Europe.[8]

Although it has other effects, the biggest concern with London smog, or sulfurous smog, is in its potential for adverse effects on human health. The smog irritates the lungs, both in chronic long-term exposures and in more acute, short-term ones, such as the various killer-smog episodes throughout history. It has not been rigorously demonstrated with scientific precision just how much of the harm is done by particulates alone, SO_2 alone, by the two working together, or by the sulfates and acid mist. There is essentially unquestionable evidence from real-life observation, however, that the overall combination causes irritation of the lungs that can kill outright in extreme cases or, more commonly, lead to progressive deterioration of the lungs, resulting in disorders such as chronic bronchitis or pulmo-nary emphysema. Hence both sulfur oxides and particulates are regarded as major pollutants, and in my opinion, are the most serious air pollution problem.

HYDROCARBONS, NITROGEN OXIDES, OXIDANTS

The three major pollutants that are identified with Los Angeles or photochemical smog are also groups of chemicals; except for nitrogen oxides, they are large, complicated groups, and they are involved in fairly complicated atmospheric chemistry. The quick mnemonic device about "hydrocarbons plus nitrogen oxides produce oxidants" is, while true, grossly simplified. Although very much is known, many of the finer details of smog photochemistry are still not understood, even after 20 years of research.

The "pollutant" nitrogen oxides include nitric oxide (NO) and nitrogen dioxide (NO_2), both gases. The nitrogen oxides are formed primarily in fuel combustion processes, not from anything in the fuel but by the chemical combination of oxygen and nitrogen from the atmosphere. This oxidation requires very high temperatures, however, which occur only in the largest, most efficient electric power plants and in the ordinary internal combustion engines we use in our cars. The total estimated national emissions of about 23 million tons per year are less than any of the other major pollutants, and nitrogen oxides is also the only major pollutant that is not clearly either a stationary- or a mobile-source problem but a 50:50 combination (Table I).

The other major ingredient of the photochemical smog, or precursor of the photochemical oxidant, to be more precise, is the class of hydrocarbons. In an air pollution context, the word "hydrocarbons" means those organic chemicals composed of only carbon and hydrogen that are gaseous at atmospheric conditions; this is much narrower than the whole group of organic chemicals, or even than the entire class of hydrocarbons, as a chemist would define them. The air pollutant nonetheless involves dozens of specific chemical species. And to make matters worse, some of the hydrocarbons are more important than others in producing smog, by virtue of being

231

more reactive in the air. For example, the single most prevalent atmospheric hydrocarbon, methane, does not participate in the least in the smog reactions. Thus when we use the term "hydrocarbons" we are glossing over a great many definitional and measurement problems. This is done not only in this brief chapter, where space might be a justification, but also in regulatory and research efforts, just as a matter of history and sheer necessity.

Hydrocarbons are essentially an automobile pollution problem; over half of the total emissions (Table I) are directly from automobiles, as unburned or incompletely burned products of the gasoline combustion, and sizable portions of the industrial process loss emissions are from petroleum refining and from gasoline distribution and retail marketing, the latter simply being gasoline vapors spilled or otherwise lost while filling tank trucks or an individual car's tank.

The third pollutant involved in the smog, photochemical oxidant, is the principal product formed by the photochemical reactions between the nitrogen oxides and the hydrocarbons. Oxidant is also a mixture of chemicals, some of them not yet identified. The major portion of the oxidant is ozone (O_3), a form of oxygen that has an extra atom. Other chemicals that are oxidants, that is, chemicals which have an oxidation reactivity greater than oxygen and enough to show up in the measuring instrument, are also included in the term "total oxidants." This implies that the definition of the pollutant depends on the instrument we use to measure it, which is, in fact, accurate, unfortunately. The next most common component of the photochemical oxidant is peroxyacetyl nitrate, or PAN, and there are many others, mostly complex organic compounds.

It is not clear which of the oxidants causes the eye-irritation symptoms of photochemical smog in humans, although it seems clear that it is not ozone, the major constituent. Both ozone and PAN are, however, the cause of much of the vegetation damage caused by the smog, and ozone is believed to be responsible for the lung-irritation effects. The visible haze is believed to be caused by organic nitrates in the form of fine particles that can scatter light.

As we noted, the chemistry of the photochemical formation of oxidants is a good bit more complicated than the simple "hydrocarbons plus nitrogen oxides produce oxidants." In fact, the vast majority of the oxidants, the ozone, is obviously formed from atmospheric oxygen, with no hydrocarbon or nitrogen in the molecule. The discussion of the chemistry of photochemical smog that follows is oversimplified but hopefully will serve to suggest some of the subtle complexities that are not unusual in chemical interactions with nature.

PHOTOCHEMISTRY

The very name "photochemical smog" indicates that the chemical reactions are driven by energy from sunlight. In an atmosphere that contains nitrogen oxides, sunlight is absorbed by nitrogen dioxide, as we see in Figure 2(a). If the absorbed energy is high enough, the NO_2 molecule dissociates into a molecule of nitric oxide (NO) and a free oxygen atom. The free oxygen is very reactive, and it quickly reacts with an ordinary oxygen atom to form a molecule of ozone. The resulting ozone is unstable, however, because it has too much energy; only if the extra energy can be dissipated in a random collision with another molecule can the ozone survive as a stable molecule. If the excess energy is not dissipated, the ozone just dissociates and leaves the free oxygen atom to try again. Even a stable ozone molecule is a pretty reactive thing, however, and it can easily react with the NO that was left from the original NO_2 dissociation; this reaction again forms NO_2 and ordinary oxygen. Thus the cycle has returned to its starting point, the only permanent change being the absorbed radiation which has gone into the kinetic energy of the extra molecule in the second reaction and which in total serves only to warm the air somewhat.

This nice, neat, little cycle by itself does not amount to photochemical smog, however; the relatively fleeting existence of the ozone amounts to far less than is actually observed on smoggy days. What happens is that the presence of organic, mainly hydrocarbon, molecules in the air can disrupt this nicely balanced cycle. Hydrocarbons become involved in several ways, reacting not only with the NO_x and ozone, but also in various ways with themselves, and with carbon monoxide, sulfur dioxide, and other constituents, although most of the latter reactions are minor.

The principal role of the emitted hydrocarbons is that they react with the free oxygen atoms to form oxygenated hydrocarbons; as in Figure 2(b), these in turn can oxidize the NO back to NO_2, in competition with the ozone's role in the third reaction discussed. If each molecule did this only once, the hydrocarbon effect would simply be to provide an alternative path for one part of the cycle. However, because they can easily form free radicals (i.e., very reactive molecules that can often perpetuate themselves by forming more free radicals as they react), the hydrocarbons manage to convert more than just one NO molecule back to NO_2; there is then a buildup of the excess ozone, the principal feature of the smog.

The oxygenated hydrocarbon free radicals also react in other ways, with other substances, and among themselves. A large number of such organic free-radical reactions have been identified, but it is not clear which are important or minor, or even whether most of them have been identified. It is these less-well-known reactions that form peroxyacetyl nitrate and also the very fine nitrogen-bearing organic particles of less than 0.5 micrometer; it is these particles, which are very effective at scattering light, that cause the milky-gray haze characteristic of photochemical smog.

CARBON MONOXIDE

The last of the major pollutants is carbon monoxide (CO), the product of the incomplete combustion of the carbon in fossil fuels. Since most fuel-burning operations try to extract the last amount of energy from their fuel, they operate with "excess air," that is, with plenty of air and oxygen, more than the chemical minimum necessary to oxidize the fuel and thus oxidize the carbon all the way to carbon dioxide (CO_2). The one big exception is the internal combustion engine in the automobile; because of its design, the conventional internal combustion engine operates with no excess air and so produces a great deal more CO than other combustion operations. Most of the 147 million tons of CO in Table I are from automobile and other internal combustion engines.

Another striking feature in Table I is that CO emissions are much the largest of the major pollutants, totaling more than the others combined. This is the basis for the oft-quoted ob-

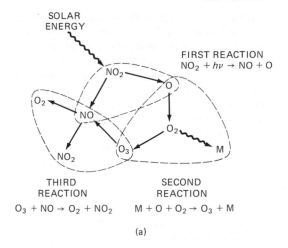

SOLAR ENERGY

FIRST REACTION
$NO_2 + h\nu \rightarrow NO + O$

THIRD REACTION
$O_3 + NO \rightarrow O_2 + NO_2$

SECOND REACTION
$M + O + O_2 \rightarrow O_3 + M$

(a)

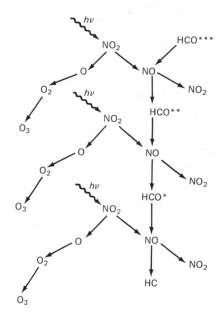

(b)

2   Diagrammatic Representation of Photochemical Reactions in Smog Formation. (a) Cycle of reactions among NO_2, NO, and O_3 without intervention by hydrocarbons. (b) With the intervention of oxygenated hydrocarbons (HCO), which can consume the NO produced by several NO_2 dissociations, the ozone is left to accumulate.

servation that the automobile is responsible for a predominant share of our air pollution. This is true, on the basis of sheer tonnage of emissions, but that is only part of the story, the balance being a matter of how much harm the pollutant does. It happens that, in general, the typical ambient concentrations of carbon monoxide, in particular the concentrations that cause adverse effects on human health, are many times higher than the typical ambient concentrations of other pollutants. When this is taken into consideration and the emissions estimates normalized by a measure of the harmfulness of each pollutant, a very different picture emerges, as seen in Table III. On this basis, sulfur dioxide, and consequently stationary-source fuel combustion, constitute the more serious problem. The reason automobile pollution has received prominent attention is partly because it has been a problem for federal, rather than state, legislation and enforcement; this has meant not only that it was effectively attacked somewhat earlier, but also that the media could follow the efforts much better than they can the more dispersed, even chaotic, efforts of states to control pollution from stationary sources.

Meteorological Influences

Turning to a more technical side of our subject, let us consider the influence of meteorology on the severity of the air pollution problem. The amount of motion and mixing in the air determines how much our pollutant emissions will be diluted and dispersed, and consequently what concentrations of pollutant we will breathe. Patterns of weather differ enough from place to place to make serious differences in the degree of pollution problems encountered. Los Angeles, which experiences photochemical smog much worse than any other location in the United States, does have a great excess of vehicle usage, but its problem is mostly due to its extremely adverse meteorological and topographical situation.

There are two quite different dimensions to the nature of meteorological dispersion of pollutants: the horizontal, where dispersion is a fairly simple function of the direction and speed of the winds, and the vertical, where mixing is a more complex function of the vertical temperature structure (i.e., of the differences in temperature in different levels or layers of the atmosphere). The direction of the wind determines which way the pollutant effluents will go, and the speed of the wind partially determines the ambient concentration of the pollution (i.e., how much pollutant there is in a given volume of air). For example, as the pollutant substance comes out of a smokestack at a constant amount per hour, it

III Relative Importance of Major Pollutants and Various Source Types

	Total emissions (%)	Health significance (%)[a]
Pollutant		
Sulfur oxides	12.9	34.6
Particulate matter	9.7	27.9
Nitrogen oxides	8.6	18.6
Hydrocarbons	13.1	17.7
Carbon monoxide	55.7	1.2
	100.0	100.0
Source		
Stationary fuel combustion	16.9	43.0
Transportation	54.5	22.2
Solid waste disposal	4.2	3.0
Industrial process losses	15.3	25.7
Agricultural burning	7.3	4.4
Miscellaneous man-made	1.8	1.7
	100.0	100.0

On a mass-emission basis, carbon monoxide and automobiles are the most important pollutants; when weighted by the amount of harm they cause, however, sulfur oxides and stationary combustion sources are seen as the major problem.
[a]Calculated by renormalizing the emissions estimates in Table I by the National Ambient Air Quality Standards in Table V.
Source: Reference 9.

is distributed into whatever air happens to be coming by at the time. If the wind speed increases, the same amount of pollutant is distributed into a larger volume of air, so the resulting concentration is lower.

Also important in diluting the emissions from a pollutant source is turbulence — the random motions in the air caused by buildings, a rough terrain, and vertical mixing. Vertical mixing is the important phenomenon in the up-and-down dimension. It is less important with respect to the plume from a particular source than it is in defining the depth of the mixing layer over an entire urban area. Because of turbulence from the surface roughness and vertical mixing, the lowest layer of the air is usually pretty thoroughly and homogeneously mixed; the deeper this layer is at any given time, the more air there is for mixing the pollution and the lower the ambient concentrations will be. The mixing layer is always at least several hundred feet deep over an urban area, because of buildings and the like, but on a nice summer afternoon the mixing layer can be as deep as several thousand feet. The reason the depth of the mixing layer changes so dramatically, particularly why it changes systematically with the time of day and the seasons, is because it depends on the vertical temperature structure, which in turn is dependent on solar radiation. We shall pursue the vertical temperature structure a little further, even though it is relatively technical, because it will lead us to a discussion of inversions, a type of special meteorological phenomenon that can dramatically increase pollutant problems and is responsible for the major killer-smog episodes.

The relationship between vertical temperature differences and vertical mixing arises from the basic physical relationship among factors of pressure, temperature, and density of a gas, such as air. The pressure and density of the air decreases with height at a standard rate determined by gravity; consequently, there is also a standard temperature gradient, called adiabatic, which serves as a sort of reference point or demarcation line. If the temperature difference becomes greater than 5.4°F per 1,000 feet, either by warming below or by cooling aloft, the air mass will become unstable, the too-light warmer air at the bottom will mix upward, and the entire air mass will become turbulent until the adiabatic relationship is again established. If, on the other hand, the temperature decrease with height is less than the adiabatic, the atmosphere will be proportionately stable and vertical mixing will be greatly reduced.

The most extreme case, where the temperature actually increases rather than decreases with height, is called a temperature inversion. In such a situation, there is essentially no vertical mixing; even the turbulence introduced by buildings and hills will be actively suppressed. Such inversion situations also tend to be accompanied by very light or no winds, so horizontal dispersion of pollution is also severely restricted. Inversions happen for a variety of reasons. On clear, cloudless nights, warmth is radiated away from the earth, which cools the earth's surface and the air nearest it; if this lowest layer of air cools enough, a *nocturnal radiation inversion* will be formed. Such an inversion breaks up in the morning when the sun again warms the earth surface and the lowest air layers. As the air continues to be warmed from the bottom up, it usually becomes warm enough to be very unstable, and in the midafternoon on sunny days the turbulent mixing may well extend to 5,000 to 8,000 feet.

Another type of inversion occurs when certain large high-pressure weather systems simply slow down and nearly stop in one place for several days. Such a situation, called a *stagnation*, most typically occurs in the fall and early winter. Because these inversions can persist day after day, they often permit air pollution levels to build up dramatically; the major killer-smog episodes have all been associated with such stagnation. The meteorological problem in Los Angeles is similar in that one of these stagnant high-pressure systems exists almost permanently through the summer in the southern Pacific coast area. Los Angeles can easily have stagnations half the time or more during the May through October smog season, whereas the worst place in the east, the Smoky Mountains, will typically have only 20 to 30 such days all year, mostly in the fall and early winter.

The weather, in addition to having its effect on pollution levels, can, in turn, be affected by air pollutants. Under certain circumstances, air pollution can induce rainfall, and carbon dioxide and fine particles are involved in the radiation balance of the entire earth; these, however, are topics discussed in Chapter 15.

Engineering Control of Pollution

In contrast to the case with water, air cannot be collected for treatment before we use it, nor is it feasible to collect our waste air in one place for treatment before discharging it back into the atmosphere. Consequently, if we are to reduce the amount of the various air pollutants we breathe, it is necessary to reduce or "control" the emissions from each individual source.

There are three general approaches to the control of an air pollution source: one may change the nature of the combustion or industrial process in such a way that the emissions are less; one can take steps to disperse very widely the pollutants emitted, so that their effects are reduced; or one can add additional equipment to remove some of the pollutant material from the effluent gases before they are released to the atmosphere. The applicability of each of these approaches, and of specific engineering options within each, differs widely among various types and sizes of sources. The greatest variety of choices is available to the various stationary fuel combustion and industrial process sources. The automobile and its internal combustion engine offer a somewhat narrowed spectrum of opportunities. Because the two types of sources are so different, we shall consider them separately, but the conceptual parallels ought not be forgotten.

CONTROLLING STATIONARY-SOURCE EMISSIONS

In the case of controlling a stationary source, the easiest thing to do is to change the process in some way so as to reduce the pollution. We do many of our activities the way we do because of decisions made decades ago, when many considerations, especially pollution considerations, were different or absent. In a complicated industrial process, many possibilities are available, ranging from better housekeeping and tighter operating practices to the adoption of totally new technologies. Most of the sizable reductions in industrial

emissions noted in Table II are due to major innovations in the process involved — the change from beehive coke ovens to by-product coke ovens, the virtual elimination of the dirtiest process for making carbon black in favor of less-polluting alternatives. Rarely, however, have these industrial processes been changed solely for pollution control reasons; more frequently, there is a sizable economic incentive also, perhaps just from eliminating the waste of raw materials, finished product, or energy that pollutant emissions often represent. In the case of fuel combustion, for example, the economic losses of fuel energy in the form of smoke and soot were a major factor in reducing such pollution by replacing old, inefficient, fuel-burning equipment during the earlier part of this century. There are at present advanced, cleaner, fuel-burning technologies, but without sizable economic advantages, they are not apt to be utilized.

The biggest stationary-source pollution problem, the sulfur oxides from fuel combustion, is also presently approached by a type of process change, a change in fuels. Because the sulfur oxides emitted result from the burning of the sulfur in the fuel, one very sensible approach to reducing this pollution is just to change to a fuel with a lower sulfur content. The sulfur content of coals range from under 1 percent to over 5 percent; the sulfur content of residual fuel oil is similar, the content of distillate fuel oil very low, and that in natural gas essentially zero (it is removed during processing, to avoid odors). In the late 1960s when regulations aimed at reducing sulfur oxide emissions became effective in major eastern cities, many fuel users did switch to lower-sulfur coal or oil, despite a somewhat higher cost. Figure 3 shows the effect in Washington, D.C., as the federal General Services Adminstration changed the fuels they used for heating government buildings. The difficulty with this otherwise-panacea solution is that there is just not enough low-sulfur fuel to go around. Low-sulfur oil is simply scarce; and the lower-sulfur coals are located in the western plains and mountains, while the need for coal is centered in the East and Midwest near the high-sulfur coal deposits. The ultimate decisions about the wisest use of our limited lower-sulfur fuel resources are thoroughly tied up with the overall question of fuel use and energy consumption (Chapter 5). During the recent petroleum shortage or "energy crisis," some major power plants that had earlier changed from coal to oil were required to change back, and in New England, where oil is used extensively for heating, air pollution regulations limiting the sulfur content of oil were suspended to permit the buttressing of oil supplies with higher-sulfur stocks.

The second general approach to pollution control is the use of very tall stacks to disperse effluents over large areas, a technique applicable primarily to very large combustion sources with high SO_2 emissions. This approach makes use of the winds to disperse the emissions widely and so reduce the concentrations, rather than to reduce them by reducing the emission quantities. The taller the stack from which the pollutants are emitted, the farther they will go and the more diluted they will become before they come back to the ground. It is possible to calculate the approximate dispersion of plumes, and such calculations indicate that, with a tall-enough stack, one would never really find measurable levels of SO_2 at ground level.

There are difficulties with this approach, however, which lead to its being seen as unacceptable by environmentalists and tolerable only in very limited situations to the Environmental Protection Agency. First, the calculated dispersion effect is necessarily a matter of average circumstances; there will unavoidably be periods of very adverse meteorological conditions when the emissions will still cause undesirable ambient levels. It is possible that the source could be taken out of operation or changed to a cleaner fuel during these very few adverse periods for less economic cost than alternative control approaches. Planning such an intermittent control strategy, however, requires much meteorological and regulatory monitoring, and it is anticipated that there would likely be considerable pressure not to follow through on promised intermittent shutdowns.

A much more fundamental difficulty with using dispersion as a control method lies in the obvious fact that it does not reduce the overall pollutant burden but merely spreads it around. Although sources closest to the source receive less immediate exposure to the concentrated emissions, receptors farther away re-

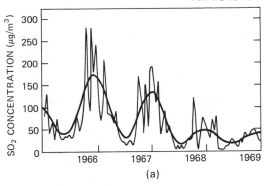

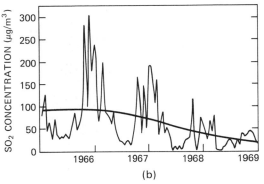

3 **SO_2 Levels in Washington, D.C.** (a) A strong seasonal pattern caused by space-heating emissions, and (b) a clear decrease as government buildings began to utilize lower-sulfur fuel.

ceive increased amounts of pollution, even at very great distances. Some portion of the increasing acid rain in the northern European countries and Scandinavia, for example, is believed to result from sulfur dioxide emitted in the United Kingdom and converted to sulfuric acid as it crosses the North Sea.[10] There is also serious concern in the United States over the conversion of atmospheric SO_2 into sulfuric acid and other sulfates. Measurements throughout the country do seem to indicate that long-distance transport and acid formation is occurring. An increase in the dispersion of the SO_2 will in no way improve this problem, as a direct decrease in sulfur dioxide emissions likely would.

If neither process changes nor tall-stack dispersion proves to be an adequate solution, the operators of a source must necessarily turn to the third general approach, the installation of some type of pollution control device or process to remove at least some of the pollutant before the exhaust gas is released to the air;

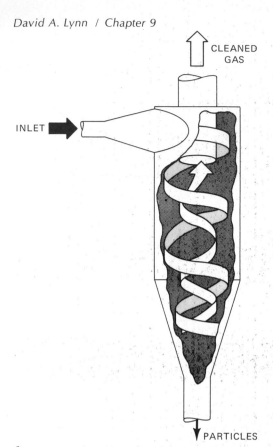

4 **Cyclone Separator.** Particles are whirled in a spiral motion to separate them by centrifugal force. Reprinted by special permission of S. J. Williamson. See Reference 11.

this is actually the most common approach, simply because the applicability of the others is often limited. To separate the pollutant substance from the stack gas, the removal process or control device must take advantage of some physical or chemical property of the pollutant; so a variety of devices may exist for the same pollutant, based on different principles of removal, let alone the wide variety prompted by the detailed engineering of capacity, size, efficiency, and so on, which differs from one installation to another.

POLLUTANT REMOVAL

Because particles are many times larger and heavier than the gases from which they need to be separated, particulate emissions are the easiest to control; some particulate control has been practiced for decades. All particulate control devices exploit the physical properties of

size and weight, so it is not surprising that the smaller the particle, the more difficult it is to control.

The easiest to collect are the heaviest particles, such as sawdust from a sawmill, which will simply settle out by gravity if the air flow is slowed by a wide place in the pipes. Separation of somewhat smaller particles is accomplished in a cyclone (Figure 4), where the airstream is swirled in a spiral motion so that the particles collect at the periphery by centrifugal force. The smallest particles, less than 2 to 3 micrometers in size, are collected with one of two types of high-efficiency collectors, electrostatic precipitators or fabric filters. Electrostatic precipitators send electrical discharges through the airstream; this ionizes the very fine particles, which are then collected by electrostatic attraction on a charged plate. Fabric filters, or baghouses, work essentially like the filter bags on household vacuum cleaners, except that fabric filter bags have not given way to paper ones. A baghouse is simply a structure with many long cylindrical bags hanging upside down; the dirty air rises up through the bags, out through the fabric filter media, and on out the top. The bags are periodically shaken to dump the collected particles into a hopper at the bottom. Although the two high-efficiency control devices are equivalent in many cases, electrostatic precipitators are somewhat more common; baghouses are both a little more effective and a little more expensive.

In the case of sulfur dioxide, all stack-gas cleaning processes are based on the fact that sulfur dioxide is an acidic gas and will react fairly readily with alkaline substances. The most common and effective processes also take advantage of the fact that SO_2 is relatively soluble, and operate with a scrubber, by means of which the stack gas and the other chemicals are brought into contact with a strong water spray. The principal differences among different processes are the chemical used, which is usually some relatively cheap, naturally occurring alkaline substance, such as lime or limestone, and the nature of the resultant sulfur product, which might be elemental sulfur, sulfuric acid, or gypsum. The processes do have different percentage-removal efficiencies, generally from 75 to 90 percent, but the primary differences among them are economic, involving not only the size and cost of the scrubbers required to get the necessary effectiveness, but also the cost of the input chemicals and the sale price or disposal cost of the output materials, which vary with time, shipping distance, and similar factors.[12]

CONTROLLING MOBILE-SOURCE EMISSIONS

The pollutants from motor vehicles are totally different from those emitted by stationary sources, and their control is similarly different. The same three general approaches still apply, at least if one stretches the concept of tall-stack dispersion to include elevated exhaust stacks on large diesel trucks. Having mentioned diesels, we should note in passing that they are not important sources of the major pollutants, although they are good sources of smoke and odors, which are more noticable and irritating (mentally) to the observer.

As seen in Table IV, the transportation pollution problem, or motor vehicle pollution problem, whatever we call it, is essentially a problem of the gasoline-powered internal combustion engine in our cars and smaller trucks; these vehicles emit the predominant share of the major pollutants, not only because there are so many of them, but also on a passenger-mile or ton-mile basis as well. There are three different mechanisms by which automobiles emit pollutants. The majority of the hydrocarbons, and all the carbon monoxide and nitrogen oxides, are emitted from the exhaust system out the tailpipe. The other emissions of hydrocarbons are gasoline that evaporates from the gas tank and carburetor, and gasoline vapors that "blow by" the pistons into the crankcase and escape from there into the air. Both sources are relatively minor and are controlled by modifying the car. The crankcase, or "blow by," emissions have been controlled for some years with the "positive crankcase ventilation" tubes and valves that used to be featured in the gasoline ads as needing cleaning by additive X. The evaporative losses from the fuel system are also controlled with some extra plumbing to collect and use the vapors; these were installed beginning with the 1971 and later-model cars.

The major problem, however, and the major challenge to the automakers' engineering talents, is the exhaust gas, formed during the combustion in the engine cylinders, and collected and discharged through the exhaust system, muffler, and tailpipe. Since both the carbon monoxide and the partially burned hydrocarbons are the product of incomplete combustion, the most obvious control approach is to modify the engine to improve the combustion efficiency, and this has in fact been the primary initial thrust of control efforts. The parameter of engine operation most critical for combustion efficiency is the air–fuel ratio, the balance of the amount of air available to burn a given amount of fuel. If the ratio is too low, there is not enough air, combustion is incomplete, and pollutant production is voluminous. Even if there is exactly the right amount of air for complete combustion according to chemical theory, some pollution will still form because of uneven mixing in the cylinders and insufficient time to complete the combustion. Stationary sources solve this by using excess air, beyond that theoretically necessary, in order to get efficient combustion. This route, however, is essentially closed to the automobile engine designer. Whereas stationary sources burn their fuel smoothly and continuously, the internal combustion en-

IV	Transportation Emissions, 1970*	Transportation mode	Nitrogen oxides	Hydrocarbons	Carbon monoxide
		Gasoline vehicles	7.8	16.5	95.7
		Diesel vehicles	1.3	0.1	0.8
		Railroads	0.1	0.1	0.1
		Vessels	0.2	0.3	1.7
		Aircraft	0.4	0.4	3.0
		Other nonhighway	1.9	2.0	9.5
			11.7	19.4	110.8

*Figures given in millions of tons per year. *Source:* Reference 5.

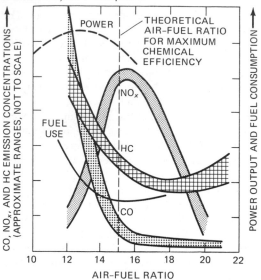

5 **Pollutant Formation in Internal Combustion Engine.** Pollution is a function of the air–fuel ratio. Operation at 19 to 20 would be best for low levels of all pollutants, but with that much extra air, the engine will stall. Prior to the advent of pollution regulations, engines typically operated near 13, to get maximum "performance." *Source:* Reference 13.

gine requires intermittent explosions, and if there is too much excess air in the air–fuel mixture, it will not ignite and explode, so the engine misses or stalls. This is a fundamental problem with conventional engines, and the only solution is to change the design fairly drastically. Such a redesign, already being marketed by Honda Motors of Japan, is the "stratified charge engine"; it solves the problem by dividing the combustion chamber into a fuel-rich zone where ignition is good and an air-rich zone where combustion can proceed efficiently.

When we add nitrogen oxides to our concerns, however, we come up against another fundamental problem with internal combustion engines. Nitrogen oxides are formed at high temperatures by combining the oxygen and nitrogen of the air itself; unfortunately, the most efficient combustion that gives lowest CO and hydrocarbon emissions also produces very high temperatures and maximum nitrogen oxides formation. As Figure 5 illustrates, *it is intrinsically impossible to achieve low emissions of all three pollutants simultaneously.* Consequently, the exhaust pollution from the au-

tomobile cannot be totally conquered by modifying the internal combustion engine for improved efficiency. Either we must make a more drastic process change, such as switch to alternative power plants or other modes of transportation, or we must utilize control devices to remove the pollutants before the exhaust leaves the tailpipe.

For the short term, it has been necessary to choose the second path, although it has been a difficult task. In contrast to sulfur oxides, which are chemically reactive and soluble in water, and particulate matter, which is very heavy compared to a gas, the pollutants from the internal combustion engine do not have any properties that make for their easy removal. The primary approach available for removing CO and hydrocarbons is to further oxidize them to CO_2 and water after they leave the engine. In some cases, this can be done to some extent with a thermal reactor; this is simply a chamber where the hot exhaust can go after it leaves the engine to be mixed with extra air and to have more time to complete the combustion process before it goes out the tailpipe. If this is inadequate, it is necessary to install a catalytic converter, a modified muffler containing catalytic chemicals that will further the oxidation process without the need for high temperatures. These devices are expensive (up to $250 per car), delicate, and perhaps unreliable. They are not yet installed on very many cars, the first ones being on some of the 1975 model cars. One positive feature of using the catalysts is that they are ruined by lead in the gasoline, so their wide usage will hasten the wider usage of lead-free gasoline and thus reduce ambient lead levels in urban areas. A negative feature along the same lines is the effect on ambient levels of sulfuric acid mist; oxidizing catalysts can oxidize not only hydrocarbons and CO, but also sulfur dioxide formed from the low levels of sulfur in the gasoline or from the ambient SO_2 levels in the air brought into the engine. For all these reasons, the total dependence on catalytic converters is not generally viewed as good, although there is little alternative in the short run if we are to meet the standards and deadlines that we have set for ourselves.

A far better approach, although it takes somewhat longer, is to act in other ways to first reduce our dependence on the individual

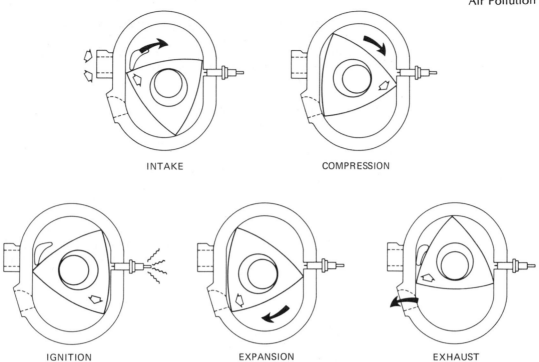

INTAKE   COMPRESSION

IGNITION   EXPANSION   EXHAUST

6 **Wankel Engine.** Utilizes novel geometry to have a standard internal combustion cycle without the reciprocating pistons. *Source:* Reference 13.

private automobile to get us about, and second, to reduce the predominant use of internal combustion engines in those individual cars that we do have to have. The latter may prove to be the easier of the two. Changing engines requires several sizable steps: (1) the engineering of the new engines themselves, (2) the assembling of a capital system to produce them in quantity, and (3) perhaps the restructuring of some of the associated aspects of our society, such as our fuel-distribution system. There are a number of totally different engines under various stages of development, and there are other, less dramatic modifications. Although it is not possible to discuss them thoroughly, we shall at least list them and see how they meet the problems of the conventional internal combustion engine.

It is possible to modify engines very slightly and run them on gaseous fuels, such as natural gas or liquefied petroleum gas. These fuels burn more easily than gasoline and create less pollution. Our "gas" stations, however, do not sell gaseous fuels, so this approach is in practice limited to fleets of vehicles, such as

buses and delivery trucks, that can utilize their own central refueling facilities. We have already mentioned the stratified-charge engine, a drastic redesign of the engine cylinders that leaves most of the rest of the car design intact. A more basic change, although still an internal combustion engine, is the Wankel engine, currently offered in Japanese Mazda cars and also licensed to General Motors. The Wankel has the same intake–compression–power–exhaust cycle as does the conventional engine, but instead of having reciprocating pistons in cylinders, it has a triangular rotor rotating in an odd-shaped housing (Figure 6). The Wankel is actually a fairly dirty engine with respect to CO and hydrocarbons, but it has a very hot exhaust and is very small for its power, making it fairly easy to control the emissions with a thermal reactor rather than a catalytic muffler.

More dramatically different engines include the diesel, steam engines, turbines, and electrically powered vehicles. The diesel has been used in cars, its liabilities being cost and weight and relatively high NO_x emissions. Its

241

advantages are fuel economy and the fact that it is a less radical departure, causing less dramatic disruption of fuel and repair facilities. The various types of steam engines and the turbine (jet) engine have the major advantage that they burn their fuel smoothly and continuously rather than in intermittent explosions; this solves most of the pollution problems, although there are serious engineering, production, and maintenance problems to be solved for them. Electric cars that use batteries are today practical and available, but they have a relatively short range between recharging and an unfavorable power-to-weight ratio that keeps speeds down. Improvement of these will apparently require a significant improvement in battery technology.

The problems of these radically different alternative power plants are in most cases an inability to match the performance of the conventional engine in one way or another; the diesel and experimental turbines lack quick power and acceleration, the battery cars lack range and speed. Some of these are "performance" characteristics that we do not really need, like the ability to peel away from a traffic light, whereas others are features that we need only because we have become accustomed to them, such as the range required for the long-distance commute produced by our suburban sprawl.

None of these difficulties, however, is apt to be the major stumbling block in the way of the widespread adoption of an alternative power plant. If the perfect replacement engine were available today, the capital facilities needed to manufacture it in quantity would still be a nearly insurmountable obstacle. The major auto manufacturers have resisted very effectively placing any major emphasis on alternative engines, and those manufacturers who have entered the field have often had to subsist on subsidies, demonstration grants, sales of off-road specialty vehicles, and so on; none of them has any great chance of truly mass production other than to sell their ideas or designs to the major manufacturers. With the alternative engines in the Japanese small cars meeting the pollution standards that Detroit says it cannot meet and gasoline shortages upsetting traditional car-buying habits, it is very difficult to guess what will happen. Ultimately some diversity of engines will no doubt

occur, as people's habits and the industry's capital policy changes slowly, and as various engines come to be used in the circumstances where they perform best.

The Social Control of Air Pollution

The various control devices and other control approaches that we have just described are a formidable arsenal of engineering answers to engineering problems. More important to the ultimate solution of the entire, overall, air pollution problem, however, is the social organization that sees to it that these engineering answers are in fact utilized. It is most unlikely that very many pollution sources would implement control technology unless there was at least the threat of compulsion; in fact, without such sanctions, much of the control technology would not even have been developed.

There are three basic levels at which our social–governmental–legal system approaches the air pollution problem. One of these, discussed elsewhere in this volume (Chapter 17), is the individual citizen's right to go to court. Although important as a legal and citizens' rights issue, and especially important in keeping government programs operating effectively, this approach alone works well only in the case of extremely gross pollution damage, caused by a readily identifiable source, and is not very effective in reducing the overall burden that we all share. Another approach, most important in the long term, is exemplified by the National Environmental Policy Act (Chapter 17); this involves building environmental considerations into other aspects of the public-policy-making process, so as to prevent problems in advance and to facilitate the adoption of long-term solutions.

The approach we wish to pursue here is that of direct governmental programs designed to clean up the pollution we already have and taking direct short-term steps to prevent more. Essentially this is done with air pollution control laws at the federal, state, and local levels of government, laws aimed directly at pollution sources and their emissions. Historically, the first air pollution laws were city smoke ordinances, beginning in Chicago in 1881; the first state law was in California in 1947, and other states followed. The first federal law, in

1955, was designed simply to help the states with money and technical assistance, but as it became increasingly apparent over the years that the states' efforts were inadequate, the federal law gradually evolved into a more stringent form. The present Clean Air Act in effect requires states to take action to clean up their pollution and provides for the federal government to step in and do it if the state will not. It also provides for federal emission standards for automobiles, research and development on health effects and control technology, and, perhaps most importantly, financial support of the states' pollution control programs.

The simplest and most direct part of the law is the control program for mobile sources. Under the Clean Air Act, EPA sets emission standards — specific pollutant emission levels that may not be exceeded — and it is then illegal to sell a new car or engine that does not meet the standards. There is a major task of certification testing involved, but the principal issue is essentially just how stringent a standard how soon. The Department of Health, Education, and Welfare, which had the federal air pollution responsibility before EPA was formed, took a generally slow and cautious approach. In 1970, in the middle of a presidential campaign, Congress concluded that the pace was too slow and legislated a speedup, requiring very strict CO and hydrocarbon standards for the 1975 model cars and very strict NO_x standards for the 1976 models. These have since been delayed one year as permitted in the law, but the automakers still claim that they cannot meet the postponed standards either. Most environmentalists, including your author, do not believe that the major manufacturers have been conscientiously trying; however, the economic impact of halting automobile production is such that EPA has necessarily recommended to Congress that further extensions be permitted, perhaps on a selective model-by-model basis.

IMPLEMENTATION OF AIR QUALITY STANDARDS

The major portion of the national air pollution control program, designed primarily for stationary sources, is a joint federal–state effort based on National Ambient Air Quality Standards (NAAQS). The ambient air quality standards, established by EPA, are those concentrations of air pollutants in the ambient air that are seen as safe, including a reasonable margin of safety. The primary NAAQS in Table V are those values believed to be protective of human health; for particulate matter and sulfur oxides, there are also secondary standards believed to provide protection against damage to materials and vegetation, although the secondary SO_2 standard is currently in the process of being deleted. The NAAQS must by law be based on an air quality criteria document, which is a compendium of the known information on the effects of a pollutant and other relevant information about the pollutant. Criteria documents are compiled by EPA and screened by advisory committees, including members of the public and representatives of industry. A criteria document is thus meant to be a scientific concensus about the ill

V   National Ambient Air Quality Standards

Pollutant	Concentration not to be exceeded
Sulfur oxides (as SO_2)	0.03 ppm (80 μg/m³) annual mean 0.14 ppm (365 μg/m³) maximum daily average
Suspended particulate matter	75 μg/m³ annual mean (geometric) 260 μg/m³ maximum daily average
Carbon monoxide	9 ppm (10 mg/m³) maximum 8-hour average 35 ppm (40 mg/m³) maximum hourly average
Photochemical oxidants (as O_3)	0.08 ppm (160 μg/m³) maximum hourly average
Reactive (nonmethane) hydrocarbons	0.24 (160 μg/m³) maximum 6–9 A.M. average
Nitrogen oxides (as NO_2)	0.05 ppm (100 μg/m³) annual mean

Primary standards, protective of human health. The "maximum" figures quoted are those levels permitted to be exceeded once per year. If the suspended particulate geometric mean of μg/m³ were to be expressed as the normal arithmetic mean, it would be somewhat higher, perhaps 05 or 90 μg/m³.

243

effects of a pollutant; and this they are. They are also, however, at least to some extent social–political documents, involving public policy choices. The very first criteria document, for sulfur oxides, which came down quite hard for a very low recommended standard, had to be withdrawn, reconsidered, and revised in the face of pressure from the coal, oil, and public utility establishments.[14] Nonetheless, it is generally felt that the present standards are quite adequately protective of the public health.

An air quality standard, however, cannot by itself lead to the installation of emission-control equipment, the use of low-sulfur oil, or any emission-reducing measure, but can only tell us, when compared with the ambient pollutant levels actually measured, whether we need to reduce our pollution and to some extent how much we might need to reduce it. Although it may be clear, as it usually is, that we are violating the air quality standard in a complex urban area, it is typically not clear which source or group of sources is the particular culprit and needs to be controlled. In order to turn the air quality standards into rules and regulations applicable to specific sources, the Clean Air Act provides that the various states shall develop "plans for the implementation of the standards." The Act also provides for deadline dates, and perhaps most importantly, provides that the federal EPA will develop and enforce such a plan if a state fails to do so.

The state implementation plans, or SIPs, thus involve the states' decisions about which sources or types of sources require stringent control, which less; the form of control required; procedures for verifying compliance; definition of a monitoring network to verify meeting the standard; designation of an agency to do it all; and so on — in effect, the entire package of what the state plans to do about the air pollution problem within its jurisdiction. The states generally approach the problem by establishing emission standards for various types of sources — limits on the amount of pollutant to be legally emitted. To be technically reasonable, these are often expressed relative to the size of the source, such as pounds of particulate emissions per 1,000 Btu of fuel energy used. Frequently, such standards are more stringent for larger sources, where greater financial and technological resources make the control of the emissions somewhat easier. An example is shown in Figure 7. The primary advantage of simple emission standards is that it leaves the source free to limit its emissions in the way that is easiest, cheapest, or whatever, from its viewpoint (Chapter 16). Even those regulations that require the use of low-sulfur fuel typically permit the burning of high-sulfur fuel if the sulfur oxides are equivalently removed from the stack effluent by a control device.

Whatever regulations that the state may adopt under the authority of the state air pollution control law, the aggregate implementation plan, once approved by EPA, also becomes enforcable by EPA under the authority of the federal Clean Air Act. This dual legal responsibility is a very effective arrangement. It permits a conscientious state agency to pursue its own pollution cleanup in its own way, but if the legal or political muscle of a particular source becomes simply too great for the agency to cope with, the much more massive power of EPA is available to help.

Lest the existence of this effective federal law imply that all is right with the world, we should hasten to add that this enforcement effort has only been in effect for a few years, that the ultimate completion of the process is not yet at hand, and that, particularly with automobiles and the power plants, social and political pressure, even at a national level, may yet force the dismemberment of the system. The present form of the law became effective in December 1970. In April 1971, EPA promulgated the NAAQS, and later that year issued guidelines for the preparation of the implementation plans. Under the time tables of the law, implementation plans for sulfur oxides and particulate matter were due first; most of them were approved in mid-1972, so the law's 3-year schedule would require that the standards be met by 1975, which is upon us. The law does provide for time extensions up to 2 years, and some have been made.

In general, the largest sources are the ones not yet appropriately controlled, and in many cases the problem is the control of sulfur oxides. As we noted, the easiest approach was to switch to lower-sulfur fuels, but the scarcity of such fuels, complicated by the energy crisis, has made massive reliance on that approach impossible. To many, including the author, the obvious answer is to install stack-gas scrubbing devices to remove SO₂ and permit the use of our vast reserves of high-sulfur coal. The electric power industry in particular, however, is resisting this approach, and less than two dozen plants, most small or medium-sized, have or are planning to install scrubbing systems. EPA and some other agencies and organizations feel that the technology is ready and should be implemented; the power companies and the Federal Power Commission disagree and urge more research and testing. To a great extent it is a circular argument; the very large power plants do not want to install scrubbers until they have been proved on very large power plants. Although it is obviously true that designing and operating a scrubbing installation is a major and difficult chemical engineering job, it is equally true that someone must be first.

TRANSPORTATION AND LAND-USE PLANNING

Although the criteria–standards–implementation plan approach was primarily designed to attack the stationary-source problem in large urban areas, the law did specify that it should also be applied to pollutants that come primarily from mobile sources. Criteria documents were prepared and ambient air-quality standards adopted for photochemical oxidants, hydrocarbons, nitrogen dioxide, and carbon monoxide. The states were thus obligated to produce implementation plans to attain and maintain the standards for these pollutants. However, the federal government had already promulgated emission standards for new motor vehicles and, except for California, which had a grandfather exemption, the states were specifically prohibited from regulating new car emissions in any way.

This redundancy was resolved by requiring implementation plans only for those areas where the air quality standards would not be met on schedule by virtue of the federal motor vehicle control program alone; projections indicated that in over 30 major metropolitan areas, even the new stringent emission standards would not be enough to meet the standards by 1975, or even by 1977, with a maximum 2-year extension. Consequently, during 1973 EPA and the states went through a pro-

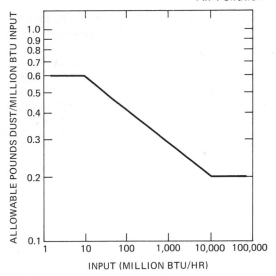

7 **Allowable Particulate Emissions from Boilers.**
The legal limit is defined as a function of size, so large boilers may emit less fuel energy per unit than smaller boilers.

cess of devising what came to be called transportation control and land-use plans, plans to implement the standards for vehicle-related pollutants by reducing vehicle usage, putting control devices on used cars as well as new ones, and so on. Although the long-term options for such control measures are not totally objectionable, including shifts to smaller cars and improved mass transit, the short-term ones that could be implemented in 2 to 4 years tended to feature such politically unpalatable measures as banning pleasure driving, compulsory car pools, and, worst of all, gasoline rationing. At one point, under court order, EPA promulgated a plan for the Los Angeles area that would have required over 80 percent gas rationing in an area with very poor public transit. In most of the affected urban areas, citizen reaction was angry — EPA and the environmentalists were seen as taking away each citizen's God-given right to drive alone in a big car. The interesting feature, in retrospect, is that within 6 months, the October war in the Middle East and the Arab oil embargo, coupled with an already very tight oil supply situation, had Americans, at least in the East, waiting in line for gasoline and buying small cars and joining carpools in large numbers.

At present, the more reasonable of the transportation measures are just beginning to

245

be implemented, the less reasonable ones are still on the books, and EPA has asked Congress for authority to extend the 1977 deadlines to avoid the worst of the difficulties.

Over the longer term, however, a great improvement could be, and many feel must be, made, not only in air quality but also in other environmental amenities, by undertaking to cause a much more fundamental change, the drastic reduction of the amount that we use in our cars. This could be done both by increasing the use of mass-transit alternatives and by redesigning our cities and our habits to require less transit of any kind. The number of cities that have any type of rapid rail transit is small, and in many even the bus service neither extends into the suburbs nor provides really good service in the central city. This is changing, however, partly as a result of pollution control requirements. San Francisco and Washington, D.C., are building the first new subway systems in decades, and Denver has just committed itself to a Personal Rapid Transit (PRT) system involving vehicles smaller and more flexible than traditional subway cars. Although an alternative is obviously necessary if we are to reduce private-vehicle usage, it need not be an expensive new transit system. Another innovative concept, currently in practice, is "dial-a-bus," a type of small transit system where small buses respond to individual requests for travel, like a taxi would, but with the cost and pollution savings of shared travel. All these provide interesting options to the private traveler in moving about the present city, as indeed does that old favorite, the car pool to work, which had been nearly eliminated by our rising affluence.

Over a still longer term, and probably to a lesser extent, it is also possible to rearrange cities to minimize or at least reduce the total amount of travel required. This can be effected primarily by long-term substantial changes in the pattern of land use, which in turn requires changes in taxation and in zoning laws and regulations. If the neighborhoods near the central business district were more livable, there would be less pressure to move to the suburbs and commute many miles; similarly, the neighborhoods near major industrial employers could attract workers into residence by cleaning up, among other things, their air pollution. Before the car came to join our society,

many people used to walk to work; today, many suburbs do not offer enough sidewalks to even walk to the corner drugstore.

On a scale larger than a single urban area, the patterns of growth and land use in recent years have proved to be a function of the transportation system, rather than the transportation system being a function of the master land-use plan. Thus it becomes necessary to attempt to shape the future pattern of land use by shaping the present growth and development of the transportation system. Together with the possible changes in the ways our society uses energy, these changes in the way that we use land resources are the ultimate solution to our pollution problem, if there is any such solution, a sort of grand and glorious ultimate process modification.

Trends in Pollution Levels

Because of the rapid pace of change in various aspects of society recently, and our increasing awareness of such changes, especially the emergence and growth of new technologies, there is a distinct interest in anticipating the future, in trying to define trends in the parameters that constitute and affect our daily lives and the environment we share. Many persons, for one reason or another, foresee a rather pessimistic future, with increasing pollution, overcrowding, resource depletion, and so on. Although I certainly concur that it is not easy to be optimistic in these times, it is also accurate and important to point out that at least in one environmental area we can anticipate for the near future decreasing trends in the emissions of the major air pollutants and, I believe, a generally improving air pollution picture. This is due, in most instances, to deliberate action that has been taken to reduce pollution.

Sulfur dioxide emissions, without deliberate intervention, could be reasonably expected to continue climbing upward with increasing energy usage. However, two trends, of different types, are operating against this. First, regulatory control of sulfur oxide emissions is forcing a decrease in the emissions per unit of energy conversion, although the current "energy crisis" has slowed this slightly, as we have noted. The second trend is more specula-

246

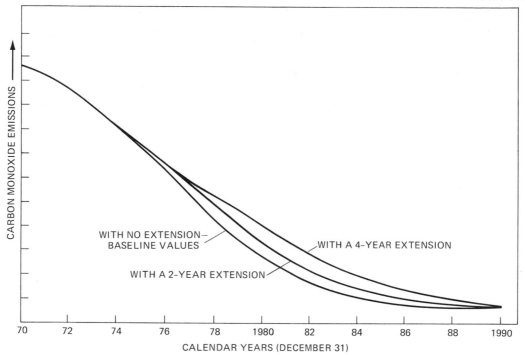

8 **Trends in Estimated Carbon Monoxide Emissions from Motor Vehicles.** Shows gradual decline until the late 1980s, whereupon the growth in vehicular usage may again cause an increase in emissions. The lowest curve is the trend calculated with presently existing new-car emission factors, including the 1-year extension already granted; the other two curves show the effect of hypothetical extensions of the present 1975 interim standards for 2 and 4 years, respectively. *Source:* Unpublished calculations courtesy of C. K. Wilcox, GCA/Technology Division, 1974.

tive: the belief, or perhaps hope, that the "energy crisis" has made enough of an impact as to make long-term energy-conservation efforts fruitful. With the possibility of various new energy sources coming into use, and the changes in the technology for using our traditional fuels, it would be foolhardy to try to anticipate the fuels and energy policies of 20 years from now with the precision needed to judge pollutant emissions. Until then, once we are past the immediate problems, the control measures currently being pursued should give us an overall decreasing trend in sulfur dioxide emissions.

The case of automobile pollution is similar — a 15- to 20-year downtrend in emissions, followed by an increase again unless permanent, fundamental changes are made. For several decades, automobile population and aggregate travel have been increasing several percent a year, somewhat faster than the human popula-

tion because of our spreading affluence. Efforts to control emissions from the conventional engine, however, have started a downturn in total vehicular emissions, as seen in Figure 8. Because it takes 10 to 15 years for the automobile population to replace itself, the present and anticipated new car emission standards are expected to maintain this decline in emissions until the early 1990s. If the internal combustion engine is then still our primary transportation mechanism, the trend will necessarily reverse itself and begin climbing with any growth in vehicle usage. Again, there is more optimism than there was 3 years ago that longer-term solutions or alternatives will be adopted; otherwise, petroleum supply may well become a far more dominant problem than pollutant emissions.

Two basic arguments make it possible for some observers to remain pessimistic despite the previous comments. First, in both cases,

247

true long-term improvement depends on society's making some fairly dramatic intelligent choices, something for which there is no real abundance of historical precedent. Resource depletion and economic considerations, however, have a very good chance of forcing such choices upon us, and although we might prefer to make them more rationally, the effect will likely be the same. The second pessimistic argument is that while it may be true that concentrated effort is reducing the major pollutants, very little is being done about others, except for technology's inventing a few new ones each year. There is a certain truth to this point; it is correct enough in its assessment of what is happening. Nonetheless, its logic is dependent to some extent on the implicit assumption that the number of pollutants, or the subtlety of their effects, or their widespread, global, nature will intrinsically make them important problems. This is not necessarily true. Although the presently "minor" pollutants are unquestionably of concern and need to be studied and controlled, it appears unlikely that any of them could reach dire proportions in the same manner as have the present major pollutants. The sulfur oxides–particulate matter problem grew to its present proportions over centuries of ignorance and resultant neglect, the automobile pollutants similarly over a few decades. In the light of current awareness, and public interest and support, I find it hard to anticipate that air pollution will be, proportionately, as serious a problem in the future as it is now.

References

1. Evelyn, J. 1661. *Fumifugium, or the Inconvenience of the Aer and Smoake of London Dissipated (together with some remedies humbly proposed)*; reprinted most recently in *The Smoake of London — Two Prophecies* (J. P. Lodge, Jr., ed.). Maxwell Reprints, Elmsford, N.Y., 1969.

2. Anon. 1969. *Air Quality Criteria for Sulfur Oxides* (Department of Health, Education, and Welfare Publ. AP-50). Government Printing Office, Washington, D.C.

3. Anon. 1969. *Air Quality Criteria for Particulate Matter* (Department of Health, Education, and Welfare Publ. AP-49). Government Printing Office, Washington, D.C.

4. Anon. 1970. *Air Quality Criteria for Photochemical Oxidants* (Department of Health, Education, and Welfare Publ. AP-63). Government Printing Office, Washington, D.C.

5. Cavender, J. H., Kircher, D. S., and Hoffman, A. J. 1973. *Nationwide Air Pollutant Emission Trends. 1940–1970* (Environmental Protection Agency Publ. AP-115). Government Printing Office, Washington, D.C.

6. Anon. 1973. *Compilation of Air Pollutant Emission Factors*, 2nd ed. (Environmental Protection Agency Publ. AP-42). Government Printing Office, Washington, D.C.

7. Anon. 1973. *The National Air Monitoring Program: Air Quality and Emissions Trends — Annual Report* (2 vols.) (Environmental Protection Agency Publ. EPA 450/1-73-001A and 001B). Government Printing Office, Washington, D.C.

8. Likens, G. E., Bormann, F. H., and Johnson, N. M. 1972. Acid rain. *Environment* 14(2): 33–40.

9. Lynn, D. A. 1975. *Air Pollution–Thrust and Response*, Addison-Wesley Publishing Company, Inc., Reading, Mass.

10. Reiquam, H. 1970. Sulfur: simulated long-range transport in the atmosphere. *Science* 170: 318.

11. Williamson, S. J. 1973. *Fundamentals of Air Pollution*. Addison-Wesley Publishing Company, Inc., Reading, Mass.

12. Anon. 1973. *Final Report of Sulfur Oxide Control Technology Assessment Panel on Projected Utilization of Stack Gas Cleaning Systems by Steam-Electric Plants* (Environmental Protection Agency Publ. APTD-1569). Government Printing Office, Washington, D.C.

13. Anon. 1970. *Control Techniques for Carbon Monoxide, Nitrogen Oxides, and Hydrocarbon Emissions from Mobile Sources* (Department of Health, Education, and Welfare Publ. AP-66). Government Printing Office, Washington, D.C.

14. Davies, J. C., III. 1970. *The Politics of Pollution*. Pegasus, Indianapolis, Ind.

Further Reading

Lynn, D. A. 1975. *Air Pollution—Thrust and Response*, Addison-Wesley Publishing Company, Inc., Reading, Mass. Broad coverage at a less technical level than Williamson or Stern.

Miller, A., and Thompson, J. C. 1970. *Elements of Meteorology*, Charles E. Merrill Publishing Company, Columbus, Ohio. A basic meteorology text.

Stern, A. C. et al. 1973. *Fundamentals of Air Pollution*, Academic Press, Inc., New York. Emphasizes traditional detailed engineering topics, although coverage is broad.

Williamson, S. J. 1973. *Fundamentals of Air Pollution*, Addison-Wesley Publishing Company, Inc., Reading, Mass. Emphasizes physical science fundamentals and meteorology.

10

Fresh Water Pollution

W. T. Edmondson   *is Professor of Zoology at the University of Washington, Seattle. He
did his undergraduate work at Yale University, and earned his doctorate there in 1942, working
under G. E. Hutchinson. During World War II he did research for the Navy at the American
Museum of Natural History and the Woods Hole Oceanographic Institution. He was a member of
the biology faculty at Harvard from 1946 to 1949. In 1949 he moved to the University of
Washington, where his research has centered on the mechanisms that control the productivity and
the populations of organisms in lakes, particularly Lake Washington. In 1973 he received the
National Academy of Sciences Cottrell Award for Environmental Quality and was elected to
membership in the Academy.*

251

Pollution of fresh waters results largely from the ease of disposal of waste there. Lakes occupy low places in the landscape and their inflowing streams are natural collecting systems from the watershed. It is easy to pour wastes into the water and let them be carried away to "somewhere else." But lakes and rivers are used for a variety of other purposes: as sources of drinking water or food (fish), for fun (boating, swimming, water skiing), for camping sites, or simply as parts of beautiful landscapes enjoyed for their existence. Each use depends upon the maintenance of a particular quality, but each of the uses has the possibility of decreasing that quality.

Rivers and lakes may be "enriched" by surface runoff carrying everything that people put into their house drains, onto their land, or disposed of on the streets: sewage, fertilizers, herbicides, insecticides, and general junk. Up to a point we have been able to get away with this because water and its biota have the capacity to absorb and break down by biological action many kinds of wastes. But some kinds of waste cannot be degraded, and any system can be overloaded. And the "somewhere else" turns out to be where somebody else is trying to live and use the water.

Waste can affect the receiving waters in a variety of ways, depending on its character. Disposal of raw sewage is obviously bad aesthetically and because of the possibility of spreading waterborne diseases, but this is only part of the problem. Treated sewage can increase biological production in such a way as to generate a great excess of undesirable types of organisms with unpleasant side effects. The structure of the natural community is distorted.

Some kinds of industrial waste contain toxic substances that kill or damage desirable forms of life, a different kind of distortion of the community. Other substances give the water a taste or odor that makes it unsuitable for drinking without extensive treatment.

Another kind of damage results from the disposal of large, resistant objects such as cans, wornout tires, and even cars. It is more than simply that the beauty of a place is destroyed. Although one can argue that concepts of natural beauty are cultural and subject to change, for people who have been brought up to value natural beauty, experiencing the sight of a lot of junk dumped into a stream can be psychologically damaging. Surely our mental health is affected by the character of the surroundings we live in and the nature of the visual experiences we have. This effect is entirely out of my field, and I will say nothing about it in this chapter, but it is a point that looms large in the minds of some limnologists as they go about their work.

If we agree that it is acceptable for man to exploit natural waters for food, transportation, and recreation, it makes sense to optimize that exploitation so as to do the least damage with the most benefit in the long run. Any exploitation inevitably will change the system, and any protective action will cost something. We are dealing with a very complex system of interacting natural and economic processes. In the past, many important actions have been taken without realization of the remote consequences, or even despite knowledge of them. As will be shown, the character of the fuel burned in one country can strongly affect the agriculture and fisheries of another. Often consequences have been evaluated only in terms of immediate costs, and actions taken to maximize monetary income without taking ac-

252

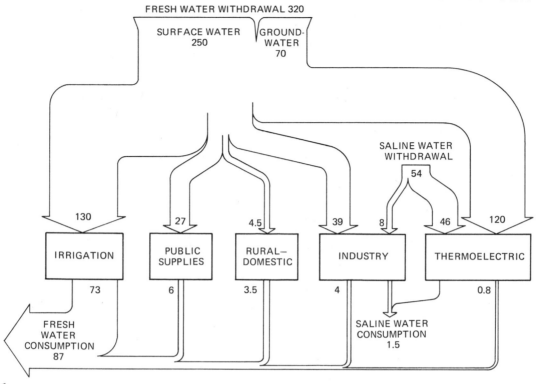

1 **Estimated Water Withdrawals and Consumption in the United States in 1970.** Billions of gallons per day. *Source:* Reference 1.

count of the fact that a corresponding cost to somebody else was being generated elsewhere (Chapter 16). One great difficulty in managing pollution has been that actions leading to maximization of profits are relatively easy to identify, but the consequences are diffuse and are difficult to identify and to measure quantitatively in terms of monetary cost.

As a result of decades of overloading the capacity of our natural waste disposal systems, we have built up a pollution debt that now must be paid off if our society is to be able to continue to function. One can make an analogy with a person running a 100-yard dash. The idea is to run as fast as possible. Under these conditions, it is physiologically impossible for a person to absorb oxygen through the lungs fast enough to keep up with the rate of consumption of oxygen in the muscles. Thus, the muscles are living anaerobically, and lactic acid accumulates from the breakdown of carbohydrate rather than being completely oxidized to carbon dioxide. This accumulation is called an *oxygen debt*. The degree of accumula-

tion is limited, and one cannot run at top speed very long. After the race, the athlete lies panting on the ground, absorbing oxygen at a rate faster than he is using it, and the extra amount largely goes to resynthesizing carbohydrate from the lactic acid; the oxygen debt is being repaid.

We have been producing wastes much faster than the natural systems can deal with them, producing an accumulation of unsatisfactory conditions. So, we must learn how to decrease the rate at which we damage the environment. The time has come to pay off the pollution debt. To do this, we have to understand exactly what the different kinds of pollution do and what the possibilities are for modification.

Large volumes of water are involved (Figure 1 and Chapter 6). In 1965 the rate of water withdrawal for public water supplies was 23.6 billion gallons per day; that is 155 gallons per person. Total use for all purposes, including irrigation, farm, and industrial, was 310 billion gallons per day, 1,600 gallons per person. Different industries have vastly different re-

253

quirements for water (Table I). The availability of adequate supplies of suitable water is often the key factor in determining the location of factories. Further, the character of the effluents from different industries is enormously varied; consequently, the complexity of the effects of industrial waste and the efforts needed to control them also vary greatly. The large numbers in Table I are meaningless in themselves. They must be judged in relation to our total water supply and our requirements. The average stream flow is 1,200 billion gallons per day. In some basins, water is used repeatedly.[3]

Many of the uses to which water is put require water of high quality, that is, with a small content of dissolved and suspended material, and lack of bacterial contamination. But some of the uses degrade the quality of the water so that it cannot be used again without treatment. A good example is the domestic use of water in houses, with the consequent production of sewage.

Eutrophication

SEWAGE AND SELF PURIFICATION OF STREAMS

Commonly, sewage has been disposed of simply by letting it run into a river without any treatment.[4] Household drains taking waste from sinks and toilets run into pipes under streets that lead to big collector sewers that in turn lead to the river. Such material is very rich in organic substances that provide nutrition for bacteria and fungi, organisms known collectively as decomposers. As the population of decomposers builds up in the river, the original material is transformed partly into the organic material of the microorganisms, but much of it is liberated to the water in a dissolved, inorganic form. For example, while some of the nitrogen in urea will be absorbed and made into bacterial protein, much will be liberated as ammonia, which

I   Industrial Waste

Industry	Waste water volume (billions of gallons)	Process water intake (billions of gallons)	Suspended solids (millions of pounds)
Food and kindred products	690	260	6,600
Meat products	99	52	640
Dairy products	58	13	230
Canned and frozen food	87	51	600
Sugar refining	220	110	5,000
All other	220	43	110
Textile mill products	140	110	—
Paper and allied products	1,900	1,300	3,000
Chemical and allied products	3,700	560	1,900
Petroleum and coal	1,300	88	460
Rubber and plastics	160	19	50
Primary metals	4,300	1,000	4,700
Blast furnaces and steel mills	3,600	870	4,300
All others	740	130	430
Machinery	150	23	50
Electrical machinery	91	28	20
Transportation equipment	240	58	—
All other manufacturing	450	190	930
All manufacturing	13,100	3,700	18,000
For comparison: sewered population of U.S.	5,300		8,800

Volume of industrial wastes before treatment, 1964.
Source: Reference 2.

may become oxidized to nitrate. Phosphorus that was originally part of organic molecules will be liberated as dissolved inorganic phosphate. The respiratory activity of the microorganisms will liberate carbon dioxide containing carbon that was originally part of carbohydrates and other carbon-containing compounds in the sewage.

These processes are a perfectly normal part of the functioning of the ecosystem. Decomposers are very important in nature because of their activities of reducing natural wastes, corpses, fallen leaves, and other by-products of life to their elemental constituents. The recycling of elements permitted by this process is essential for the continued growth of plants, the primary producers (Chapter 1).

All these processes take time as the water moves downstream, which establishes a series of zones below the input of raw sewage depending upon the stage reached in the series of processes by the time the water passes a given point. As in most biological matters, these zones are not clearcut regions with sharp boundaries but blend into each other in a gradual manner [4] (Figure 2).

Immediately below the outfall is the *zone of degradation*, where stream water mixes with sewage. With the development of increasing numbers of decomposers, the water enters the *zone of active decomposition*, where the concentrations of organic materials decrease with a corresponding increase of inorganic by-products. If the load of sewage is great enough, the organisms use oxygen faster than it can diffuse it through the surface of the stream, and the water becomes devoid of dissolved oxygen. This *septic zone* is characterized by the absence of animals and the presence of hydrogen sulfide (H_2S), which smells like rotten eggs. Eventually, enough of the organic substrates are consumed that the bacterial activity slows down and oxygen reappears, being stirred in from the air as the water moves along. Next, *zone of recovery* processes take place that change the water back toward its original condition. The dense population of bacteria forms an excellent food supply for a variety of protozoans and small invertebrates. Some of the bottom-dwelling invertebrates are able to live with little oxygen, and sludge worms (Tubificidae) appear in great numbers early in the recovery zone. Further down-

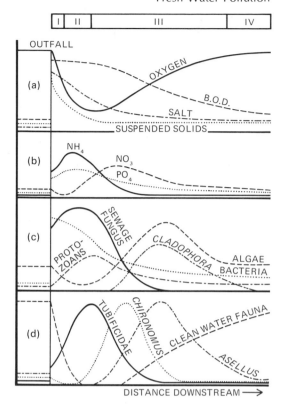

2 **Stream Below Sewage Outfall.** Changes in the condition of a stream below an outfall of raw sewage. (a) and (b) show mostly chemical changes; (c) shows changes in the abundance of microorganisms and (d) shows changes in the abundance of invertebrate animals. The zones described in the text are shown at the top. *Source:* Reference 2.

stream, after more aeration, invertebrates with higher requirements for oxygen increase in numbers.

With the relatively high concentrations of carbon dioxide, phosphate, and other nutrients now present, algae can grow, and luxuriant streamers of filamentous algae attached to stones may develop. By their photosynthetic activity they contribute dissolved oxygen to the water. These algae provide cover and food for invertebrates, so the variety and abundance of animals increases downstream in this zone. Free-floating microscopic algae may develop in large numbers also, providing food for those animals adapted to them, especially in slowly moving water.

Eventually, all the decomposible material that entered the stream has been degraded by biological action, and the water is restored to something close to its original condition, the *clean water zone*. This series of processes has been described as the "self-purification of streams."

Obviously, the discussion just presented concerns the effects only of materials that are degradable by biological action. Some industrial wastes contain highly toxic materials that kill everything and prevent proper action of the natural self-purification processes. Such wastes can damage streams as far downstream as it takes for the material to be diluted below biologically effective concentrations. These wastes will be discussed later in the chapter.

It is clear that the length of river occupied by the affected zone (zones of degradation and active decomposition, septic zone, and recovery zone) will depend largely upon the rate of addition of sewage and the rate of flow of the river. The establishment of a village on a stream early in this century may have produced an affected zone that was relatively short, and this amount of degradation might have been acceptable. But as that village grew to a city, the rate of sewage input would have increased in proportion to the population size and activity, and the length of the affected zone must have increased downstream accordingly. This would present a problem for the next city downstream, which originally stood in a zone of clean water. As the city upstream grew, an increasingly long stretch of river would have been affected, and the downstream city would find its water deteriorating. If the water was used for drinking, the degree of filtration and treatment would have had to be increased. In the meantime, the lower city would be adding its own sewage, so the entire stretch between the two cities and some distance below would be affected. Entire streams have been turned into essentially open sewers by this kind of overloading.

Many cities have been built on lakes and have used the lakes as recipients of their raw sewage. The biological consequences are similar to those in the river, but they are not laid out in such linear order. The surface water of lakes is likely to be blown around and mixed so actively by wind that definite zones are not likely to be seen except in special circumstances. The production of a large quantity of bacteria feeds the animals, and the liberation of nutrients increases algal production throughout the lake.

SEWAGE TREATMENT TECHNIQUES

When it became clear in the early part of this century that streams and lakes were becoming overloaded and that very unsatisfactory conditions would develop, various processes for treating sewage to reduce its unpleasant effects were invented. For example, a typical modern two-stage biological treatment plant operates as follows. Large pieces of solid material are screened out, and preliminary settling in large tanks permits settling of finer material as a sludge that can be removed. The remaining material is led into large tanks, where air is bubbled through it. Under these conditions, bacteria and fungi multiply, decomposing the organic materials. Another settling stage permits removal of much of the organisms as sludge. The clear liquid above the sludge (the supernatant) is then chlorinated to kill bacteria and is discharged into whatever water is convenient.[5]

Obviously, what this kind of treatment does is to carry through the normal ecosystem decomposer function in a confined place under controlled conditions. Although it might be thought that treatment would solve the sewage pollution problem, it has solved only part. Disposal of effluent from normal secondary treatment plants can generate problems because most nutrients are released in dissolved inorganic form. Although much of the carbon dioxide is lost during the aeration process, the effluent contains much more phosphorus, nitrogen, and some other elements than do most natural waters. Of these, phosphorus is especially enriched, relative to other elements. Effluent from secondary treatment typically contains about 8 milligrams per liter of phosphate-phosphorus and 20 of inorganic nitrogen, including ammonia.[5] Although natural values vary greatly with place and time, typical values for an unproductive lake would be, respectively, 0.01 and 0.10 milligram per liter; that is, the sewage has 800 and 200 times the natural concentration. The annual human release to sewage is about 1.4 pounds of phosphorus and 11 pounds of nitrogen per person per year. Added to this is phosphate from detergents, which accounts for roughly half the phosphate content of sewage. This means that the effluent is a very rich fertilizer which supports the growth of algae and other photosynthetic organisms. It might be thought to be favorable to increase the biological productivity of a lake by fertilization, and up to a point it can be. But this system, like any other, can be overloaded, and there can be very unpleasant consequences of enrichment with treated sewage effluent.[6,7]

The great increase in the supply of dissolved nutrients increases the production of photosynthetic organisms, and the abundance of the microscopic algae in the open water (phytoplankton) increases. This makes the water cloudy, so that one cannot see very far into it, interfering with its use for drinking and recreation. Other interesting effects magnify the unpleasantness of overproduction of algae. One is that the kinds of algae that grow most abundantly under these conditions (certain species of the blue-green algae) tend to float when the weather is calm, and they concentrate into scums at the surface. Under the sun, they die, and large quantities of decaying algae float around or blow downwind and pile up on beaches. The odor is unpleasant, and if the water is to be used for drinking it requires elaborate treatment. Such conditions have been described as "masses of decaying algae . . ." which "look like human excrement and smell exactly like odors from a foul and neglected pigsty."

Another effect is that the particular kinds of blue-green algae that form the bulk of the population appear not to be used effectively as food by the small invertebrates in the open water (zooplankton). Thus, predatory control, which is important in some communities, is at a minimum.

Planktonic algae are not the only ones that respond to fertilization. The filamentous ones that grow attached to stones and docks and boats also thrive and may form dense growths which break away, float around, and wash up on the beach. They contribute to the odor problem, and their presence in masses is unpleasant for swimmers. The alga *Cladophora* has been a nuisance in some of the Great Lakes.

Finally, some of the blue-green algae produce substances toxic to animals. Cattle frequently have died after drinking from a pond with a decaying bloom of such algae. This is unlikely to be a frequent problem for people, but some are sensitive and develop skin rashes or vomiting after swimming in lakes with abundant blue-green algae.

LIFE HISTORY OF LAKES

Before showing how this kind of problem has been handled, it may be useful to consider the normal changes that a lake goes through in its life.[8] Lakes originate from various geological processes, but many of the most familiar lakes in inhabited regions of the temperate zone owe their origin or present form largely to the action of glaciers. Many of the small lakes in the temperate zone originated as blocks of ice left as the glacier melted back at the end of the Ice Age. Typically, such a lake becomes populated by organisms of various kinds that can be transported by wind, by migrating birds, or by their own movements. Algae start growing in the open water, and seeds of water plants germinate on the bottom in shallow water and form stands of underwater vegetation. Animals become established. Eventually a complete aquatic community is developed. In the meantime, similar events are taking place on land in the drainage area around the lake. A cover of vegetation develops which reduces erosion of the developing soil. Before this time, erosion of the land may have delivered large quantities of nutritive elements to the lake. While development of a cover of vegetation may reduce erosion, it does not stop the inflow of nutrients. The plants tap the chemical resources of the soil and make available nutritive elements that are released when they decompose.

As long as the lake is relatively deep, it enters a long period of equilibrium during which the biological productivity and the abundance of plants and animals match the input of nutritive material from the drainage area. During this time, the lake gradually becomes shallower as the bottom sediment builds up, but as long as there is a rather large volume of water that is deeper than plants can grow, the lake remains in something like a steady state. The steady state may become disturbed by landslides, changes of climate, fires, or other major events that affect the input of material or sunshine into the lake. During this time the lake is stratified in summer, with a distinct layer of warm water floating on top. The warm layer is stirred by the wind, but it seals off the rest of the lake from contact with air.

In time the lake becomes so shallow that the relation between the shore region and the deep-water region are changed in an important way. The area of bottom that can be inhabited by plants increases greatly, so that a larger proportion of the area of the lake is inhabited by massive, long-lived plants that provide cover for animals and surface for attachment of algae and animals. When the lake becomes shallow enough, it is stirred to the bottom by wind all summer instead of being stratified. This changes some of the chemical processes that affect the productivity of plants. The net result is that, toward the end of its life, a lake develops much larger quantities of organisms per unit area than it did earlier. Eventually it becomes filled entirely and covered by land vegetation. When a lake is very shallow at the time of origin, it very quickly goes into the last, weedy stage.

Before the lake becomes too shallow to stratify, important changes take place in the deep water during summer. Because of the increased consumption of oxygen in the deep water by decomposing organisms, the dissolved oxygen may become depleted, rendering that part of the lake uninhabitable by fish. Even a deep lake may lose the oxygen from the deep water if the productivity is increased by sewage enrichment. Typically, enriched lakes have lost their deep-water populations of salmonid fish, even though coarse fish may thrive in the upper, shallow layers.

Lakes that have a large supply of nutrients are called *eutrophic* and those with small supplies *oligotrophic*. Eutrophic lakes usually support a high rate of biological production and produce dense populations of organisms. The term "eutrophication" is usually used to indicate an increase in the rate of supply of nutrients. The term "eutrophic" is sometimes used to describe lakes that produce dense growths of plants or algae. This variety of use to cover different concepts with the same words has led to some confusion, but the basic ideas are pretty simple.[9]

Changes in the productivity and abundance of organisms as a result of enrichment with sewage have been observed in many lakes.[6] Some of the best-studied cases are in central Europe, where limnology was well advanced in the late part of the nineteenth century. The lake at Zurich in Switzerland is a very well known and studied example. In 1898, the algal population abruptly changed, and by 1910 the valued deep-water coregonid fish were in trouble. A more recent example of eutrophication by treated sewage in the United States is especially interesting because it was brought under control by enlightened citizens' action long before the extent of our pollution problems was well understood in the country.

THE LAKE WASHINGTON STORY

The city of Seattle lies between Puget Sound and the west side of Lake Washington.[8] Early in the twentieth century, the lake was used for disposal of raw sewage, and unsatisfactory conditions developed. In the early 1930s, the sewage was diverted to Puget Sound, and for a few years the pollution of the lake was reduced. However, Seattle was expanding and smaller towns around the lake were growing. In 1941, a two-stage biological sewage treatment plant was established on the lake, and by 1954, 10 such plants had been built. An additional one was built on one of the inlets to the lake in 1959. At the maximum, the inflow of treated sewage effluent was about 20 million gallons per day. In addition, some of the smaller streams were contaminated with drainage from septic tanks. Studies of the lake in 1933,

3   **Changes in Lake Washington.** Concentrations of two nutrients are shown as means for the months January through March, when they are highest. Phosphate-phosphorus and nitrate-nitrogen are given as micrograms of the element per liter. The abundance of algae in summer is indicated by the mean chlorophyll content for the months July and August. The transparency of the lake, strongly affected by the abundance of algae, is shown by the depth to which a 25-cm white disk can be seen; note that the scale has zero at the top, so the farther down the value, the clearer the lake. The vertical bars show the range of values observed in the summer, July and August, and the cross bar is the mean. The relative amount of treated sewage entering the lake is shown as a percentage of the maximum rated capacity of the treatment plants, 20 million gallons per day.

Note that the lake began to respond as soon as part of the sewage, about one-third, was diverted. The concentrations of phosphate stopped increasing and began to decrease when the sewage was half-diverted. Nitrate was not changed as much, relatively.

258

1950, and 1952 showed increases in algae and nutrient content (Figure 3). In 1955, a conspicuous growth of the alga *Oscillatoria rubescens* developed. This event attracted attention because this species occurred early in the process of deterioration of a number of European lakes, including the lake at Zurich. Thus, it seemed to be a vanguard of pollutional deterioration, and Lake Washington appeared to be responding to enrichment with sewage effluent. It was exhibiting perfectly normal behavior when it began its increase in the abundance of algae, and it was possible to predict with considerable confidence what would be

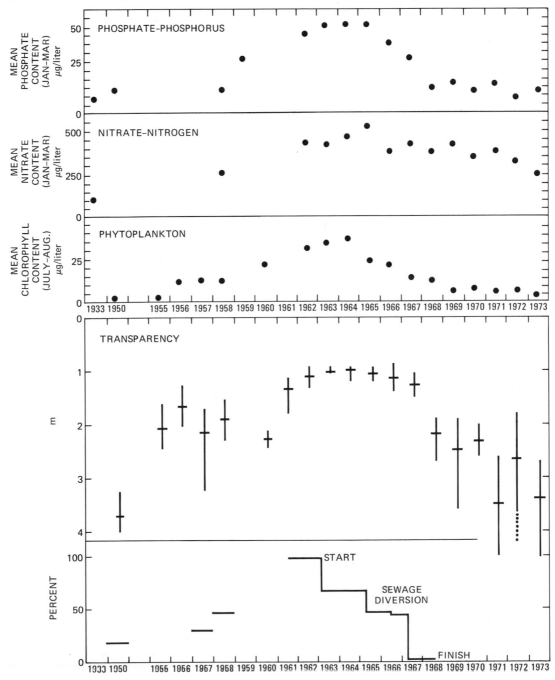

the consequences of continued or increasing enrichment.

Public concern had been growing about the sewage situation in the entire Seattle metropolitan area. In 1955, the mayor of Seattle appointed a Metropolitan Problem Advisory Committee to study, among other things, these sewage conditions. The obvious deterioration of Lake Washington and the rather clear-cut predictions that could be made about its future conditions gave focus to the public concern. At the same time, it was recognized that unsatisfactory conditions existed in Puget Sound and that a broadly based, coordinated program was necessary.

As a result of the Committee's action, a campaign was organized by public-minded citizens' groups to develop a governmental organization to handle the problem (Municipality to Metropolitan Seattle, or "Metro"). An active informational campaign was carried out, using to a large extent information about the actual deterioration of the lake and predictions about its future.

After a certain amount of difficulty, the public voted for the authorization of Metro in 1958. A sewage diversion project was started in 1963 and completed in 1968. Much of the sewage that formerly entered Lake Washington is taken to a large primary treatment plant that has its outfall very deep in Puget Sound and far from shore, where strong tidal action dilutes and disperses the effluent. The rest is given secondary treatment and discharged into a river that leads to Puget Sound. These changes are not expected to cause algal problems in the Sound because of the special chemical and hydrological conditions there (see below).

With the first diversion of about one-fourth of the sewage, deterioration of Lake Washington slowed, and further diversions were promptly followed by more improvement, as measured by increased transparency of the water and decreased amounts of phosphorus and algae (Figure 3). During late summer of 1969, some of the deep water had more oxygen than in 1933, phosphate was nearly down to 1950 concentrations, and the summer transparency, as judged by the depth to which a white disk can be seen, was 2.5 times as great as in 1963.[8] In the summer of 1971, the maximum transparency was 4.5 m, more than

had been observed in 1950. It is important to realize that action was taken before the lake had deteriorated very far relative to the well-known lakes in Europe and the Midwest of this country. The condition of the lake had changed conspicuously enough that there was no doubt about it, but action was taken early in the process. It is clear that Lake Washington responded promptly and sensitively to the increase and decrease in nutrient income. Lake Washington should not be regarded as an unusual lake; many lakes would respond just as sensitively. Many of the results of the Lake Washington study can be generalized and used for evaluating other enrichment situations.

A point of particular interest has to do with the relative importance of phosphorus, nitrogen, and carbon, a matter about which there has been much debate. In Lake Washington, phosphorus has decreased much more than nitrogen or carbon dioxide. The abundance of algae has decreased in very close relation to phosphate, not in relation to nitrate or carbon dioxide.[10] This suggests that, in similar lakes, limitation of the amount of concentrated sources of phosphorus reaching the lake can be beneficial. Considerable attention has been given to methods for reducing the phosphorus content of sewage effluent.

Puget Sound, filled with upwelled water from the Pacific Ocean, is a naturally productive body of water, and it is valued for the resulting rich fishery. Concentrations of nutrients are higher than they were in Lake Washington at its richest, and they are hardly ever exhausted from the bulk of the Sound. Conditions do not permit the development of algal nuisances of the type that disturb people about eutrophicated lakes. In much of the Sound, algal populations are not able to build up to the point of exhausting the nutrients, because at times of high tides, the water is mixed so deeply that the algae do not experience enough light to keep growing. Thus the extra load of nutrients from Lake Washington is unable to contribute to nuisance conditions. The water of Puget Sound is mixed so thoroughly and exchanged so frequently with the ocean by tidal action that sewage effluent cannot accumulate as discrete masses. Thus, the nutrient enrichment problem of Lake Washington was not sent to Puget Sound.

Furthermore, part of the Metro operation was to collect 70 million gallons of raw sewage per day that had been entering at sea level and put it through a big treatment plant that has its outfall in deep water far off shore. This made a pronounced improvement in conditions along the Seattle waterfront.

Of course, any body of water can be affected by enough contamination. Care and caution should continuously be used. Particular attention should be given to new developments of potentially toxic household or industrial products which are disposed of through sewers. Puget Sound receives large volumes of municipal wastes, and the possibility of damage must always be considered and preventive action taken.

In fact, Metro has legal authority to refuse to accept waste with concentration of toxic materials higher than specified limits, and industries that generate such wastes must treat them or dispose of them elsewhere than into the public sewer. This point is worth emphasizing because many proposals for the control of pollution are based on *removal* of undesirable materials. In view of the difficulties of disposal that are generated, it makes sense instead to limit the *input* of such materials to sewage when possible.

OTHER WAYS TO AVOID EUTROPHICATION

The solution of the Lake Washington problem was complete diversion of sewage effluent to Puget Sound, where it could not create the problems that it was creating in the lake. Such a solution is not available to most communities. Usually, in contrast to the Lake Washington situation, such a move would just send the algal problem somewhere else. For example, in 1936 effluent was moved downstream from Lake Monona, Wisconsin, to Lake Waubesa, which promptly started to make nuisance conditions.[7] Therefore, the question arises of what else can be done to avoid eutrophicational damage to lakes.

Cases are known in which lakes have deteriorated under the influence of wastes that were rich in phosphorus but not in carbon dioxide or nitrogen, and in one lake, removal of the effluent was followed by prompt recovery of the lake. Since about half the phosphorus in modern sewage originates in household detergents, a distinct reduction in the phosphorus input to lakes that receive sewage would be made by stopping the use of phosphorus detergents. In some cases this could be enough of a reduction in phosphate load to make a distinct improvement in the lake, but not if the human population were large enough that the physiologically produced phosphate were enough to generate nuisance conditions.

Problems of eutrophication by sewage existed before detergents were invented, but detergents have magnified the effect of sewage. Thus, an important difference in the phosphorus income could be made by changing the formulation of detergents, but it is not a simple action to take. People want to clean their dishes and clothes effectively, so an effective substitute must be found that will not produce its own problems. Arguments have been made against a change on the basis that the proposed substitutes do not clean well, have bad biological effects, are hazardous to young children, or would eutrophicate the coastal waters of the ocean. Unfortunately, many of the people who have argued in favor of keeping phosphorus detergents have felt it necessary to deny the importance of phosphorus in the control of productivity in many lakes, and have presented incorrect or misleading statements that have interfered with formulation of rational decisions.[9] Although a reduction in detergent phosphorus could be helpful to many lakes, it would not be a universal solution to all eutrophication problems, and other techniques are being considered.

In recent years improved methods have been developed for advanced treatment of sewage, to remove phosphate on a mass scale. The question is whether removal or reduction of the phosphate content of sewage will be adequate. This is a field in which considerable controversy exists, and different problems have been mixed up.

On the one hand, if we discuss the ecological requirements for production of dense populations of algae, we have to mention that they are affected by the amounts of carbon dioxide, nitrogen, phosphorus, and light; the temperature; and a number of other conditions, too. To explain why a given population exists requires a lengthy study. If, on the other hand, we are concerned with what actions we

261

need to take to improve a lake that is enriched with sewage, the problem becomes simpler. In theory, if we limit any essential element enough, production will be checked, but only a few elements are available for control. For chemical reasons, nitrogen is difficult to eliminate from sewage. Much carbon dioxide is eliminated in secondary treatment, but there are vast reservoirs in the atmosphere, and hard waters especially have bicarbonate that can support plant growth. The influential element that is most easily eliminated from sewage is phosphorus.

The proposal to control eutrophication by removal of phosphorus from the effluent has been very attractive. It can be regarded as part of a program to regard wastes as potential sources of valuable materials, and to recycle them. It would be advantageous to reclaim phosphorus from sewage and use it again for fertilizer. Unfortunately, this is not a simple solution. Most systems under consideration involve the use of chemicals to precipitate phosphate, and widespread adoption of such systems would impose serious demands for materials. For instance, Clair N. Sawyer has pointed out that to treat the sewage of the Washington, D.C., area by one particular process would require 8 percent of the current U.S. production of alum. The systems also require additional pumping of effluent, which would require more energy.

Further, precipitation of phosphate produces a sludge that presents difficult problems of disposal. It would be advantageous to use the sludge for fertilizer, thus recycling the phosphorus, but some sludges are unsuitable because they also have concentrated toxic metals such as vanadium and mercury, and pesticides. The largest single cost of sewage treatment is in dealing with the residual sludge.[5] It can take up to 50 percent of the capital and operating costs. The quantities involved are great; Chicago alone produces 900 tons daily on a dry weight basis.

Another phosphorus-removal system that is under test has considerable promise because it minimizes the use of additional chemicals and reconstruction of treatment plants. It involves increasing the absorption of phosphate by organisms in the treatment process by manipulating the oxygen supply.[11]

The foregoing discussion has applied to the disposal of household wastes in a very dilute form, where material from toilets and sinks is mixed into a large volume of water. It is much easier to handle concentrated materials, and radically different methods of disposal are being studied. In effect, the wastes are collected in a concentrated form and converted to compost. Problems of the control of disease organisms must be solved, and a considerable change in public attitudes would be required to make this type of system workable.

Not all lakes that generate nuisance conditions do so simply because of overproduction of floating algae. Some are in the last stages of ecological succession and have filled in to the point where they can support a dense growth of rooted vegetation over much of the bottom. Diversion of effluents from such a lake may make little improvement, but large improvements have been made by cutting the vegetation and dredging out a few feet of the soft sediments.[12] Green Lake in Seattle was further improved by increasing the flow of nutrient-poor water through it.[7]

AGRICULTURAL DRAINAGE

Human sewage is not the only kind that makes problems. Animals in feedlots produce wastes, and some of the material drains into waterways. That which does not offers large difficulties of disposal. Domestic animals excrete from 17 to 45 pounds of phosphorus per thousand pounds of animal per year. The production of animal wastes in the Potomac Basin is almost six times that of the human population there.[2] In areas of big-scale cattle and chicken farming, disposal of animal wastes far outweighs the human sewage problem.

Nor is sewage the only source of concentrated nutrients. In some regions much agricultural fertilizer gets into lakes because it is spread on frozen ground or snow, a very wasteful practice. Such conditions seem especially common in the midwest of the United States.[7]

Fortunately, phosphorus is generally more tightly bound in soil than is nitrogen, and water draining agricultural land does not carry a load of phosphorus proportional to its application, relative to nitrogen. In general, fertilized lawns are relatively unimportant sources of phosphorus. Nevertheless, wasteful techniques of application should be avoided, especially on lawns and gardens around small lakes.

A remarkable difficulty has occurred in some regions, where excess nitrate originating from septic tank or agricultural drainage has gotten into well water. Nitrate in drinking water becomes reduced to nitrite in the stomach and combines with hemoglobin, reducing the oxygen capacity of the blood. When the concentration of nitrate exceeds about 38 milligrams per liter, the condition can be fatal to infants and cattle, although the Public Health Service standard for drinking water is 45 milligrams per liter.[5]

This problem has become serious in southern California, which has a long history of pumping water for irrigation, and where drinking water comes from wells replenished by groundwater. Agriculture in southern California at first depended on pumped groundwater for irrigation. As pumping continued over the years, the level of the water table dropped lower and lower, being lowered by as much as 170 feet over a wide area. At the same time, fertilizers were applied and the irrigation water leached nitrate into the soil. In recent years, southern California has been using water imported from farther north, causing a change in the water balance so that the water table is rising again. When water comes in contact with the layers that have been enriched with nitrate, the nitrate is moved into wells used for drinking water. Children of farm laborers have suffered from nitrite poisoning.

SEDIMENT

When forests are cut or land is kept in a disturbed state, for example, by repeated plowing without precautions, large quantities of silt are washed into rivers and lakes (Chapter 1). The effect is multifold; part of the damaging effect results from smothering the bottom community, covering spawning beds of fish, and from making the water so turbid that light cannot penetrate to support plants. In addition, the increased rate of inwash of eroded materials greatly hastens the extinction of small lakes. This is particularly serious in small lakes

that are near extinction through natural aging. But even large lakes can be affected. Two and a half million tons of silt enter Lake Erie annually. Further, when erosion is from heavily fertilized agricultural land, some of the fertilizer that was intended to increase the production of useful land plants goes to eutrophication instead.

Silt appears to be more of a problem in the semiarid regions of the west in North America than in the forested areas of the more humid zones, but local activities can produce unsightly conditions anywhere.

Other kinds of activity can increase the rate of filling of lakes. A spectacular case that attracted much attention in the 1970s involved the disposal of finely divided mineral waste from iron ore (taconite) processing. Reserve Mining Company was putting 67,000 tons per day into Lake Superior, one of the world's major lakes, which was until recently hardly affected by pollution.[7] Legal action was started to cause the company to treat the waste or to go to land disposal because of the prospects of several kinds of damage to the lake. The problem became acute in 1973 when it was found that the waste also contained asbestos fibers, a serious health hazard in air pollution. Although asbestos in water has not been demonstrated to be a hazard, the fact that several cities get their drinking water from Lake Superior caused concern and led to a court order to close the plant. The action was contested, and recently the company was allowed to continue operation.

This case appears to be a good example of the situation with much industrial pollution. The position of the company appears to be that it would cost too much to install settling tanks or other processes, because profits and competitive position would be impaired. On the other hand, it can be argued that treating the wastes should be part of the cost of doing business and reflected in the price of the product (Chapter 16). As it is, a debt is being generated in the form of eventual deterioration of the lake. One very objectionable argument that is made repeatedly by defenders of this kind of industrial action is to say, in effect, "Nobody has been killed yet." It appears that we must generate a real disaster before it is worthwhile to take protective action. But long

263

before real disasters develop, very unsatisfactory conditions exist.

Toxic Wastes

Many industrial wastes contain material that is toxic to organisms and not only interferes with the self-purification processes in streams but kills the native fauna as well. Some metal treatment processes produce an effluent that contains cyanide. A spill of this material, accidental or otherwise, can kill the fish for many miles downstream. Although most pollution laws specify disposal of this kind of waste in some other way, trouble continues to be had with it in various places.

A number of materials have the property of being concentrated from the environment by organisms and becoming additionally concentrated from food by the higher members of the food chain (Chapter 1). That is, a substance is absorbed from food and built into the tissues of the feeding animal. As time goes on, the concentration of the substance in the tissues increases and may reach a level at which it interferes with physiological functions. The most famous case of this sort of problem is with DDT, which has given great trouble both in terrestrial and aquatic communities. Carnivorous fish can develop such high concentrations in their tissues that they are affected and become unsuitable for human food. The successful introduction of the coho salmon into the Great Lakes was met with enormous enthusiasm, but disappointment ensued when the sale and consumption of the fish was banned because of their high DDT content. On land, DDT interferes with formation of eggshells by predatory birds, and several species have been affected. The peregrine falcon has been eliminated from some areas (see Chapter 13). *Homo sapiens* can tolerate higher tissue concentrations without symptoms of damage than many of the species he wants to encourage.

Some metals have the same property of becoming concentrated in organisms. Rather suddenly, mercury has become recognized as a major health problem in Japan, Sweden, and all parts of the United States. Many inorganic and organic compounds of mercury are soluble and can be widely dispersed in natural waters. Mercury is toxic. During prolonged exposure to mercury compounds, even at low concentration, mercury accumulates progressively in the nervous system, impairs nervous function, and eventually causes death. The phrase "mad as a hatter" was derived in the nineteenth century from the behavior of hat makers, who were in the habit of holding felt in their mouths while fashioning it and were thus poisoned by mercury compounds used to treat the felt. Even low mercury levels in mothers can lead to abnormalities in their babies.

Mercury enters the environment from natural and industrial sources. In the past 20 years its use has increased greatly in industry, where it is used, for example, as a catalyst in the production of plastics and as a fungicide in pulp mills. Untreated effluents from these industries can raise the concentration of dissolved mercury compounds in the water manyfold. Metallic mercury is used for electrodes in the production of chlorine and sodium hydroxide, and much of this metal has been discarded into the Great Lakes. In Lake Erie and associated waters, the fish contain so much mercury (1.3 ppm; the legal limit is 0.5) that the sale of fish was banned, and fishing was prohibited in Lake St. Clair.

In agriculture, organic compounds of mercury are used on seeds as fungicides. Eating such seeds has led to the death of many seed-eating birds and the accidental death of children. Some of the mercury is absorbed by the plants that grow from the seed, increasing the mercury content of food. Some of the remainder is leached from the soil into rivers and lakes. This has been a major problem in Sweden, where mercury is still increasing in fresh waters by leaching from the soil, even though agricultural use of mercury was restricted in 1966.

In lakes, various processes result in the accumulation of mercury in animals. One is the usual food-chain effect. Some is lost to the sediments in the form of dead organisms, but part of the sedimented material is taken up by invertebrate scavengers and part is released by bacterial action. Bacteria can produce methyl mercury, even from metallic mercury, which is soluble and is absorbed directly by fish.

Lead is another metal that is being spread

around the landscape, including lakes and streams.[13] Lead is liberated not only by industries but in the exhaust of automobiles and boats that use leaded gasoline. The vegetation close to major highways has significantly more lead than that a few hundred feet away. Lead deposited on roadways can easily be washed into the nearest watercourses by rain. Symptoms of lead poisoning in man include depression and impaired mental function. While the levels of lead in food animals have not been as high as that of some other metals, lead seems to be one more part of the burden of toxic substances that we are carrying around in us, each contributing proportionately to the damage of our mental and physical functions.

A very prevalent product of industry is waste oil. Oil spilled over a lake or stream can damage plants and animals and interfere with swimming and drinking. Major oil-spill accidents are not the only source of oil problems on lakes. Most of the oil added to the fuel of outboard motors is not burned and contributes to the oil slicks that are becoming prominent on some recreational lakes. Disposal of waste oil is difficult, but putting it down the sewer is a very poor solution. The oil can be broken down by microorganisms, but it takes time, and in the meantime, considerable damage can result.

Oil spills from wells and tankers are mainly a marine problem, but the large lakes are being increasingly affected. Undesirable amounts of oil are liberated even in routine shipping and transfer operations, not just as a result of major accidents. One ordinarily does not think of rivers as being fire hazards, but the Cuyahoga River in Ohio is now famous for an occasion on which a spill of flammable material caught fire and damaged bridges.

MINE DRAINAGE

Drainage from coal mines presents a rather special but widespread problem. Coal deposits contain iron sulfide, which reacts with oxygen in water to form sulfuric acid and iron hydroxide, which precipitates, making the substrate unsuitable for normal stream animals. The strong acidity of the water prevents the occurrence of the normal stream flora and fauna. Thousands of miles of streams have deterio-

rated under the influence of coal mine wastes, particularly in areas of strip mining.

RADIOACTIVE ISOTOPES

The problem of environmental contamination with radioactivity from fallout, nuclear accidents, or normal loss from industrial use of atomic energy is widely known. Freshwater organisms concentrate some of the isotopes, such as strontium, vastly in excess of natural amounts. A special feature of fresh waters is their susceptibility to contamination by leaking underground storage tanks of radioactive wastes. At present, the technology of storage appears not to be advanced enough for us to dismiss the possibility of difficult future problems (Chapter 12).

Thermal Pollution

Thermal pollution simply means the addition of hot water to a stream or lake.[14,15] Steam and nuclear power plants need water for cooling, especially the latter, which produce 60 percent more heat per unit of electricity than coal plants. The water is usually taken from a stream or lake, passed through the cooling system, and returned at a temperature higher by 5 or 10°C. The manufacturing industry used over 9,000 billion gallons of cooling water in 1964, and electrical generating plants used almost 42,000 billion gallons. With the projected increase in nuclear power, the requirements for cooling water are growing greatly, and in heavily industrial regions, all the available water could be used for cooling. The cost of cooling increases with the degree of cooling. The difference in construction cost between permitting a 10°F increase in the temperature of cooling water or a 20°F increase in 514 steam plants operating in 1965 would be $132 million.[2]

There is much controversy about the extent of damage caused by thermal pollution. Some animals are killed outright by the hottest water, but with mixing, relatively little water is above lethal temperature. That does not mean that it is harmless. It has been suggested, for instance, that spread of a bacterial disease of salmon in the Columbia River (columnaris disease) is facilitated by a slight elevation in temperature. Thus, there is the possibility of an interaction of factors, so that a degree of heating, not in itself damaging, may facilitate a damaging condition caused by something else.

The major damage is probably caused not by heat, but by mechanical injury to the small fish and invertebrates swept through the cooling tubes at high speed, most of which are killed. Several plants along a river could impose a very significant mortality on the populations and could eliminate some species of fish. On the other hand, fish may congregate near a warm-water outfall, and in doing so may be able to grow better during cold weather. This matter is still under study and the situation is not clear. Some people are optimistic that the hot water can be used beneficially, either industrially or to increase productivity of useful fish. However, the temperature adaptations of animals are results of a long evolutionary history. It is unlikely that a major change in environmental conditions can be met without affecting the populations. Tropical rivers and hot springs have existed for a long time and populations have adapted to their thermal regime. Doubtless, life can continue in heated rivers. The question is whether the resulting changes will be acceptable. The possibility exists that different genotypes will be selected in the newly warmed places, and we may soon be seeing interesting experiments in natural selection.

Is Lake Erie Dead?

Lake Erie has become the focus of much attention. Without question it is a very badly polluted lake, receiving all the kinds of wastes discussed above and probably more.[5] It has been heavily eutrophicated and the productivity increased despite the inflow of toxic wastes. It appears that toxic damage is somewhat localized, and for hydrological reasons, the greatest deterioration is limited to shallow inshore waters. The open part of the lake continues to be inhabited by plankton, bottom fauna, and several species of fish. The species that were most desirable a few decades ago, such as the lake trout and sauger, are no longer present, but Lake Erie produces about

50 million pounds of fish per year, which is about half the entire production of the Great Lakes.[7] The fishery in the mid-1960s was dominated by yellow perch, smelt, and sheepshead. That pollution has been an important influence in Lake Erie is suggested by the fact that some species of fish disappeared from the more polluted western end earlier than from the eastern end.

We should investigate the condition of a lake that can properly be described as "dead." Surely, a dead lake would be one that is sterile, devoid of life. It seems anomalous to call a lake "dead" when it is swarming with living things. This does not mean that the lake is in good condition, but it is anything but dead. To call Lake Erie "dead" carries certain implications that confuse one's thinking about how to improve it. After all, death is irreversible, but eutrophication is not. The sight of dead fish washed up on the shore of a lake should not mislead anybody. One cannot have dead fish without having live ones first. Of course, this is not to say that a massive die-off is a good thing.

Lake Erie has been a naturally productive lake during all of recorded history. It was famous in the early part of this century for producing great swarms of mayflies that flew all over, became crushed on street car tracks, and made a nuisance by their abundance in the wrong place. These animals spend most of their life as immature aquatic stages in the lake and the vast yield to the air of adults was a sign of the great productivity of the lake. The mayflies no longer swarm because they cannot tolerate the low-oxygen conditions that now exist, but the bottom of the lake supports many sludge worms (*Tubifex*) and other invertebrates.

Lake Erie is but one of the Great Lakes, all of which are generating some concern. Lake Erie is the smallest in volume and the most heavily polluted, so it has deteriorated the most, but all show some signs of effect. There has been a massive decrease in the abundance of desirable species of fish, but not every decrease can be attributed to pollution. Some decreases have resulted from the introduction of predatory, nongame fish when the Erie and Welland canals opened, and some have been the result of overfishing. For example, the decline of the lake trout in Lake Michigan by 1953 was importantly influenced by predation by the sea lamprey.[16]

Some accounts of eutrophication have mentioned "the point of no return," suggesting that a lake could be damaged so much that it could not recover, no matter what remedial efforts were made. In particular, some people have expressed fear that Lake Erie is beyond repair. I am convinced that it is not. I think that limitation of the input of sewage phosphorus could make a large and noticeable improvement in its condition. The question has been raised whether considerable improvement in lakes can be made by eliminating the use of detergents with phosphate. Whether this, rather than removal of phosphate from effluent, would have enough effect on Lake Erie is difficult to judge. It would seem reasonable to do a certain amount of experimentation on a large scale. In any case, to reduce the loading of concentrated phosphorus on eutrophicated lakes makes sense. Phosphate limitation would not be enough by itself to restore Lake Erie to a fully satisfactory condition. Two other operations would be required to get the lake back to something like its original condition: reduction of the input of concentrated toxic industrial wastes and cleaning up of the junk that has been permitted to accumulate in bays, inlets, and shallows. Attempts are being made to improve all these conditions.

Although recovery might be slow in coming after a lake is relieved of its burden of pollution, very few situations exist in which some improvement could not be expected. Even Lake Erie could be restored to its original mayfly-rich condition. For a lake really to be "killed" would require either that it receive so much toxic waste that nothing could live or that it be filled in to the point where it no longer existed as a lake.

The real question concerns not whether a lake is "dead" but the fate of dissolved nutrients and toxic materials. Many factors, such as the rate of replenishment of the lake's water and the amount of nutrients retained in the sediments, must be taken into account. To calculate the magnitude of the effect of dissolved materials very exactly in many cases is beyond our present abilities. To make the attempt on some well-selected cases would increase our understanding.

Air Pollution and Water Pollution

Rain is not simply distilled water. It brings down with it particles and dissolved materials that have been entrapped or absorbed from the air. One of the outstanding problems generated by air pollutants dissolved in rain is caused by sulfur dioxide produced by the burning of fuel oil rich in sulfur. Sulfur dioxide dissolved in water becomes oxidized to form sulfuric acid; over 5 milligrams per liter of sulfur can be found in rain in industrial regions.[5] In regions of heavy use of sulfurous fuels, the rain can be distinctly acid, with a pH value less than 4.0. This has been especially well studied in Europe, where, since 1955, the rain has shown a distinct tendency to become more acid, with values higher than pH 6.0 becoming rare, and values less than 5 becoming prevalent. While the rain is relatively dilute sulfuric acid, it can have pronounced effects, especially when it forms the water of streams and lakes in regions where, for geological reasons, there is little calcium bicarbonate to neutralize the acid, that is, where the natural waters have little buffering capacity.

The pH of the lakes of southeastern Sweden has been decreasing over the past two decades, and in recent years pH values as low as 3.8 have been observed. The average change has been 0.4 pH unit in the past 20 years. It is no mere coincidence that unusual numbers of dead salmonid fish are now seen in these lakes, and some lakes have become unsuitable for fishing.

Further damage may arise from leaching of calcium from the soil by acid rain, leading to poor growth of trees and crops. This problem crosses national boundaries; much of the sulfuric acid falling in Sweden originated in the industrial midlands of England (Chapter 1). While the acidified rain does much damage by corroding metals, not the least damage being done is the dissolution of marble and limestone buildings and works of art. The effect has been especially serious in Venice. This city, an art museum in itself, has been irrevocably dam-

267

aged by the erosion of statues and buildings. Some very important ties with our artistic past no longer exist in recognizable form.[17]

Pollutants as Resources

A productive approach to the pollution problem may be to regard the pollutants as resources to be recycled. Indeed, many of the pollutants are scarce and valuable materials. The facts of thermodynamics require that to recover these resources from dilute form will cost more than from concentrated form, such as ore or even industrial waste. But without such a program, other kinds of costs will be generated. At one time, the cheap and easy way to dispose of used mercury seemed to be to throw it into lakes, but the consequences in Lake St. Clair were awesome. Thus removal of such toxic materials from wastes is necessary, and the most effective way is to keep it out of sewage in the first place. Thus, the possibilities of recycling trace metals would be increased by keeping industrial wastes segregated from household wastes.

As for nutrients, one way to recycle them is to put them into a body of water large enough to accept them without being damaged, where

useful biological production would be increased. The most likely type of community to benefit from this kind of influence is the inshore water of the ocean in places where circulation is great enough that noxious conditions would not be generated. Some bad experiences have been had with sea disposal of sewage. In some cases the disposal of raw sewage has generated large volumes of water without dissolved oxygen. In other cases, inadequate circulation permitted the overproduction of algae, as in estuaries along Long Island Sound that were enriched with drainage from duck farms. Nevertheless, the richest fishing areas of the world are places where the upwelling of nutrient-rich water provides adequate nutrition for the phytoplankton. This effect could be simulated by sewage nutrients in those places with adequate access to sea disposal and where the content of toxic materials can, in fact, be limited. In some areas, a more controlled fish farming based on use of sewage might be practicable.[18]

Another technique for productive use of sewage nutrients is land disposal where the growth of terrestrial vegetation is increased by spraying sewage effluent. A major difficulty here is that the rate of flow of sewage may be much larger than the absorptive capacity of the

II Waste Treatment Requirements

Regions	Total plant required	Value of plant in place	Additional investment required
North Atlantic	814.0	575.5	238.5
Southeast	276.1	208.0	68.1
Great Lakes	973.4	784.2	189.2
Ohio	658.5	526.7	131.8
Tennessee	80.4	47.8	32.6
Upper Mississippi	205.1	149.9	55.2
Lower Mississippi	230.1	144.8	85.3
Missouri	88.2	64.2	24.0
Arkansas-White-Red	49.2	33.0	16.2
Western Gulf	286.8	168.9	117.9
Colorado/Great	25.9	17.0	8.9
Pacific Northwest[a]	167.6	121.1	46.5
California[b]	143.3	105.6	37.7
	3,998.6	2,946.7	1,051.9

Regional distribution of waste treatment requirements, 1968, in millions of 1968 dollars.
[a]Includes Alaska.
[b]Includes Hawaii.
Source: Reference 2.

available land. The soil can be saturated, so that the surface runoff to streams and lakes is rich enough to cause eutrophication.

No single technique will be universally applicable, but by a combination of controlling what goes into sewage and where and how the sewage is disposed, considerable improvements can be made. In each area, an appropriate mixture of techniques will have to be adopted to meet the particular environmental situation.

Magnitude of the Cleanup

It is obvious that we have many problems, but we understand enough to be able to deal with them very effectively. Before starting, it is necessary to find out how big the problems are, for much of the difficulty is financial: How much are people willing to spend, once they clearly see the consequences of failing to clean up?

It has been estimated that of the total population of the United States, as much as 40 percent has less than adequate municipal treatment of domestic sewage, or none at all. To develop adequate facilities would require roughly $8 billion for capital costs alone, and this would provide no more than secondary treatment, thus not really solving the eutrophication problem in much of the country.

Industrial wastes in 1964 amounted to 13,100 billion gallons containing 18,000 million pounds of suspended solids, as compared to 5,300 billion gallons of domestic sewage containing 8,800 million pounds of solids.[2] Studies of the problem show that all parts of the country need attention (Table II) and that a wide variety of industries is involved (Table III).

A recent estimate by the Council on Environmental Quality suggests that more than $120 billion should be spent during the period 1972–1981 for the control of water pollution. Only a small fraction of that is covered by the current rate of appropriation, and part of that has been impounded. Thus, the rate of expenditure necessary to control water pollution is similar to that of the Vietnam war. Such estimates are difficult to make with any precision because of incomplete information about the

III Cost of Industrial Waste Treatment Improvement

Industry	Annual investment to reduce existing requirement	Total investment to reduce waste treatment requirements and meet growth needs				
		1969	1970	1971	1972	1973
Food and kindred products	43.9	63.2	65.4	69.9	70.0	69.9
Meat products	7.0	10.1	11.2	11.2	11.7	11.6
Dairy products	4.6	5.1	5.7	5.5	5.5	5.5
Canned and frozen foods	6.7	11.4	12.4	12.6	12.9	13.0
Sugar refining	13.5	19.3	18.4	22.6	21.4	21.5
All other	12.1	17.3	17.7	18.0	18.5	18.3
Textile mill products	5.3	9.8	10.9	11.1	11.0	11.6
Paper and allied products	15.1	19.1	25.5	26.0	26.4	27.0
Chemical and allied products	56.0	75.7	76.9	77.7	79.4	77.9
Petroleum and coal	15.4	15.4	18.1	30.5	31.7	32.1
Rubber and plastics	6.2	7.0	7.9	7.1	7.2	7.1
Primary metals	29.9	83.6	91.3	93.3	96.2	97.8
Blast furnaces and steel mills	19.6	52.4	59.1	60.1	63.0	63.0
All other	10.3	31.2	32.2	33.2	34.2	34.8
Machinery	5.0	6.9	6.9	7.1	7.1	7.3
Electrical machinery	1.7	3.6	3.8	3.8	4.0	4.1
Transportation equipment	8.3	11.7	11.9	12.2	12.1	12.3
All other manufacturing	23.5	32.3	32.6	33.0	33.5	33.8
All manufacturers	210.3	328.3	351.6	371.7	378.6	380.9

Annual investment required to reduce the existing industrial waste treatment deficiency in 5 years, in millions of 1968 dollars.
Source: Reference 2.

needs in some areas and because it is hard to predict the effect of some recent legislation on costs. In any case, most of the estimates are based on plans for secondary treatment of sewage, which was seen to be inadequate to prevent the deterioration of lakes, as with Lake Washington.

A major development in the control of water pollution was the 1972 Federal Water Pollution Control Act Amendment, which set as a goal the zero discharge of pollutants by 1985. This legislation has caused considerable controversy, largely on the basis of the disproportion between costs and benefits when the goal is complete removal of all pollutants from all waters. Partial treatment is much cheaper and in many cases would restore acceptable conditions. It has been difficult to identify ways to meet the goals without abolishing cities. The difficulties probably will cause increased scrutiny of methods of keeping toxic materials out of effluent.

The rate of cleaning up has been far below the rate of increase of pollution. For an effective cleanup, it will be necessary to make some major readjustments in national priorities. Since production of pollutional effects is linked to population, it is hard to see how in the long run environmental problems can be controlled without population control. Some improvements can be made by reducing per capita quantities of pollutants, but the possibilities are limited, both by limitations of the physical principles involved and by the expectations people have for the requirements of a satisfactory life.

References

1. Murray, C. R., and Reeves, E. B. 1972. *Estimated Use of Water in the United States in 1970* (U.S. Geol. Survey Circ. 676). Government Printing Office, Washington, D.C.

2. Federal Water Pollution Control Administration. 1968. *The Cost of Clean Water*, 4 vols. Government Printing Office, Washington, D.C.

3. Murray, C. R. 1968. Estimated use of water in the United States, 1965 (U.S. Geol. Survey Circ. 556). Government Printing Office, Washington, D.C.

4. Hynes, H. B. N. 1963. *The Biology of Polluted Waters*. Liverpool University Press, Liverpool, England.

5. American Chemical Society. 1969. *Cleaning Our Environment. The Chemical Basis for Action*. American Chemical Society, Washington, D.C. (Supplement, 1971).

6. Edmondson, W. T. 1968. Water quality management and lake eutrophication: the Lake Washington case. In *Water Resources Management and Public Policy*. University of Washington Press, Seattle.

7. National Academy of Sciences. 1969. *Eutrophication: Causes, Consequences, Correctives*. National Academy of Sciences–National Research Council, Washington, D.C.

8. Edmondson, W. T. 1973. Lake Washington. In *Environmental Quality and Water Development* (C. R. Goldman, J. McEvoy III, and P. J. Richerson, eds.). W. H. Freeman and Company, San Francisco. pp. 281–298.

9. Edmondson, W. T. 1974. Review of *The Environmental Handbook. Limnol. Oceanog. 19:* 369–375.

10. Edmondson, W. T. 1970. Phosphorus, nitrogen and algae in Lake Washington after diversion of sewage. *Science 169:* 690–691.

11. Levin, G. V., Topol, G. J., Tarnay, A. G., and Sanworth, R. B. 1972. Pilot-plant tests of a phosphate removal process. *J. Water Pollution Contr. Fed. 44:* 1940–1954.

12. Bjork, S. 1972. Swedish lake restoration program gets results. *Ambio. 1:* 153–165.

13. Patterson, C. C. 1965. Contaminated and natural lead environments of man. *Arch. Environ. Health 11:* 344–360.

14. Clark, J. R. 1969. Thermal pollution and aquatic life. *Sci. Amer. 220* (3): 18–27.

15. Krenkel, P. F., and Parker, F. L. 1969. *Biological Aspects of Thermal Pollution*. Vanderbilt University Press, Nashville, Tenn.

16. Smith, S. H. 1968. Species succession and fishery exploitation in the Great Lakes. *J. Fish. Res. Bd. Canada 25:* 667–693.

17. Sargeant, W. 1968. The crumbling stones of Venice. *The New Yorker*, Nov. 18.

18. Ryther, J. H., Dunstan, W. M., Tenore, K. R., and Huguenin, J. E. 1972. Controlled eutrophication — increasing food production from the sea by recycling human wastes. *BioScience 22:* 144–152.

270

Further Reading

Council on Environmental Quality. 1970–1974. *Annual Report*. Government Printing Office, Washington, D.C. These reports contain detailed information on kinds and quantities of pollutants and the progress being made in controlling them.

Likens, G. (ed.). 1972. *Nutrients and Eutrophication*. American Society of Limnology and Oceanography Special Symposia 1.

McCaul, J., and J. Crossland. 1974. *Water Pollution*. Harcourt Brace Jovanovich, Inc., New York. 206 pp.

National Water Commission. 1973. *New Directions in U.S. Water Policy*. Government Printing Office, Washington, D.C. 197 pp.

President's Science Advisory Committee, Environmental Pollution Panel. 1965. *Restoring the Quality of our Environment*. Government Printing Office, Washington, D.C. 316 pp.

White, G. F. 1969. *Strategies of American Water Management*. University of Michigan Press, Ann Arbor, Mich. 155 pp.

Wolman, M. G. 1971. The nation's rivers. *Science 174:* 905–918.

11
Marine Pollution

Edward D. Goldberg   *is Professor of Chemistry at the Scripps Institution of Oceanography, La Jolla, California. He received his doctorate from the University of Chicago in 1949. His subsequent work has been primarily with the chemistry of the marine ecosystem and with the radiometric dating of sediments. He has worked with national and international organizations on marine pollution problems and recommended courses of action to protect oceanic resources.*

Kathe K. Bertine   *is an Assistant Professor of Geology at San Diego State University, San Diego, California. She received her doctorate from Yale University in 1970. Since then, her principal research has been concerned with trace metal pollutants in sediments, organisms, glaciers and water.*

273

Because of the mind of man, materials are mobilized about the surface of the earth in amounts that are now beginning to approach those of natural processes. Man was recognized as a growing geologic force by the Russian geochemist Vernadsky,[1] who introduced the term *noösphere* as the geosphere incorporating the activities of man's thinking.

Annually the world's population moves about 3 billion tons of materials about the planet in its use of minerals, plants, and animals. In addition, a slightly larger amount of fossil fuels (coals, oils, and natural gases) are combusted, releasing waste products to the atmosphere. Nearly all of these discards have little, if any, effect upon those characteristics of our environment that give quality to human life, be it man's health, the health of other living systems, or the many attributes of our inanimate surroundings. A few substances, however, have polluted parts of the earth, disrupting the normal course of events. Our concerns in this chapter will cover the impact of such pollutants upon the marine environment and upon the organisms that utilize its resources, including man.

The resources of the sea include its fish and shellfish for human consumption as food; its mineral resources, including sand, gravel, phosphorites, and the ferromanganese minerals; its shipping lanes; its recreational areas; and finally waste space for the discards of human society. The use of any one of these resources may have adverse effects upon another. For example, the mining of sand and gravel from coastal areas can destroy fish spawning grounds; the release of oil from tankers can soil beaches; mercury released to the oceans can be taken up by fish subsequently eaten by man, and are responsible for his illness or death. The problem, then, is to bring into harmony our multiple uses of the oceans.

The Oceans and Pollution in General

CHARACTERISTICS OF THE MARINE SYSTEM

The world ocean covers about 71 percent of the earth's surface to an average depth of slightly less than 4 kilometers. Because of the extent of the ocean, the utilization of the seas as a reservoir for the wastes of man has seemed very attractive. But when we discard a material in the ocean, we should know the consequences of our action, if at all possible. Whether a pollutant remains in the water, in the sediment, or concentrates in biological materials is a function of its chemistry. Depending upon the pollutant, it may be more dangerous in the sediment, water, or biota.

Dissolved chemicals are introduced to the oceans by rivers, winds, or glaciers. They remain dissolved in the water for time periods (residence times) extending from hundreds of years to millions of years before being accommodated in the sediments or before destruction (Table I). Such estimates of the length of time that a dissolved chemical remains in the water are derived from a steady-state model of the oceans in which the amount of a chemical introduced per year is balanced by an equal amount removed to the sediments or destroyed. In such a scheme, the ocean concentrations never change. Processes other than sedimentation that remove chemicals from seawater include radioactive decay and photochemical decomposition or destruction by liv-

274

I Chemical Characteristics of Seawater

Element		Concentration (μg/liter)	Residence time (years)	Element		Concentration (μg/liter)	Residence time (years)
Hydrogen	H	1.1×10^8	—	Silver	Ag	0.3	4×10^4
Helium	He	7×10^{-3}	—	Cadmium	Cd	0.1	—
Lithium	Li	1.7×10^2	2.3×10^6	Indium	In	<20	—
Beryllium	Be	6×10^{-4}	—	Tin	Sn	0.8	—
Boron	B	4.5×10^3	1.8×10^7	Antimony	Sb	0.3	7,000
Carbon	C	2.8×10^4	—	Tellurium	Te	—	—
Nitrogen	N	1.5×10^4	—	Iodine	I	60	4×10^5
Oxygen	O	8.8×10^8	—	Xenon	Xe	5×10^{-2}	—
Fluorine	F	1.3×10^3	5.2×10^5	Cesium	Cs	0.3	6×10^5
Neon	Ne	0.12	—	Barium	Ba	20	4×10^4
Sodium	Na	1.1×10^7	6.8×10^7	Lanthanum	La	3×10^{-3}	6×10^2
Magnesium	Mg	1.3×10^6	1.2×10^7	Cerium	Ce	1×10^{-3}	—
Aluminum	Al	1	1.0×10^2	Praseodymium	Pr	0.6×10^{-3}	—
Silicon	Si	3×10^3	1.8×10^4	Neodymium	Nd	3×10^{-3}	—
Phosphorus	P	90	1.8×10^5	Samarium	Sm	0.5×10^{-3}	—
Sulfur	S	9.0×10^5	—	Europium	Eu	0.1×10^{-3}	—
Chlorine	Cl	1.9×10^7	1×10^8	Gadolinium	Gd	0.7×10^{-3}	—
Argon	Ar	4.5×10^2	—	Terbium	Tb	1.4×10^{-3}	—
Potassium	K	3.9×10^5	7×10^6	Dysprosium	Dy	0.9×10^{-3}	—
Calcium	Ca	4.1×10^5	1.0×10^6	Holmium	Ho	0.2×10^{-3}	—
Scandium	Sc	$<4 \times 10^{-3}$	$<4 \times 10^4$	Erbium	Er	0.9×10^{-3}	—
Titanium	Ti	1	1.3×10^4	Thulium	Tm	0.2×10^{-3}	—
Vanadium	V	2	8.0×10^4	Ytterbium	Yb	0.8×10^{-3}	—
Chromium	Cr	0.5	2.0×10^4	Lutetium	Lu	0.1×10^{-3}	—
Manganese	Mn	2	1.0×10^4	Hafnium	Hf	$<8 \times 10^{-3}$	—
Iron	Fe	3	2.0×10^2	Tantalum	Ta	$<3 \times 10^{-3}$	—
Cobalt	Co	0.4	1.6×10^5	Tungsten	W	0.1	1.2×10^5
Nickel	Ni	7	9.0×10^4	Rhenium	Re	0.008	—
Copper	Cu	3	2×10^4	Osmium	Os	—	—
Zinc	Zn	10	2×10^4	Iridium	Ir	—	—
Gallium	Ga	3×10^{-2}	1×10^4	Platinum	Pt	—	—
Germanium	Ge	7×10^{-2}	—	Gold	Au	1×10^{-2}	2×10^5
Arsenic	As	2.6	5×10^4	Mercury	Hg	0.2	8×10^4
Selenium	Se	9×10^{-2}	2×10^4	Thallium	Tl	<0.1	—
Bromine	Br	6.7×10^4	1×10^8	Lead	Pb	0.03	4×10^2
Krypton	Kr	0.2	—	Bismuth	Bi	0.02	—
Rubidium	Rb	1.2×10^2	5×10^6	Polonium	Po	—	—
Strontium	Sr	8×10^3	4×10^6	Astatine	At	—	—
Yttrium	Y	1×10^{-3}	—	Radon	Rn	6×10^{-13}	—
Zirconium	Zr	3×10^{-2}	—	Radium	Ra	1×10^{-7}	—
Niobium	Nb	0.01	—	Actinium	Ac	—	—
Molybdenum	Mo	10	7×10^5	Thorium	Th	$<5 \times 10^{-4}$	<200
Ruthenium	Ru	—	—	Protactinium	Pa	2.0×10^{-6}	—
Rhodium	Rh	—	—	Uranium	U	3	3×10^6
Palladium	Pd	—	—				

One microgram in 1 liter is a concentration of 1 part per billion. Notice that residence times vary from 100 years for aluminum to 19 million years for chlorine.
Source: Reference 2.

ing organisms such as bacteria or yeasts. The more highly reactive an element is, the shorter is its residence time.

There are other features of the steady-state model that are important for pollution studies. For example, a pollutant introduced in a constant amount per unit time to the oceans will build up to a steady-state (maximum) concentration in a time equal to about four times its residence period. Also, for polluting substances that have counterparts in the oceans today, the residence times of the counterparts allow predictions of the buildup of the pollutant, if estimates of its future productions can be made.

Clearly the steady-state model is an oversimplification. It assumes that there is a complete mixing of a chemical within the ocean in times that are short with respect to its residence time. This may not be a valid assumption for chemicals whose residence times are less than the times for oceanic mixing processes, which take place over periods of centuries. The important concept to be gained from residence times is the recognition that the oceanic time scales for the retention of materials is huge compared with other systems. Whereas rivers maintain introduced materials for periods of days to months, lakes for decades to centuries, the oceans can hold on to its dissolved chemicals for periods up to hundreds of millions of years. Substances injected by man today may be seen for many, many generations into the future.

A second important characteristic of the

marine system involves those biological and geological processes that result in the extraction of dissolved chemicals from seawater into both living and nonliving solids, called *bioconcentration* when it occurs in organisms (Chapter 1). The concentration factor (the ratio of the concentration of the chemical in the whole fresh organism, or in the mineral to that of its concentration in seawater) can reach values of 1 million or more.

Marine organisms not only accumulate unusual levels of heavy metals in their bodies, but they display remarkable specificities for particular substances. For example, it has been known for about a half a century that some species of tunicates extract vanadium from seawater where its concentration is extremely small. Other tunicates concentrate niobium, whose seawater content is one-tenth less than vanadium. Still other tunicates can accumulate neither element. No tunicate has yet been observed that can dramatically amass both.

Some species of oysters are enriched in zinc, some seaweeds in ruthenium, some sea grasses in beryllium. Enough examples of enrichment are now known so that many marine scientists feel that for any given chemical in the oceans, whether naturally occurring or introduced by man, there will be at least one species of organism capable of remarkably concentrating it.

A second form of assimilation of species from seawater involves the intake of particles, such as iron hydroxide particles of micron and submicron size and materials sorbed onto them. Filter-feeding organisms, such as clams and oysters, have the ability to accumulate iron hydroxide particles and thus any chemicals, pollutant or natural, associated with them.

TRANSPORT PATHS OF POLLUTANTS TO THE OCEANS

The movement of materials from the continents takes place naturally through three transport paths: rivers, winds, and glaciers. Man has added two others, ships and outfalls (pipes that discharge domestic and industrial wastes). The entry of materials to the marine environment from rivers, outfalls, or glaciers usually results in local dispersion while wind systems are capable of transport across continents and oceans. For example, the radioactive

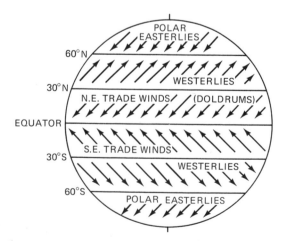

1   **Planetary Winds of the Earth.**

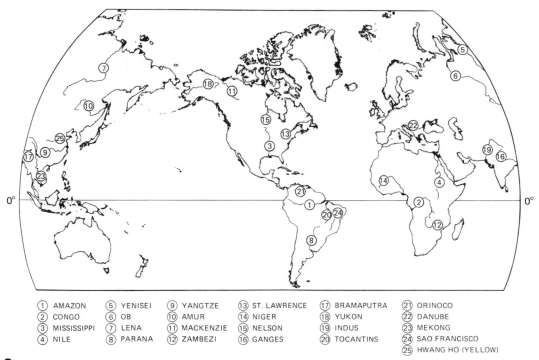

①	AMAZON	⑤	YENISEI	⑨	YANGTZE	⑬	ST. LAWRENCE	⑰	BRAMAPUTRA	㉑ ORINOCO
②	CONGO	⑥	OB	⑩	AMUR	⑭	NIGER	⑱	YUKON	㉒ DANUBE
③	MISSISSIPPI	⑦	LENA	⑪	MACKENZIE	⑮	NELSON	⑲	INDUS	㉓ MEKONG
④	NILE	⑧	PARANA	⑫	ZAMBEZI	⑯	GANGES	⑳	TOCANTINS	㉔ SAO FRANCISCO
										㉕ HWANG HO (YELLOW)

2   Principal Rivers of the World. Ranked on the basis of annual discharge.

debris introduced to the troposphere from the detonation of a Chinese nuclear device at Lop Nor (40°N and 90°E) in May 1965 circled the world in about 3 weeks. Radioactive fallout in rain was measured in Tokyo, Japan (36°N, 140°E), and Fayetteville, Arkansas (36°N, 94°W).

The main *wind systems* of the lower atmosphere tend to carry continentally introduced materials along lines of latitude (i.e., east and west). The directions and velocities of these winds vary and there can be a spreading of air masses both to the north and to the south. However, in general, a pollutant introduced into one hemisphere will remain in that hemisphere.

The three principal wind systems (Figure 1) are:

1. The trades, easterly winds that prevail between 30°S and 30°N, with their intensities decreasing toward the equator. Reversals in direction can take place in the upper troposphere, except in equatorial regions, where the wind directions are always easterly.
2. The midlatitudinal westerlies, the jet streams, which occur between 30°N and 70°N and between 30°S and 65°S, often with greater intensities in the upper troposphere than the lower troposphere.
3. The polar easterlies, winds between 70°N and 90°N and between 65°S and 90°S. Their intensities decrease with height and at about 3 kilometers they reverse their directions.

In addition, there are the continental monsoons, such as those associated with India, whose movements are determined by continent–ocean temperature differences. They can reverse their direction from summer to winter.

The two major *rivers* of the world, based upon annual discharge, the Amazon and Congo, drain into the equatorial Atlantic. Their drainage basins do not encompass highly industrialized societies. Only five of the first 25 major rivers, ranked as above, drain into the Pacific (Figure 2). Thus, the Atlantic Ocean receives the bulk of river discharge.

Rivers draining areas with advanced technologies show higher levels of pollution than those flowing through the developing nations.

277

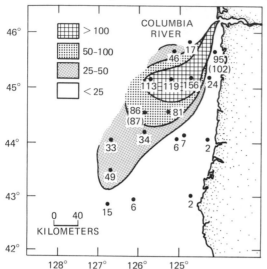

3 **Columbia River Plume as Defined by the Chromium 51 Activity of Surface Seawater.** Counts per minute per 100 liters of water. Copyright 1965 by the American Association for the Advancement of Science. See Reference 4.

For example, the rivers of Europe and North America have significantly higher sulfate contents than do those of Asia, Africa, and South America.[3] Most of the man-generated sulfate has its source in the combustion of fossil fuels, which releases sulfur dioxide to the atmosphere. This gas is subsequently oxidized to sulfur trioxide, which interacts with water to form sulfuric acid. Subsequent rains (Chapter 1) wash out this sulfur in the form of sulfate, which later is carried by the rivers to the oceans.

Radioactive wastes from the coolant waters of nuclear reactors, introduced into rivers, have provided tracers that have allowed us to follow the penetration of both the particulate and dissolved phases of the rivers into the marine environment. Radioactive chromium ^{51}Cr showed the entry of Columbia River water into the Pacific Ocean for distances of 350 kilometers (Figure 3). This nuclear isotope exists primarily in the dissolved state in seawater and is an effective tracer of the water mass. On the other hand, the radioactive nuclides zinc 65 and cobalt 60 are predominantly taken up by the sediments of the river system and are useful in following sediment movements. Upon entering the oceans, these two isotopes

stay attached to the solids, and they have shown that Columbia River sediments entering the Pacific move northward at rates of 12 to 30 kilometers per year along the shelf and 2.5 to 10 kilometers per year westward away from the coast.

Glaciers are responsible for the transport of the solids they pick up in the polar regions to latitudes up to 50° in either hemisphere. The greater bulk of solid movement by glaciers takes place in the Antarctic. As yet, glacier transport has not been associated with any pollutants.

Discharges from *ships* result in the transportation of materials from the continents to the oceans on both a global and a regional scale. The shipping lanes of the world have a surface soiling of oil and tars introduced both consciously and inadvertently from their traffic. Further, the dumping of materials in coastal areas is increasing with the greater use of materials by an enlarging world population. For example, in the United States, intentional dumping off coastal areas quadrupled between 1949 and 1968. About 80 percent of these discharges were dredge spoils; industrial and domestic wastes each accounted for about 10 percent of the total.

The nearly 9 million tons of waste solids (excluding rubbish and flotable debris) dumped annually in the coastal ocean near New York City appears to be the largest single sediment source from North America entering the Atlantic Ocean. Another 5 million tons of waste is discharged into the outer Thames Estuary, originating with the inhabitants of London. The sediments of the sea floor maintain a record of these dumpings. The deposits of the dumping area off New York contain a vast number of substances that have passed through human society: plastics, petroleum products, synthetic organic chemicals, and heavy metals. One of the ubiquitous ingredients found in the English sediments near disposal areas were the husks of tomato seeds, highly resistant to degradation in these environments and thus useful as a tracer of the dispersion of dumping.

The particles released from sewer *outfalls* appear to be disseminated over distances up to 10 or so kilometers from the injection sites. Heavy metals such as mercury and lead, attached to particles, as well as microorganisms,

278

have been used to trace the extent of travel of material from outfalls.

MAJOR INJECTION SITES OF OCEAN POLLUTANTS

The gross national product (GNP) is related to the flow of materials through a society as a result of its industrial, agricultural, and domestic activities. Leaks of such substances to the environment, whether unintentional or deliberate, constitute the nation's pollution. Thus, GNP proves a measure of man-generated material fluxes from the continents to the oceans. Countries bordering on the world ocean clearly have a better chance of contributing to its pollution than those that are inland, although atmospheric injections from the latter may eventually fall out into seawaters.

An examination of the nations with the highest GNPs indicates that the principal injection of pollutants into the environment take place in the midlatitudes of the Northern Hemisphere (Figure 4). The 10 countries with the highest GNPs are within this zone (the

numbers in parentheses are annual GNPs in billions of dollars, for 1969–1971): United States (1,050), USSR (290), Japan (200), Federal Republic of Germany (213), France (148), Great Britain (103), Canada (92), Italy (83), Democratic Republic of Germany (70), and China (60). Their combined value is $2,300 billion. The next 11 countries have a total GNP of $343 billion; of these, only Australia, Brazil, and Argentina are in the Southern Hemisphere: India (43), Australia (39), Brazil (35), Spain (32), Poland (31), The Netherlands (31), Sweden (30), Belgium (29), Mexico (28), Switzerland (26), and Argentina (19). Thus, the combination of the latitudinal zonation of man's atmospheric releases and the westerly flowing winds of the jet streams provides a mechanism for the generation of global pollution problems in the oceans of the Northern Hemisphere.

The ratio of the GNP to the area of a country gives an approximate, but perhaps useful, measure of the potential pollution that it is experiencing.[5] A type of country that may encounter serious marine problems was identified by this ratio — islands. Two countries

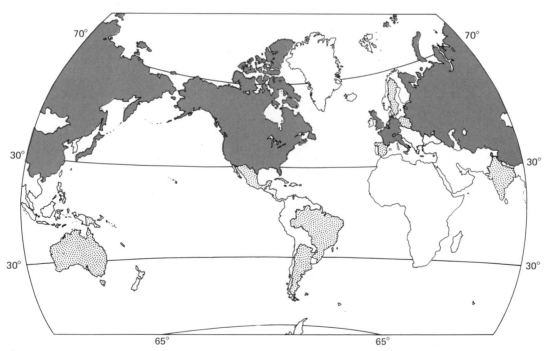

4   **Pollution Band.** The paths of the prevailing westerly winds are enclosed by the latitudinal lines between 30°N and 70°N and between 30°S and 65°S. Countries with GNPs greater than $60 billion per year are heavily shaded; those with GNPs between $19 and $43 billion per year are dotted; and those with less than $19 billion per year are clear.

279

with the highest values of the ratio (Hong Kong and Singapore) are densely inhabited and are undergoing vigorous economic and population growth. Their coastal areas already have significant pollution problems.

MATERIAL FLUXES TO THE OCEANS

The geological agency involving the mind of man, the noösphere, is mobilizing materials in the sedimentary cycle to within an order of magnitude of natural movements (Table II). Rivers are the primary natural means by which materials are moved, bringing annually from the continents to the oceans somewhat over 20 billion tons of dissolved and solid phases. The wastes of society (excluding the combustion products of fossil fuel) are about 3 billion tons per year or, on a volume basis, a cube 1 kilometer on a side. A substantial amount of these discards eventually reach the ocean system.

The atmosphere receives naturally occurring solid phases from continental weathering and volcanic debris, as well as injections of sea salt from the oceans. Calculations for particles that can be transported by wind systems over great distances (in general under 10 micrometers in diameter) have yielded ranges or upper limits for atmospheric fluxes. It appears that the fluxes of sea salts (0.3 billion ton annually) and continental rock and soil particles (0.1 to 0.5 billion ton annually) are about a factor of 10 higher than man's emissions from industrial activity (0.012 billion ton annually) and from fossil-fuel combustion (0.025 billion ton annually). The small particles introduced by man arise mainly from fossil-fuel combustion, in the crushed stone industry, the iron and steel industry, and the cement industry. The smaller particles are emitted mainly from activities that involve combustion, condensation, and vaporization rather than from mechanical processes. The most abundant types of particles are oxides of silicon, iron, aluminum, calcium, and magnesium and calcium carbonate.

Landscape alteration by man has altered the intensity of the sedimentary cycle and consequently the delivery of solid phases by the winds and rivers to the oceans. The transformation of forests to croplands and grazing areas increases the erosion rate and presumably the delivery of solid phases to the oceans approximately tenfold. During the construction of roads, the erosion rate usually increases 10 to 100 times. With one-fourth of the United States involved in such processes its erosion rate may now be larger than in the past by a factor of three or more.

Specific Pollutants

The potentially polluting types of societal wastes that have been injected into the oceans

II   Some Fluxes in the Major Sedimentary Cycle

Material	Geosphere receiving material	Flux (millions of tons per year)
Suspended river solids	Oceans	18,000
Dissolved river solids	Oceans	3,900
Continental rock and soil particles	Atmosphere	100–500
Sea salt	Atmosphere	300
Volcanic debris	Stratosphere	3.6
Volcanic debris	Atmosphere	<150
Wastes of society (excluding fossil fuels)	Hydrosphere, lithosphere atmosphere	3,000
Wastes dumped from ships	Oceans	140
Carbon and fly ash from fossil-fuel combustion	Atmosphere	25
Industrial particulates		
total	Atmosphere	54
<2 μm		12

Source: Adapted, in part, from Reference 6.

can be divided conveniently into six major categories: (1) heavy metals, (2) halogenated hydrocarbons, (3) petroleum, (4) carbon dioxide, (5) artificial radioactivities, and (6) litter.

HEAVY METALS

Some metals are being injected by man into the major sedimentary cycle in amounts that are comparable to those of natural weathering processes. For example, the combustion of fossil fuels (coals, oils, and natural gases) introduces a whole suite of heavy metals into the atmosphere and subsequently into the ocean (Table III) at rates sometimes comparable to those of river discharge. Similarly, the production of cement, essentially the roasting of a 1:1 mixture of shale and limestone, mobilizes materials to the air, again at rates approaching those of natural processes. Since the principal sites for such activities are in the midlatitudes of the Northern Hemisphere, changes in the heavy-metal compositions of natural waters and air will be most evident at these latitudes. The higher sulfate levels in rivers of industrialized countries, as compared to those in developing nations, are attributed to fossil-fuel burning.

Besides these entries from dispersed sources, there are regional or local sites of injection, such as domestic or industrial sewer outfalls. In the case of lead, whose man-caused entries are of the same order as natural fluxes to the ocean system, it appears that storm and river runoff, sewer discharge, and atmospheric washout all introduce comparable quantities to the Los Angeles Basin. On the other hand, the introduction of mercury to Minimata Bay, Japan, an activity that resulted in an epidemic poisoning, resulted solely from the discharge of wastes from a single chemical plant.

The world mining of *mercury*, somewhat under 10,000 tons per year, is small in comparison with the natural degassing of mercury from the interior of the earth, estimated to be somewhere between 25 and 150 kilotons annually. Thus, one would expect to see man-moved mercury only in localized situations where there is a massive input from an essentially point source.

Two tragic poisoning episodes in Japan focused the attention of environmental scientists upon the problems involved with promiscuous releases of mercury to inshore marine waters. In 1953 the first cases of severe neurological disorders were observed among the residents of Minimata Bay in southwestern Kyushu. By the end of 1970, 111 cases of the disease had been diagnosed and 41 deaths had been recorded. The victims were primarily from the fishing community. Cats and rats living in the region also succumbed to the affliction. All victims had eaten either fish or shellfish from Minimata Bay. In 1965, a second outbreak occurred, in Niigata, Japan, far removed from Minimata Bay; of 26 cases of neurological disorders, 5 deaths occurred.

In 1959, mercury was found to be in high concentrations in seafood recovered from the Minimata Bay region, and in 1963 the active agent causing the disease was identified as methyl mercury chloride. The source of the toxin was the Minimata Factory, a large chemical industry involved in the manufacture of polyvinyl chloride resin and in the production of octanol and dioctol phthalate, using acetaldehyde as the primary starting material. The wastes from these processes, discharged into Minimata Bay, were found to contain methyl mercury chloride. This organic mercury compound was bioaccumulated by the fish and shellfish, whose consumption in large quantities produced the disease. Two factors combined to create these tragedies: first, there was a production of organic compounds containing mercury in the wastes, and second, there was a marked accumulation of methyl mercury chloride in the marine biota.

The groundwork has thus been laid for the importance of methylated compounds in the environment. Following the Minimata Bay disaster, Swedish scientists established that the principal form of mercury in fish was methyl mercury. Subsequent investigations indicated that microorganisms are capable of producing methyl mercury and dimethyl mercury from inorganic mercury compounds. Vitamin B_{12} was implicated in the synthesis of methyl mercury chloride and of dimethyl mercury. Thus, the injected inorganic forms of mercury to the marine environment can be converted, under the proper conditions, into the toxic methyl mercury.

Lead is the one heavy metal whose open ocean concentrations have been measurably altered by man's intervention into the

III   Comparison of Amounts of Elements Mobilized into the Atmosphere

Element	Symbol	Man-made (fossil-fuel) mobilization (thousands of tons per year)			Natural mobilization[a] (thousands of tons per year)	
		Coal	Oil	Total	River flow	Sediments
Lithium	Li	9			110	12
Beryllium	Be	0.41	0.00006	0.41		5.6
Boron	B	10.5	0.0003	10.5	360	
Sodium	Na	280	0.33	280	230,000	57,000
Magnesium	Mg	280	0.02	280	148,000	42,000
Aluminum	Al	1,400	0.08	1,400	14,000	140,000
Phosphorus	P	70			720	
Sulfur	S	2,800	550	3,400	140,000	
Chlorine	Cl	140			280,000	
Potassium	K	140			83,000	48,000
Calcium	Ca	1,400	0.82	1,400	540,000	70,000
Scandium	Sc	0.7	0.0002	0.7	0.14	10
Titanium	Ti	70	0.02	70	108	9,000
Vanadium	V	3.5	8.2	12	32	280
Chromium	Cr	1.4	0.05	1.5	36	200
Manganese	Mn	7	0.02	7	250	2,000
Iron	Fe	1,400	0.41	1,400	24,000	100,000
Cobalt	Co	0.7	0.03	0.7	7.2	8
Nickel	Ni	2.1	1.6	3.7	11	160
Copper	Cu	2.1	0.023	2.1	250	80
Zinc	Zn	7	0.04	7	720	80
Gallium	Ga	1	0.002	1	3	30
Germanium	Ge	0.7	0.0002	0.7		12
Arsenic	As	0.7	0.002	0.7	72	
Selenium	Se	0.42	0.03	0.45	7.2	
Rubidium	Rb	14			36	600
Strontium	Sr	70	0.02	70	1,800	600
Yttrium	Y	1.4	0.0002	1.4	25	60
Molybdenum	Mo	0.7	1.6	2.3	36	28
Silver	Ag	0.07	0.00002	0.07	11	0.03
Cadmium	Cd		0.002			0.5
Tin	Sn	0.28	0.002	0.28		11
Barium	Ba	70	0.02	70	360	500
Lanthanum	La	1.4	0.0008	1.4	7.2	40
Cerium	Ce	1.6	0.002	1.6	2.2	90
Praseodymium	Pr	0.31			1.1	11
Neodymium	Nd	0.65			7.2	50
Samarium	Sm	0.22			1.1	13
Europium	Eu	0.1			0.25	2.1
Gadolinium	Gd	0.22			1.4	13
Terbium	Tb	0.042			0.29	
Holmium	Ho	0.042			0.36	2.3
Erbium	Er	0.085	0.0002	0.085	1.8	5.0
Thulium	Tm	0.014			0.32	0.4
Ytterbium	Yb	0.07			1.8	5.3
Lutetium	Lu	0.01			0.29	1.5
Rhenium	Re	0.007				0.001
Mercury	Hg	0.0017	1.6	1.6	2.5	1.0
Lead	Pb	3.5	0.05	3.6	110	21
Bismuth	Bi	0.75				0.6
Uranium	U	0.14	0.001	0.14	11	8

[a]Two different techniques were employed to calculate the weathering fluxes of the metals. One was based upon sedimentation and one upon river flow. The differences between the two techniques usually reflect the inadequacies of the existing data.
Source: Reference 7.

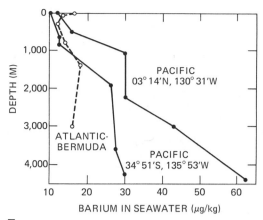 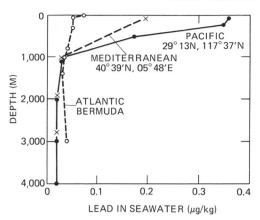

5 **Barium and Lead Concentration Profiles in Ocean Waters.** The graphs show that, whereas the natural concentration of barium (in micrograms per kilogram) increases with depth, the natural pattern for lead has been altered by man so that concentration is much higher in surface waters than deeper in the oceans. *Source:* Reference 8.

sedimentary cycle. The principal dispersive process involves the combustion of lead alkyls in internal combustion engines, where these chemicals act as antiknock agents. The natural river fluxes of lead to the oceans are only several times higher than the rate of lead alkyl combustion (Table IV). The lead issuing from the exhaust pipes of engines is in the form of small particles (usually in the micron or submicron range). This atmospherically injected lead later returns to the earth in rain or dry fallout, over half probably being deposited in the vicinity of the injection site.

An insight into the impact of this lead alkyl burning on the marine environment may be found through a comparison of the profiles of lead and barium concentrations as a function of depth in ocean waters. Both of these elements have similar marine chemical behaviors, but barium is not mobilized in large quantities by man. Where the concentrations of bar-

ium show that barium increases with depth (or in some cases is nearly constant with depth), the concentrations of lead are often much higher in surface waters (Figure 5), reflecting recent inputs from man's activities. The present average lead concentrations today in Northern Hemisphere coastal surface waters (0.07 microgram per kilogram) are significantly higher than those estimated to have occurred before the use of lead alkyl (about 0.01 to 0.02 microgram per kilogram).

The history of lead pollution in areas of high industrial activity can be recorded in coastal sediments. Lead is accumulated by marine organisms whose death, consumption by other organisms, or metabolic waste products result in a net transfer of lead from surface to deeper waters or sediments. The deposits off the coast of southern California, whose layers can be dated either by varving or by radiometric methods, show two significant patterns (Fig-

IV Comparison of Lead Introduced into the Environment

	Lead (millions of tons per year)
Man-made	
World lead production, 1966	3.5
Northern Hemisphere production	3.1
Lead burned as alkyls	0.31
Natural	
River input of soluble lead to marine environment	0.24
River input of particulate lead to marine environment	0.50

283

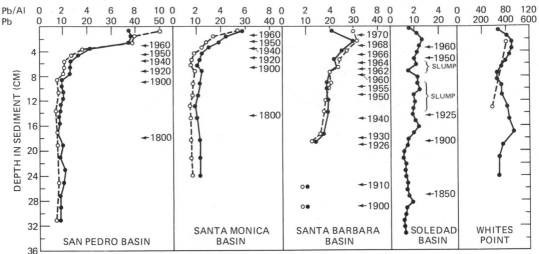

6 **Lead Concentrations and Lead/Aluminum Ratios.** Lead concentration is given in parts per million (solid circles) and lead/aluminum ratios × 10⁻⁴ (open circles); sediments were from the Santa Monica, San Pedro, and Santa Barbara basins off southern California, from the Soledad Basin off the coast of Baja California, and from a site near Whites Point, Los Angeles County, outfall. Note that the lead concentration from Whites Point is 10 to 40 times as high as that in other areas, that the basin off Baja has low and unchanged lead concentrations, and that the evidence for increased lead in the past 25 years is especially strong in the San Pedro and Santa Monica basins. *Source:* Reference 9.

ure 6). First, there are evident increased rates of lead accumulation beginning in the late 1940s, about 25 years after the introduction of lead alkyls into gasoline, especially in San Pedro and Santa Monica basins. Second, there are much higher values of lead (a 10- to 40-fold increase) found in sediments adjacent to a sewer outfall (White's Point) than in those deposits much farther removed. These changes in lead concentrations in seawater and sediments is more an indication that man's activities can measurably alter the composition of the oceans than any cause for immediate alarm about danger to man or the marine biota. There is no evidence at the present time that these increased lead concentrations are hazardous to living organisms.

Other than lead, no *other heavy metals*, excluding artificially produced radionuclides, have had their open-ocean concentrations demonstrably affected by the activities of man. On the other hand, in coastal areas, lagoons and estuaries, the sediments, organisms, and waters often show enhanced levels of some heavy metals which have been used and released by human society. For example, the heavy metals lead, chromium, cadmium, zinc,

copper, mercury, silver, vanadium, and molybdenum have accumulated at higher rates in the coastal sediments off Los Angeles during the past several decades than they had previously. Many of these elements appear to be transported by winds from industrial areas. The fluxes of many of these metals to the sediments in recent times are similar to those found for atmospheric precipitation collected at nearby Santa Catalina Island. In the New York Bight, the concentrations of chromium, copper, lead, nickel, and zinc in the surface layers of sediments collected near waste disposal areas were 10 to 100 times greater than those in adjacent uncontaminated deposits.

Cadmium appears to be several orders of magnitude higher in filtered inshore waters off the British coast than in the open ocean.[10] Coastal waters have tenths to units of micrograms of cadmium per liter; the open ocean, hundreths of micrograms. In this area, oysters have the highest cadmium content in areas of high industrial discharge. Curiously, crabs from these same areas have very low cadmium levels.

It appears that heavy metal pollution around the British Isles is most intense in the Irish Sea

284

region. Here, seawaters and some organisms (limpets and *Porphyra*) have generally higher levels of zinc, iron, manganese, copper, nickel, lead, silver, and cadmium than do other coastal zones. However, in both polluted and unpolluted areas there has been little change in the 10 years previous to 1970 in the concentrations of these elements in their waters.[11] As is the case with lead, there is no evidence at present which links any of these metals to disruptions of marine ecosystems.

Of the thousands of organic compounds synthesized industrially, those that contain halogen atoms (fluorine, chlorine, bromine, or iodine) have received the closest scrutiny as potential threats to living resources of the sea. Two groups of compounds are of concern: (1) the low-molecular-weight chemicals, usually containing one or two carbon atoms, which are used as intermediates in chemical synthesis, as aerosol propellants, refrigerants, and cleaners; and (2) the higher-molecular-weight compounds, containing aromatic rings, which include the pesticides DDT, dieldrin, and aldrin and the industrial chemicals, the polychlorinated biphenyls. These substances have the following characteristics: high production levels, well-documented toxicities to living systems, and persistence in air and water. In addition, all reach the marine environment primarily through atmospheric transport.

The continental and marine airs and Atlantic surface waters have been found to contain a number of *low-molecular-weight chemicals*: chloroform, carbon tetrachloride, trichlorethylene, perchlorethylene, methyl iodide, dichlorodifluormethane, and trichlorfluormethane, as well as other unidentified halogen-containing species. The atmospheric concentrations are of the order of nannograms per cubic meter (10^{-9} g/m^3) and the surface ocean waters contain these chemicals in the range of nanograms per liter. The fluorinated species (those containing fluorine) and the chloroethylenes are definitely of human origin; the methyl iodide is apparently produced by marine algae. There are no obvious marine sources for either chloroform or carbon tetrachloride, whose concentrations in Northern and Southern Hemispheric waters are similar. Their origins are yet to be determined.

We already have evidence that this group of compounds pose a hazard. For example, mix

tures of short-chained hydrocarbons, waste by-products from the manufacture of vinyl chloride, were dumped into the North Sea and have been implicated in the deaths of plankton and fish. Furthermore, some of these low-molecular-weight halogenated hydrocarbons may inhibit naturally occurring fermentation processes. There is therefore the haunting possibility that environmental concentrations of these low-molecular-weight hydrocarbons could attain such levels that fermentation processes in oceanic areas might be diminished, an event that might affect related biological activities.

One of the most troublesome of the *high-molecular-weight chemicals* is DDT. It and its degradation products DDE and DDD have been dispersed about the marine environment since the end of World War II, following the use of DDT as a pesticide in agriculture and in the control of malaria. Although there is no clear evidence of direct threat to human health from its use, the development of resistance in target organisms and the decimation of nontarget organisms necessary to community balance in ecosystems (Chapter 14) have resulted in ' restrictions or bans upon its use by a number of developed nations. World production figures of DDT have never been assembled, and most studies of DDT in the global environment have involved extrapolations of U.S. production data. In the late 1960s, the annual world production was estimated to be of the order of 100,000 tons per year and up to that time about 2 million tons of DDT appeared to have been manufactured.

DDT and its residues DDE and DDD are transferred from the continents to the oceans primarily through atmospheric transport following its application in agriculture or public health activities such as malaria control or through vaporization from plants and soils. These compounds have been found in atmospheric dusts, in rainwater, and as vapor in the air. Quantification of the rates of transfer are difficult to make, primarily because so few data on environmental concentrations have been collected. On the basis of what limited data are available, rivers appear to contribute several orders of magnitude less DDT to the oceans than does rain. Thus, whereas atmospheric fluxes may have had a maximum value of about 24,000 tons per year (before restric tions upon its use were enacted), the maximum estimate for river transport is 3,700 tons per year.

An additional piece of evidence that long-range atmospheric transport predominates over any local discharges comes from a survey of marine organisms from polluted and unpolluted coasts off many countries that were analyzed for DDT residues in the period 1969–1971. No significant differences in DDT residue concentrations between polluted and unpolluted waters for samples of the same species of fish and shellfish were found — a most surprising observation. It appears that the concentration of DDT in the U.S. marine environment has declined significantly since DDT use was curtailed in the United States. Over the period 1965–1972, a U.S. monitoring program was carried out for estuarine mollusks, primarily the eastern oyster *Crassostrea virginica*.[12] Over 8,000 samples were analyzed and DDT residues were found in 63 percent of the samples. In only 38 samples (0.5 percent) did the residues exceed 1 ppm, and these were from samples collected in drainage basins in California, Florida, and Texas which have intensive adjacent agricultural activity. During 1965–1972 there was a pronounced decline, about 55 percent, in the number of samples containing DDT residues in excess of 0.1 part per million and a 44 percent increase in the number of samples showing no DDT residues. These concentrations are sufficiently low that there is no discernible threat to the health of the mollusks.

Both DDT and its residues and the PCBs (see below) have been implicated in the reproductive failures of marine birds (Chapter 13). In birds, with high concentrations of halogenated hydrocarbons there is abnormal calcium metabolism and a consequential thinning of eggshells, which results in fewer live births. There appears to be an inverse relationship, for example, between the eggshell thickness in brown pelican eggs from the United States and the DDE contents in their eggs. Eggshell thinning became evident at the parts-per-million level of DDE in the egg.[13] On the other hand, during the same investigation it was found that the herring gull showed no thinning at the 1 ppm level and only 11 percent at the 80-ppm level. Thus, thinning occurs at different concentrations for different species of birds.

The brown pelican population of Anacapa Island off the California coast allegedly experienced severe population failures in 1969–1972 through eggshell thinning and the inability of the eggs to hatch through their friability.[14] The source of the DDT residues was from a factory in Los Angeles which discharged 200 to 500 kilograms of DDT daily into the Los Angeles sewage system and hence into the sea. The DDT residues were accumulated in marine organisms, which were subsequently eaten by the birds.

The polychlorinated piphenyls (PCBs) are perhaps the most widespread and most abundant of the synthetic organic chemicals in the oceans. Unlike DDT and its degradation products, these compounds, or possibly the contaminants in them introduced during manufacture (the chlorinated dibenzodioxins or dibenzofurans), have been implicated in diseases of man and animals.

The PCBs were introduced as industrial chemicals in the early 1930s and have been used as dielectric fluids in capacitors and transformers, in hydraulic fluids (especially in systems involving higher temperatures), and in aircraft as heat-transfer fluids. In addition, they have been employed as plasticizers and resin extenders in adhesives, sealants, paints, and printing ink.

PCBs were first measured in the environment during the late 1960s. Through concern about their threat to living systems, the sole U.S. producer in 1970 restricted their sale to uses in closed systems (i.e., those where escape to the environment was minimal). Before this action, about 60 percent of industrial uses were in essentially closed-system applications, electrical and heat-transfer equipment, 25 percent for plasticizers, 10 percent for hydraulic fluids and lubricants, and less than 5 percent for such applications as surface coatings, adhesives, printing ink, and pesticide extenders.

It is difficult to arrive at a world production rate for PCBs. The compounds are or have been manufactured in the United States, Japan, Spain, Italy, France, England, West Germany, East Germany, and the Soviet Union. A first approximation for the time before 1970 might be a trebling of the U.S. rate, which would give a value of about 100,000 tons per year. Major transport routes to the ocean include the atmosphere and domestic and industrial sewer outfalls and rivers. The annual amount escaping to the environment is estimated to be 28,000 tons, or about 80 percent of the 1970 production of PCBs in North America. It is portioned in the following way[15]: 22,000 tons, disposed of in incinerators, dumps, and sanitary landfills; perhaps 3,000 tons destroyed by burning; 4,000 to 5,000 tons, leaks and disposal of hydraulic fluids, lubricants, heat-transfer fluids, and transformer oils; and 1,000 to 2,000 tons, evaporization of plasticizers. The annual environmental fluxes in 1970 are estimated from these figures as: 1,000 to 2,000 tons into the atmosphere; 4,000 to 5,000 tons into fresh and coastal waters; and 18 tons into dumps and landfills.

Of the PCBs entering the U.S. atmosphere, the greater portion will be deposited within 2 or 3 days onto land and coastal waters near the urban sites of entry. The terrestrial fallout is estimated to be 1,000 to 2,000 tons per year with about 0.5 thousand ton entering the marine environment from the atmosphere. The amount of PCBs transported to the ocean in solution or attached to particles is estimated as 0.2 thousand ton. The few measurements of PCBs in seawaters, made in 1972 localized to the North Atlantic Ocean, have confirmed the major role played by the atmospheric transport of PCBs, because no relationship between PCB concentration and proximity to land was observed. PCB concentrations decrease with depth in the oceans, but there are measurable levels even at a depth of 3,000 m.

An extensive study of halogenated hydrocarbons in the organisms of the North and South Atlantic[16] provides an understanding of their behavior in the food web. First, the levels in the organisms in general reflect the levels measured in the waters in which they live. Low values in PCBs in the organisms from the Sargasso Sea were associated with low values in the water itself compared to other parts of the Atlantic. PCB levels in open ocean plankton are of the same levels as in plankton collected in coastal areas and harbors. Second, even though more PCBs are released into coastal waters from sewage runoff and from

the atmosphere, they are probably removed to the sediments more rapidly there by sorption, a process that is less intensive in the open ocean. Third, PCB concentrations in North and South Atlantic plankton were the same, a result that is not explainable at the present time, as more PCB is used in Northern Hemisphere, and atmospheric transport is primarily east-west.

Somewhat surprising was their observation that there is no biomagnification of PCBs going up the food chain on either a total-weight or lipid (fat)-weight basis. Plankton had the highest average concentrations, and flying fish, which feed on plankton, carried orders of magnitude less PCB and DDT than did the plankton. It is thought that organisms get halogenated hydrocarbons from the water across their permeable surfaces; this may explain why concentrations do not increase up food chains.

Finally, the differences between the behavior of PCB and that of DDT are interesting. DDT is much less common than PCBs. The ratio of DDT to PCB has been measured in ocean waters of the North Atlantic and appears to be less than 0.05.[17] The ratio in plankton samples was always less than 0.03.[16] By contrast, more DDT is produced than PCB: U.S. production ratios (DDT/PCB) in the 1960s were somewhere between 2 and 4, with reduced usage of both PCBs and DDT beginning in the 1970s. The most reasonable explanation for the lower marine values of the ratio is that the residence time of DDT and its degradation products is much less than that of PCBs, most probably as a consequence of microbial or inorganic, perhaps photochemical, decomposition reactions. On the basis that PCBs will remain in ocean waters for longer periods of time, they are of greater concern as toxic agents.

Laboratory investigations of the toxicities of PCBs and DDT residues to marine organisms indicate that effects are only observed at much higher concentrations than are presently observed in the oceans. For example, the growth of marine algae is inhibited by both groups of compounds, but the critical amounts vary from species to species. Effects were noted with concentrations of tens of parts per billion. Juvenile shrimp cultivated in running seawater and exposed to 5 parts per billion of a PCB suffered a 72 percent mortality after 20 days. Juvenile blue crabs were more resistant to such levels; only 1 of 20 crabs died in a 20-day experiment.

PETROLEUM

Petroleum hydrocarbons are visible pollutants that have entered the ocean system. Their unpleasant effects include the soiling of beaches, the staining of surface waters with films and tarballs, and the death of marine organisms and seabirds. There is an invisible penetration also. Marine organisms now carry body burdens of man-mobilized petroleum hydrocarbons. The effects, if any, of these materials upon their life processes have not yet been ascertained. The increased demands for energy will result in greater injections of petroleum to the oceans.

In 1969, 1.8 billion tons of petroleum were mined from the earth and 1.2 billion tons, or 65 percent, were transported in ocean-going vessels. The losses from activities involving petroleum production, petrochemical manufacture, and the movement of these materials to the oceans are estimated to be about 2.1 million tons per year (Table V). Shipping operations are responsible for about 60 percent of the total. These chronic discharges from ships include deliberate and accidental discharges at sea, the flushing of oil tanks, and terminal transfer losses. Episodic releases, such as the spectacular breakup of the *Torrey Canyon* off the coast of Great Britain, can introduce large amounts of petroleum to the oceans. In this case 118,000 tons were lost.

Offshore oil production is estimated to introduce about 100,000 tons per year. The Santa Barbara, California, blowout injected about 11,000 tons of oil into the oceans. As offshore oil drilling expands in the future, a potentially higher injection rate can be expected, assuming no improvements in control measures to reduce losses.

Refinery operations and municipal and industrial wastes bring about 0.3 million ton per year of petroleum to the oceans. It is interesting to note that about 5 percent of the total oil production is involved in the manufacture of petrochemicals; the remainder is used as fuel.

The transport of oil to the sea via the atmosphere and river and urban runoff is estimated at 0.6 and 1.9 million tons per year, respectively. Both values are based on indirect evidence and may be changed substantially as more observations of petroleum products in the atmosphere and rivers are made.

A value of 0.6 million ton per year for natural seepages of petroleum into the ocean comes from an extrapolation of observations on several localized investigations,[19] in particular the Santa Barbara Channel. Thus, today the mobilization of petroleum into the oceans by man appears to be substantially greater than natural fluxes. Both of these flows have changed with time. For instance, during World War II, about 4 million tons of oil entered the

V Involvement of Petroleum with the Marine Environment

World oil production (1969)	1.82	billion tons/year
Oil transport by tanker (1969)	1.18	billion tons/year
Direct injections into marine environment through man's activities		
Tanker operations	0.5	million tons/year
Other ship operations	0.5	million tons/year
Offshore oil production	0.1	million tons/year
Accidental spills	0.2	million tons/year
Refinery operations	0.3	million tons/year
Industrial and automotive wastes	0.45	million tons/year
	2.1	million tons/year
Torrey Canyon discharge	0.118	million tons
Santa Barbara blowout	0.003–0.011	million tons
Natural seepage into ocean	0.6	million tons/year

Reprinted by permission of the M.I.T. Press, Cambridge, Mass. See Reference 18.

oceans through the sinking of tankers, probably a substantial increase over the chronic losses preceding these military operations.

Alterations to the composition of petroleum in the oceans may take place through a variety of mechanisms: solution, surface-film formation, evaporation, oxidation, sedimentation, or degradation by organisms. Of particular importance are those processes that result in the breakdown of petroleum. There has been a general sense that petroleum is entirely biodegradable and that bacteria are primarily responsible for its decomposition in the oceans. This concept has been critically examined recently and perhaps is not valid. Laboratory experiments of microbial degradation of oil have yielded conflicting results, and extrapolations to the marine system can be misleading. At present we can say that some marine bacteria are capable of degrading some components of petroleum but that the rates and extents in the natural situation are poorly known.

On the other hand, petroleum products persist in marine sediments for at least years. Extensive studies have been conducted of a 1969 spill of fuel oil into Buzzards Bay, Massachusetts.[20] The disappearance of oils from the sediments took place both by microbial degradation and through solution, with the latter being quantitatively more important. Some naturally produced hydrocarbons have been preserved in bottom sediments for geologic time periods, and there is no reason to expect that man-generated petroleum hydrocarbons will not be able to persist there for similar lengths of time.

Oil spills such as the *Torrey Canyon* accident near Lands End, England, in 1967 have caused mass extinctions of benthic fauna in the immediate vicinity of the spill, probably by smothering the animals. Furthermore, attempted corrective measured to contain spilled oils such as applying detergents to inactivate and disperse the oils have sometimes been more injurious to living organisms than the petroleum. The faunal assemblages of such areas generally take several years to recover.

Marine organisms now have body burdens of petroleum in the parts-per-million range on a wet-weight basis. The effects of such concentrations upon their metabolism have not yet been determined. In addition, there may be other types of interactions with life in the sea that warrant assessment. For example, certain petroleum components may interfere with the processes of chemoreception through the blocking of the detection organs. Pheromones (chemicals produced by organisms to communicate information) can be masked by the presence of petroleum in the water. In the case of lobsters, these pheromones are exuded by the female to initiate copulation with the male.

CARBON DIOXIDE AND FOSSIL-FUEL COMBUSTION

The combustion of fossil fuel is introducing into the atmosphere and subsequently into the oceans not only such gases as carbon dioxide, carbon monoxide, nitrogen oxides, sulfur dioxide, and hydrocarbons but also many metals as volatile and particulate substances. Many of the chemicals are being produced at rates approaching or even exceeding natural fluxes. For example, the amount of carbon dioxide introduced into the atmosphere in 1967 by energy production, 13 billion tons, is one-sixth of the amount of carbon dioxide utilized in photosynthesis on land and sea, 76 billion tons. Natural sulfur dioxide in the atmosphere is presumed to be produced by the oxidation of hydrogen sulfide, which enters from swampy regions at a rate of about 100 million tons per year. At the present time fossil-fuel burning appears to be introducing an equal amount of sulfur dioxide. About the same amount of lead is transferred annually from land to the oceans as a consequence of the use of tetraethyllead as an antiknock additive in gasolines as has been added to the marine environment through rivers. The oceans receive a substantial fraction of the fossil-fuel combustion products. Since transport from the continents takes place primarily through the atmosphere, transfer processes involve washout, dry fallout, or gas-exchange reactions between seawater and the atmosphere. Although the surface sea waters are clearly a sink for many of the combustion products, others of these species enter the atmosphere following production in the oceans. For example, carbon monoxide and some low molecular weight hydrocarbons, including methane, appear to be produced in the oceanic biological cycles and often become supersaturated in surface waters.

Although carbon dioxide is a minor constituent of the atmosphere (about 300 parts per million by volume) and of the oceans (about 50 parts per million by weight), it plays key roles in the radiation balance of the atmosphere, in sedimentation processes in the ocean, and in plant productivity both on land and in the oceans. A knowledge of the partition of the man-generated carbon dioxide between these three reservoirs — the oceans, the atmosphere, and the biosphere — is essential for forecasting changes in climate, in primary productivity of plants, and in sedimentary processes (Chapter 15).

A predictive model for the accumulation of man-mobilized carbon dioxide in various reservoirs has been formulated.[21] This model is used to foretell the environmental levels of carbon dioxide in the environment, assuming an annual increase in the combustion of fossil fuels of 4 percent up to 1979 and of 3.5 percent between 1980 and 1999. The values of prediction and observation were normalized to each other in 1958 with an air concentration of 313 ppm at the Mauna Loa Observatory. This station in Hawaii has been measuring carbon dioxide concentrations in the atmosphere since 1958. Figure 3 in Chapter 15 indicates that predicted and observed values were in accord from 1958 to 1970.

VI Partition of Fossil Fuel $C_{12}O_2$, 1970, According to Model Calculations

	Carbon	
	Tens of billions of tons	Percent
Atmosphere	5.6	55
Stratosphere	4.8	47
Troposphere	0.8	8
Biosphere	1.5	15
Long-term land	1.3	13
Short-term land	0.2	2
Marine	0	0
Oceans	3.0	30
Mixed layer	2.5	25
Deep layer	0.5	5
	10.1	100

Source: Reference 21.

The partition for man-made carbon dioxide between the various reservoirs is indicated in Table VI. In 1970 more than half still remained in the atmosphere. Thirty percent is in the ocean and 15 percent in the terrestrial biosphere. The marine biosphere has accumulated an insignificant amount of the gas.

RADIOACTIVITY

The artificially produced radioactive isotopes constituted the first collection of substances recognized by the scientific community as providing a challenge to the resources of the sea. The initial concerns involved the dumping of radioactive wastes produced in nuclear reactors into the oceans, common international property. The mood of one of the first large-scale meetings of scientists confronting the potential dangers of radioactive releases was summarized as follows[22]:

> The use of the sea for waste disposal, in particular, can jeopardize the other resources and hence should be done cautiously. . . . The large areas of uncertainty respecting the physical, chemical and biological processes in the sea lead to restrictions on what can now be regarded as safe practices. . . . If the sea is to be seriously considered as a dumping ground for any large fraction of the fission products that will be produced even within the next ten years, it is urgently necessary to learn about these processes to provide a basis for engineering estimates.

In addition, these scientists were aware that an understanding of the distribution of materials and predictions of future disseminations could only come about if an adequate bookkeeping of environmental dispositions were available:

> From the standpoint both of research and property control of this new kind of pollution careful records should be maintained of the kinds, quantities, and physical and chemical status of all radioisotopes introduced into the seas together with the detailed data on locations, depths and modes of introduction. This can probably best be done by national agencies reporting to an international record center.

Environmental scientists directed authorities responsible for the management of radioactive wastes to minimize high-level releases to the marine environment or to the atmosphere. Acceptable levels for the waters, plants, and animals of the sea were developed on the basis of potential dangers to man through his ingestion of foods or through direct exposure. Discharges of nuclear wastes have been regulated and dumpings of high-level radioactivities to the ocean have been prohibited by the actions of nations. The management of nuclear debris (excluding that released in the testing of weapons) has become so effective that there probably will be a reduction in the amount of radioactive material reaching marine environments per unit of fuel burned and reprocessed.[11]

Today, the greatest contribution to the artificial radioactive burden of the oceans arises from nuclear devices (Table VII). The total artificial radioactivity in 1970 was less than 1 percent of the total natural radioactivity, due primarily to potassium 40, and the great bulk of this was tritium from nuclear explosions. Weapons testing contributed more than 100 times as much radioactivity as did nuclear energy programs. Controlled discharges to the oceans of radioactive wastes from outfalls and from nuclear-powered vessels have sometimes resulted in high local concentrations of radioactive nuclides, but atmospheric fallout of matter introduced by bomb detonations has been relatively uniformly dispersed along lines of latitude except in the immediate area of the explosion site.

Through 1968 there had been 470 nuclear explosions carried out by the United States, the Soviet Union, the United Kingdom, France, and China. The oceans received a substantial part of the resultant radioactivity except for those experiments conducted underground or in outer space. Underwater detona-

tions introduce radioactive chemicals to a rather localized area initially, although subsequent mixing spreads the nuclides over vast distances. Many radionuclides from surface or air detonations enter the stratosphere, where they may remain for periods of several years before returning to the earth's surface. Usually, the fallout from the stratosphere takes place in the hemisphere of introduction, although stratospheric mixing processes do allow for interhemispheric transfers. Removal of materials from the stratosphere to the troposphere usually takes place at the upper and middle latitudes of either hemisphere in the late winter and early spring.

The consequences of the planned discharges of liquid radioactive wastes from nuclear power stations are being monitored in surveillance programs of the United Kingdom.[23] The largest inputs of radioactivity occur at the Windscale installation, where in 1971, 300,000 curies of alpha radiation and 6,000 curies of beta radiation were authorized for discharge. Only about half of these amounts were actually utilized in the disposal operations.

When we use the *critical-pathways approach* (see the section, "The Oceans as Waste Space"), three significant exposures of man to these ionizing radiations became apparent, two internal and one external. In the former case, the principal concern involved the consumption of laverbread, made from the seaweed *Porphyra*, which had accumulated the radioisotope ruthenium 106. Although in the preparation of the foodstuff, the contaminated ingredients are usually diluted with uncontaminated ones, a fail-safe premise is made that only undiluted *Porphyra* is used. On this

VII Total Inventory of Artificial Radionuclides Introduced into the World Oceans, 1970

	Curies
Nuclear explosions (worldwide distribution)	
Fission products (exclusive of tritium)	$2-6 \times 10^8$
Tritium	10^9
Reactors and reprocessing of fuel (restricted local distribution)	
Fission and activation products (exclusive of tritium)	3×10^5
Tritium	3×10^5
Total artificial radioactivity	10^9
Total natural potassium 40	5×10^{11}

Source. Reference 24.

basis, the exposed group of people receive but 6 percent of the permissible dose recommended by the ICRP.

A second problem area involves the eating of fish that have accumulated the radionuclides cesium 134, cesium 137, and, in smaller amounts, ruthenium 106. There has been an increased intake by the British public of these radioactivities in recent years. Exposure is still small, however, an estimated 2 to 3 percent of the ICRP recommended dose limit.

External exposure to ionizing radiation could result from the sorption of radionuclides on the coastline sediments in the vicinity of the power station. Most of the contamination around Windscale is low and is only evident in areas where there is an accumulation of fine sediments, such as in harbors and estuaries. The radioactivity on sand beaches is usually very low. The most highly known exposed individual, a salmon fisherman, was estimated to have received only 11 percent of the recommended ICRP dose. In the Thames region of England, where smaller amounts of radioactivity are discharged from nuclear power reactors, the critical pathway to man is not through fish or direct exposure but through drinking water.

For the forthcoming years, the radioactive chemicals that appear to pose the more formidable problems encompass the transuranics, the nuclear fuels, and the products produced from them through the capture of such particles as neutrons. Plutonium will be one of the important participants in the production of nuclear power for the next several decades and it is a very toxic substance. It has become one of the most studied toxicologically of the hazardous elements because of worries about its potential inhalation or ingestion by man.

Today, the largest flux of plutonium to the environment results from fallout from nuclear weapons tests in the atmosphere. However, future employment of the element as a fuel in nuclear power reactors, as heat sources in thermoelectric power devices, and in weapons can in principle introduce intolerable amounts to the environment. There are many possible nonmilitary ways for plutonium to be lost to our surroundings. Accidents in nuclear power plants or with power devices will occur. For example, the primary source of plutonium 238 in the environment today resulted from the unintentional burnup of a power device in a rocket sent to outer space.

Both uranium and plutonium fuels must be treated chemically every year or so to remove unwanted chemical species that build up with time. For the world, it is estimated that in 1980 about 200 tons of plutonium will be transported to and from reprocessing plants to maintain power reactors and weapons in operative conditions. Microgram quantities of plutonium, if inhaled, are toxic. It is easily possible to imagine ways in which man could be exposed to plutonium if it escapes to the oceans, and we must hope that this will be prevented.

MARINE LITTER

The oceans are being visibly soiled by the discard of trash. Beaches, sea surfaces, and the sea floors carry a burden of litter, man-fabricated materials promiscuously released to the environment. These substances are introduced from rivers, dumping of wastes, domestic and industrial sewer outfalls, ship discards, and atmospheric fallout. An estimated flux of around 6 million tons per year, of which about 75 percent originates from ships at sea, has recently been made. Typical composition of wastes discharged from ships is: food, 43.0 percent; paper, 26.0; metal, 7.0; cloth, 4.0; glass, 4.0; plastics, 3.5; rubber, 0.5; and other, 16.0.

The presence of litter in seascapes creates an aesthetic insult, and this appears to be the primary impact upon our society. Occasionally, such objects as plastic sheets become caught in the water intakes of ships and impede their normal flows. Ropes and wire have become entangled in the propellors of ships and can inhibit normal propulsion. This litter is being incorporated in the biosphere. Of 33,000 fur seals examined in St. Paul Island, Alaska, 81 had bits of net caught on them, 6 had ropes and wire, 52 had plastic bands about their necks and 76 had net marks on their flesh. Elastic bands have been found in the guts of puffins, birds that feed upon marine organisms. Apparently, these pieces of rubber were mistaken for food.

Plastic spherules, often polystyrene used in the fabrication of plastic ware, have been observed in the guts of fish and zooplankton and in New England coastal waters. These polystyrene spherules may contain PCBs, which they presumably extracted from seawater.

The litter also provides artificial substrates for a variety of organisms, from algae to barnacles. These artifacts with their adhering tenants can be transferred over vast distances of water by ocean currents. There also exists the possibility that new ecological niches are being created which can enhance biological productivity.

The Oceans as Waste Space

The vast volume of the oceans constitutes an important resource that can be utilized to accept a part of the wastes of man. The most effective utilization of it as waste space with minimal undesirable effects upon other resources is a formidable challenge to those responsible for the management of the marine environment. Rational control programs to minimize risks to man's well being, to the vitality of other life systems, and to the amenities provided by the oceans should involve the following steps:

1. Identification of any substances that might jeopardize marine resources.
2. Determination of acceptable levels in marine waters.
3. Determination of major discharge sites and routes to the oceans.
4. Imposition of regulations to manage releases to maintain acceptable levels in seawaters.
5. Surveillance programs to ascertain that the acceptable levels are not exceeded.

Ways of carrying out such strategies of management have been formulated and utilized by those involved in the discharge of artificially produced radionuclides to the oceans. Here, the concerns have primarily involved hazards to human health through ingestion of or exposure to ionizing radiation. Each one of the above steps is fraught with difficulties, yet, in the case of radioactivity, such tactics have allowed reasonable quantities of radioactive wastes to be accommodated in the marine systems.

Any assessment of what constitutes a potential hazard to the oceans from among the thousands of materials utilized in man's activities is usually based upon short exposure times to organisms or ingestions of large amounts of any of these materials. For human health, permissible body levels and hence dosages are often extrapolated from the results of experiments on laboratory animals. The effects of long-term, low-level exposures, which are more important, are much more difficult to discover because experimental costs are so high (Chapter 8).

Before release of a substance to the environment, predictions of its distribution within air, water, organisms, and sediments are most desirable. These estimates are of great importance (1) in ascertaining what regulations on releases to the oceans appear reasonable before its first use, and (2) in directing those responsible for monitoring its distribution in the environment to the materials that should be assayed. These initial estimates can later be refined through the results of the surveillance programs.

Since there are economic limitations to the numbers of samples that can be analyzed in a monitoring program, a strategy based upon the more important routes or reservoirs in which environmental impacts are possible has been devised and is known as the *critical-pathways approach*. Here, it is assumed that the number of pathways from the ocean to man through ingestion or direct exposure to a toxin is usually limited to a small number, say two or three. Thus, the surveillance programs can be limited to these routes rather than becoming involved with a more extensive monitoring effort. For example, in the discharge of radioactive wastes, the isotope ruthenium 106 is concentrated in the seaweed *Porphyra*, which is eaten in the form of a pudding (laverbread) in south Wales. This single route to man allows an effective surveillance program. Similarly, the mercury levels in edible fish can often exceed those defined as acceptable. Protection against the unwanted effects of this heavy metal can be sought through the assay of fish.

The acceptable levels of radioisotopes have been developed by the International Commission on Radiological Protection (ICRP). For each radioisotope, depending upon the nature of its radiation and on its average concentrations in parts of the body, permissible levels can be defined. For example, ruthenium causes greatest damage in the gastrointestinal tract. Therefore, the maximum permissible levels are based on its content in the gastrointestinal tract. On the other hand, the ingestion of methyl mercury results in impairment of the nervous system. Methyl mercury concentrations in hair and/or in blood have been used as measures of acceptable body levels resulting from the ingestion of this chemical.

PREDICTING OCEAN POLLUTANTS

We have identified two contrasting approaches to the management of waste materials that can potentially enter the marine environment. In one approach, an estimate is made of the amount of a given substance that can be accommodated in the oceans without sacrificing resources or jeopardizing human health. Such estimates, continuously subject to revision as new information about permissible environmental levels evolves, provide the basis for surveillance programs. As maximum acceptable levels are approached, leakages to the environment can be reduced. This has been the strategy in the management of radioactive wastes.

The second approach develops as a reaction to catastrophic episodes when, as a result of an intolerable injection of a material into the environment, a disaster occurs. As a consequence, regulatory actions are formulated to control production, use, or leakage of the offending material. For example, controls upon mercury discharges and upon the consumption of seafoods with mercury levels higher than considered safe were imposed following the Minimata Bay epidemic of mercury poisoning.

The chemical properties of the substance and its degradation products can be of use in predicting its bioaccumulation. High levels of PCBs are often found in marine organisms along with high levels of DDT residues in Northern Hemisphere waters. Both accumulate in the lipid phases.

Determination of the amount of a chemical that is toxic to man or to other living organisms is perhaps the most difficult information to acquire. The levels of lethal action may be of less importance to long-range management problems than the levels of risk associated with low-level, long-time exposures.

Conclusion

All the world's ocean waters bear the signature of man's technology through an accommodation of his wastes. Of the thousands of chemicals that man has dispersed about his surroundings, very few have so far posed threats to marine resources. The protection of the environment from undesirable impacts of new substances or from increased usage of existing materials requires predictive abilities. How will a given material react with the various components of the ocean system? How much of the material can the oceans carry before a resource is jeopardized? This ability to foretell the behavior of substances in the ocean depend upon a knowledge of the fundamental processes naturally occurring there, and as a greater variety of materials are used by world society, we will need more and more knowledge of environmental chemistry and biology.

References

1. Vernadsky, W. I. 1945. The biosphere and the noösphere. *Amer. Scientist 33:* 1–12.
2. *Anon.* 1971. *Radioactivity in the Marine Environment (RIME).* National Academy of Sciences, Washington, D.C.
3. Berner, R. A. 1971. Worldwide sulfur pollution of rivers. *J. Geophys. Res. 76:* 6597.
4. Osterberg, C., Cutshall, N., and Cronin, J. 1965. Chromium-51 as a radioactive tracer of Columbia River water at sea. *Science 150:* 1584–1587.
5. Goldberg, E. D., and Bertine, K. K. 1971. GNP/area ratio as a measure of national polution. *Marine Pollution Bull. 2:*94.
6. Goldberg, E. D. 1972. Man's role in the major sedimentary cycle. In *The Changing Chemistry of the Oceans.* (D. Dyrssen and D. Jagner, eds.). Almqvist & Wiksell Förlag AB, Stockholm, pp. 267–288.
7. Bertine, K. K., and Goldberg, E. D. 1971. Fossil fuel combustion and the major sedimentary cycle. *Science 173:*233–235.
8. Chow, T. J., and Patterson, C. C. 1966. Concentration profiles of barium and lead

in Atlantic waters off Bermuda. *Earth Planetary Sci. Letters 1:* 397–400.

9. Chow, T. J., Bruland, K. W., Bertine, K., Soutar, A., Koide, M., and Goldberg, E. D. 1973. Lead pollution: records in southern California coastal sediments. *Science 181:* 551–552.

10. Preston, A. 1973. Cadmium in the marine environment of the United Kingdom. *Marine Pollution Bull. 4:*105–107.

11. Preston, A., Jefferies, D. F., Dutton, J. W. R., Harvey, B. R., and Steele, A. K. 1972. British Isles coastal waters: the concentrations of selected heavy metals in sea water, suspended matter and biological indicators —a pilot survey. *Environ. Pollution 3:*69–82.

12. Butler, P. A. 1973. Organochlorine residues in estuarine mollusks, 1965–72— National Pesticide Monitoring Program. *Pesticides Monitoring J. 6:* 238–362.

13. Blus, L. J., Gish, C. D., Belisle, A. A., and Prouty, R. M. 1972. Logarithmic relationship of DDE residues to eggshell thinning. *Nature 235:* 376–377.

14. Risebrough, R. 1972. Cited in "Birds and pollution," an article in *Nature 240:* 248.

15. Nisbet, I. C. T., and Sarofim, A. F. 1972. Rates and routes of transport of PCBs in the environment. *Environ. Health Perspectives 1:* 21–38.

16. Harvey, G. R., Miklas, H. P., Bowen, V. T., and Steinhauer, W. G. Observations on the distribution of chlorinated hydrocarbons in Atlantic Ocean organisms. *J. Marine Res. 32:*103–118.

17. Harvey, G. R., Steinhauer, W. G., and Teal, J. M. 1973. Polychlorobiphenyls in North Atlantic Ocean water. *Science 180:* 643–644.

18. Matthews, W. H. (ed.). 1970. *Man's Impact on the Global Environment.* SCEP. The MIT Press, Cambridge, Mass.

19. Wilson, R. D. 1973. Estimate of annual input of petroleum to the marine environment from offshore production operations. Background volume for NAS Workshop on Inputs, Fates and Effects of Petroleum in the Marine Environment. Arlie, Va.

20. Blumer, M., and Sass, J. 1972. Oil pollution: persistence and degradation of spilled fuel oil. *Science 176:* 1120–1122.

21. Machta, L. 1972. The role of oceans and biosphere in the carbon dioxide cycle. In *The Changing Chemistry of the Oceans* (D. Dyrssen and D. Jagner, eds.). Almqvist & Wiksell Förlag AB, Stockholm. pp. 121–145.

22. National Academy of Sciences–National Research Council. 1957. *The Effects of Atomic Radiation on Oceanography and Fisheries* (NAS–NRC Publ. 551). The Academy, Washington, D.C.

23. Mitchell, N. T. 1973. *Radioactivity in Surface and Coastal Waters of the British Isles, 1971* (Ministry of Agriculture, Fisheries and Food, Radiobiological Lab. Tech. Rept. FRL 9). Lowestoft, England.

24. Preston, A., Fukai, R., Volchok, H., Yamagata, N., and Dutton, J. 1971. Radioactivity. In *Seminar on Methods of Detection, Measurement and Monitoring of Pollutants in the Marine Environment.* FAO, Rome.

Editor's Commentary— Chapter 12

RADIATION has been singled out for separate discussion in this chapter by Earl Cook because it is the most hazardous pollutant that exists, and it illustrates very well the dilemma we face in evaluating trade-offs when we examine our energy alternatives (Chapter 5). No other environmental issue has raised so many fears or sparked such fierce debate as the safety of the nuclear power industry. At the center of the debate lies real uncertainty about safety (see Chapter 19). The nuclear power industry does have a good safety record in comparison with other industries, but its mistakes are potentially very costly in human suffering. Furthermore, many of the calculations concerning reactor safety are just that—theoretical calculations based on computer models of reactors, rather than being based on real experiments, which of course we cannot do very easily. Furthermore, the Atomic Energy Commission (AEC) has not always been frank about potential hazards, so one has a sense of unease even when assurances of safety are given.

The whole issue of nuclear safety is by no means resolved, and there are still a number of nuclear physicists and students of safety in the nuclear power industry who feel, for example, that the breeder reactor program should not be undertaken under currently foreseeable circumstances. (The breeder reactor creates radioactive material as well as ''burning'' it, and has been proposed as the mainstay of nuclear power in the intermediate future.) The problem is so serious that Dr. Cook suggests that a national referendum is called for.

Perhaps the most serious potential problems of an enlarged nuclear power industry are: the handling and storage of high-level wastes; the build-up of radiation in the environment and its accumulation in food-chains leading to man; and nuclear blackmail by highly organized gangsters. Dr. Cook believes, however, that we could prevent the accidental release of radioactivity to the environment if we were willing to pay the cost, and that in fact the cost is not prohibitive. The same, of course, is not true of nuclear blasts for digging canals and mining gas, where release of radioactivity to the biosphere is a virtual certainty. Whatever the potential hazards and costs are, Dr. Cook draws a highly significant conclusion, namely that those exposed to the risks and costs—that is, the public—should make the final decisions.

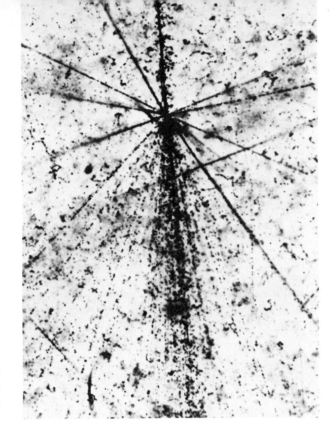

Star pattern of ionized particles is produced on film when high-energy radiation strikes an atom. Photo by Air Force Cambridge Research Laboratories.

12
Ionizing Radiation

Earl Cook *is Professor of Geology and Geography at Texas A & M University. Born in 1920, he received his doctorate in geology in 1954 from the University of Washington. In a career that has combined engineering, science and administration, he has had considerable opportunity to observe the impact of technology on environment. As staff officer of one of the committees of the National Academy of Sciences, he has investigated the problems of protecting the Earth from radioactive pollution. Currently he is studying the process of decision-making about environmental problems.*

Ionizing radiation is a natural environmental hazard whose potential for physiological (somatic) and genetic damage has been augmented greatly by man's discovery of atomic power. With advances in technology, the beneficial uses of man-produced radiation, especially in the production of electric power, are being continually multiplied, but so are man's chances of exposure to harmful radiation. The long hazard life of some radioisotopes (hundreds to thousands of years), the lack of knowledge of the effects of low chronic radiation doses, and the fact that dangerous isotopes introduced into the biosphere at low and perhaps harmless levels may be selectively concentrated by biological processes in man's food chain make adequate environmental protection a matter for cautious balancing of benefits and risks.

There are many kinds of radiation, but only radioactivity, cosmic rays, and cathode-ray tubes produce radiation capable of tearing electrons away from, or *ionizing*, atoms with which they interact. When the atoms disrupted by ionization form part of living tissue, serious biologic damage may occur, ranging from reduced life expectancy to almost immediate death; cancer and leukemia may be induced; and genetic damage to the species may arise from inherited defects that result from increased mutation rates. Rational management of man-produced radiation, so that its benefits can be enjoyed without a long-range cost in human misery and death from environmental pollution, is a supreme test of man's decision-making capability.

Before discussing radiation as an environmental hazard, we need to review some basic facts of nuclear and health physics. The structure of atoms and the nature of isotopes of elements are described in Chapter 5.

Radiation and Its Effects on Man

RADIOACTIVITY

Of the more than 320 isotopes that exist in nature, about 60 are radioactive. In addition, man has learned how to create, through nuclear fission, radioactive isotopes not found in nature; about 200 of these "artificial" radioisotopes have been identified.

Radioactivity involves a spontaneous nuclear change that results in the formation of a new element. The nuclear change takes place by emission of high-energy particles; a radioactive material is said to *decay* by disintegration. Each isotope decays at its own particular rate regardless of temperature, pressure, or chemical environment. *Allowing isotopes to decay naturally is the only practical means of eliminating their radioactivity.* The time required for any given radioisotope to decay to half of its original quantity is called its *half-life*. Half-lives range from microseconds to billions of years.

Radiation exposure or dosage is measured by the *rem, rad,* or *roentgen*. The most useful of these for our purposes is the rem, which measures the extent of biological injury of a given type that would result from the absorption of nuclear radiation.

All common matter on earth is radioactive in some degree. Organisms on the earth are subjected continuously to a low level of radiation from radioactivity within their tissues, from the immediate environment, and from cosmic rays. The principal internal radiation source is potassium 40 of the bones; the main external natural sources are cosmic rays from outer space and gamma rays from the earth's crust. The range of estimated average annual gonadal radiation doses (radiation on the sex cells) to the general population from natural sources is given in Table I (a millirem, mrem, is 0.001 rem). Total dosage is 0.081 to 0.172 rem per year (Table I).

Total radiation doses from environmental sources vary with altitude, because cosmic rays penetrate the thinner atmosphere of high places more readily than the heavier atmosphere of sea level; they vary with latitude because of the differences in travel paths of cosmic rays through the atmosphere; they vary with the nature of the underlying and adjacent rocks because granitic rocks contain more radioactive thorium, uranium, and potassium than other kinds of rocks; they vary with height above the land surface because of dilution of radon (a radioactive gas common to the three most important natural radioactive decay series); and they vary with the built environment because radon gas from natural building materials can become concentrated in closed spaces.

In the open ocean at the equator the average total radiation dosage is only 53 mrem per year, but at mile-high Denver near granitic rocks, the background is 170 mrem per year, and on a 20,000-foot granite peak at 55°N latitude, background rises to 560 mrem per year. Travelers in high-flying aircraft and astronauts in space vehicles are exposed to even larger radiation dosages (from cosmic rays).

As we shall see later, radiation protection guides or standards are set in *dose rates*, a fraction of a rem per week or month for radiation workers, or per year for the general population. It may be useful to keep in mind, as we discuss levels of radiation, that a guideline of the Federal Radiation Council sets 5 rem per generation (30 years), or about 170 mrem per year, as the maximum permissible gonadal dose for the general public, excluding background or natural radiation and medical irradiation. This dose rate represents substantially more than a doubling of the average background dose, which is a little over 100 mrem per year.

A little-known "natural" source of radiation to which man has exposed himself is radium in groundwater. In several areas of the United States deep wells have penetrated geologic formations that contain water whose radium content exceeds the maximum permissible concentration in drinking water as set by the Public Health Service.[2] Some of these wells are used as sources of municipal and domestic water; others are used mainly for stock, irrigation, and industry. Like uranium mines, these high-radium wells show that man can greatly increase his exposure to "natural" radiation, apart from his ability to produce "artificial" radiation.

Man-produced ionizing radiation includes x-rays from x-ray machines and television sets and the radiation from the products of nuclear fission. In nuclear fission, bombardment of a fissionable isotope by neutrons splits its nucleus and produces an enormous release of energy as well as a wide variety of fission products, almost all of which are radioactive. Some of the principal fission products are not only extremely radioactive but have half-lives as long as 33 years. One important fission product, plutonium, has a half-life of 24,360 years. If we use the commonly accepted guideline of 20 half-lives to safety, such isotopes will be dangerous for periods of 600 to half a million years!

In terms of environmental pollution, the most important fact about nuclear fission is that the fissioning of 1,000 grams of a fissile isotope produces 999 grams of waste materials (fission products and neutrons) which are either radioactive or capable of making inert material radioactive.

It is important to remember that man's additions to natural radioactivity have both inten-

I	Average Annual Gonadal Radiation Dose	
Source		**Annual dose (mrem)**
External sources		
Cosmic rays from space		32–73
Gamma rays from crust		25–75
Internal sources		
Potassium 40		19
Radium 226		3
Carbon 14		2
		81–172

Source: Reference 1.

sity and time factors. The most important artificial radioisotopes in radioactive wastes from nuclear power plants have half-lives of 30–33 years and hazard lives of 600–700 years. These are very long times in terms of man's generations, population shifts, and political changes. Furthermore, continued production of radioactive materials will extend the environmental hazard indefinitely.

It is also important to recognize the environmental distinction between x-radiation and radiation from radioactive isotopes. X-rays add to the radiation dose and to the biological hazard of the individuals exposed to them, but they cannot raise the level of environmental radiation, for they *cease to be propagated* when their energy source is turned off, and they attenuate or fade very rapidly in passing through air. Radioactive isotopes, on the contrary, add to the environmental or natural radiation level, because they endure and continuously emit radiation as they decay (Table II).

Finally, it should be noted that biological damage from radioactivity depends not only on the strength and quantity of the radioactive emissions, but also on the location of the radioactive source in relation to the body. A speck of plutonium on the skin may be blown or washed off without much harm, but in the bloodstream that speck is fatal. Uranium may be handled without danger from direct contact, but inhaled uranium dust can cause cancer (as may inhalation of air that has been in contact with uranium, for it will contain radon, a gaseous product of the radioactive decay of uranium).

II Average Annual Whole-Body Dose Rates in the United States

Source	Annual dose (mrem)
Environmental	
Natural	102
Global fallout	4
Nuclear power	0.003
Medical	
Diagnostic	72[a]
Other	3.8
	182

[a]Based on abdominal dose.
Source: Reference 3.

SOMATIC EFFECTS OF IONIZING RADIATION

In discussing biological damage from ionizing radiation, we must distinguish *somatic* damage from *genetic* damage. Somatic refers to the body cells of an animal, as distinguished from the germ cells. Somatic damage is limited to the individual receiving an injury. Injury to the germ cells, on the other hand, may be propagated through succeeding generations, resulting in inherited or genetic defects that lessen the survival potential of countless individuals. The differences between the two kinds of damage are crucial and the distinction should be kept in mind at all times.

In terms of energy delivered to a living cell, ionizing radiation is by far the most potent of all agents, physical or chemical; for example, about 100 million times more energy has to be introduced as cyanide to produce an equally deadly effect.[4]

Ionizing radiation disrupts chemical bonds in the molecules of living cells that absorb energy from the incident rays. New chemical bonds, if established, may be in a changed configuration that may not be compatible with normal life processes. Because the living cell contains giant molecules whose structure is vital to their function, the rupture and rearrangement of bonds may have far-reaching consequences.

Depending on the nature of the exposure, radiation injury takes many forms, ranging from small and long-delayed effects to short-term lethal effects. In the individual, the biologic damage ranges from reduced life expectancy through cancer and leukemia to death.

Different parts of the body vary in their response to radiation. The gonads are especially radiosensitive and the eyes only slightly less so. Human individuals appear to get less sensitive to radiation as they grow older. The embryo in utero is highly sensitive. It has been suggested that this sensitivity reflects the percentage of body cells undergoing cell division, and that these are the cells most susceptible to cancer induction; consequently, irradiation early in life is much more serious in increasing cancer occurrence than irradiation later in life.

The biologic effects of radiation are known from the following studies:

1. Studies of the survivors of Hiroshima and Nagasaki by the Atomic Bomb Casualty Commission.
2. Observation of victims of accidental high acute radiation exposure.
3. Study of patients treated by radiation for nonmalignant diseases.
4. Study of the occurrence of lung cancer in uranium miners.
5. Study of leukemia and other cancer in children whose mothers received irradiation during the pregnancy.
6. Many experimental studies on animals subjected to various doses, dose rates, and kinds of radiation.

The most acute damage is radiation sickness; it ranges from short-term effects from which the individual recovers fully, to death within minutes. Radiation sickness is the direct consequence of a nuclear accident or of warfare. The gross effects of very high radiation doses are given in Table III.

Cancer of the skin, marrow, bone, lung, and thyroid gland, as well as leukemia, can be induced by radiation. There is a latency period of 5–20 years after exposure before the cancer appears. Leukemia is perhaps the most likely form of malignancy; its occurrence in survivors of Hiroshima has been correlated with nearness to the center of the explosion.[5] Leukemia shows up much more quickly in an exposed population than do other types of cancer; the Atomic Bomb Casualty Commission reported in 1971 that other cancers were beginning to be seen in the Hiroshima and Nagasaki survivors 25 years after the bombs were dropped. Several studies suggest that very small exposures to radiation before conception or during pregnancy may increase by 50 percent the child's risk of leukemia.

Some radioisotopes show a tendency to become concentrated in a specific biologic environment. For example, there are bone seekers, all of which are known to be carcinogenic; plutonium is both an extremely toxic element and a bone seeker. Other bone seekers are radium 226 (tenaciously retained, found in bones 25 to 35 years after exposure), strontium 90, barium 137, thorium 232, and actinium 227.

Finally, general physiological aspects, such as growth, development, and aging, are af-fected. In Hiroshima child survivors, as radiation exposure increased, there were small but statistically significant decreases in body measurements at all age levels and in growth rate at postpubertal age levels. Animal experimentation has shown that ionizing radiation can induce a shortening of life span attributable to no specific disease but to an accelerated occurrence of disease in general. In Hiroshima and Nagasaki survivors, a general increase in mortality, exclusive of death from cancer, has been found.

GENETIC EFFECTS OF IONIZING RADIATION

A subtle and serious consequence of some radiation injuries is genetic transmission of physiological defects following an increase in mutation rates. Thus, the species may suffer genetic damage from inherited defects. A mutation is a chemical or physical accident that changes the composition of a gene. Mutations induced by radiation are no different from spontaneous ones. Indeed, geneticists believe that 5 to 12 percent of the "spontaneous" mutations are caused by background radiation (Table II).

The effects produced by mutations are mostly harmful. Some have an immediate effect, some lie hidden for many generations. Mutations are responsible for a large part of human premature death, illness, and misery.

III Effects of High Radiation Doses

Dose (rem)	Effect
100,000	Death in minutes
10,000	Death in hours
1,000	Death in days
700	Death for 90 percent within months but 10 percent survive
200	Death for 10 percent within many months; 90 percent survive
100	No deaths, but chances of cancer and other forms of reduced life expectancy greatly increased; can induce permanent sterility in females, 2- to 3-year sterility in males

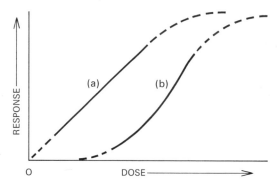

1   **Dose–Response Curves.** (a) Nonthreshold curve: holds for genetic damage and probably for some cancer induction. (b) Threshold curve: holds for radiation sickness and possibly for delayed somatic damage.

DOSE-RESPONSE CHARACTERISTICS

Determination of the biological risk of any toxic substance requires knowledge of the relation between the dose and the corresponding response. It also requires a search for a possible threshold level of exposure (Chapter 8).

Some toxic substances can be taken in small quantities without detectable effects. As the quantity or dose is increased, the most susceptible individuals begin to show an effect. This may or may not be a threshold dose, the dose below which there is no effect. It may simply be the dose below which actual effects are not detectable, because they are widely dispersed in a large population and cannot be related to a single cause, or because they are concealed as genetic defects that may not become visible for several generations. If greater doses increase the observed effect more than proportionately, there is probably a threshold. But if the curve relating response to dose is a straight line, any exposure probably has some effect (curve a, Figure 1).

IV   **Respiratory Cancer Deaths in White Uranium Miners of the Colorado Plateau**

	Smokers	Nonsmokers
Person-years	26,392	9,047
Cancers expected[a]	15.5	0.5
Cancers observed	60	2

[a]If uranium miners react to smoking like everyone else and if the observed group of nonsmokers has the normal incidence of cancer in the U.S. male white population.
Source: Reference 7.

It is known that a given dose causes more somatic damage when absorbed in a single brief exposure than when absorbed in multiple exposures or in an exposure protracted at a low dose rate. In other words, there appears to be a threshold dose below which there is no permanent damage, probably because the physiological system can repair tissue damage from background (and a little more) radiation at about the same rate at which damage occurs. If the total dose is prolonged at a low rate rather than concentrated in time, the system has time for recovery and replacement of damaged and dying cells. If the dose is concentrated or if it falls above the threshold, repair is no longer possible and permanent injury results.

Whether or not there is a threshold for somatic effects that are delayed is not known, and at the present time there is great disagreement over the question of cancer induction by prolonged or chronic exposure to low levels of radiation. There may be a threshold level below which radiation has no carcinogenic effect. However, similar carcinogens such as ultraviolet light show no threshold effect[6] (there is a linear relation between dose and response down to very low dose levels) and their effects are cumulative. The most reasonable assumption is that all doses of radiation, however small, can cause cancer in some fraction of the population.

There is also argument over the way in which radiation adds to the incidence of cancer produced by other sources. Some scientists assume a fixed risk per rem that should be *added* to the "natural" incidence of any particular form of cancer to calculate the carcinogenic effect of radiation, whereas others point to evidence such as that now available on uranium miners of the Colorado Plateau to support their contention that radiation acts with other carcinogens to *multiply* the number of cancers induced. The fact that leaps from these simple statistics is that radiation exposure caused the risk of both smokers and nonsmokers to multiply four times instead of just adding a fixed risk or number of deaths to each category.

There is no threshold below which radiation is genetically harmless. Radiation-induced mutations are known to be induced at a rate proportional to the total amount of radiation received (i.e., the dose–response curve is linear), regardless of the duration or time distribution of the exposure. The genetic effects of radiation doses are cumulative both in the individual and in his offspring; an increased mutation rate in a descendant of irradiated parents will be multiplied in proportion to the dose the descendant receives during his lifetime. *Perhaps the most important hazard related to radiation pollution is this snowballing genetic effect:* an increased radiation level that produces little adverse effect in the first few generations may, if maintained through those generations, produce disastrous effects, even extinction, in far-distant generations.

BACKGROUND RADIATION: BENIGN OR MALIGN?

Although many people, including some scientists and engineers, speak of background or the "natural level" of radiation as if it were benign or harmless to man, a steady background dose of ionizing radiation which cannot be reduced may be the cause not only of a fraction of the "spontaneous" mutations in man but also of a certain very low incidence of cancer.

The assumption of harmlessness of background radiation is based on the belief that man, over thousands of years, has adapted himself physiologically to life in a sea of radiation and that his body can repair damage from natural dosage levels. This benign assumption is questioned more and more as linear response curves based on animal experiments and on studies of humans exposed to chronic or continued radiation doses are extended to lower and lower dose levels.

Although no one actually claims that a rem of man-made radiation is more dangerous than a rem of natural radiation, such an attitude seems implicit in radiation protection guides that define permissible limits of exposure *above* background. It is difficult to understand why

280 mrem per year, exclusive of medical irradiation, is unsafe or "unacceptable" for a man who lives in New York City but safe or "acceptable" for a man who lives in Denver. The rationale for such standards appears to be that permissible limits expressed in total exposure would tend to restrict nuclear industry to coastal areas and to deprive the higher parts of the country of their fair share of new industrial development.

At least for genetic effects, and probably for long-term somatic effects, man-made radioactivity released to the environment is simply an increase in a hazard already present (Table V).

BIOLOGICAL CONCENTRATION PROCESSES

Although maximum permissible concentrations for radioactivity in air and water have been set to protect the public from harmful biological effects of gaseous and liquid effluents from nuclear reactors, fuel-reprocessing plants, and nuclear explosions, these standards do not take into account certain biological processes that may concentrate radioactivity in the food chain of man (Chapter 1).

Just as DDT and other pesticides may be concentrated in the living tissue of fish and birds, so may radioisotopes be concentrated to levels thousands of times greater than their concentrations in the air and water surrounding the living organisms. Below the Hanford, Washington, plant of the AEC, for example, eggs of ducks and geese have been found to contain radiophosphorus in concentrations 200,000 times greater than the effluent solution.

The interception and retention of fallout by plants leads to the rapid injection of radioactive material into terrestrial food chains. In the days of bomb-test fallout it was discovered that Eskimos of the Arctic were being subjected to a unique hazard because Arctic lichens collect radioisotopes from the air and con-

V   Summary of Health Effects

Effect	Response	Threshold?
Radiation sickness (somatic damage)	Immediate, cumulative, multiplicative (synergistic)	Yes
Cancer of various sorts	Delayed	Probably no
Genetic effects	Immediate (to germ cells) and inherited	No

centrate them. Caribou feed on the lichen and retain the radioisotopes in their bodies, with further concentration. Finally, the Eskimos eat the caribou, ingesting the radioisotopes, which then become selectively concentrated in various parts of their bodies.

Biological concentration is a limiting factor on the level of radioactive waste discharged into the sea at the Windscale works in England. Extensive studies have shown that the persons liable to the greatest exposure from this discharge are those who eat a special type of bread made from edible seaweed of the area. Accordingly, the discharges have been lowered to protect this bread-eating group. In the United States, on the other hand, when it was discovered that certain Tennessee mountaineers were eating fish cakes made from the meat and bones of fish taken from the Clinch River, into which low-level wastes from Oak Ridge are discharged, the Atomic Energy Commission (AEC) attempted to "educate" the mountaineers to remove the bones (in which certain radioisotopes are concentrated) before making their cakes.

Studies of iodine 131 and strontium 89 from fallout from Project Sedan (a Plowshare cratering test in Nevada) showed that desert plants collected these radioisotopes and that they were further highly concentrated in the thyroids of jackrabbits that feed on the plants.[8] After deposition on pasturage, iodine 131 can make its way swiftly into cow's milk and thence into the human body; unacceptable concentrations were found in milk in southwest Utah after Nevada bomb tests in 1953. Although iodine 131 has a short radioactive half-life (8 days), it has a *biological half-time* of about 3 months, because it is quickly taken up by the thyroid gland and is eliminated slowly. Biological half-time is the time required for the amount of a particular element in the body to decrease to half its initial mass through elimination by natural biological processes. It is important to understand also that the biological effectiveness of a radioisotope is inversely related to its radioactive half-life: a given mass of a radioisotope of short half-life will decay and emit particles at a greater rate than the same mass of another isotope which has a longer half-life. Consequently, the radioisotope with the shorter half-life is more biologically "effective" or damaging. The radioiso-

topes representing the greatest internal hazard are those with a relatively short radioactive half-life and a comparatively long biological half-time. However, any radioisotope with a long biological half-time, including those with long radioactive half-lives, can be a serious internal hazard; two of these are radium 226 and plutonium 239.

Man-Made Radiation Hazards

URANIUM MINING AND MILLING HAZARDS

The highest natural background radiation levels on the earth's surface are found over deposits of uranium ore. Man can greatly augment his dosage from this natural source by sinking mines into the uranium ore bodies, where he will be exposed to dose rates of 5,600 mrem (5.6 rem) per year or greater, some 60 times the surface background at New York City. In addition, the miner will inhale radon gas, a decay product of uranium; the radioactive daughters of radon are solid and will probably lodge in his lungs, where their potential for damage is much greater than on his skin.

The hazard of the uranium mine environment was first revealed in the late 1930s by a study of pitchblende miners of the Erzgebirge in Germany and Czechoslovakia. Pitchblende is a mineral source of radium as well as uranium; it was then mined for its radium content. About 15 years after the start of pitchblende mining in this area of central Europe, the miners began to die of lung cancer. Ultimately, about 50 percent perished from this disease, and 80 percent of the remainder died from other lung diseases. Miners in other kinds of mines showed a much lower incidence of lung cancer. Beyond statistical doubt, the high lung cancer mortality rate was related to the pitchblende ore and to the inhalation of radon by the miners.

Despite the ominous evidence from the Erzgebirge and the documentation and interpretation of that evidence in the scientific literature in the United States, when uranium mining started in the United States in the 1940s, standards of individual exposure and mine ventilation were not set high enough to prevent a recurrence of the tragedy. In 1955, U.S. Public

304

Health Service tests on air samples from 75 U.S. uranium mines showed that most far exceeded standards set by the International Commission on Radiological Protection (ICRP). A 1967 study[7] of 3,414 U.S. uranium miners indicated that 46 of them had already died of lung cancer because of their mine exposure to radon and suggested that more of the studied group would yet have their lives shortened by their mining experience. Charles C. Johnson, Jr., head of the U.S. Consumer Protection and Environmental Health Service, has said that "of the 6000 men who have been uranium miners, an estimated 600–1100 will die of lung cancer within the next 20 years because of radiation exposure on the job." In 1967, more than 20 years after the start of mining, exposure standards for U.S. uranium miners finally were set at levels that *may* be adequate for protection. Both the AEC and the U.S. Department of Labor now have permissible exposure limits for uranium miners; the fact that the Labor Department's standard is more than three times as stringent as that of the AEC reflects a continuing argument over the shape of the lower end of the dose–response curve.

Another hazard belatedly recognized involves the finely ground waste material, called tailings, from the uranium ore-processing mills. In the upper Colorado River basin, these tailings are known to contain a large amount of radium 226 as well as significant quantitites of thorium 230 and lead 210, all of which are radioactive.

Radium 226 is a significant environmental hazard. It is readily leached from the tailings by rainwater and enters streams in solution. It is highly toxic, having the lowest maximum permissible concentration in drinking water of any of the 264 isotopes (natural and man-made) considered in the ICRP standards. It has a long half-life (1,622 years), remaining hazardous for thousands of years. It decays to radon gas, which can concentrate in buildings built on tailings used as landfill.

Radium leached from tailings has raised radium concentrations, at some places and times far above permissible levels, throughout thousands of miles of the Colorado River system, and has contaminated groundwater reservoirs. In places where tailings have been used as convenient landfill, buildings can contain hazardous concentrations of radon gas. Investigations, begun in 1966 when the hazard was first detected, by mid-1971 had identified more than 5,000 homes and commercial buildings in the Colorado Plateau region which had been built on uranium tailings and which, as a consequence, contained anomalously high concentrations of radon.

NUCLEAR WAR AND REACTOR ACCIDENTS

The greatest potential for worldwide radioactive pollution of the environment lies in nuclear war and related weapons testing. There seems little reason to discuss this disastrous potential in detail here. Man has developed the ability to render the entire surface of this planet uninhabitable. Although a nuclear war probably would stop somewhere short of annihilation of the human species, environmental conditions for the survivors would be so changed that the content of this chapter would be of no more than historical interest to them.

The greatest potential for local radioactive pollution of the environment lies in the possibility of reactor accidents (excepting catastrophic releases of stored high-level wastes by flooding or earthquake, or a conflagration at a plutonium plant; these sites are many times fewer in number than reactor sites, but each has more releasable radioactivity than any single reactor). No explosion like that of an atom bomb can occur in a nuclear reactor of the type in use today. However, several types of major accidents are possible. Any of these may result in melting or physical disruption of the reactor core and possible release of some of its great load of radioactivity to the local environment. This type of accident is assumed to be highly unlikely, because of the elaborate and comprehensive safety design requirements for nuclear power plants. On the other hand, the disastrous consequences of such an accident and the fact that it can occur (reactor safety design does not consider sabotage or warfare, for example) has caused commercial insurance companies to refuse to assume the risk inherent in a commercial power reactor; under the provisions of a law known as the Price–Anderson Act, this risk is jointly "assumed" by the Federal Treasury and the exposed public. How bad such an accident might be is indicated by a 1957 report of the Atomic Energy Commission on *The Theoretical Possibilities and Consequences of Major Accidents in Large Nuclear Power Plants*, which concluded that a single major accident might result in 3,400 deaths at distances up to 15 miles, 43,000 injuries at distances up to 45 miles, and property damage of as much as $7 billion. Present estimates of damage, based on the assumptions of the 1957 report, might be much greater because of the larger power plants now being constructed. Although nuclear experts speak of the "exceedingly low" possibility of a serious nuclear power plant accident, the Joint Committee on Atomic Energy in 1965 admitted that there had not been "sufficient operating experience to form an adequate judgment of risk." Although reactors continue to be licensed on the assumption of "no undue risks to the public," the Price–Anderson Act in effect recognizes an undue (or at least unknown) risk to industry in the operation of the same reactors, for it indemnifies the utilities and equipment manufacturers against a risk neither they nor the commercial insurance companies are willing to assume.[9]

Reactors are built with an emergency core-cooling system to protect against venting as a result of an accident; without such a cooling system, the heat in the core, say as the consequence of a leak or rupture in the primary coolant system, could increase enormously within a few minutes, melt the reactor and its containment vessels, and spread death and destruction over a wide area. The first full-scale test of the emergency core-cooling system used in the present U.S. power reactors is now scheduled for 1974 or 1975 (a striking example of putting promotion ahead of protection). To achieve economies of scale, power reactors have been growing in size, and therefore in their internal burden of radioactivity. The power density (a measure of heat intensity) of reactor cores also has been increasing. To keep power-transmission losses low, power plants are sited as close to centers of power use (cities) as the regulatory authorities will allow. Large reactors, high power densities, urban siting, and a lack of certainty about the proper functioning of the emergency system combine to create what appears to be a formula for disaster which fully justifies the AEC's stubborn insistence on licensing com-

mercial reactors as "experimental." It became known only in 1972 that many nuclear safety experts favor a moratorium on nuclear-power increases until safety research catches up with construction and licensing of power reactors. The director of nuclear safety at Oak Ridge National Laboratory has said that "no one really knows what will happen in a reactor core in the event of a loss-of-coolant accident."[10]

A kind of reactor accident that is more likely than an accident in a commercial power reactor is the loss of a nuclear-powered vessel at sea. Indeed, such an accident has already happened. The U.S. nuclear-powered submarine Thresher, lost in the North Atlantic in 1963, released a large quantity of radioactivity to the marine environment.[11] The potential for radioactive pollution of the oceans during a major war is very great, *even if nuclear weapons are not used.*

MEDICAL IRRADIATION

As early as 1900 a physician described the "irritating" effects of x-rays. The first death from a radiation-induced tumor occurred in 1904. Then reports began to appear in the medical literature describing mentally retarded children with small heads born of mothers who had received radiotherapy during early pregnancy. Over the ensuing years it was noted that x-ray workers, including physicists and physicians, had a much higher incidence of skin cancer than could be expected from random occurrence. During the same period many dentists developed cancer of those fingers used to hold dental x-ray films in the mouths of patients. A 1944 report showed that leukemia was reported 1.7 times more often as a cause of death among all U.S. physicians than among the general population of adult white males. An increase in leukemia incidence has been verified among patients treated by x-rays for ankylosing spondulitis, an arthritic condition of the back. A 1965 study of 877 Nova Scotia women who received numerous fluoroscopic examinations while being treated for tuberculosis showed clear evidence of radiation-induced breast cancer at approximately 24 times the normal incidence. Recent studies show that children born after *in utero* radiation of embryos for diagnostic purposes have about 50 percent more danger of developing leukemia and other forms of cancer than do other children.

Although the use of x-rays in medicine is still expanding, there is a strong movement within the profession to reduce radiation exposure from diagnostic x-rays. Not only are techniques available for grossly reducing the radiation exposure, but many doctors now use x-rays only when they judge the patient would be in more danger without the exposure than with it. Medical irradiation gives the average individual in the United States a yearly dose of about 76 mrem.

X-RAYS FROM TELEVISION SETS

Any electronic tube operating at a potential above a few thousand volts may be a source of x-radiation. Such sources include oscillographs, electron microscopes, and television tubes. Of these the home television tube is of greatest interest because a high percentage of the population is involved.

Television tubes in general are designed so that the maximum emission at a distance of 5 cm from the tube surface is less than 0.5 mrem per hour. On the basis of an average yearly viewing time of 1,000 hours, and correction factors for viewing distances of 100 cm (children) and 200 cm (adults), it has been calculated that the average yearly gonadal skin dose for such viewers would be 40 mrem (100 cm) and 10 mrem (200 cm).[12]

In 1967, many color television sets were found to be emitting excessive radiation, generally from the sides, top, bottom, or back, and in 1969 the U.S. Bureau of Radiological Health warned viewers of color television against the hazard of sitting closer than 6 feet from the screen and of exposing themselves for long periods to the sides or back of an operating set. By 1973, however, many manufacturers of television sets had switched from electronic tubes to solid-state devices that do not produce x-radiation, and the new sets with tubes were better shielded.[13]

RELEASES FROM NUCLEAR EXPLOSIVES

Fallout from atmospheric tests of atomic explosives prior to the Limited Nuclear Test Ban Treaty of 1963 raised background radiation levels throughout the Northern Hemisphere, most dramatically in areas downwind from the Nevada Test Site. Averaged over the United States, however, the fallout during the years of maximum testing has been calculated at about 30 mrem per year, well below the guide of 170 mrem per year. The guidelines for "permissible" exposures, however, must not be interpreted as no-risk limits. Any exposure to ionizing radiation will produce some shortening of human lives in the exposed population. It has been calculated that the atmospheric bomb tests of 1945–1963 ultimately will cause 400,000 "statistical deaths," equivalent to the shortening by 1 year of 28 million lives.[14] This calculation alone justifies concern over atmospheric tests carried out in recent years by China and France, neither a signatory to the test-ban treaty.

Since 1963 the danger of environmental contamination from nuclear explosives in the United States has come from underground weapons tests and from underground nuclear explosions of the Plowshare program. Plowshare is a program of the AEC to promote the peaceful use of nuclear explosives in a wide range of applications, from the creation of underground storage facilities to the blasting of artificial harbors and canals. Although the potential engineering applications of nuclear energy are vast, the potential for environmental contamination is also great. Unlike the man-made radiation of nuclear reactors and their wastes, which can, with proper care in design and practice, be kept out of the biologic environment, the man-made radiation of an underground nuclear explosion is created in that environment and cannot be kept out of it. Cratering explosions release 10 percent of their radioactivity directly to the atmosphere and produce fallout characteristic of atmospheric weapons tests. Some of the 90 percent of the radioactivity that remains in the ground may be leached and transported by moving groundwater to points of potential water use. Completely contained underground nuclear detonations create a similar danger of groundwater contamination and, in addition, may involve deliberate releases to the environment such as occurs in the flaring of radioactive natural gas after a shot designed to stimulate its production from geologic formations of low permeability. Three nuclear stimulation tests have already been carried out, and in 1974 or 1975 a sequence of five nuclear explosions in one borehole will produce the largest cavern yet made in the natural-gas stimulation tests. To produce significant amounts of gas, it is expected that about 370 nuclear explosions each year will be required!

Although environmental contamination from one properly placed underground nuclear blast might be nothing to worry about, the contamination potential of hundreds or thousands of such blasts within a relatively small area is causing concern. The Bureau of Natural Gas of the Federal Power Commission has stated: "In order to substantially increase natural gas availability . . . thousands of nuclear devices will have to be detonated . . . such large-scale application might not gain public acceptance."

Natural gas "stimulated" by Plowshare blasts will itself be too radioactive for use; therefore, it is planned to add it slowly to gas distribution systems so that it will be diluted to acceptable or permissible levels by "clean" gas from other sources. If this occurs, an important decision will have been made for the ultimate consumers of the mixed gas by government experts; these experts will have made a judgment that the expectable benefits exceed the risk, and very likely will not have asked the opinion of those who will be exposed to that risk.

REACTOR WASTES

A fission reactor produces a great quantity of radioactivity as waste. This radioactivity is generated in the core of the reactor that houses the fissionable fuel. (Figure 2 shows three kinds of fission reactors.) Reactor fuel consists of uranium in metallic, oxide, or carbide form, fashioned into rods or plates put together into what are called fuel elements. The elements are encased in zirconium-alloy or stainless-steel cladding. The kinetic energy of fission is converted to heat which is picked up by water or gas (air or carbon dioxide) circulating among the fuel elements. The heat is used to generate steam, either directly (boiling-water reactor) or indirectly (pressurized-water or gas-cooled reactors), and the steam drives a turbine generator that produces electricity.

Only a small portion of the material in the fuel elements is actually burned up at the end of a fuel cycle. Much of it is not fissionable, and for technical reasons not all of the fissionable portion can be used or burned. Conse-

quently, when it comes time for the fuel elements to be removed from the reactor core (about 20 percent are removed and replaced annually), there remains in them some highly valuable unspent fuel which it is economically important to recover. This recovery is done at fuel-reprocessing plants. However, by the time that the fuel elements are removed, they have become highly radioactive and extremely dangerous. The cladding and the unburned contents not only absorb most of the radioactive fission products but, although previously inert, become radioactive themselves under the neutron bombardment that takes place during fission. By far the greater part of the radioactive waste of a nuclear reactor is contained in the spent fuel elements. If the elements are properly handled, none of this radioactivity is released to the environment at the power-plant site.

During operation of the reactor, radioactivity also builds up in the primary coolant (water or gas), in two ways. Fission products escape from the fuel elements by diffusion or through defects in the metal cladding. In addition, radioactive products, especially tritium (^{3}H or T), are formed in the water by neutron activation. Water (H_2O) in which one of the ordinary hydrogen ions (H) is replaced by a tritium ion (T) is called *tritiated water* (HTO). Chemically and physically (except for its radioactivity) tritiated water is very similar to ordinary water, and it is not possible to separate the tritium by conventional waste treatment processes. It is common practice to release tritiated water, after dilution, by uncontaminated water, into the ground at reactor sites. Tritium has a relatively long half-life (12.36 years) and as tritiated water can become a serious environmental hazard.

Another hazardous radioisotope that has been difficult to manage in reactors is krypton 85. On a laboratory scale, it is possible to remove krypton from gas streams, but until recently no commercially feasible technique had been developed, and krypton 85 is released to the atmosphere at present reactor sites.

Radioactive iodine (two of the iodine isotopes are radioactive), an extremely noxious waste, was released as a gas in considerable quantities into the atmosphere from the early commercial reactors, but reactors now under

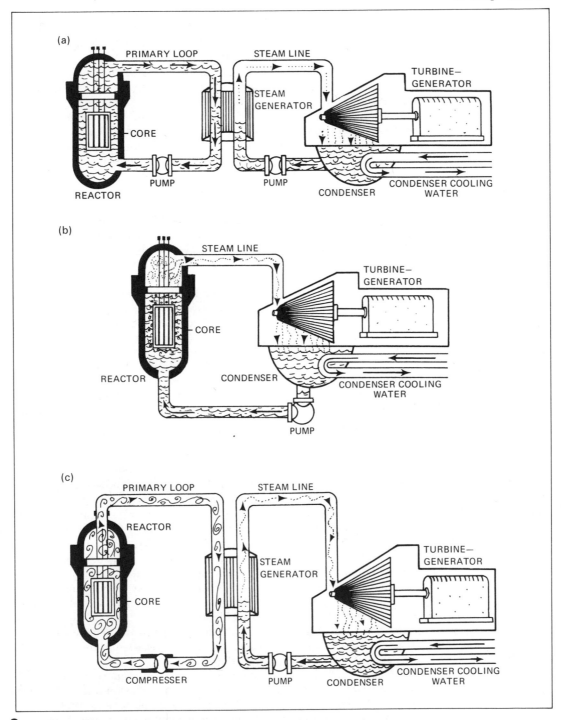

2 **Fission Power Reactors.** Examples of the three types of reactors in commercial use. Pressurized water reactors: Yankee Atomic Electric Company, Rowe, Massachusetts, and Consolidated Edison Company, Indian Point, New York. Boiling water reactors: Commonwealth Edison Company, Dresden Nuclear Power Station (near Chicago, Illinois). Gas-cooled reactors: Hinkley Point, Somerset, England.

construction are designed to remove well over 99 percent of the produced iodine from the gaseous effluent. Iodine 131, much the more abundant of the two isotopes, has a half-life of only 8 days. If it can be trapped, concentrated, and contained for a period of 150 days or so, it will decay to an innocuous level; on the other hand, because it becomes concentrated by organisms, this short half-life is not the protection that it might appear to be.

Here it is important to note that a continuous advance in the technology of managing the radioactive wastes of reactors appears to have culminated in a zero-release design that is technically feasible and economically practical. In the new system, krypton 85 will be concentrated and stored. Tritium will be fed back into the reactor system and, in effect, stored there. Adoption of a zero-release reactor system would focus the radioactive waste disposal problem of the nuclear power industry at the fuel-reprocessing plants and at the permanent repository for high-level solidified wastes.

WASTES FROM FUEL REPROCESSING

A fuel-reprocessing plant is a chemical plant where the fuel elements are dissolved in nitric acid, and the depleted fissionable fuel is recovered. Until 1966, the AEC did all fuel reprocessing at its own plants in South Carolina, Idaho, and Washington. Then the first private reprocessing facility was established at West Valley, about 30 miles south of Buffalo, New York. Another has been completed at Morris, Illinois, and several others are planned. Each fuel-reprocessing plant is designed to serve a local or regional cluster of reactors. As more power reactors come into operation, the demand will increase for more and larger reprocessing plants.

When the spent fuel elements are removed from the core of a reactor they are highly radioactive and must be handled with great care. The spent elements are stored at the reactor site for a period of about 150 days, during which time their radioactivity is greatly reduced by decay. They are then transported by truck or railroad in large lead and steel casks weighing as much as 70 tons to a fuel-reprocessing plant, where all the radioactive waste materials are extracted. After extraction,

some of the radioactive wastes are in liquid form; others are gaseous.

Most of the radioactivity is contained in high-level liquid wastes. Recently published AEC procedures for handling this very dangerous waste, which contains strontium 90 and cesium 137 with hazard lives of 600 to 700 years, call for storage at the reprocessing plant for a period not to exceed 10 years, conversion to solid form, and shipment to a designated federal repository. If adequate safeguards are maintained during storage, conversion, and shipment of this high-level waste, and if its "ultimate disposal" is in a deep salt formation from which it could be retrieved if necessary, it should constitute no significant environmental hazard. On the other hand, location of such a plant near any large city increases the hazard inherent in the possibility of the accidental rupture of a high-level waste storage tank. The new fuel-reprocessing plant at Morris, Illinois, avoids the liquid-storage hazard by solidifying its high-level liquid waste as it is produced.

THE BREEDER POWER PLANT

Introduction of a breeder power plant (Chapter 5), although it will extend known reserves of nuclear fuel 60 to 100 times, will increase the problems of thermal pollution, accident containment, and waste handling. Not only must the breeder reactor be large to be economical, but its core will contain more radioactivity per unit of power output than does the core of a reactor of the present generation. In addition, more plutonium will be recovered from the spent fuel and reprocessed into new fuel elements. Plutonium is an almost incredibly toxic substance: one-millionth of a gram injected into the skin of a mouse has caused cancer; a similar amount injected into the bloodstream of dogs has caused a substantial incidence of bone cancer; a few hundred-thousandths of a gram inhaled by each of 24 beagles at Hanford resulted in the deaths of 22 of them from lung cancer at an average of half their normal life-span. But plutonium is the most desired of the fission fuels. By many students of the national energy problem, the breeder power plant is considered a necessary and large contributor to the energy needs of the country. The hazards it represents, however, are so formidable that the equivalent of a national referendum on the question of proceeding with its development appears called for. In 1973 the National Resources Defense Council and the Scientists' Institute for Public Information won an important lawsuit against the AEC, which must now prepare a full environmental-impact statement for the liquid-metal fast-breeder reactor research-and-development program. The basis will thus be laid for a full and public assessment of the risks and benefits inherent in developing the breeder reactor. The potential plutonium peril is social as well as environmental; it includes the chilling prospect of blackmail on a scale never before possible, by persons using stolen plutonium to construct what has been called a "basement" hydrogen bomb. It would take about 10 pounds of plutonium to make a bomb; by the year 2000, the annual U.S. production of plutonium may be 100 tons and rapidly increasing; sometime thereafter it might be quite possible to divert enough to make a blackmail bomb.

The two justifications for the present heavy investment in breeder development are (1) we shall soon run out of low-cost fuel for the present generation of nonbreeder nuclear plants, and (2) other countries will get a technological lead on the United States if we do not push the breeder design along, and our balance of trade will suffer.

The first point has been countered by studies which show that nonbreeder plants could stand a fuel-price increase of three to five times the present price of uranium, at which levels there would be ample supplies to carry a rapidly growing nuclear power industry to the year 2020 or beyond.[15] The economic part of the second argument has been made to appear doubtful by those who question the economic benefits claimed for the breeder.[16]

Radioactive Waste Storage and Disposal

We have seen that radioactive wastes are generated in nuclear reactors and associated fuel-reprocessing plants. Although research, experimental, and production reactors (those which produce such fissionable isotopes as plutonium 239) produce wastes that represent local environmental hazards, the great bulk of the radioactive wastes of the future will come from commercial power reactors.

In early 1974 there were 44 nuclear power plants licensed to operate in the United States; 54 more were under construction and another 109 were on order. It will be seen from the map (Figure 3) that there is a strong concentration of present and planned U.S. nuclear power plants in the eastern part of the nation. This concentration is but crudely accordant with population distribution; the capacity distribution of all plants indicated on the map (and four more to be built by the Tennessee Valley Authority, sites not yet selected) is as follows: Southeast, 36 percent; Northeast, 27; North Central, 22; Pacific Coast, 9; South Central, 5; Rocky Mountains, 0.2. This distribution appears to reflect regional differences (1) in availability and cost of fossil-fuel alternatives and (2) in public acceptance of nuclear power. The AEC has predicted that the installed nuclear electric capacity will rise 100-fold, from 7,000 megawatts in 1970 to 734,000 megawatts in the year 2000.

Storage of radioactive wastes implies that the material is retrievable; disposal, that it is not. Methods of storage and disposal fundamentally depend on the biological harmfulness of the radioisotopes involved, their hazard lives, their concentrations, and the physical state of the waste material (solid, liquid, or gas). Other factors that may affect specific practices are economy, expediency, and faith in printed standards and in the capacity or willingness of the environment to cooperate.

Although most present methods for storage and disposal of radioactive wastes appear satisfactory in terms of existing quantities and present knowledge, some will certainly not be acceptable for the much higher quantities of the future and may not be acceptable even in terms of present quantities as our knowledge of the biological risks improves.

A common classification of current disposal methods for fluid (liquid and gaseous) wastes is the following: concentrate and contain; delay and decay; dilute and disperse.

Concentrate and contain applies to treatment of the high-level (tens to thousands of curies per gallon) liquid wastes produced at the fuel-processing plants. (The curie is a unit of quantity of radioactivity, originally defined as

311

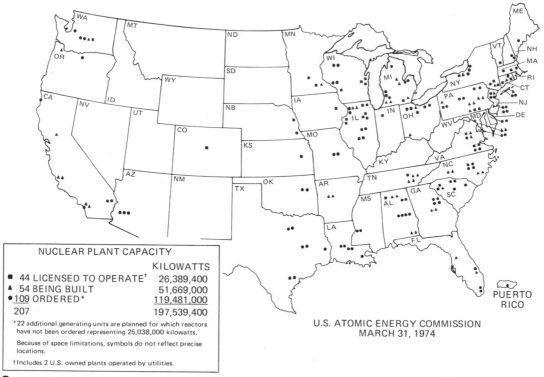

NUCLEAR PLANT CAPACITY

KILOWATTS

■ 44 LICENSED TO OPERATE†   26,389,400
▲ 54 BEING BUILT   51,669,000
● 109 ORDERED*   <u>119,481,000</u>
207   197,539,400

*22 additional generating units are planned for which reactors have not been ordered representing 25,038,000 kilowatts.

Because of space limitations, symbols do not reflect precise locations.

†Includes 2 U.S. owned plants operated by utilities.

U.S. ATOMIC ENERGY COMMISSION
MARCH 31, 1974

3   **Nuclear Power Reactors in the United States.**

the radioactivity of 1 gram of radium.) These wastes must be stored, for they are too dangerous to be placed where they could not be retrieved. So far, no method of permanent storage has been adopted, although several methods have been under research and development by the AEC for a number of years. Almost 80 million gallons of hot, high-level wastes, concentrated by evaporation, are stored in about 200 steel-and-concrete tanks at AEC reprocessing sites (140 of these are at the Hanford, Washington, site). Some of the waste has been there for over 25 years; all of it will be hazardous for at least 600 years. The tanks must be strong, corrosion-resistant, leak-protected, and provided with a constant cooling mechanism. As the tanks age, the hazard from leakage becomes serious. Several tanks at Hanford and at Savannah River have sprung leaks during the past 10 years; in at least one instance about 100,000 gallons of high-level waste leaked into the underlying gravel. The most promising procedures for permanent storage would convert the wastes to solids in the form of glass or ceramic slugs, after which

they could be interred in salt formations deep underground or in concrete-and-metal bins at or near the earth's surface. In either method, it is critically important to keep the radioactivity isolated from circulating groundwater and thus from the biological environment. Underground storage in salt is preferable, because salt beds are highly impervious to groundwater flow, act plastically to seal any fractures that form in them, are insulated from the effects of rain and snowfall, and underlie about 400,000 square miles of the United States. The salt mine near Lyons, Kansas, selected by the AEC as the first national high-level solidified-waste repository, was abandoned as a repository because active solution mining was going on within 1,500 feet of the site and because of uncertainty over the location of old drill holes that would threaten the integrity of the repository.

Delay and decay applies to treatment of most of the radioactive gaseous wastes of the nuclear-power industry, for they are mainly composed of short-lived isotopes. The gases may be compressed and stored in tanks until the radioisotopes, such as iodine 131, decay to innocuous levels; but iodine 131 needs to be stored for less than 6 months to be harmless, whereas krypton 85 has a hazard life closer to 50 years.

In some places intermediate-level (from a microcurie to a few curies per gallon) liquid wastes are discharged directly into the ground. Such discharge is based upon a judgment that the hydrologic factors, the ion-exchange properties of the soil, and local population distribution will allow sufficient time for the radioactivity to decay to nonhazard levels as the waste moves slowly through the soil; the fundamental assumption that these factors will remain constant throughout the hazard life of the waste material is open to serious question; the practice itself is a temporary expedient with an inherent long-term hazard potential, and should be terminated.

At Oak Ridge a unique delay-and-decay disposal method for intermediate-level wastes involves injecting them as a slurry into hydraulically induced fractures in shale at depths of 700–1,000 feet, where the slurry "sets" into a concretelike solid. This method is safe only for locations below the level of circulating potable groundwater and where vertical fractures in the overlying rock are essentially nonexistent; such locations are probably rare.

Dilute and disperse applies to low-level (less than a microcurie per gallon) wastes when diluted by water or air and discharged into the environment. In many cases the most *expedient* method of dealing with a radioactive gas is to discharge it to the atmosphere from a high stack and thus to dilute the radioactivity to a "permissible" level.

It should be emphasized that any disposal of radioactive waste to the biosphere or to any geological environment where there is a risk of contaminating the biosphere is not a safe practice. However, it has been and it continues to be done; although not safe, it is at present "permissible."

The tremendous volume of the ocean might seem to make it an ideal medium for the dilute-and-disperse technique of radioactive waste disposal (Chapter 11). According to recommendations of the International Atomic Energy Agency, low-level and intermediate-level wastes may be discharged into the ocean under controlled and specified conditions. As in all releases to the biosphere, however, there is widespread concern for the unknown hazard potential in the ability of certain organisms of man's food chain to concentrate radioactivity.

Solid radioactive waste (contaminated objects, not solidified liquid waste) is buried. If high level, it is stored in vaults above the water table. If lower level, it may be buried in relatively shallow trenches at sites where calculated groundwater movement and ion-exchange capacity of the soil make it certain that decay and dilution will reduce the radioactivity to permissible levels by the time the radioisotopes reach places where the water is available for human use.

In 1966, the Committee on Geologic Aspects of Radioactive Waste Disposal of the National Academy of Sciences issued the following warning:

> The current practices of disposing of intermediate and low-level liquid wastes and all manner of solid wastes directly into the ground above or in freshwater zones, although momentarily safe, will lead in the long run to a serious fouling of man's environment. Such methods represent a concept of easy disposal that has had and will continue to have great appeal to operators, but we fear that continuation of the practices eventually will create hazards that will be extremely difficult to eliminate.[17]

The same committee pointed to the necessity of taking a long-range view of waste storage and disposal. Not only are some present environmental disposal sites and practices unsuited for disposal of the much greater quantities of wastes that will be produced by the growing nuclear power industry, but they may be subject to serious hydrologic changes during the 600 years or more of the hazard life of some of the long-lived isotopes. In arid regions such as central Washington and southeast Idaho, irrigation projects and dams can change hydrologic conditions from safe to unsafe within a few years.

SAFE UNDERGROUND STORAGE AND DISPOSAL

Short-lived radioisotopes can be contained or delayed until they decay to safe levels, in artificial containers or in surface ponds and soil layers sufficiently distant or isolated from points of water use. Long-lived radioisotopes, however, pose a different problem. Eventually the present practice of discharging long-lived radioisotopes to the environment, even in dilute concentrations, must be stopped, for the hazards of buildup and biological reconcentration in the environment are too great. Geologic and hydrologic research carried out over the past decade indicates that there is an alternative, that long-lived radioisotopes can be effectively isolated from the biological environment during their hazard lives. All indicated safe methods involve intelligent use of the underground, nonbiologic environment.[17,18]

High-level liquid wastes can be solidified and placed in safe permanent storage in natural salt formations, at sites proved to contain no openings that would permit or threaten the entrance of water. Economic studies at Oak Ridge indicate that the cost for the safer, permanent storage of solidified wastes in salt mines is about the same as that for perpetual tank storage. Intermediate- and low-level liquid wastes can be injected through deep wells into deep geologic basins in tectonically stable portions of the continent where water movement is extremely slow, on the order of a few feet a year.[18]

A permeable geologic formation (e.g., sandstone), from which upward migration of fluids is prevented by an impervious overlying formation (e.g., shale) and in which fluid motion is, say, only 3 feet per year, is almost ideal for disposal of liquid radioactive waste. Tritium, which takes perhaps 75 years to decay to harmlessness, would in that time move only 225 feet from the injection well; cesium 137 and strontium 90 would move not more than 600 yards away from the well before becoming innocuous.

Solid wastes containing long-lived radioisotopes could be placed in permanent underground storage in an environment isolated from circulating groundwater. The concept of one or two national burial grounds for long-hazard wastes has much to commend it; salt-mine interment of solidified high-level liquid wastes could be supplemented by a desert cemetery for contaminated solid wastes.

It should not be assumed, from these statements, that finding suitable underground storage and disposal sites is easy, or that just about any place will do. Suitable geologic environments for radioactive waste disposal are relatively rare. But they exist, and in such size as to contain readily all the radioactive wastes of the foreseeable future. Before any subsurface storage or disposal operations are undertaken, the underground environment needs to be investigated thoroughly so that the risk of contaminating the biosphere can be made vanishingly small. To be avoided are earthquake zones and places where the rocks are fractured and might leak radioactivity into formations from which water is drawn into the biosphere. Any subsurface injection of waste materials, radioactive or not, should be carried out in accordance with the Federal Water Quality Administration's 1970 *Policy on Disposal of Wastes by Subsurface Injection*, which requires the following conditions:

1. Demonstration that subsurface injection is the best available disposal alternative.
2. Preinjection tests sufficient for prediction of the fate of the wastes.
3. Evidence that such an injection will not constitute an environmental hazard.
4. Best practicable pretreatment of the wastes.
5. Best possible design and construction of the injection system.
6. Provision for adequate and continuous monitoring of the operation.
7. Provision for securing the environment if the disposal operation is discontinued.

SITING OF FUEL-REPROCESSING PLANTS

None of the AEC plants which produce large quantities of radioactive wastes, from Savannah River in the southeast to Hanford in the northwest, appears to have been located with safe waste disposal in mind. The main factors in site selection seem to have been distance from existing centers of population and availability of abundant dependable supplies of fresh water for reactor cooling. As it turned out, none of these plants was located near a favorable area for permanent disposal. As a consequence, economic and operational expediency has led to storage and disposal practices at these sites which are only temporarily safe and which must ultimately be abandoned in favor of safer methods.

This failure to foresee the importance of good geologic (and meteorologic) conditions for disposal of low- and intermediate-level wastes is now being repeated in the siting of commercial fuel-reprocessing plants. Site testing for underground disposal potential has been given little consideration in the location of the first three commercial plants.

Disposal of low-level liquid and gaseous wastes is (or should be) the critical problem in siting commercial reprocessing plants.[19] The high-level liquid wastes represent a minor siting factor, for they are to be stored at the reprocessing plant for not more than 10 years, after which they will be solidified and transferred to one of a limited number of federal repositories which will provide "permanent isolation of the wastes from man's biological environment."

Most of the waste tritium of the nuclear power industry is released at the reprocessing plant. Tritium as tritiated water is readily incorporated into living tissue. To keep tritium from the biologic environment requires containment of tritiated water in closed-cycle recirculating systems, and disposal of that which is released in deep geologic formations in which the contained water moves so slowly that the tritium will decay before entering the biological environment. Such disposal requires that the reprocessing plant be located over geologic formations which have been proved by test wells to be suitable for safe disposal of liquid radioactive wastes. An alternative, but probably more expensive way of handling excess tritium would be to separate it from tritiated water by a distillation process, condense it (^3H is a gas), and contain it until decay.

Fission-product gases such as krypton 85 and xenon 133 are released in significant amounts at fuel-reprocessing plants. Recent Oak Ridge studies show that these gases will (or should) impose the most severe limitations on siting of reprocessing plants. If releases of krypton 85 to the atmosphere were to continue at the present rate per megawatt, by the year 2060 the radiation exposure of the general public from this radioisotope *alone* could equal more than half the permissible dose.[20] Isolation of these chemically inert gases from the biological environment during their hazard life requires either concentration and surface containment or injection into geologic formations in which they will be trapped or delayed and dispersed in their rise to the surface.

If, instead of being contained or disposed of underground, low-level liquid and gaseous radioactive wastes are to be deliberately released to the environment from fuel-reprocessing plants, the plants should be located in remote areas, suggested in a 1968 report by the energy policy staff of the Office of Science and Technology,[21] which went on to point out that there is a great deal of flexibility in siting a reprocessing plant because of the relatively low cost of transporting the small tonnage of solids involved. But of the first two commercial reprocessing plants, one is only 30 miles from Buffalo and the other is 60 miles from Chicago.

"Low-level" liquid waste from the western New York plant (West Valley), which approaches maximum permissible concentrations some distance *below* its point of discharge into a surface stream, eventually finds its way to Lake Erie. Because of continued great difficulty in meeting effluent standards, the plant operators in 1970 asked permission to put in a deep (7,000-foot) waste-disposal well but were turned down by the AEC. The site appears well suited for deep-well disposal (although this was not a factor in its selection), and the alternative disposal into surface streams cannot — or should not — long continue without serious objection being raised; consequently, the sustained official reluctance to test the underground for disposal capabilities is difficult to understand.

The Decision Matrix

The three principal peacetime sources of potential environmental pollution from ionizing radiation are (1) nuclear reactors, (2) nuclear fuel-reprocessing plants, and (3) Plowshare blasts.

Public concern about possible radioactive pollution has been expressed recently in a number of confrontations with planners or operators of nuclear projects. Plans to excavate harbors in Alaska and Australia and to blast out an underground storage cavern in Pennsylvania with nuclear explosives have been abandoned because of public protests. Opposition to nuclear power plants in California, Minnesota, New York, and elsewhere has caused abandonment of sites, modification of designs, and delays in construction or opera-

315

tion. For purposes of illustration, three such confrontations, each representing one of the main potential sources of pollution, will be described briefly.

MONTICELLO NUCLEAR POWER PLANT

In 1967, the Atomic Energy Commission issued a construction permit to the Northern States Power Company for a 545-megawatt boiling-water reactor and power plant to be built near Monticello, Minnesota, on the Mississippi about 35 miles upstream from the point where St. Paul and Minneapolis extract their water. When the power company applied for a waste-disposal permit from the newly created Minnesota Pollution Control Agency (MPCA), it ran into trouble. Citizens led by University of Minnesota scientists criticized the projected discharge of low-level radioactive wastes to the environment, even at levels considerably below those considered by the AEC as acceptable; questioned the effluent standards set by the AEC; and accused the AEC of a conflict of interest because it is required to promote the use of nuclear power as well as protect the public.

The power company countered these charges by arguing that standards stricter than those of the AEC were unnecessary, that liquid discharges below the Monticello plant would contain less radioactivity than domestic tap water, and that a person living beside the plant would receive about half the radiation an average American receives from watching television.

In February 1969, the MPCA, acting upon the recommendations of a consultant who pointed out that the AEC standards neglect the problem of multiple sources of radioactive pollution, issued a waste-disposal permit limiting radioactive discharges from the reactor to levels much lower than those permitted by the AEC.

In August 1969, its plant almost completed, Northern States Power Company brought suit against the state pollution agency, challenging its authority to set radioactive-discharge standards. A few days later Governor LeVander of Minnesota received the unanimous endorsement of the National Governors' Conference in support of the principle that state governments should have independent authority to protect their residents from environmental dangers, even in fields such as radioactivity, where federal agencies such as the AEC claim exclusive jurisdiction. Later, 17 states entered the law suit as friends of the court in support of Minnesota's position. In 1971, the court ruled against Minnesota in a decision subsequently upheld by the U.S. Supreme Court.

RADIOACTIVE WASTE DISPOSAL IN IDAHO

A wide range of activity levels is represented in liquid wastes produced, largely from fuel reprocessing, at the National Reactor Testing Station (NRTS) located on the Snake River Plain near Idaho Falls, Idaho. For years, both low-level and intermediate-level liquid wastes have been discharged into the ground at NRTS, despite the fact that the site lies above one of the nation's great underground sources of fresh water. Reliance has been placed on the considerable depth to the water table, on the ability of the soil to capture radioisotopes by ion exchange, on the aridity of the region, and on the distance from the disposal site to places where groundwater is available for use by the public. Solid waste materials, also representing a wide range of hazards, are buried in the same environment. Plutonium-contaminated wastes, highly hazardous in perpetuity (half-life: 24,360 years), are given shallow burial in steel drums on the same assumption under which the liquid wastes are discharged to the environment.

Although in 1960 and again in 1966 a committee of the National Academy of Sciences had expressed concern about "overconfidence" in the protection afforded by aridity at both NRTS and the Hanford works,[17] this did not become public knowledge until 1970, when in response to questioning by Senator Frank Church of Idaho, the committee report was made public. Senator Church's action reflected the expressed concern of some Idaho citizens who live "downstream" from NRTS in terms of groundwater flow. At the request of the Senator, four federal agencies investigated and recommended substantial changes in the disposal program. The Federal Water Quality Administration, for example, urged the AEC to stop burying contaminated solid waste at NRTS and to halt the discharge of tritiated water into the ground above the Snake River Plain groundwater reservoir (aquifer); FWQA recommended subsurface, deep-well injection of the tritium into a pretested, slow-flushing formation below the aquifer.

PROJECT RULISON

In 1969, a public storm blew up in Colorado over the AEC's Project Rulison. In partnership with the Austral Oil Company, the AEC announced plans for a 40-kiloton nuclear explosion deep below the scenic surface of western Colorado, in hope of stimulating natural-gas production and recovery from the enclosing low-permeability Mesaverde Formation of the Rulison Gas Field.

Citizens concerned about possible detrimental effects to the environment (contamination of groundwater, induced earthquakes, surface and atmospheric contamination from subsequent flaring of the radioactive gas) found that they had no right even to a hearing, under a law that gives full decision authority to the AEC.

The Colorado Open Spaces Council and the American Civil Liberties Union thereupon sued the AEC in an attempt to halt Project Rulison. Two federal courts upheld the AEC's right to proceed with the test, subject only to the condition that no reentry and flaring take place until at least 6 months after the shot. The blast took place on September 10, 1969. Another shot (Rio Blanco), again vigorously protested, took place in June 1973. No radioactivity was vented, no earthquake induced by either shot. Critics of the project were far from reassured that all would be well, however; their concern went beyond Rulison and Rio Blanco, to the potential environmental effects of the hundreds or even thousands of such shots that may be required to develop the field.

Recurrent themes in these and other public confrontations with the proponents of nuclear energy projects are the questions: What is safe? How are protection standards set? If these standards involve acceptance of risk in return for benefits, why does the concerned and exposed public find it so difficult to participate in the acceptance or assumption of the risk?

We shall briefly consider each of these questions in turn.

WHAT IS SAFE?

Safety from the effects of ionizing radiation is difficult to define. Absolute safety does not exist, because some genetic damage is caused by any sustained level of radiation. Therefore, we can speak only of relative safety, the determination of which depends on a knowledge of the risks inherent in various levels and durations and kinds of exposure to ionizing radiation. But we do not have the fundamental information necessary to calculate risks at low levels of chronic exposure; we do not have enough experience with reactors to be able to calculate the risk of serious accidents; we do not know enough about the concentrating and transporting mechanisms for radioisotopes in the biosphere to be able to forecast the ecologic risks of man-made radiation; and we have very imperfect means of estimating how technologic, demographic, political, and even geologic events of the future may serve to compound (or perhaps alleviate) present risks.

At relatively high levels of both acute and chronic exposure, we are beginning to know some of the risks of ionizing radiation fairly well. But these are risks of somatic damage to individuals exposed under rather special environmental conditions, such as the children who grow from irradiated embryos, the uranium miners of the Colorado Plateau, and the survivors of Hiroshima. What we need to know is how much we shall have to pay in shortened lives and human misery for each increment of radioactivity that is deliberately released to the biological environment. Although such calculations of human costs have been made, they are based on unproved assumptions and vary widely. Estimates of the number of deaths from radiation-induced cancer and leukemia to be expected in the United States from sustained radiation exposure at the guideline dose rate of 170 mrem per year range from 160 per year[22] to 32,000 per year.[23]

The United States in 1970 had an environmental level of man-made radiation that was only a small fraction of the guideline level of 170 mrem per year above background and medical dosage; of the approximately 200

317

mrem per year average gonadal exposure in the United States, less than 5 mrem was attributable to nuclear energy; most of the rest was from background and medical irradiation. The human cost of nuclear energy, in other words, is not yet very high. But since a human price is paid for any increase in the environmental level of ionizing radiation there can be no truly safe level, only levels that are deemed acceptable in terms of the benefits and costs involved.

HOW THE STANDARDS HAVE BEEN SET

Protection standards have been set to represent "acceptable" tradeoffs between calculated benefits and the incalculable risks. Because the risks are not calculable, standards have been set at levels where the consequent biological damage will be difficult if not impossible to distinguish or measure. The standards have been set by technical experts who determine their public acceptability without recourse to the trial of public review and debate.

The two main standard-setting organizations whose guidelines are used in the United States are the International Commission on Radiological Protection (ICRP) and the Federal Radiation Council (FRC). The ICRP states that "the problem in practice is to limit the radiation dose to that which involves a risk which is not unacceptable to the individual or to the population at large" and that "long-term effects should be a major preoccupation of the collective conscience."

In its first report (1960) the Federal Radiation Council, established the year before to advise the President on radiation standards, expressed a similar philosophy:

Fundamentally, setting basic radiation protection standards involves passing judgment on the extent of the possible health hazards society is willing to accept in order to realize the unknown benefits of radiation. There should be no manmade radiation exposure without the expectation of benefit resulting from such exposure.

The FRC has also stated its conviction that "exposure from radiation should result from a real determination of its necessity." However, in 1962–1963, during great public concern about high iodine 131 levels in milk from cows fed on grass contaminated by fallout from Nevada weapons tests, levels that exceeded FRC guidelines, the FRC (1) declared that its protection guides did not apply to nuclear fallout, (2) suggested that action by the State of Utah in diverting contaminated milk from the market was reprehensible because it deprived children of milk needed to avoid malnutrition, and (3) raised its guideline for radioactive iodine by 20 times, all without discussing "the expectation of benefit" involved!

The permissible dose of 5 rem per person per generation (30 years) exclusive of natural background and medical exposures appears to have been selected because it was thought that a doubling of the natural background level would involve neither serious genetic risk nor measurable somatic effects. Both groups, however, have repeatedly pointed to the lack of knowledge of the biological effects of low chronic radiation doses.

AEC regulations are based on FRC and ICRP guidelines. They are designed to protect both workers and public exposed to man-made radioactivity and are set in terms of dose rates and maximum permissible concentrations. They are not designed to limit total amounts of radioactivity released to the environment by any single plant or project or by any combination of plants or projects. They are not designed to protect against slow buildup in the environment of long-lived radioisotopes or against biological concentration in the food chain.

WHO ACCEPTS THE RISK?

The risks to the exposed public are inherent in radiation-protection standards. The experts who set the standards, by their own testimony, have weighed the possible health hazards "acceptable to society" against the expected benefits to be gained from the use of radiation. They have accepted the risk for the public. Many critics of the present regulatory and licensing system charge that the setting of protection standards, as well as the siting of a nuclear power plant or the decision to implement a Plowshare project, involve moral and political judgments made by persons whose limitations of competence and responsibility should restrict them to scientific and engineering judgments.

The furor over fallout in the 1950s and early 1960s and the present concern about the siting of nuclear power plants and underground nuclear explosions may stem as much from frustration over exclusion of the affected public from the decision process as from worry about the biological harm of the resultant radiation. Atomic experts grow exasperated with citizens who fuss about a risk which the experts feel they can demonstrate is many times less than that of driving a car on a public highway. But risks imposed are much more difficult for some people to accept than risks voluntarily assumed. It appears that most if not all radioactive wastes developed from use of U.S. nuclear fuel will, as a matter of U.S. policy, be returned to the United States for disposal. The added risk to the U.S. environment from adoption of such a policy was accepted without asking the public.

DUAL ROLES OF THE AEC

Atomic energy was developed for military purposes under wartime restrictions. Under those circumstances, it was natural that the development of applications as well as the protection of those exposed to radiation hazard should be vested in the same organization. When, in 1947, responsibility for development of a national atomic energy program was given by Congress to the AEC, it still seemed appropriate, in view of the continuing importance to the nation of the military applications of nuclear energy, and the fact that almost all the nuclear technologists in the country worked for the AEC, that the Commission be given sole authority to review reactor designs and sites and to license reactors.

The fallout experience of the western states in the 1950s and early 1960s, however, raised some interesting questions about the dual promotion–protection role of the AEC. Perhaps in defensive reaction to mounting criticism, the Atomic Energy Act of 1954 explicitly confirmed the intent of Congress to remove jurisdiction over radiation hazard from the states and vest it solely in the AEC. Although a 1959 amendment to the act deleted the preemption statement, the AEC continues to exercise this jurisdiction. In 1969, the state of Minnesota, as noted earlier, challenged the AEC by prescribing stricter standards for effluent from the Monticello reactor, but lost the case.

The conflict of interest inherent in the dual role of the Commission to promote the applications of nuclear energy and to protect the public from its hazardous effects has been repeatedly pointed out, as has the fact that every modern nation in the world that has a commercial nuclear power program, except the United States, vests the promotional and regulatory functions in separate agencies. The transfer in 1971 of some AEC functions to the Environmental Protection Agency did not eliminate the conflict of interest, since the regulatory functions remain vested in the AEC.

A significant change in the decision system, if not in the structure of decision making in the nuclear power field, came about in 1971 as a result of a court decision which held that the AEC was required by the provisions of the National Environmental Policy Act of 1969 to consider the adverse environmental impacts of licensing specific nuclear facilities. Following this decision in the Calvert Cliffs case, the AEC for the first time allowed the introduction of environmental-impact testimony in hearings on permit applications for nuclear power plants. The Commission has, however, chosen to interpret the court decision as *not* requiring consideration of adverse effects in the fuel cycle except in the vicinity of the proposed power plant; the increased environmental risk due to the increased need for mining, milling, waste transport, fuel reprocessing, and ultimate waste storage or disposal represented by the proposed power plant cannot be discussed "for the record" by intervenors in a licensing hearing.

WHERE DO WE GO FROM HERE?

Present standards for effluents from nuclear facilities and projects, as pointed out earlier, are designed to protect workers and the nearby public from the effects of radiation released by that power plant or project. There is nothing built into such standards that will prevent a slow buildup of radioactivity in the environment as nuclear facilities and projects proliferate, and there is nothing in them that will protect the public from the ill effects of ingesting radioisotopes that have been biologically concentrated in man's food chain.

It is not technologically necessary to release any radioactivity to the human environment from either nuclear reactors or fuel-reprocessing plants. Releases are made for reasons of cost. It is technologically difficult to capture the noble gases like krypton 85 from the gaseous waste and almost, if not actually impossible, to recover tritium from the liquid waste. But these technical difficulties do not mean that these harmful substances have to be released to the human environment. Given proper containment design and adequate siting relative to geologic and hydrologic disposal criteria, even krypton and tritium could be kept out of the biosphere. These substances will set an upper limit to the development of nuclear power only if their release to the environment is allowed to continue at, or anywhere near, the present release rate per unit of installed energy capacity (per megawatt, for example).

Any radioactive release to the biological environment, the National Academy of Sciences Committee on Resources and Man recommended in 1969,[24] should be monitored by an agency truly independent of the organization promoting the activity causing the release. (In 1971, the monitoring function was transferred from the AEC to the EPA.) Such release should be analyzed in terms of national and global environmental consequences, not just their effects on the health of the nearby human populations. Both the data and the analyses should be made available for public information and independent review.

There is usually not just one way of reaching a social goal. More often there are several, each involving a different set of costs and consequences, and each affecting in some manner the march toward other social goals.

Meeting the increasing demand for electric power at a cost that will allow wide participation in its benefits is a social goal. Restoring and maintaining a healthy and clean environment is a social goal. Assuring our nation of a continuing supply of the mineral sources of energy (coal, oil, and natural gas, uranium, and perhaps thorium) is a social goal. Optimum realization of these three goals requires a thorough knowledge of available resources and of the available technological alternatives, and it demands a continuing comparison of the benefits to be derived, and of the costs and risks to the public and to the environment inherent in each technological alternative. We have good means of obtaining the needed knowledge and of making the necessary calcu-

lations. We have poor means of comparing alternatives and of making reasoned public choices among them.

We need improved means for public participation in the choices that have to be made in our use of energy. We want cheap, dependable power. We want a clean, healthy environment. We cannot have the best of both (cheapest power and cleanest environment). Be it power from fossil fuels or radioactive materials, we have to decide how clean an environment, and what reduction of risk to ourselves and to our descendants, we want at what price in the increased cost of power (Chapter 5).

At the present level of environmental releases and human risk, the economic cost of environmental protection is not great. It has been estimated that the cost of radioactive waste containment, treatment, and disposal may run as low as 1.0 to 1.5 percent of the capital and operating costs of the large nuclear plants now being designed. Westinghouse has stated that a zero-release power-plant design would add less than 1 percent of the cost of a 1,000-megawatt plant. Eliminating the accident risk by placing the reactors securely underground would increase the capital cost of a large plant by only a few percent[25,26] and should add almost nothing to the operating costs. Placing power reactors in remote places like Hudson Bay and transmitting the power over high-voltage direct-current lines to urban areas would entail both increased capital costs and transmission losses but not of a size that an affluent society could not absorb without difficulty. More important, siting on Hudson Bay might turn the gross impediment of thermal discharge into an advantage. Even in urban locations, the hot water from power plants can be used to heat buildings and homes; little consideration appears to have been given to this sort of alternative in site planning.

It will be imperative to site nuclear fuel-reprocessing and fuel-element fabrication plants so as to minimize the transport of plutonium in the age of breeder power plants we may be about to enter. All the plants of a single fuel-cycle system should be planned simultaneously; but our present decision system does not allow such a comprehensive and rational approach to national problems.

320

In blasting canals and harbors with nuclear explosives, as well as in underground stimulation of natural gas production by nuclear blasts, there is no zero-release alternative to consider. Some radioactivity will always be released to the human environment by a Plowshare project. Consequently, consideration of technological alternatives and comparison of costs and consequences become even more important.

IN SUMMARY

There is no absolutely safe level of ionizing radiation. Any man-made addition to the natural level increases the potential for biological harm. Consequently, the use of radiation and nuclear-energy materials that release radioactivity to the environment involves a balancing of benefits and risks. Reducing the risk means reducing the benefits (if only by increasing the cost of the product). Choice of benefit/risk ratios involves political and moral decisions, not scientific and technical ones.

There is no technical necessity for release to the human environment of radioactivity from nuclear reactors or fuel-processing plants. Not only can reactors be made as clean as one may wish to pay for, but they can be made as safe as one is willing to pay for. Although reactor design engineers feel that modern reactors are very safe, there will always remain some risk of accident, and a reactor accident could be catastrophic. Edward Teller has said: "In principle, nuclear reactors are dangerous . . . [they] belong underground."[27] An underground reactor (Sweden, Belgium, and Switzerland already have them) without radioactive effluents could be both safe and clean, in almost absolute terms.

Putting nuclear power plants in the ocean, frequently cited as an alternative to underground siting, has some strong disadvantages: the very large threat to the marine environment represented by a reactor accident or a waste-transport accident; the additional costs of transmitting power to onshore cities; probable higher construction costs than for underground siting; higher maintenance costs because of a corrosive and stormy environment; and naked exposure to attack in the event of war.

Present radiation standards have little relation to environmental contamination, because they do not take into account radioactivity buildup in the environment from multiple sources, and they do not protect against biological concentration of radioactivity in the food chain. They appear to represent levels of exposure below which the harmful effects of man-made ionizing radiation are at present impossible to distinguish.

The rapid growth of the nuclear power industry, the promotion of large seacoast agro-industrial complexes powered by huge reactors, and the pollution potential of Plowshare projects presage a great increase in the quantity of radioactive wastes that must be guarded against and disposed of. Present practices which depend on dilution and dispersal of radioactive wastes cannot be continued without a significant buildup of radioactivity in the environment.

Decisions made today on human exposure to radiation and on releases of radioactivity to the environment will be of great concern to posterity. High-level wastes from nuclear fuel-processing plants have a hazard life of 600 years or more. Plutonium has a half-life of 24,360 years and its hazard life may extend beyond the duration of our species. Even at fairly low levels of exposure, radiation-induced mutations may, through a score of generations, wreak more havoc than a nuclear war. Any buildup of long-lived radioisotopes in the environment will, for practical purposes, be irreversible. It is this long-lived lethality that makes man-made radiation not just another environmental hazard but the ultimate environmental hazard.

Even a very small amount of risk becomes a certainty over a long period of time. Somewhere, sometime, if present practices and decision mechanisms persist, there will be a catastrophic reactor accident. The consequences of such an accident will be unlike those of an earthquake or of conventional warfare. They will be more like the postblast effects of Hiroshima, expanding malignantly down the generations. Somewhere, sometime, after the proliferation of breeder reactors, a political fanatic, a madman, or a distraught nation will be able to assemble one or more hydrogen bombs from stolen plutonium and to use them for social if not physical debilitation. Somewhere, sometime, the *perpetual* surveillance of high-level radioactive wastes will break down

321

and a large amount of radioactivity will enter the biosphere.

The possibility of massive, irrevocable environmental contamination and genetic damage to the human race is real. Never have the risks and benefits of further development of nuclear power been presented for public review and national decision. Allen Kneese, a well-known resource economist, in recent remarks to the AEC,[28] urged that this momentous question be reviewed thoroughly in Congress and that an "explicit decision . . . be made by the entire Congress as to whether the risks are worth the benefits." It is not too late for such a review, and a negative answer is not impracticable.

References

1. Federal Radiation Council. 1960. *Report 1.* Government Printing Office, Washington, D.C.
2. Scott, R. C. 1963. Radium in natural waters in the United States. In *Radioecology.* Van Nostrand Reinhold, New York. pp. 237–240.
3. National Academy of Sciences. 1972. *The Effects on Populations of Exposure to Low Levels of Ionizing Radiation.* The Academy, Washington, D.C.
4. Pollard, E. C. 1969. The biological action of ionizing radiation. *Amer. Scientist 57:* 206–236.
5. Miller, R. W. 1969. Delayed radiation effects in atomic-bomb survivors. *Science 166:* 569–574.
6. Blum, H. F. 1959. Environmental radiation and cancer. *Science 130:* 1545–1547.
7. Lundin, F. E., Jr., et al. 1969. Mortality of uranium miners in relation to radiation exposure, hard-rock mining and cigarette smoking — 1950 through September 1967. *Health Phys. 16:* 571–578.
8. Martin, W. E. 1965. Interception and retention of fallout by desert shrubs. *Health Phys. 11:* 1341–1354.
9. Green, H. P. 1968. "Reasonable assurance" of "no undue risk." *Scientist and Citizen 10:* 128–140.
10. Gillette, R., 1972. Nuclear safety (III): critics charge conflicts of interest. *Science 177:* 970. See also *Science 177:* 330–331, 771–776, 1080 — 1082.
11. Polikarpov, G. G. 1966. *Radioecology of Aquatic Organisms.* Van Nostrand Reinhold, New York.
12. Braestrup, C. B., and Mooney, R. T. 1959. X-ray emission from television sets. *Science 130:* 1071–1074.
13. Consumers Union. 1974. TV and X radiation. *Consumer Rept. 39:* 2.
14. Metzger, H. P. 1972. *The Atomic Establishment.* Simon & Schuster, Inc., New York.
15. Holdren, J. P. 1974. *Uranium availability and the breeder decision* (Calif. Inst. Techn. Environ. Quality Lab. *Memorandum 8*). California Institute of Technology, Pasadena, Calif.
16. Cochran, T. B. 1974. *The Liquid Metal Fast Breeder Reactor.* The Johns Hopkins Press, Baltimore, Md.
17. National Academy of Sciences. 1966. Unpublished report to the Atomic Energy Commission by the Committee on Geologic Aspects of Radioactive Waste Disposal. On open file, National Research Council, Washington, D.C. Reprinted in *Underground Uses of Nuclear Energy.* Hearings of the Subcommittee on Air and Water Pollution of the Committee on Public Works, U.S. Senate, Nov. 18–20, 1969, pp. 461–512.
18. Galley, J. E. 1968. Economic potential of geologic basins and reservoir strata. In *Subsurface Disposal in Geologic Basins* (Amer. Assoc. Petrol. Geol. Mem. 10). AAPG, pp. 1–10.
19. De Laguna, W. 1968. Importance of deep permeable formations in location of a large nuclear-fuel reprocessing plant. In *Subsurface Disposal in Geologic Basins* (Amer. Assoc. Petrol Geol. Mem. 10). AAPG, pp. 21–31.
20. Coleman, L. R., and Liberace, R. 1965. Nuclear power production and estimated krypton-85 levels. *Radiological Health Data and Reports.* pp. 615–621.
21. Executive Office of the President. 1968. *Considerations Affecting Steam Power Plant Site Selection.* Office of Science and Technology, National Science Foundation, Washington, D.C.

22. Storer, J. 1969. Comments on manuscript "Low Dose Radiation, Chromosomes, and Cancer" by J. W. Gofman and A. R. Tamplin. In *Environmental Effects of Producing Electric Power*. Hearings of the Joint Committee on Atomic Energy, Pt. I, pp. 653–654, 91st Congress, Oct. 28–Nov. 7, 1969.

23. Gofman, J. W., and Tamplin, A. R. 1970. The cancer-leukemia risk from FRC guideline radiation based upon ICRP publications. Hearings of the Joint Committee on Atomic Energy, Pt. II, 91st Congress, Feb. 20, 1970.

24. Hubbert, M. K. 1969. Energy resources. In *Resources and Man*. W. H. Freeman and Company, San Francisco.

25. Bernell, L., and Lindbo, T. 1965. Tests of air leakage in rock for underground reactor containment. *Nuclear Safety 6:* 267–272.

26. Rogers, F. C. 1971. Underground nuclear power plants. *Bull. Atomic Scientists 27:* 8, 38–41, 51.

27. Teller, E. 1965. Energy from oil and from the nucleus. *J. Petrol. Technol. 17:* 505–508.

28. Kneese, A. V. 1973. The Faustian bargain. In *Resources, no. 44*. Resources for the Future, Inc., Washington, D.C. pp. 1–5.

Further Reading

Cember, H. 1969. *Introduction to Health Physics*. Pergamon Press, Inc., Oxford. 422 pp.

Glasstone, S. 1967. *Sourcebook on Atomic Energy*, 3rd ed. Van Nostrand Reinhold, New York. 883 pp.

Lewis, R. S. 1972. *The Nuclear-Power Rebellion*. The Viking Press, New York, 313 pp.

National Academy of Sciences, Advisory Committee on the Biological Effections of Ionizing Radiations. 1972. *The Effects on Populations of Exposure to Low Levels of Ionizing Radiation*. The Academy, Washington, D.C. 217 pp.

Parker, F. L. 1969. Status of radioactive waste disposal in USA. *J. Sanitary Eng. Div., Amer. Soc. Civil Engr. 95:* 439–464.

Upton, A. C. 1969. *Radiation Injury*. University of Chicago Press, Chicago. 126 pp.

Editor's Commentary— Chapters 13 and 14

THE NEXT TWO CHAPTERS, on pesticides and alternative pest control techniques, form a unit that first describes the problems created by modern pest control methods and then suggests solutions. They thus deal in more detail with points raised in the chapter on food supply (Chapter 3). Some of the institutional questions concerning pesticides and pest control are discussed in Chapter 19, and some ideas relevant to pest control are presented in Chapter 1.

Pesticides are now sufficiently notorious that I need not provide a summary here of the issues, which are discussed in the next two chapters. Rather, I will note a few points that are sometimes overlooked in a first analysis of the ''pesticide problem''. First of all, to a great extent the pesticide problem is really an insecticide problem. Although herbicides are used on the same scale as insecticides, their known deleterious environmental effects are minor by comparison. They do sometimes kill non-target plants, their use sometimes increases soil erosion and subsequent siltation of freshwater systems, they have polluted freshwater lakes, a few may be a health hazard (though this is contested), and in agriculture they tend to cause one weed problem to be replaced by another. But at least so far they have not been plagued by the sorts of problems that beset insecticides (fungicides and rodenticides also have some associated problems, but are used only on a small scale).

Second, although most people probably think of interference with bird reproduction and the disruption of ecosystems when they think of insecticides, perhaps the major problem with insecticides is that they frequently fail to do their job because insect pests become resistant to them (whether this will occur with weeds and herbicides remains to be seen). Thus, the long-term dependability of some parts of agricultural production is placed in jeopardy.

Third, in the United States the complexion of the insecticide problem is changing rapidly. ''Insecticide'' has meant DDT in the public mind for so long that one is apt not to realize that different types of hazards are coming to dominate this area. In fact, the use of DDT was rapidly declining towards the end of the 1960s, as a response to the development of resistance to it by major pests. The banning of DDT in the United States simply hurried along the end of this process. However, DDT has largely been replaced by much more highly toxic compounds (e.g. methyl parathion) so that the main danger is now to the poorly protected and badly organized farm laborer.

These two chapters lead to two clear conclusions. First, pesticides epitomize our tendency to throw around large quantities of chemicals whose environmental and even health effects we do not well understand and whose movement through the environment we have not been able to trace adequately. Second, it is quite likely that we can develop feasible alternative pest control techniques to solve most of the problems for which we now use insecticides. But both problems need more intensive research than they have received so far—and particularly pest control in the tropics needs much more study if we are to avoid the disasters we have caused in temperate agriculture.

324

Biplane dusts sulfur onto a vineyard to retard the growth of mildew on grape vines. EPA—Documerica photo by Gene Daniels.

13
Pesticides

Robert L. Rudd   *is a Professor of Zoology at the University of California, Davis. Born in 1921, he earned his doctorate in 1953 at the University of California, Berkeley. He has spent most of his academic career at Davis, but has lectured at many institutions and conferences in the United States and abroad. He is the author of* Pesticides and the Living Landscape *(University of Wisconsin Press). His current research interests include pesticide kinetics, the tropical rain forest as a three-dimensional ecosystem, and the activity rhythms and speciation mechanisms of shrews.*

The presence of novel chemicals in natural systems is a consequence of technological man. The appearance of such synthetic chemicals and the effects that they produce have no precedent in organic evolution. Most pesticides as we know them today have no natural counterparts; thirty years ago most did not exist. Accelerated use of organic pesticidal compounds occurred during the years after World War II. The residues of many pesticidal compounds, most notably DDT, are distributed worldwide. One DDT metabolite, DDE, may be the commonest and most widely distributed synthetic chemical on the globe. DDE can now be found in biological tissues from the open ocean to the polar ice caps, in airborne dusts over cities, and in plants, animals, and waters in remote forests and mountains.[1,2]

Pesticides are biologically active chemicals. They must be highly so to achieve the intended purpose of pest control. The spectrum of activity is, however, sufficiently broad in most instances to justify the most applicable word, *biocide*. In sum, therefore, pesticides are widely distributed by natural means; but, in contrast to natural materials such as dusts, they retain a good part of their biocidal activity.

To the ecologist, normal pesticide use poses two major difficulties. First, pesticides are biologically active and regularly strike at non-target organisms as well as their intended targets. Normally, both biological effects and the judgments rendered about these effects are short term. Second, necessarily viewed in a long-term perspective are effects attributable chiefly to the residues of pesticides, whose biological consequences cannot be easily forecast. This second category of effects is more serious. Not only are these effects unpredictable, but they act at temporal, biological, and spatial distances whose routes of transfer and action are poorly known.

This second set of problems can be enumerated as follows:

1. Many pesticides persist and we cannot dispose of them.
2. They may cause unintended effects in place; usually these are population phenomena (e.g., resistance, faunal displacement).
3. They may occur at considerable distances from points of origin. Clearly, fluid transport systems (air and water flows) are chiefly responsible for dispersal.
4. Following dispersal of residues, differential magnification (concentration) in biological systems may cause unintended and unexpected results.

One may be able to categorize the problems associated with the use of pesticides broadly and simply. In so doing, however, it is equally easy to lose sight of three elementary facts:

1. There exists already a great body of information dealing with pesticide technology.
2. Pesticides, that is, chemicals used in pest control, are only a part of the more general problem of pest control. Too many scientists and technologists have obscured the fact that the control of offending animal and plant populations is a *biological* problem, not primarily a chemical one.
3. Whatever our knowledge, the additional fact is that we are all largely unaware of the longer-term consequences of continued pesticide use. We have in progress a global

experiment whose conclusions have yet to be fully foreseen.

This chapter attempts to provide a general understanding of the two questions: (1) why chemical control is needed and (2) what the general environmental consequences of chemical use are. Neither question should be overlooked in assessing the importance of pesticide use. Throughout this chapter I shall stress the environmental effects. But it is important to understand first why pesticides are used.

Competitors with Man

Most people recognize only a few dozen kinds of animals or plants that affect — either help or hinder — them directly. They may be cereal grains and fruits, game species for recreation, domestic animals, plants and animals providing fibers for clothing, and a few that are moral or political symbols. Pests are a special category of animals and plants that affect man. Although most people believe that the word "pest" is clearly definable, it is important to realize that the word has no biological meaning. A pest may very well be a hazard to us because we have made it so. For example, the Colorado potato beetle originally, like most insects, was not a pest at all. It fed on wild members of the family Solanaceae, to which the potato and nightshade belong. The movements of people and, more particularly, changes in agricultural practices, allowed this normally benign beetle to increase its numbers and its range and, particularly in Europe, to cause great damage.

When the numbers of an offending animal or plant are reduced to levels that no longer cause an important hazard or inconvenience, we may not properly call it a pest. Reduction to such low numbers is the object of pest control, in which chemical agents figure prominently.

The competition between man and living things takes many forms — not all competitors are pests. We shall discuss five categories of living things that are in competition with man.

DIRECT AND IMMEDIATE HAZARDS
One direct and immediate hazard to man are vertebrate animals that attack human beings.

"Pest" is not quite strong enough to describe the poisonous snake that injects a fatal dose of venom. The same is true of predatory mammals. Whether hazardous to man or his livestock, tigers, bears, lions, and other large animals are no longer a general "pest" problem. Traps, guns, and poisons render likely their reduction in numbers or their extinction.

Also included in this category are diseases and parasites of man. Direct chemical therapy is standard medical practice in treating irritations or illnesses that result.

DIRECT BUT DELAYED HAZARDS
Many infectious microorganisms find their way to human beings in the bodies of other living animals. Such disease hosts are called *vectors*. The best known examples are mosquitoes that carry the sporozoan parasite *Plasmodium*, which causes malaria in human beings. The control of vector-caused disease is approached in two ways. The first is drug therapy when disease symptoms manifest themselves in the human individual. The second is with attempts to reduce or eliminate the vector population that carries the disease. At the present time chemical pesticides are very widely used for that purpose. Scientists who work on vector-control problems come from many disciplines; they include epidemiologists, entomologists, and ecologists, to name a few. The importance of their work is obvious. Still, general environmental contamination (notably by DDT) is a frequent consequence of their efforts.

DIRECT HAZARDS TO DOMESTICATED ANIMALS
Almost all domesticated animals of basic importance to man are birds or mammals. Unfortunately, these warm-blooded animals are subject to the same kinds of hazards as man. Predatory animals are controlled through trapping, shooting (aided by bounty systems), or poisoning. In the United States sheep and poultry now bear the brunt of attacks by predators. Although frequently economically unjustifiable, large-scale poisoning programs continue. Compound 1080 and strychnine are the most widely used chemicals. Fear of predatory mammals is a deeply ingrained human characteristic. We have, we might say, institutionalized an overreaction to that fear in

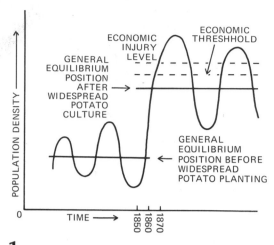

1 **Growth of the Colorado Potato Beetle as a Pest.** Economic injury levels in relation to numbers of a pest species and man's agricultural enterprise. The example here is the Colorado potato beetle, which became a pest as agriculture expanded in the nineteenth century.

continuing poison programs on the scale we do.

Disease and parasites also plague domestic animals. Veterinarians depend primarily on therapeutic chemicals for the treatment of a wide variety of afflictions. Systemic insecticides, for example, are commonly used against bot fly infestations in the skin of cattle.

DIRECT HAZARDS TO COMMODITY PRODUCTION

No agricultural or forest crop can be grown in an insect- and disease-free environment. Animal and plant pests consume a significant part of commodity production. The efficient producer wants to keep pest losses to a minimum. There are many ways of attempting this, but pesticides are one of the most important. The widest variety and volume of pesticide chemicals are used in agriculture, and the great magnitude of environmental effects and the current concern stem largely from this source. The use of pesticides is clearly encouraged by economic considerations. Some scientists say (rather generously, I think) that, generally, $1 spent on pesticides results in $3 of increased yield. In some specialty crops chemical dependence is so great that, given present production methods, the crops could not be grown economically without pesticides. The yield increment attributable to pesticide use in all commodities is translated into economic terms.

The general case can be illustrated as in Figure 1. Economic density or injury level simply means the level of population at which pest damage becomes too serious to overlook. Below the threshold level the grower's efforts in pest control would not return full value to him. Actual economic injury is rarely calculated; the *belief* that injury will occur is more frequently the stimulus to apply pesticides. The extent of pest control operations is staggering. In California, for example, major control programs are waged against some 200 pest species annually. The approximate cost of these efforts is abour $250,000,000 per year. A good part of this cost is for pesticides.

NUISANCES ONLY

Throughout the world there are many kinds of interfering animals and plants that pose no serious economic or health hazard. These nuisance species range from fleas to elephants, from algae to trees. Human life is more comfortable without direct impingement of these nuisance species. Where we can, we try to control them. Pesticides are frequently used, particularly in urban areas, to attempt control. The control used is normally sporadic and localized. Systematic effort, such as that characterizing commodity production, rarely occurs. The result is that general nuisance abatement has not been too successful.

The Creation of Pests

Man's every action affects the biological world in ways often disguised and unknown. A few examples: Kentucky bluegrass originally came from Europe. The grasslands of California may look wild and natural, but most species of grasses date from early Spanish occupation. The house sparrow and the starling, now so common in the United States, came from only a few birds introduced less than a century ago. Baboons have become serious crop pests in Africa, possibly because the demand for leopard skins has reduced the number of predators.

The foregoing examples illustrate the range of species, not necessarily pests, whose ecological relationships have been altered by man.

Pests are "made" in the same fashion, in an altered environment. Basically, pests important to man are created in four ways:

1. Certain species of plants and animals are selected by known evolutionary processes to become more numerous and successful.
2. Many thousands of species of plants and animals are moved from one place on earth to another. Many species, such as crop plants, are intentionally moved, but most are carried by man accidentally.
3. A reduction in biological diversity often takes place, sometimes intentionally and sometimes accidentally.
4. Living things may adapt to changed environments and, in the case of pest species, "biological weeds" are produced.

The first point can be dismissed quickly because it is obvious; yet it is, of course, profoundly important, as it concerns the entire history of agriculture (Chapter 3). Essentially, most crop species were developed very early in the history of man. Modern agriculture, classified solely by the plants it uses, is neolithic agriculture. Mechanization and chemicalization began on a large scale in the nineteenth century. On the basis of these two technological changes, traditional monocultural (single-species) plantings could be enlarged. For a century ever-expanding monocultures allowed by essentially industrial methods have produced a variety of social changes, a rural–urban shift being the most prominent. But in crop production itself, yield increase became increasingly dependent on mechanical and chemical tools. The result has been an increasingly precarious ecological pattern in most basic crops. Large-scale monoculture produces in an accelerating cycle more pest numbers that must be held down by increasing the use of pesticides (or other pest control means). A large part of the present concern for uncontrolled pesticide residues in the environment stems from the successes of modern technology-based agriculture.

Biological translocations are the natural product of species dispersal. Yet, geologically speaking, the process was slow enough to result in a clear identity for each of the zoogeographical regions of the world. The biotic isolation that has characterized much of the earth's history has recently undergone dramatic change. Much pest control, and the need to use pesticides, comes about because of man-carried animals and plants introduced into areas favorable for their multiplication.

Whether man-carried or naturally dispersed, pest species had to find in their new sites an environment that was favorably reconstructed, that is, modified by man for the pest's advantage before it could become established. A series of environmental "bridges" is often required. The cotton boll weevil — nemesis of cotton growers and beloved by folk singers — originally lived in Mexico. Not until the barriers of southwestern deserts had been bridged by irrigation schemes and agricultural plantings did the weevil gain entrance into the United States. Its presence is clearly felt. No other insect species is sprayed or dusted with insecticides so often. The Colorado potato beetle (Figure 1) moved eastward in the United States on a "potato bridge," whence it moved to Europe by other means.

Most species are transported accidentally. There are now some 5,000 species of insect pests worldwide. Adding all other kinds of pests, plant and animal, brings the list to 10,000 species. And these are only the economically important ones. The exchange of organisms continues daily; without too much exaggeration, one can say that, for major crop types, the one-time continental isolation no longer exists. One may speak of citrus, cotton, or wheat pests and be reasonably sure that they are the same kinds of arthropods whether from Israel, Australia, South Africa, or the United States. The worldwide exchange of pests became nearly complete before stringent quarantine was enforced.

A pest in one land is usually not a pest in its native land. When it is carried to an alien shore, it usually leaves behind its predators and parasites, thus allowing its numbers to increase greatly in its new habitat. As indicated earlier, many environmental conditions created by man favor pests. The economic need to plant single species in large stands abets the problem by favoring the dispersal and food supply of the pest species. *Reduction in biological diversity* is a normal consequence of clearing and planting. Further reduction can also be planned for special reasons. Still further reduction, with inherently catastrophic

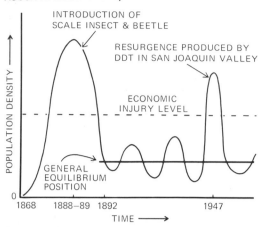

2 **History of a Pest Species.** A scale insect of California citrus orchards and effect of the ladybird beetle successfully introduced for its control. Note particularly the resurgence of pest numbers when DDT came into use following World War II (see the text).

instability, can be the unplanned consequence of many pest control procedures, particularly those depending on chemical pesticides.

The total biological changes that follow breaking of virgin ground could not have concerned pioneers too much. Crop production was the goal and unintended pest production the likely consequence. The two aspects have only in recent years been associated and documented. The best-studied example to illustrate that the simple ploughing of land for planting reduces the biological diversity and produces pests comes from the recent massive agricultural plantings in the virgin steppe of Kazakhstan in the USSR. Insect fauna from both original and modified (wheat-field) environments were identified, counted, and studied. These were the general conclusions: (1) there were more than twice as many species in the virgin steppe than in the wheat fields, and (2) the density of insect populations in the wheat fields was almost double that of the steppe. The number of dominant and constant species in the steppe was twice that of wheat fields, but in the steppe they comprised only half the fauna, whereas in wheat they amount to 94 percent. Therefore, there was a greater variety of species in the population of the virgin steppe and a better numerical balance between the species. In the wheat fields a few species became dominant. Finally, most of the dominant species in the wheat fields were important pests.

The foregoing changes took place in an amazingly short time, only 2 years. This pattern is followed wherever man clears, cultivates, and plants. This reduction in biological diversity, sometimes called *artificial simplification*, is therefore a normal, although not a totally desirable, thing.

The process can be reversed in certain instances, thereby requiring less pesticidal control. A specific instance is the use of strip harvesting in California alfalfa fields. Three strips of alfalfa 120 feet wide are cut at different times. The uncut strip provides a reservoir of pest enemies immediately available to the regenerating, newly cut strip. Actual counts of insect predators and parasites from alfalfa fields in which both regular and strip cutting were practiced were made by University of California scientists. Three groups of arthropod insects illustrate the important difference, as shown in Table I.

There were almost four times as many controlling insects where diversity was maintained by strip farming than where it was not. The expenditure for pest control was correspondingly reduced. Moreover, the yield of alfalfa hay was almost 15 percent greater under the strip-farming system.

Pesticides are used largely to combat the disturbances arising from a reduced and unstable biological community. It is unfortunate that unknowingly we make the problems of both production and pollution worse with pest control chemicals.

Figure 2 illustrates a simple case of a control problem aggravated by pesticides. The ladybird, or vedalia beetle, has been introduced into California from Australia in the nineteenth century to control a scale insect pest of oranges. The predaceous beetle quickly

I **Arthropods in Alfalfa Fields**

Natural enemies	Regular farming	Strip farming
Ladybird beetles		
Adults	46,000	205,000
Larvae	11,000	232,000
Parasitic wasps	70,000	287,000
Predatory spiders	105,000	1,094,000

Average number per acre.

established itself and for over 60 years kept the scale in check. In 1946 the newly discovered DDT found its way into use in citrus groves. Very quickly the susceptible beetle became rarer, and, as quickly, the scale insect again became a major pest. The withdrawal of DDT restored the "natural balance," but 3 years were required in most citrus orchards before control was again satisfactory. (Yet, strangely, in 1974 the Environmental Protection Agency received requests for reinstatement of DDT usage in control of certain citrus pests.)

Continued pesticide use not only influences the numbers in single species but can affect the composition of the entire animal community. As an illustration, Cornell University scientists made a comparison between soil organisms from orchards frequently treated with pesticides and those from nearby untreated orchards. There were several times more organisms in the treated orchards, but the biomass (total living weight) was less. Continued treatment resulted in more of fewer kinds, and these weighed less. Larger predatory forms, particularly, were reduced in the treated area. In fact, predatory organisms were seven times more abundant in untreated soil. The opportunity for natural control is clearly much greater in the untreated area. More disturbing was the discovery that there was a downward shift in the trophic level in the treated orchard. This shift simply means that repeated pesticide treatment resulted in vastly greater numbers of plant-feeding forms. The presence of these chemicals is another form of environmental pollution producing results very similar to those described for the Russian steppe.

The simplest definition of a *weed* is that it is a plant in the wrong place. The word "weed," like the word "pest," biologically has no meaning. The idea of weeds, however, can be extended very easily to any living thing that changes its natural relationships as a result of man's activities. Logically, animals can be thought of as weeds as well as plants. Whether animal or plant, they occur where we do not want them, in numbers that are competitive. They are alien to the area where they are considered as pests, and almost always they are aided by man's activities. They also maintain higher densities in disturbed areas than they did in their native habitats. They

may have compensatory mechanisms such as increased reproductive rate to counter efforts to control their numbers. Pests are mainly biological weeds created by man-made environments. The necessity to use pesticides is most frequently a crude attempt to correct an ecologically unbalanced agricultural ecosystem.

Chemicals in the Control of Pests

The great variety of chemicals and their uses dictates against a full description.[1-3] About 10,000 commercial pesticidal formulations are registered in the United States. Only a few dozen basic pesticide chemicals are commonly used (Table II). Many thousands of chemical compounds have been screened for their biocidal capabilities. But the magnitude of this screening effort marks a fundamental change in the use of pesticides. The era of synthetic organic chemicals in pest control began with the discovery of the insecticidal capacity of DDT in 1941, coupled with wartime development of the nerve poisons (organic phosphorus compounds) in Germany. Synthetic compounds have dominated pest control for more than a quarter century. Their heavy and widespread use during this period is one of the major changes from neolithic agriculture. A host of technologists, ranging from developmental chemists in industry to agricultural extension advisors, have arisen to accommodate this new development. Unfortunately, the emphasis has been largely on chemical development and application, a view leading to both narrowness and controversy.

The ecologist or environmentalist must take a point of view that is expanded in both time and space beyond that of the usual pesticide technologist. At once he must consider the entire living community and the physical factors on a global scale that influence the distribution of pesticides. The broadest range of pest control methodology is outlined in Figure 3. This chapter concerns itself only with chemical pesticides, and more particularly with new synthetic materials widely used. Chapter 14 emphasizes bioenvironmental alternatives.

Two general themes underlie the ecologist's criticisms of present pesticide application practice. These are lack of selectivity and lack of controllability. All environmental problems, whether they are the immediate difficulties of the pest control practitioner or the broadest problems of the social humanist, stem from these two lacks.

Lack of selectivity, together with basic *pesticide kinetics*, or transfer, are illustrated in Figure 4. Later figures illustrate the range of residue uncontrollability and subsequent ecological effects. Figure 4 describes well how broad-spectrum pesticides are applied. The announced intention of many pesticide technologists is selectivity. The fact is that very little success toward that aim has been achieved in practice, whatever the character of the chemical types used.

I have indicated the general themes and viewpoints associated with chemical pest control. Technical aspects cannot be reviewed in detail in this chapter (see Further Reading). I have elected to discuss generally two important aspects of chemical use. One is mode of action, using common chemical classes and well-known pesticides involved in contention within them. The second is the concept of toxicity versus hazard as normally viewed by the toxicologist and pest control advisor. The last section of this chapter is essentially an expansion of this concept to the social arena.

Mode of Action of Pesticides

A chemical can enter an animal's body by three routes: ingestion, by being eaten; respiration, by being breathed; dermal assimilation, by absorption through the skin or body covering. These are referred to, respectively, by

II Basic Pesticide Chemicals

Chemical group or action	Examples
Insecticides and Acaricides	
Inorganic	
Arsenicals	Lead arsenate
Copper-bearing	Copper sulfate
Organic, naturally occurring	
Nicotine alkaloids	Nicotine sulfate
Pyrethroids	Pyrethrum
Rotenoids	Rotenone
Organic, synthetic	
Chlorinated hydrocarbon compounds	Aldrin, benzene, hexachloride, DDD, DDT, endrin, heptachlor, methoxychlor, ovatran, toxaphene
Organic phosphorus compounds	DDVP, malathion, parathion, Phosdrin, schradan, TEPP, Systox
Carbamates	Isodan, pyrolan, Sevin
Fungicides	
Mercurials	Mercuric chloride, "organics"
Dithiocarbamates	Nabam, ziram
Others	Captan
Herbicides	
Contact toxicity	Sodium arsenite, "oils"
Translocated (hormones)	2,4-D, 2,4,5-T, dalapon
Soil sterilants	Borates, chlorates
Soil fumigants	Methyl bromide, Vapam
Rodenticides (Mammal Poisons)	
Anticoagulants	Pival, warfarin
Immediate action	Endrin, phosphorus, sodium fluoroacetate (1080), strychnine, thallium
Other Vertebrate Targets	
Birds	Strychnine, TEPP
Fishes	Rotenone, toxaphene

332

toxicologists, as oral, respiratory, and dermal routes of entry. The same terms are used when the toxicity of a chemical is being examined. Thus we have oral toxicity of a particular chemical, dermal toxicity, and so on. The description of both route of entry and comparisons of toxicity by different routes are not simply conveniences. The basic character of the chemical and its technical formulation, as well as the structure of the pest, interact to assure that one route of entry is favored over another. For example, most DDT formulations have a low dermal toxicity. They normally must be eaten before becoming effective. The "nerve gas" parathion, on the other hand, has a very high dermal and respiratory toxicity. The rodenticide strychnine is simply not hazardous unless eaten. The same is true of the sodium arsenite in ant baits.

Further decisions about the routes of entry must be made according to the group of animals and its particular structure. The chitinous exoskeleton of insects is relatively impervious

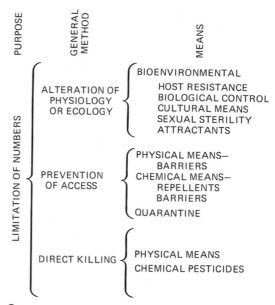

3 **Pest Control Methods.** A general outline of the varied ways by which pests can be controlled. Chemical pesticides are only one means.

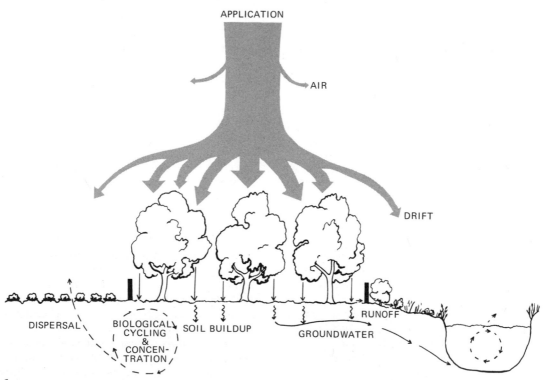

4 **Pesticide Kinetics.** The general case of persistent pesticide kinetics. A large fraction of spray goes to the intended area. A sizeable fraction drifts onto unintended surfaces. A portion is carried into the atmosphere. Surviving residues at the surface may leach into water or enter the food chain, where biological concentration begins.

to most common chemicals. This is certainly not true of mammals. A chemical may be highly toxic by one route of entry to one group of organisms and hardly toxic at all by the same route in another group. The structure and physiology of all animal groups likely to be exposed must be known so as to assess both pest control efficiency and hazard. For example, a small concentration of DDT or toxaphene in a lake will not directly harm a swimming man even if he gulps a fair amount of water. Yet that same concentration will be very toxic to the gill-breathers — fishes and invertebrates — of the lake. Examples of this sort are legion and make responsible pest control a very complicated affair. Methylparathion probably presents the greatest human hazard. It is extremely toxic, both orally and dermally. More methylparathion is used on U.S. farms than any insecticide except toxaphene.

With some pesticides, and particularly with their actions on certain groups of organisms, contact with part of the body surface is all that is needed for a toxic reaction. The surface may be skin, gills, the alveoli of the lungs, or the intestinal wall. The pesticide barely enters the body. For example, DDT affects the gills of fishes directly by immobilizing the blood supply, hence suffocating the affected fish. But with most pesticides, the chemical must not only contact and penetrate a part of the body surface, it must also be circulated to the part of the body most sensitive to its action. The circulating pesticide by itself does not differ at this stage from any of the hundreds of other compounds carried by the blood and lymphatic system. Inevitably, a sensitive area, usually a part of the nervous system, will be reached and, if the concentration is high enough, a toxic reaction can be expected.

The physiological target of a toxic chemical is usually some phase of nervous control. It may be central, that is, the brain or the spinal cord may be affected. These central nervous system poisons are exemplified by the chlorinated hydrocarbon insecticides such as DDT (Table II). Toxicity may be manifest in the peripheral nervous system. This is the action of the organophosphates. Their pest control ability rests on "shunting out" the nerves that control body muscles. They do this by destroying the enzyme cholinesterase that is necessary for nervous transmission at the nerve ends. Or, as a third kind of attack on nerves, the nerve supply of a vital organ may be destroyed or disrupted. The rodenticide sodium fluoroacetate (1080), for example, kills by disrupting nervous control of the heart.

The vital system that the pesticide seeks may not be directly a part of the nervous system. The rodenticide warfarin, for example, acts to lower the prothrombin (a clotting factor) level in circulating blood. In time, blood loses its important ability to clot. The affected animal hemorrhages internally and dies. Another example is the chemosterilant Metepa. Its purpose is to slow or prevent the production of sperm. Although the animal carrying the chemosterilant may not be noticeably poisoned, it has no ability to reproduce. The effect, given time, is to lower population numbers in the next pest generation. Reducing numbers is the chief purpose of pesticides. The time involved, organs and systems affected, and efficiency in accomplishing this purpose differ according to the type of chemical.

Of the wide range of herbicidal chemicals available, by far the greatest use is made of growth accelerators. The best known of these are 2,4-D [(2,4-dichlorophenoxy)acetic acid] and 2,4,5-T. The sprayed material passes through the phloem of a plant to an area where food is being utilized. When in responsive plant tissue, 2,4-D apparently acts rather similarly to the natural growth hormone, 3-indoleacetic acid. This hormone encourages the cell wall to take up water, with resulting elongation of the cell. In the normal plant cell, various mechanisms operate to control the rate of growth, including the destruction of the growth hormone when too much is present. The plant reacts to 2,4-D as it does to indoleacetic acid. Yet it is unable to oxidize or in any way inactivate the herbicide. Hence, unlimited growth continues until the plant dies. 2,4-D acts chiefly on dicotyledonous (broad-leaved) plants. Although broad spectrum in one sense, it can be used as a selective herbicide to remove weeds from fields of cereal grain, all monocotyledons. Moreover, the volatility and solubility of 2,4-D can be modified by formulation; hence, the herbicide can be used in a variety of specific ways. One illustration is its use as a defoliator, producing "partial death" by removing leaves but, at least for a short time, not killing the plant. Cotton fields may be defoliated for easier mechanical harvest. Or, as a military weapon, entire stretches of tropical jungle have recently been sprayed with 2,4,5-T to enable the discovery of men and installations (producing, among other effects, a great deal of social concern).

The second group of herbicides inhibits growth, perhaps to the point of death, and does so by interference with the cell machinery. The action may be on enzyme formation or in disturbance of an oxidative reaction. These herbicides produce biochemical lesions. The chemical nature of herbicides that cause this type of effect is extremely varied, ranging from copper sulfate and the arsenical compounds to totally new organic products recently synthesized. Many are used, particularly as fungicides, on seed or in soil. Some, such as dinitrocresol (DNOC), can, according to formulation, be used as herbicides, fungicides, or insecticides.

I have termed the third group "species enhancers." In a real sense this is not a chemical but an ecological category. Many chemicals may do the job. Sometimes they are referred to as selective soil sterilants or preemergence herbicides. These chemicals are normally applied to soil and their action is on seeds or newly sprouting and rooting plants. They remove unwanted plants and make more likely survival and good growth of desirable kinds. Careful selection of chemical type ensures that result. How the favored kind is able to respond in this way is more of an unknown. It is known that many plants, particularly common, vigorous "weedy" kinds, exude chemicals from their root systems that prevent the establishment of roots of other individuals. The diffusates may reach many inches, sometimes feet, away from the parent root. Any herbicide that removes such a plant then produces an effect on the survival of another plant that is not directly a chemical effect. In the context of community ecology, we can conclude that the effect is a partial floral sterility.

Whether herbicide or insecticide, and regardless of the mode of action, almost every application made is characterized by the two criticisms of the ecologist indicated earlier: lack of selectivity and lack of controllability.

Toxicity Versus Hazard

Toxicity is the inherent capacity of a chemical to harm. Hazard is the total risk taken when the chemical is used practically. Almost any chemical is toxic if taken in too large a dose. But relatively few are genuinely hazardous. Yet it does no practical good to say that table salt or water can be toxic. Concern should be with those chemicals that are known to be poisons. In this way all pesticides are toxic. The purpose of these chemicals is to harm something living. That this something is called a pest makes no difference. Almost any pesticide sufficiently active physiologically to kill a pest can harm some other living organism. This fact brings us to hazard. Very few application techniques can now be considered thoroughly safe as far as potential hazard to people and nontarget organisms under all conditions of use is concerned. The same applies to medicinal chemicals prescribed by physicians. Therefore, a degree of hazard, or risk, is inherent in almost every pest control operation. The controversial aspect of pest control chiefly stems from the word "risk." Rather, it stems from the fact that people do not agree on what that risk is. Inevitably, the question of particular interests and values enters the risk equation. A judgment on practical use, a conclusion on the degree of risk, must always be made.

Since most pests occur in large numbers, the effective toxicity of a pesticide is usually described in the percentage reduction of numbers in a pest population. We use a death measure for those chemicals that kill directly. The normal expression of toxicity is the lethal dosage or LD. Total death to a test population is expressed as LD_{100}. The most reliable figure is that amount of chemical formulation that will kill one-half of a test population, LD_{50}. Other indexes may be used as well. The effective dosage (ED) must be defined according to desired effect. Frequently, death is not the goal of measurement. It may be, perhaps, that amount of chemical that produces a noticeable rise in heart beat in one-half of an experimental population (hence ED_{50}). Special problems require special measures. In fishes, for example, the standard measurement of toxicity is the median tolerance limit following 96 hours of exposure. This is symbolized 96-hr TL_M and has about the same meaning as the LD_{50}, except that a temporal factor is added.

Temporal effects of toxic chemicals are sometimes broadly indicated by the terms "acute" and "chronic." *Acute* is usually taken to mean immediate toxic reactions from a single exposure. *Chronic* is even less precise. Exposure may be single or continuing, but the effects may appear either over or only after a long period of time. With the discovery that subtle effects may be long delayed following exposure to poisons (Chapter 8), tests may now run as long as 2 years. The distinctions between the two temporal descriptions may be imprecise.

The actual amounts of toxicants are normally expressed in fractions of a million by weight or volume. Parts per million (ppm) is the way it is usually seen. Recently, however, the greater sensitivity of chemical detection procedures has often allowed us to express contamination ratios in parts per billion (ppb). This expression is particularly useful in the analysis of aquatic environments, where even the slightest traces of pesticides may have important biological effects.

Risk decisions must be made. Toxicity quantified in the ways just described has become the normal measurement by which hazard is assessed. Both success in pest control and contamination levels in environment and living tissue are judged relative to an *acceptable threshold* amount. The environmental monitoring schemes of public health agencies and the food-sampling programs of the Food and Drug Administration have established safety margins based on the threshold concept. As shown later, *the concept has little validity in ecosystems*. The true extent of risk in the broad sense is rapidly becoming apparent. The basic themes of this section illustrate why risk determinations in the recent past have been entirely too narrowly based.

The hard reality of risk determination regarding human health is that it is based almost entirely on toxicological and marketing considerations. Varying field conditions and differing social and traditional values play small roles in determining use in the United States. Yet the incidence of accidental death or injury is large and mounting.[4]

Only in the State of California does the hazard to human beings from agricultural chemicals approach proper assessment (although judged to be incomplete). In the 20-year period 1951–1970, there were 36 fatalities directly due to these chemicals. Illness and morbidity reports are much higher annually and have shown a steady rise in the past few years. These changes arise from both increases in chemical use and from the kinds of chemicals used. In 1969, for example, 32 percent of all cases were caused by organophosphate poisoning. In 1970, 22 percent of total illness was ascribed to organophosphorus compounds; chlorinated hydrocarbons accounted for 6 percent; and all herbicides contributed 14 percent. A doubling of cases of systemic poisonings occurred in 1973. The majority of such cases occur among farm laborers, many of whom are poorly educated and who often do not report lesser illnesses to doctors. Socioeconomic factors (e.g., union activities) are clearly involved. In a way one should be surprised that poisoning is not even commoner. Approximately 6 million pounds each of all chlorinated hydrocarbon and organophosphate pesticides are applied in California every year. DDT usage was reduced some 75 percent in 1970 and 1971. Quite likely, the compensatory increase in the use of organophosphate chemicals explains the recent rapid increase in number of systemic poisonings among field workers.

Occupational exposure among field workers is not only common but full of unknowns. Why is it, for example, that peach and strawberry pickers commonly show signs of phosphate poisoning after parathion applications in which all regulations have been followed? Recent work suggests that parathion on foliage converts to paraoxon, which is more stable and 10 times as toxic as parathion. Yet this fact is not yet recognized in regulatory practices! The herbicide 2,4,5-T is probably teratogenic (causes birth defects). But it has been used widely in Vietnam and is commonly used in the United States, even by the U.S. Department of Agriculture.

Newspaper accounts of accidental poisonings are now commonplace. Official accounts are often paradoxical. For example, the World Health Organization recently reported that in a survey of 200,000 workers occupationally exposed to DDT, not a single human death was attributable to the chemical. Yet that organization regularly issues a *Circular on Toxicity*, which reports hundreds of annual instances of death and morbidity that result from DDT and other pesticides.

Nor are paradoxes simple. Thousands of rural people in Iraq, Guatemala, and Pakistan became ill, some fatally, from eating bread made from seed wheat that had been treated with methyl mercury fungicides. The seed grain was not intended for human consumption and was dyed red for identification. Most came from out of the country in massive aid or trade programs. All sacks had warning labels, but what use are these to those poor who cannot read in their own language, much less in the language of the donor? This complicated picture reached its most dramatic expression in Iraq in late 1971 and early 1972.[5] Thousands of tons of treated and dyed seed wheat from the United States and Mexico were distributed throughout the Iraqi provinces. Within 5 months over 6,500 people had been hospitalized; over 500 died. Those not hospitalized or whose deaths were not attributed to methyl mercury could not be counted. Bread was the source (but the warnings existed!). Mortality rates were relatively constant (about 7 percent) for age and sex grouping except for pregnant women. In this case hospital admissions records showed a mortality rate of 45 percent. And what of those yet to be born? Mercury demonstrably affects the fetus *in utero*. How many more thousands have been affected?

As a further example of paradox, why has Japan totally banned the use of parathion and tetraethyl pyrophosphate (TEPP) when these and similar materials have increased in use threefold in the United States in the last 10 years? And the ultimate example of paradox: Why do we "allow" residual amounts of chlorinated hydrocarbons in human milk considerably in excess of those legally allowed in dairy products?

The risk of pesticide poisoning to human beings is normally judged by acute toxicity. The presence of residues in human tissues is now universal, ranging from an average low of about 3.0 ppm in Alaskan Eskimos to about 25.0 ppm in India. The average is about 12 ppm in the United States.[2,6] No one is fully confident that these insidiously acquired residue loads are without effect. A study in Florida suggests a relationship between the concentration of DDT (and metabolites) and liver tumors, cirrhosis, and hypertension. For example, in terminal cases of liver malignancy, the mean *p,p'*-DDE (dichlorodiphenylethane) concentration in fat was three times normal, and seven times as high in liver. Race and socioeconomic status are other elements in the risk equation. Southern blacks, for example, particularly children, consistently show higher residue levels than southern whites.[2,6,7]

I have already alluded to some symptoms occurring in human beings who have received pathological amounts of organophosphate and mercuric pesticides. Following is a condensed accounting of human (mammalian) toxic responses to the major classes of pesticides.[3] Clinical terms are used only when unavoidable or when particularly significant.

BOTANICALS

Rotenone and pyrethroids: very low toxicity; kidney and respiratory effects in severe cases. Nicotine compounds: highly toxic but not now widely used; convulsions, cardiac irregularity, and coma in severe cases.

CHLORINATED HYDROCARBONS

For the entire group of chlorinated hydrocarbons: low to moderate acute toxicity; most obvious target the central nervous systems (CNS); incoordination and tremors characteristic in severe cases. Heptachlor, aldrin, and dieldrin (cyclodienes generally) show CNS disturbance and parasympathetic failures. Later stages show results of lipid storage and release (as in liver damage). DDT and DDD considerably less toxic in acute exposure; symptoms as described. Problem is storage and buildup in lipid tissues. System-wide gonadal and endocrine effects.

MERCURIALS

Organic mercury (Hg) is more toxic than inorganic mercury, particularly alkyl Hg. Alkyls convert to methyl Hg under anaerobic conditions; stored in living tissues as methyl Hg. CNS symptoms appear first (incoordination, parathesias, tremors), followed by muscular atrophy and mental instability.

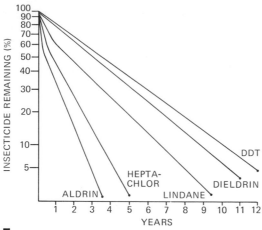

5 **Pesticide Disappearance Rates.** The normal method of showing persistent pesticide disappearance rates in soil. Note that total "loss" is expected and predicted on essentially a straight-line basis. The representation, common though it is, is ecologically naive. First, in both inorganic and organic environments, soils are among the poorest media for showing insecticide survival. Second, no indication is given of the survival and enhancement of residues in living systems.

CARBAMATES

Normal dimethyl carbamates are strong cholinesterase (ChE) inhibitors. Parasympathetic system symptoms because excess acetylcholine produced. Bradycardia and gastrointestinal hypermobility (diarrhea and vomiting); tremors and muscle seizures follow. Mental aberrations in chronic exposure (sustained low ChE levels).

ORGANOPHOSPHATES

Organophosphates are all extremely toxic, although (academically) some, as TEPP, more so. Absorbed through all routes of entry. Action and symptoms as for carbamates. Unusually hazardous on direct exposure, but rarely chronic poisoning. Parathion converts in fields to paraoxon (more toxic). Guthion and malathion least toxic of widely used organophosphates.

RODENTICIDES

There are no common features among rodenticides; common rodenticides are not chemically related. The three most widely used are compound 1080 (sodium fluoroacetate), cyanide, and strychnine. Compound 1080 is extremely toxic; it disturbs citric acid metabolism, with resulting cardiac depression, fibrillation, and peripheral nervous system effects. Cyanide (as sodium cyanide in "coyote-getters") is extremely toxic; it produces respiratory system failure (suffocation). Strychnine is also extremely toxic; it affects all parts of the CNS, resulting in asphyxia and dyspnea. Convulsions are a prominent feature. There are no antidotes for mammal poisons.

HERBICIDES

Herbicides consist of many chemically unrelated materials. Their acute toxicity to man is generally very low (exception, arsenicals, not now widely used). Chlorophenoxy compounds (2,4-D, 2,4,5-T) produce a mild irritation on exposed areas, gastrointestinal effects if taken internally. Attempts to commit suicide using these compounds produces only severe poisoning. 2,4,5-T has been shown to be teratogenic on chronic exposure.

Some Chemicals Survive

The chemicals used in pest control are valuable only when they intimately intrude into the pest's environment. Normally in agriculture the major portion is directed toward the leaf surfaces of plants. In addition, pesticides, depending on the purpose, may be placed on walls, water surfaces, or the ground surface. Very rarely is the method of use so precise that only the minimal area required is actually treated. More often a much larger area is treated simply because there is no choice. Moreover, these pesticides normally drift or drip or leach into other areas nearby, so that the area covered by the spray or dust is increased.

The actual fraction of pesticide reaching the target surface and remaining there is small. A significant part vaporizes into the atmosphere either before it reaches the crop or shortly thereafter, as the spray is drying. In one orchard where DDT was used for 20 years, measurements of DDT in soil, plants, and runoff accounted for only 50 percent of the total dosage. The remaining half probably entered the atmosphere at or shortly following spraying. In another study, a maximum of 38 percent of DDT applied to a cornfield by air

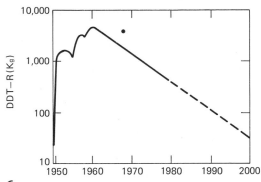

6 Model Derivation of Total DDT-R. DDT-R available from all residue types in mud and silt of the Clear Lake ecosystem. High points represent three intentional applications of DDT, added amounts known. The datum point represents estimated DDT-R in 1968 from field-collected data. The dashed line represents our own uncertainty about long-term linear projections. *Source:* Reference 10.

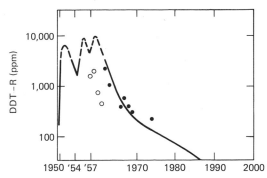

7 Total Residue Concentration in a Single Modular Compartment. Shown here is the mathematically derived "curve" and actual field-derived data describing the DDT-R load in the fat of western grebes (see Figure 9). The dashed portion is computer-derived but meaningless. No western grebe can survive with more than 2,500 ppm of DDT-R in its tissues. Circles and dots are averages of tissue analyses. Circles represent residue figures in years in which die-offs due to DDT-R would significantly disrupt our predictions based on mathematical modeling. *Source:* Reference 10.

actually ended up on the corn plants or in the soil.[6]

The chemical mixture, wherever it does fall, however, forms a *deposit* whose presence is necessary for pest control. This deposit does not normally keep its original chemical character for very long. Change comes about as the deposit is acted upon by living systems and by heat, light, and water. The remainder of this transformed deposit is called a *pesticide residue*. This remaining portion will contain reduced amounts of the original toxic substance, physically and biologically transformed chemical derivatives of the original substance, and parts of the original solvent or dilutent carriers.

All pesticides in common use produce residues that survive for noticeable periods on foliage or soils. It may be only a day or two, as for TEPP, or perhaps a couple of weeks, as for parathion. Or it may be for a very long time, many years, as with DDT. Chlorinated hydrocarbon insecticides have the greatest staying power. Figure 5 illustrates the average survival in soil of five common pesticides of this group, taking into account the environmental variables in different locations.[8] Probably the upper limit to residue survival is shown in Table III. This study was atypical in the sense that all factors contributing to the decomposition of residues were minimized. In this situation survival times ran several times longer than in the average case. The clear "winner" is

DDT, 39 percent of which survived in original form for 17 years.[9] Figures 6 and 7, which are derived from mathematical models of my research with pesticide dispersal in an aquatic ecosystem, illustrate both the "disappearance rate" and future probable quantities of residues in that system. The term "DDT-R" indicates the summation of all DDT-like residues present. Figure 6 shows total amounts in mud and silt, and Figure 7 shows the amount in one bird population, the western grebe.[10]

The period of persistence is important information. There are two ways of viewing

III Longevity of Chlorinated Hydrocarbons

Years elapsed	Pesticide	% Remaining
14	Aldrin	40
	Chlordane	40
	Endrin	41
	Heptachlor	16
	BHC	10
	Toxaphene	45
15	Aldrin	28
	Dieldrin	31
17	DDT	39

The maximum longevity of chlorinated hydrocarbon insecticides in soils. Survival of residues in this instance is maximal because of experimental design.

Source: Reference 6.

339

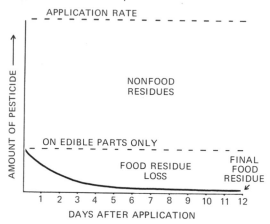

8 Pesticide Residues on Food. The normal manner of representation of surviving pesticide residues on human foodstuffs. Important as this approach is for human welfare, it, too, is ecologically naive. The implication that pesticide residues no longer are significant because "tolerance limits" are not exceeded on foodstuffs overlooks the ecologic fate of the much larger remainder.

such information. One relates to the original purpose: the length of time that residual protection from pests can be expected. The other is the judgment of hazard. In this instance, the persistence period indicates how soon before harvest or before food consumption the chemical can be applied. It indicates, too, how likely is the chance of hazard from toxic residues to living things other than man and the animals or plants that he has set out to control. For these reasons a great deal of research effort has gone into determining how long a pesticide chemical survives in various environmental situations.

For practical purposes those chemicals that survive for only a few days, or even weeks, can be termed *nonpersistent*. Those that survive for longer periods, months or years, are called *persistent*. Both use and hazard depend on how long and in what form pesticide chemicals survive. A nonpersistent chemical not only "disappears" in a short space of time but has no opportunity to disperse widely. On the other hand, a sizable fraction of the original deposit of a persistent chemical remains. These residues become scattered so that they are no longer found only on the area where applied. A small amount predictably will have been carried outside the general area to some

distant point. The ways in which residues are dispersed are many and varied. Natural forces, both physical and biological, are responsible. Residue dispersal is clearly important practically and has led to a new field of study called *pesticide kinetics*.[11]

Most chemicals now used in pest control are synthetics; that is, they are made, not found in nature. Man must decide where and when to use these chemicals. Since synthetics are usually toxic to many living things, including man, in sufficient amount, not only must pesticide kinetics be understood but practical steps to avoid excessive contamination of our own environment, particularly of our own foodstuffs, must be taken. The way in which food technologists view residues as they apply to human foods is illustrated in Figure 8.

Persistent chemicals in foods are a small fraction of the total amount applied in the agricultural environment. Moreover, the greater fraction on foodstuffs degrades within a few days, and the processing and marketing phases allow still more to degrade or to be removed. Human concern is the "final food residue." This quantity of the original toxic material present in foods cannot be allowed to exceed certain values (tolerances) judged to be within safety limits. There have been many arguments about these final residue values. What exactly is safe? On the one hand, food production is vital and yields of foodstuffs are increased by the use of pesticide chemicals. On the other hand, there are many unknowns in the judgment. Much scientific effort goes into the determination of what the maximum permissible residue values will be. There remains constant concern about the correctness of judgments. New data often force changes. There are also some people who, on personal, religious, or philosophical grounds, argue that no traces at all of any of these chemicals should be allowed in foods.

Unfortunately, the problem of residues does not stop with human food. The ecologist must attend to the much greater amounts of residues that do not remain on foods. A quick inspection of Figure 8 shows that the application rate of pesticides is many times larger than the maximum possible contamination of "edible parts only." Domestic animals and wildlife also eat. It is possible to build up residues in soil to such levels that they might

340

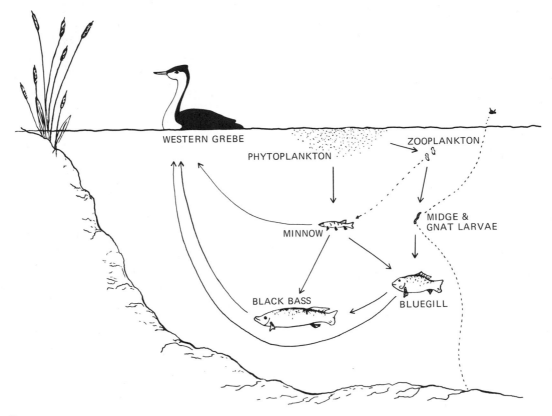

9 **Aquatic Food Chains.** A simplified representation of aquatic food chains at Clear Lake, California. Slow release of persistent pesticide residues from the surrounding watershed and the lake bottom enables biological "capturing," storage, and magnification in the food chain. As a general rule, it is the last link in the food chain (the ultimate consumer) that shows toxic symptoms of concentrated residues. Figures 6, 7, and 9 through 11 should be viewed as successive attempts to refine research analysis.

harm plants, become phytotoxic, or that they might once again pose residue problems for us.[12]

Pesticide Distribution

Both physical and biological transfer systems have been imputed to account for the spatial distribution of residues. In general, the transfer of residues in physical systems can be directly measured. There is a variety of data which demonstrate that residues occur in aerial dust and smog, rainfall, in air far at sea, in ground and flowing waters, and in soils at different depths and of different types. No one has yet conducted a full study of pesticide

kinetics in natural ecosystems. Our studies at Clear Lake, California, are an attempt to do this in an ecological microcosm. (See Figures 6, 7, and 9–11) for a sequential development of this idea.) Currently, we are in the process of applying this same type of analysis to a major agricultural ecosystem, the San Joaquin Valley, California.

Generally, fluid movements — air and water — are responsible for residue transfers over long distances. Vaporization and codistillation with water are the primary entries into air. Slowly we are acquiring the belief that larger fractions (up to 50 percent) than previously thought enter the atmosphere in this way. Pesticide residues in water tend to absorb differentially to suspended (particularly organic matter) particles. In both air and water the

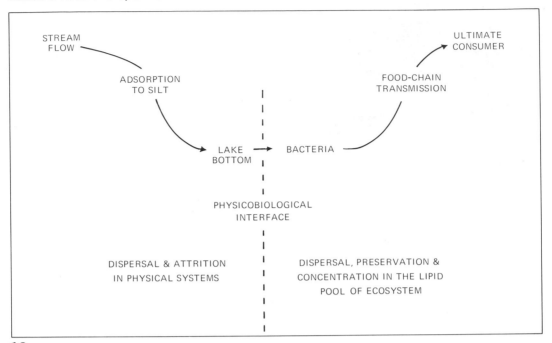

10 **Pesticide Kinetics in Lakes.** Schematic representation of pesticide kinetics in lacustrine ecosystems. Residues disperse and degrade in the physical component but are entrapped, survive, and their amount enhanced in the biological segment. The relatively inert lipid fraction amounts to an ecological preservative for these residue "fossils."

smallest particles seem to carry the greatest residue burden because of their abundance.[6]

The occurrence of residues in air and water at points distant from sites of application is now established. DDT (also DDE and DDD) occurred in atmospheric dust over the Barbados Islands in the amount of 41 ppb (Figure 12).[13] A more recent study in the Sargasso Sea confirms the phenomenon but adds that the gaseous component of these compounds (also PCBs and chlordane) in the atmosphere was two orders of magnitude higher than previously reported for particulate matter.[14] DDT in dust over Pittsburgh in the latter half of 1964 averaged 0.25 part per trillion (ppt).[15] Residues in British rainwater in 1966 and 1967 showed measurable amounts of BHC (lindane), dieldrin, and DDT. Even in the remote Shetland Islands, where pesticides are rarely used, total residues measured 229 ppt.[6] At this station as well as others in the United Kingdom, it is postulated that pesticides were being carried in the air over thousands of miles of ocean waters, in this case in the prevailing westerly winds from North America. On the basis of an average concentration of 170 ppt of all residues at all stations in Britain, one can calculate that 1 inch of rainfall would deposit 1 ton of residues. Or, in other terms, with an average rainfall of over 40 inches per year, Great Britain receives over four times the amount of pesticide residues that the Mississippi dumps into the Gulf of Mexico annually. The estimated maximum amount of residues of DDT and metabolites in Antarctic snow and ice is over 2,400 tons.[16] Almost 2 tons of residues per year enter San Francisco Bay from the Sacramento and San Joaquin rivers.[17] Even so, this amount is scarcely sufficient to explain the almost total contamination of the biota from the confluence of these rivers and in the ultimate marine environment. Air transport must be the major disseminator of residues. How else, as an example, can one explain that all frogs examined (several hundred) in the Sierra Nevada, including those above 5,000 feet, are contaminated with DDT products (average 3.19 ppm)?[18] And in a totally different

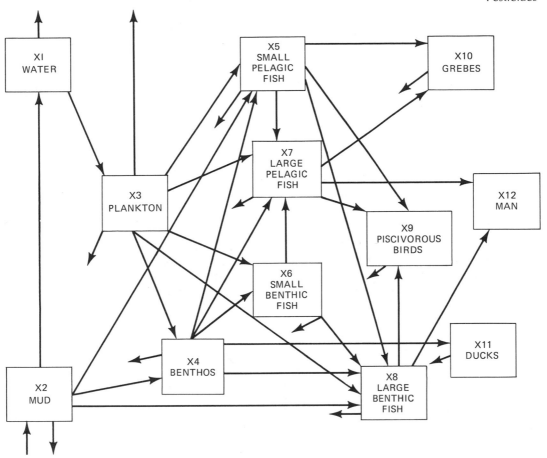

11 **Graphic Compartment Model of the Clear Lake Ecosystem.** Small arrows indicate backflow to mud. The small arrows from water and plankton represent outflow. The OW symbols describe open-water fish. X^n refers to labeled compartments in the mathematical model. *Source:* Reference 10.

ecological environment, how can one explain average levels of DDT and metabolites in the blubber of gray and sperm whales amounting to 0.36 and 6.0 ppm, respectively?[19] The differential can be explained rationally because sperm whales eat larger organisms, higher up the food chain. But the presence of residues in migratory mammals from the open ocean is more difficult to account for. Atmospheric fallout must be the source.

Biological mechanisms in residue transfer are often inferred. Descriptions of amounts in tissue are, of course, easily made. But to assess transfer mechanisms, a good knowledge of the life histories of implicated species and of trophodynamics in ecological communities is required. Too often, the basic ecological infor-

mation needed for full assessment is not available. Schematic representation of the biological and physicobiological systems involved in these assessments is provided in Figures 9 through 11. These models have been developed from my own detailed studies at Clear Lake, but they have a general applicability to freshwater ecosystems anywhere. With appropriate substitution of ecological information, the same pathways in an open-ocean environment can be illustrated (Figure 13).[13,20]

Pesticide residues in animal tissues have their source in an environment which normally contains traces of residues. Initial entry into the animal food chain is generally by means of contaminated foods. This source must be essentially the only way that terres-

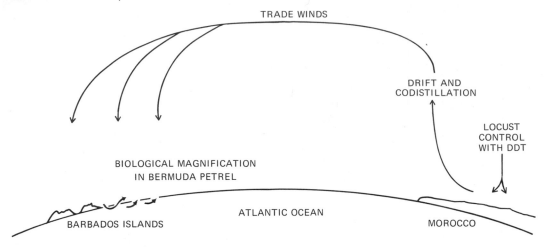

12 **Transport of Pesticide Residues.** Aerial transport of residues and subsequent biological magnification reaching toxic levels in an ultimate consumer. This example is extreme but describes generally how residues enter biological tissues in areas remote from points of application. *Source:* References 15, 16.

trial animals acquire tissue residues. Contaminated foods are also major sources of tissue residues among aquatic animals, but are particularly important as secondary links in the food chain. Initial entry into aquatic food chains may derive from the tendency of pesticides to absorb on suspended particulate matter which is consumed by filter-feeding organisms known to be environmental concentrators. There is some evidence as well that living organisms in environments in which

RESIDUE

ECOLOGICAL
TARGET

ECOLOGICAL
TRAP

BIOLOGICAL CONCENTRATION
IN PLANKTON

13 **Pesticide Kinetics in the Ocean.** One means of ecological "entrapment" and confinement of pesticide residues. Airborne particles of pesticide striking a water surface do not disperse below the "skin." Rapid uptake by planktonic "environmental concentrators" ensures spatial confinement of residues to the top biological layers in which residue magnification occurs.

sublethal amounts of residue are present may acquire residues directly through the gills and skin.

Two serious areas of ignorance remain. The first concerns the physicobiological interface. How precisely are residual transfers made from an inorganic environment to a biological community? The second is the amount and nature of return residues to a biological community following the death of contaminated tissues. In short, what is the nature of biological recycling of pesticide residues? This question rests on the observation that residues simply do not degrade in biological environments at the rate predicted from physical evidence (Figure 5, and discussion in Reference 21).

Proof of Environmental Contamination

The argument persists that recent environmental contaminants cannot be responsible for some of the major biological effects that have been observed. When the existence of DDT residues in Antarctic wildlife was first announced in 1965, the standard response of chemical producers and users was that chromatographic analyses were artifacts, due either to naturally occurring chemicals or to incompetence of analysts. Either defense has

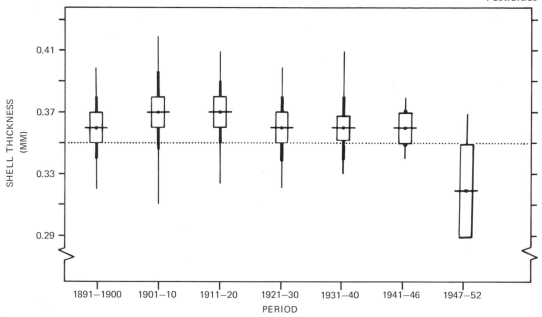

14 **Eggshell Thickness Versus Time.** Thickness of 614 peregrine falcon eggshells collected since 1891. The dotted horizontal line is the midpoint between the 95 percent confidence limits of the 1947–1952 group compared to all prior groups. The heavy horizontal lines are mean values; the open rectangle represents 95 percent confidence limits for each group. *Source:* Reference 25.

been shown to be foolish. Yet it is true that the major evidence linking biological effect in the field of environmental contamination is correlative, not experimental. Ecologists regularly depend on correlations among events to reach conclusions; experimentalists seldom do.

I offer here two examples (there are now many others) of residual occurrences linked with major biological events whose discovery was first made by field biologists and later confirmed by experimental work. The first concerns mercury in Swedish fish and wildlife. Mercury in Sweden derives primarily from point sources, that is, industrial effluents on streams, but agricultural use of mercuric seed dressings is a known major general source. Studies of fish and birds first showed high levels of mercury in tissues (over 1 ppm). A series of investigations elucidated the general problem and led to stringent industrial controls in 1966 and to the aboliton of organic mercury as a seed treatment in 1967.[22] Sweden does not have naturally occurring mercury. Confirmation of the relatively recent nature of environmental contamination by mercury was provided by a study that de-

scribes the mercury content in feathers of 11 species of birds in museums collected over a century.[23] In all species mercury was relatively constant in the period 1840–1940. Beginning in the 1940s, mercury content increased 10- to 20-fold. As expected on biological grounds, highest levels (up to 30 ppm) were in fish-eating birds and the lowest in seed-eating birds.

The second illustration links the eggshell thickness of peregrine and other falcons with time.[24-26] Thinning of eggshells due to DDE, dieldrin, mercury, and PCBs (polychlorinated biphenyls) has now been experimentally established.[27-30] Before such confirmation, correlations of eggshell thickness with time strongly suggested a cause–effect relationship. The "coincidence," now shown in several species of raptorial and fish-eating birds, is clearly shown in Figure 14.[31-36] The remarkable (and statistically significant) decline in shell thickness in the 1947–1952 period correlates well with the first general use of DDT. In my judgment final confirmation has come from Peakall's work, in which he determined DDE contents in the eggs of peregrine falcons col-

345

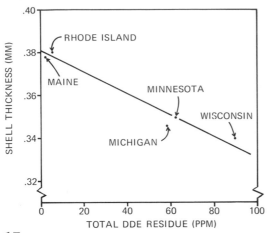

15 **Eggshell Thickness Versus DDE Concentration.** Variation in eggshell thickness and DDE concentrations in herring gulls collected from several locations in 1967. Only at the highest residual concentration (Wisconsin) were reproductive effects of pesticide contamination observable. *Source:* Reference 25.

lected over many years.[37] He was able to show that at least as early as 1948, sufficient DDE was present in eggs to cause eggshell thinning. The recent exhaustive review of this subject by Cooke[29] also leads to no other conclusion. The conclusion is still being challenged.[38,39] But, as Peakall[37,40] wisely notes, species effects are differential — not all bird species respond in the same way to the same residue loads. For example, it is now well known that gallinaceous (e.g., chickens, pheasants) and raptorial birds do not show the same "thinning" response given the same residue contaminations. Possibly this differential response is "explained" by unlike metabolic and excretory rates, as has recently been suggested.[45] Still, the phenomenon in carnivorous birds is marked and follows about this sequence of beginning dates[29]:

1. In Britain: peregrine, sparrowhawk, and golden eagle, 1945–1950; kestrel, 1946; merlin, 1951; hobby, 1952.
2. In Sweden and Finland: osprey, 1947–1949.
3. In California: brown pelican, 1968; peregrine, 1947–1952.

All the above results have been carefully studied; other prospective examples are less thoroughly reported. The relationship between shell thinning and DDE (sometimes PCBs) levels is striking.

That the observable effect is related to environmental foci of residue concentration is shown in Figure 15. It is shown that eggshell thickness (in herring gulls) is a function of DDE concentration, which in turn is a measure of local environmental contamination. Only in Wisconsin were reproductive effects observable. The herring gull colony there lived on the highly polluted shores of Lake Michigan, in which all fishes have also been shown to be heavily contaminated with residues.[41]

Biological Magnification

The diversity of responses of living things to environmental contaminants is probably beyond knowing. Because residues of many types persist, are distributed widely, and are incorporated into biological systems, the most important response to environmental contamination is that of biological magnification.[1,42]

"Capturing" existing residues occurs either by living environmental concentrators or by food gatherers in a biological chain. Magnification or concentration of captured residues may occur by three methods:

1. *Physiological concentration* occurs when residues are stored and accumulated by particular tissues within the bodies of individual organisms. A nearly straight-line relationship exists between lipid or fatty fractions of a tissue and the amount of residues contained.
2. *Biological concentration* occurs in those instances in which residues are picked up directly through the skin or respiratory surface. This method is rather common in aquatic environments.
3. *Trophic concentration*, or food-chain concentration, occurs simply because one organism depends upon another for something to eat. Stored residues, of course, accompany ingested foods and are assimilated and stored in the predator, which, in turn, if eaten, passes along a larger amount of residues.

All three methods of concentration lead to

magnified quantities of residues in secondary consumers in food chains. The hazard, needless to say, is greatest for the ultimate consumers.

Too frequently, environmental contamination as measured in soil and water is dismissed as negligible because residual values are usually very low. Overlooked is the incredible ability of natural mechanisms to concentrate residues. In an Ohio marsh, for example, DDT labeled with radioactive carbon concentrated 3000-fold in algae within only 3 days of application. Studies by the U.S. Fish and Wildlife Service show that many filter-feeding organisms have an unusual capacity to pick up and store residues from the environment around them. Oysters, for example, can concentrate DDT residues 70,000-fold after only 1 month of exposure.[22] A modular laboratory "ecosystem" consisting of a seven-element food chain treated with ^{14}C-labeled DDT at an equivalent of 1 pound per acre demonstrates well the concentration abilities of both individual organisms and trophic sequences.[43,44] Radiolabeled DDT was accumulated in mosquito larvae, snails, and fish as DDE, DDD, and DDT and concentrated 10,000- to 100,000-fold within about 1 month's time.

Differences in food habits which reflect positions in food chains are easily discoverable by looking at residual levels of contamination. In Maine forests, for example, the litter fauna was sampled for pesticide residues over a period of 9 years.[45] In this instance, the major emphasis was on residues in areas sprayed only once with DDT at 1.0 pound per acre. Some areas, however, were treated at the same rate as much as three times over a period of years. Whether the application was single or multiple, persistent residues occurred in all fauna examined. Table IV compares mice and voles with shrews. The persistence of residues is striking. It took 9 years for mice and voles to approach the residue levels that occur in nontreated areas. In areas treated three times, values remained five- to sixfold above untreated areas. Carnivorous shrews, on the other hand, showed values ranging from 10- to 40-fold those of the herbivores. Probably about 15 years would be required, after only a single DDT treatment, before residue levels in shrew tissues could return to normal. The effects of higher residue level are not known. Where

multiple treatment occurs, those levels would remain high, having effects not only on the individual organisms, but more likely harming animals that prey upon them. Local extinction of both predator and prey are likely consequences.

Entire ecosystems can be contaminated with residues. Trophic concentration has now been documented in many instances. In my own studies at Clear Lake, California, the magnitude of concentrations of DDD in tissues over that originally occurring in water was shown to be as follows: plankton, 265-fold; small fishes, 500-fold; predaceous fishes, 85,000-fold; predaceous birds (grebes), 80,000-fold.[1,11,21]

A similar pattern of residual concentration is shown in studies by University of Wisconsin scientists working on Lake Michigan.[41] The following listing indicates residue levels of chlorinated hydrocarbon insecticides in a trophic sequence: bottom sediments, 0.0085 ppm; small invertebrates, 0.41 ppm; fishes, 3.0 to 8.0 ppm; herring gulls, 3,177 ppm.

The most complete community study yet made of residual contamination centered on a salt marsh bordering Long Island Sound.[46] Arrangement of residue levels of DDT in order of increasing value showed a clear progression related to both size of organism and trophic level. Larger organisms and higher carnivores had greater concentrations than smaller organisms at lower trophic levels. Total residues ranged through three orders of magnitude, from 0.04 ppm in plankton to 75 ppm in a ring-billed gull. Shrimp contained 0.16 ppm;

IV DDT Residues in Small Mammals

Years since single treatment	Mice and voles (ppm)	Shrews (ppm)
0	1.06	15.58
2–3	0.07	0.72
3–4	0.08	2.50
5–6	0.10	1.43
6–7	0.05	1.81
8–9	0.04	1.88
Untreated	0.03	0.30
Multiple treatment (3 times)	0.17	4.77

Total residues of DDT and metabolites in small mammals collected from plots in Maine forests with variable treatment history. Mean values (ppm) only are shown.
Source: Reference 21.

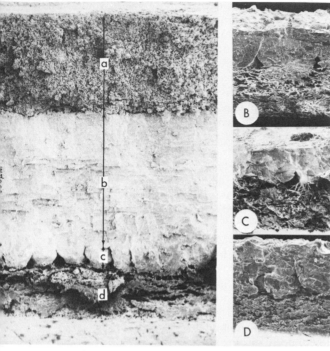

16 **Radial Sections of Eggshells of Pelicans.** The larger photograph illustrates "normal" construction of eggshell; the three smaller photographs represent high levels of organochlorine residues. Note the lack of palisade layers in those individuals (B, C, D) with high residue loads. *Source:* Reference 31.

eels, 0.28 ppm; marsh insects, 0.30 ppm; and predaceous fishes, 2.07 ppm. In general, concentrations of DDT were 10 to 100 times as high in carnivorous birds as in the fishes on which they fed. Actual values of DDT residues in water were about 0.00005 ppm. Based on this value, birds near the top of the food chain have concentrations of DDT residues 1 million times greater than the concentration in water. This example, among others, suggests that DDT residues are moving through the biological and chemical cycles of the earth at concentrations that are having far-reaching, yet little-known effects on ecological systems.

Most notably in raptorial and fish-eating birds, declines in numbers seem to be clearly related to the presence of pesticide residues. In our continuing work at Clear Lake, high residue levels are clearly correlated with inhibition of reproductive success in western grebes. The rare Bermuda petrel clearly shows these

residue effects on reproduction.[47] Yet, more dramatically, the total failure of brown pelicans to breed successfully on the coastal islands of California in 1969 and 1970 is probably attributable to biological magnification of pesticide residues "trapped" and preserved in marine ecosystems.[48] In 1970 only one young pelican among 552 nests studied survived to fledge. Regulations severely limiting DDT use and the reduction of contaminated effluent from a coastal factory manufacturing DDT have had some beneficial results since 1970. By 1972 residue loads in offshore fishes had dropped three- to fourfold, and eightfold in pelican eggs. These reductions were reflected in increased fledging success among pelicans. On Anacapa Island, for example, 247 pairs successfully produced 34 young in 1973, an increase, to be sure, but a figure much lower than normal production of young.[49] The appearance of normal and "thinned" eggshells in pelicans related to DDT-R content is shown in Figure 16.[31] Like pelicans, fish-eating mammals seem to respond to high DDE levels with reproductive inhibition.[50] Premature partus in sea lions on California's Channel Islands since 1968 has been observed. Although premature pups are born alive, whether late or early, all die within a few days. Those born later in the pupping season lack motor coordination and full respiratory development. The correlation with DDT-R levels seems marked. The blubber of full-term adult sea lions averaged 103.2 ppm DDT-R, whereas the concentration in parturient females giving birth to premature young averaged 824.4 ppm, eight times as much. Confusing the picture, PCB values showed similar disparities, averaging 17.1 ppm in the former case, 112.4 in the latter. Also, concentrations of mercury in the livers of a small sample of premature-bearing females seem inordinately high (34 to 68 ppm). Thus, several causes may be operating in addition to DDT, including lead, mercury, and PCB. Although the source of these materials is not pest control, the effects, alone or in concert, may well be the same.

In broad terms, regulatory restrictions can have the effect of reducing environmental residues and their unfortunate consequences. The pelican example is an illustration. Additional concrete examples are hard to find.

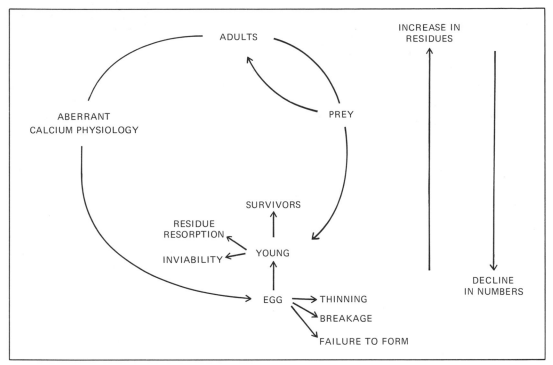

17 **The Cycle That Causes Thin Eggshells.** A simplified diagram that illustrates the physiological mechanisms by which declines in numbers of raptorial and fish-eating birds might be induced. Numbers decline in these "ultimate consumers" as pesticide residues increase. The breakdown of estrogens affects the circulating calcium level in the blood, in turn affecting both adults and unhatched eggs (see References 25 and 35).

There is the suggestion that prairie falcons in Colorado produced more young in 1972 than in 1967–1968.[24] DDE in eggs in the latter sample averaged 5 ppm with eggshells 14 percent thinner than in pre-1941 samples, whereas in 1972 eggshells were only 10 percent thinner and ranged between 2 and 3 ppm per egg.

Species decline, that is, decline in numbers and distribution, presumably caused by ubiquitous residues, has been observed in many species of animals. The evidence is strongest in raptorial and fish-eating birds, representing in ecological terms ultimate consumers, the ends of food chains. Most concern lies with the peregrine falcon, whose decline in numbers correlates very closely with expanding use of DDT (Figure 14). Several studies focus on observations and mechanisms to explain the decline in numbers. Young birds have died; clutch size has been reduced; eggshell size and thickness has decreased; and egg breaking by adults has increased. These changes have been

linked in part to aberrant calcium metabolism. DDT and DDE (as well as PCB) have been shown to be powerful inducers, at very low residual levels, of liver enzymes that degrade estradiol.[35,40] Estrogenic hormones in female birds play an important role in calcium metabolism. Insecticide residues, uncontrolled in environment and concentrated in animal tissues, can now be shown to have altered the normal hormonal balance in wild species, leading unfortunately to species-wide effects. Some 10 families of birds now seem to be affected in this manner (Figure 17). How far these effects ultimately extend in birds or other animal groups is not known. Importantly illustrated by this action is that the threshold concept of toxic effects has to be discarded. Residues not only act by accumulation ecologically or physiologically to yield toxic effects, but they can also act in *trace* amounts to result in species disaster. Only recently has this subtlety of contaminant effects been appreciated.

349

The Future of Pesticide Production

The production and use of pesticides continues to grow throughout the world. Synthetic organic pesticide production in the United States continues to increase at about 15 percent per year. Total sales for 1975 are estimated at $3 billion annually. Rates of increase in different chemical categories are not equal. Herbicide sales, for example, rose 271 percent in the period 1963–1969. This increase is more than double that for all other pesticidal classes. Herbicidal sales alone were probably $1.35 billion in the United States in 1974. The United States produces 50 to 75 percent (dependent on chemical class) of all pesticides and regularly exports about one-third of its production. Following the United States in volume of production worldwide are West Germany, Japan, France, and the United Kingdom. The journal *International Pest Control* in late 1973 estimated a world production of chemical pesticides of 2 million tons annually with a manufacturer's value of about $2.25 billion. Annual demand for pesticides worldwide is increasing at about 10 percent per year. By 1982 total sales (at the manufacturer's level) should reach almost $5 billion.

Use of pesticides on many crops has increased dramatically in a 25-year span. Corn, for example, required neither insecticides nor herbicides for its production in 1945. In 1950 insecticides applied were 0.1 pound per acre. In 1970, 1.0 pound per acre each of insecticides and herbicides were applied. Corn production trebled during this period.[54]

Both production and export of DDT have declined since 1967, but increases in other chlorinated hydrocarbon and organophosphate insecticides have more than compensated for that decline. On the world market the dominant pesticides in use are insecticides and fungicides, and this domination is expected to continue indefinitely. The conclusion that an environmentalist may clearly draw is that synthetic pesticidal use will increase worldwide, with attendant ecological hazards.

A Concluding Commentary

Pesticide chemicals will continue to be used in pest control, and they will continue to cause

effects in place. The broad-spectrum activity of these chemicals ensures that nontarget organisms will be affected as well as intended targets. Some effects and alternative methods that might lessen or eliminate these effects are considered in Chapter 14.

But the second category of major effects described at the beginning of this chapter is of more serious concern to the environmentalist. Surviving pesticide residues are at once insidious, uncontrollable, and unpredictable in their action. The only conclusion to draw is that all pesticides whose toxic effects cannot be confined to the site of application must be banned. Sufficient risks remain even with such a restriction in use. These risks are properly the responsibility of the pest control practitioner and range from hazard to the farm worker to pesticide resistance in pest species. Even with total banning of persistent pesticides, sufficient amounts of residues will remain in the earth's ecosystems to ensure ecological disruptions for decades. Estimates of these amounts range upward into the millions of pounds. These estimates are, at best, guesses. The actual amounts make no real difference, however. Examples of biological disruption are now sufficiently documented to establish that existing uncontrolled residues, whatever their amounts, are seriously damaging.

Further evidence of the subtlety with which pesticide contaminants can act is provided by an example which, if established as generally occurring, could have staggering ecological consequences. It rests on an important question: How do insecticide-resistant vertebrates affect the organisms that feed upon them? Recent work at Mississippi State University provides a first answer.[52] In experimental studies, 95 percent of the vertebrate predators (11 species) died after each had consumed one endrin-resistant mosquito fish (*Gambusia*). In other terms, resistant mosquito fish tolerate endrin residues sufficient to kill potential predators several hundred times their own weight. These were laboratory studies. No one can yet say that the effects occur in nature, but, even if locally applicable, resulting ecological disruptions could be catastrophic.

In effect, with pesticidal chemicals we have introduced an evolutionarily new, density-independent factor into the environment. We have argued the need for these chemicals in pest control. It is now necessary to argue for stringent pesticide control, to depart from the major emphasis on pests alone. Pesticides are now uncontrolled. We need to recognize that pest control is basically an ecological, not a chemical, problem. The dizzying changes induced into the earth's ecosystem in the last two or three decades constitute an experiment out of hand, whose consequences now seem ecologically catastrophic.

Both optimistic and pessimistic approaches are possible in recent pronouncements relating to pesticide use.[53] The wisest approach is to take all factors affected by chemical pest control — economics, international relations, environmental and health risks, and societal aspirations — into consideration. Pimentel states this case well in his proposed "systems management."[54] Limited thinking — particularly a return to old patterns, having learned little from empirical experience — is currently expressed in many ways. The decision to spray DDT at 0.75 pound per acre over 650,000 acres on forests in the Pacific Northwest in 1974 is clearly retrogressive. Equally clearly, the Environmental Protection Agency made its decision on political grounds only.

Restoring control requires perspective and judgment. Sufficient facts are available. No one need dodge behind the welter of data points — of LD_{50}s, TL_Ms, and chemical formulas. Neither pest control nor the uncontrolled effects of pesticides is toxicology simply extrapolated. The scientist must not only seek facts; he must judge facts in entire systems, including those within the human community. Ecologists are obligated to serve this human expectation. Through them the primary goal, the reestablishment of a *working control system*, can be achieved. Persistent pesticides, along with many more waste products of technological man, cannot be allowed to continue to work against such a controlling system.

References

1. Rudd, R. L. 1964. *Pesticides and the Living Landscape*. University of Wisconsin Press, Madison, Wis.

2. U.S. Department of Health, Education, and Welfare. 1969. *Report of the Secretary's Commission on Pesticides and their Relationship to Environmental Health*. Government Printing Office, Washington, D.C.

3. Brown, R. L. 1966. *Pesticides in Clinical Practice*. Charles C Thomas, Publisher, Springfield, Ill.

4. California Department of Public Health, Bureau of Occupational Health and Environmental Epidemiology. *Annual Reports*. Sacramento, Calif.

5. Bakir, F., Damluji, S. F., Amin-Zaki, L., Murtadha, M., Khalidi, A., Al-Rawi, N. Y., Tikriti, S., Dhahir, H. I., Clarkson, T. W., Smith, J. C., and Doherty, R. A. 1973. Methylmercury poisoning in Iraq. *Science 181:* 230–241.

6. Frost, J. 1969. Earth, air, water. *Environment II:* 14–28, 31–33.

7. Davies, J. E., Edmundson, W. F., Schneider, N. J., and Cassady, J. C. 1968. Problems of prevalence of pesticide residues in humans. *Pesticide Monitoring J. 2:* 80–85.

8. Crosby, D. G. 1964. Intentional removal of pesticide residues. In *Research in Pesticides* (C. O. Chichester, ed.) Academic Press, Inc., New York. pp. 213–222.

9. Nash, R. G., and Woolson, C. A. 1967. Persistence of chlorinated hydrocarbon insecticides in soils. *Science 157:* 924–927.

10. Craig, R. B., and Rudd, R. L. 1974. The ecosystem approach to toxic chemicals in the biosphere. In *Survival in Toxic Environments* (M. Khan, Ed.) Academic Press, Inc., New York.

11. Rudd, R. L., and Herman, S. G. 1972. Ecosystemic transferral of pesticide residues in an aquatic environment. In *Environmental Toxicology of Pesticides* (F. Matsumura, G. M. Boush, and T. Misato, eds.). Academic Press, Inc., New York. pp. 471–485.

12. Bevenue, A., and Kawano, Y. 1971. Pesticides residues, tolerances, and the law (USA). *Residue Rev. 35:* 103–149.

13. Risebrough, R. W., Huggett, R. J., Griffin, J. J., and Goldberg, E. D. 1968. Pesticides: transatlantic movements in the northeast trades. *Science 159:* 1233–1236.

14. Bidleman, T. F., and Olney, C. E. 1974. Chlorinated hydrocarbons in the Sargasso Sea atmosphere and surface water. *Science 183:* 516–518.

15. Cohen, J., and Pinkerton, C. 1966. Wide

351

spread translocation of pesticides by air transport and rain-out. In *Organic Pesticides in the Environment* (Advan. Chem. Ser. 60). American Chemical Society, Washington, D.C. pp. 163–176.

16. Peterle, T. J. 1969. DDT in Antarctic snow. *Nature 224:* 620.

17. Bailey, T. E., and Hannum, J. R. 1967. Distribution of pesticides in California. *J. Sanitary Eng. Div., Amer. Soc. Civil Engr., Proc. 5510:* 27–43.

18. Cory, L., Fjeld, P., and Serat, W. 1970. Distribution patterns of DDT residues in the Sierra Nevada Mountains. *Pesticides Monitoring J. 3:* 204–210.

19. Wolman, A. A., and Wilson, A. J. 1970. Occurrence of pesticides in whales. *Pesticides Monitoring J. 4:* 8–10.

20. Risebrough, R. W., Reiche, P., Peakall, D., Herman, S., and Kirven, M. 1968. Polychlorinated biphenyls in the global system. *Nature 220:* 1098–1102.

21. Herman, S. G., Garrett, R. L., and Rudd, R. L. 1969. Pesticides and the western grebe. In *Chemical Fall-out* (M. M. Miller and G. G. Berg, eds.). Charles C Thomas, Publisher, Springfield, Ill. pp. 24–53.

22. Miller, M. W., and Berg, G. G. (eds.). 1969. *Chemical Fallout — Current Research on Persistent Pesticides*. Charles C Thomas, Publisher, Springfield, Ill.

23. Berg, W., Johnels, A., Plantin, L. O., Sjöstrand, B., and Westermark, T. 1966. Mercury content in feathers of Swedish birds for the last 100 years. *Oikos 17:* 71.

24. Enderson, J. H., and P. H. Wrege. 1973. DDE residues and eggshell thickness in prairie falcons. *J. Wildlife Management 37:* 476–478.

25. Hickey, J. J., and Anderson, D. W. 1968. Chlorinated hydrocarbons and eggshell changes in raptorial and fish-eating birds. *Science 162:* 271–273.

26. Hickey, J. J. (ed.). 1969. *Peregrine Falcon Populations — Their Biology and Decline*. University of Wisconsin Press, Madison, Wis.

27. Bitman, J., Cecil, H. C., Harris, J. J., and Fries, G. F. 1969. DDT induces a decrease in eggshell calcium. *Nature 224:* 44–46.

28. Porter, R. D., and Wiemeyer, S. N. 1969. Dieldrin and DDT: effects on sparrow hawk eggshells and reproduction. *Science 165:* 199–200.

29. Cooke, A. S. 1973. Shell thinning in avian eggs by environmental pollutants. *Environ. Pollution 4:* 85–152.

30. Peakall, D. B., and Lineer, J. L. 1972. Methyl mercury: its effect on eggshell thickness. *Bull. Environ. Contamination Toxicol., 8:* 89–90.

31. Garrett, R. L. 1973. Effect on technical DDT on eggshell structure in Coturnix quail (*Coturnix coturnix*) and brown pelicans (*Pelicanus occidentalis*). Unpublished Ph.D. dissertation. University of California, Davis, Calif.

32. Lincer, J. L., and Peakall, D. B. 1970. Metabolic effects of polychlorinated biphenyls in the American kestrel. *Nature 228:* 773.

33. MacLane, M. A. R., and Hall, L. C. 1972. DDE thins screech owl eggshells. *Bull. Environ. Contamination Toxicol. 8:* 65–68.

34. Odsjo, T. 1971. Klorerade kolvaten och aggskalsfortunning hos fiskgjuse. *Fauna och Flora (Sweden) 66:* 90–100.

35. Peakall, D. B. 1967. Pesticide-induced enzyme breakdown of steroids in birds. *Nature 216:* 205.

36. Stickel, L. F. 1972. Biological data on PCBs in animals other than man. In *Polychlorinated Biphenyls and the Environment*. Interdepartmental Task Force on PCBs, National Technical Information Service, Springfield, Va. pp. 158–172.

37. Peakall, D. B. 1973. DDE: its presence in peregrine eggs in 1948. *Science 183:* 673–674.

38. Gunn, D. L. 1972. Dilemmas in conservation for applied biologists. *Ann. Appl. Biol. 72:* 105–127.

39. Hazeltine, W. 1972. Disagreements on why brown pelican eggs are thin. *Nature 239:* 410–411.

40. Peakall, D. B. 1970. Pesticides and the reproduction of birds. *Sci. Amer. 222:* 72–78.

41. Hickey, J. J., Keith, J. A., and Coon, F. B. 1966. An exploration of pesticides in a Lake Michigan ecosystem. *J. Appl. Ecol.* 3(suppl.): 141–145.

42. Kenaga, E. E. 1972. Factors related to bioconcentration of pesticides. In *Environ-*

mental Toxicology of Pesticides (F. Matsumura, G. M. Boush, and T. Misato, eds.). Academic Press, Inc., New York. pp. 198–228.

43. Kapoor, I. P., Metcalf, R. L., Mystrom, R. F., and Sangha, G. K. 1970. Comparative metabolism of methoxychlor, methiochlor and DDT in mouse, insects and fish in a model ecosystem. *J. Agr. Food Chem. 18:* 1145–1152.

44. Metcalf, R. L., Sangha, G. K., and Kapoor, I. P. 1971. Model ecosystem for the evaluation of pesticide biodegradability and ecological magnification. *Environ. Sci. Technol. 5:* 709–713.

45. Moriarty, F. 1972. Pollutants and food chains. *New Scientist*, Mar. 16: 594–596.

46. Woodwell, G. M., Wurster, C. F., and Isaacson, P. A. 1967. DDT residues in an east coast estuary: a case of biological concentration of a persistent insecticide. *Science 156:* 821–824.

47. Wurster, C. F., and Wingate, D. 1968. DDT residues and declining population in the Bermuda petrel. *Science 159:* 979–981.

48. Risebrough, R. W., Sibley, F. C., and Kurven, M. N. 1971. Reproductive failure of the brown pelican on Anacapa Island in 1969. *Amer. Birds 25:* 8–9.

49. Anderson, D. W. 1974. U.S. Fish and Wildlife Service. Unpublished reports.

50. DeLong, R. L., Gilmartin, W. G., and Simpson, J. G. 1973. Premature births in California sea lions: association with high organochlorine pollutant residue levels. *Science 181:* 1168–1170.

51. Pimentel, D. 1972. Pesticides, pollution and food supply. Environmental Biology, Department of Entomology, Cornell University, Ithaca, N.Y.

52. Rosato, P., and Ferguson, D. E. 1968. Toxicity of endrin-resistant mosquito fish to eleven species of vertebrates. *BioScience 18:* 783–784.

53. Borland, N. E. 1972. Mankind and civilization at another crossroad: in balance with nature — a biological myth. *BioScience 22:* 41–44.

54. Pimentel, D., Hurd, L. E., Belotti, A. C., Forster, M. J., Oka, I. N., Sholes, O. D., and Whitman, R. J. 1973. Food production and the energy crisis. *Science 182:* 443–448.

Further Reading

Carson, R. 1962. *Silent Spring*. Houghton Mifflin Company, Boston. 368 pp.

Graham, F. 1970. *Since Silent Spring*. Houghton Mifflin Company, Boston. 333 pp.

Matsumura, F., Boush, G. M., and Misato, T. 1972. *Environmental Toxicology of Pesticides*. Academic Press, Inc., New York. 637 pp.

Moore, N. W. 1967. A synopsis of the pesticide problem. *Advan. Ecol. Res. 4:* 75–129.

Napa Valley vineyard and the nearly microscopic wasp parasite (inset) that
is capable of controlling the population level of the destructive grape
leafhopper. Photos by F. E. Skinner.

14

Better Methods of Pest Control

Gordon R. Conway   *is a Research Fellow in the Environmental Resource Management
Research Unit at the Imperial College of Science and Technology in London. He received his B.S.
in zoology from the University College of North Wales and then studied agriculture at Cambridge
and at the University of the West Indies in Trinidad. From 1961 to 1966 he was entomologist at
the Agricultural Research Centre in North Borneo, responsible for controlling "everything from
elephants to red spider mites." He then went to the University of California at Davis where he
earned a doctorate in biomathematics. Conway writes: "I was dissatisfied with the empirical
approach to pest control I had been following in Borneo. I wanted to see to what extent
mathematics and the techniques of systems analysis could provide a better way of tackling pest
problems." In recent years he has served as a consultant on environmental problems in the
developing countries.*

355

Chapter 13 was concerned with the pollutant effects of modern organic pesticides. As we saw, these compounds pose serious hazards to wild animals and plants and to man. By itself this is quite sufficient justification for seeking alternatives to pesticides or, at least, trying to minimize their use. But pesticides are not only pollutants; they now also commonly fail to control pests.

Furthermore, these two aspects of the pesticide problem are related to each other. The particular characteristics of many modern pesticides which produce the failures in pest control are also partly responsible for the wider environmental problems. In this chapter we shall look first at the pattern of pesticide failure and analyze some of the underlying causes to show this connection. As a next step we shall attempt to define the features of an approach to pest control which is better both in terms of being more efficient and of producing fewer hazards in the environment. The rest of the chapter will be devoted to reviewing, on this basis, alternatives that have been or are being developed.

The emphasis will be on the problems of insect pest control. As we have seen, insecticides tend to be the worst environmental polluters, and most of the failures of pest control have been associated with their use. However, many of the arguments that will be developed also apply to other pests, such as fungi, nematodes, and weeds.

The Problem

THE EARLY SUCCESSES

In the early years of modern pesticide use, during and immediately after World War II, the successes were frequently spectacular. Many pests of crops and of man and his domestic animals were controlled to a degree that had been impossible before. In 1944, a major outbreak in Naples of typhus, which is carried by lice, was ended by treating over 3 million people with DDT. A few years later, DDT and a related organochlorine insecticide, dieldrin, were being widely and successfully used to control the mosquito vectors of malaria. Death rates from this disease in Madagascar and Ceylon were almost halved in the first 2 years of DDT spraying campaigns.

In agriculture there were similar success stories. For instance, in the 4 years prior to 1944, over 15 percent of American apple crops were lost annually to the ravages of the codling moth, but in the 4 subsequent years, use of DDT brought this loss down to 4 percent. As they became available, other organic insecticides worked with even greater success. And along with the new insecticides came new acaricides for mite control, nematocides for nematode control, fungicides for controlling fungus diseases, and herbicides for controlling weeds. Table I shows examples of the kind of gains that were achieved on a variety of crops in the United States from this "pesticide revolution."

There were failures, though, from the beginning, and by the middle 1950s poor control and worsening pest problems began to show up in an increasing number of situations. Insecticides and acaricides were by far the worst offenders, but some other forms of pesticides were also involved.

THE PATTERN OF FAILURE — COTTON

The pattern of pesticide failure is well illustrated by the history of pest control on cotton.

Over 100 species of insect and spider mite attack cotton in the United States. But only a few of these are major pests in the sense that if not controlled they would, year after year, cause serious damage to the crop. Over most of the cotton-growing acreage in the United States the boll weevil (*Anthonomus grandis*) is the major pest, and before the use of organic pesticides it caused considerable losses of cotton. Pests of lesser importance are the American bollworm (*Heliothis zea*) (a caterpillar), the cotton fleahopper, the cotton leafworm, and the cotton aphid. In 1943, DDT became available and was found to control the bollworm and the fleahopper but not the boll weevil or the other pests. Two years later, however, benzene hexachloride (BHC), another organochlorine insecticide, was found to be effective against all the pests that DDT did not control. In the years that followed, cotton pest control came to rely very heavily on these two compounds and on a number of other organochlorine insecticides. Although resistance by cotton aphid to BHC and by leafworm to toxaphene began to show up quite soon, in the early 1950s growers and entomologists felt confident that cotton pests were well under control.

In 1955 this confidence was shattered by the development of high resistance to a range of organochlorine insecticides by the boll weevil in the lower Mississippi Valley. This proved to be the beginning of the development of widespread resistance in nearly all the other major cotton pests. The change was made to organophosphorus insecticides and then to carbamates, but in each case it was not long before resistance occurred. As a report of a panel of the President's Science Advisory Committee has stated[2]: "By the end of the 1963 season, almost every major cotton pest species contained local populations that had developed resistance to one or more of the chlorinated hydrocarbons, organic phosphorus or carbamate insecticides, or mixtures of chlorinated hydrocarbons. Moreover, strains have developed in the laboratory that are resistant to all of these." In addition to this resistance effect, the heavy insecticide spraying had severely depleted the numbers of beneficial parasites and predators, and upsurges occurred of many pests that had hitherto been considered of secondary or minor importance. The American bollworm and other related bollworms, spider mites, and aphids were among the species which became considerably more important than they had been.

THE CANETE VALLEY STORY

A similar pattern of events occurred on cotton in other countries. To emphasize the shared, global nature of the problem, it is worth recounting the recent history of cotton growing in a small valley in Peru. The valley is called the Canete, and its story is a classic in the history of pest control.[3] The valley is situated in the arid coastal zone of Peru, one of 40 such valleys separated by stretches of desert. Each valley contains a river which provides irrigation water for the crops. In effect, each of these valleys is a small isolated agroecosystem. The Canete Valley originally grew sugar cane but in the 1920s shifted to cotton; today some 15,000 hectares, about two-thirds of the valley, is under that crop. The cotton growers are very technologically advanced; they have their own association and support their own experiment station. Cultivation of the crop is mechanized and many advanced agronomic techniques have been

I   **Pesticide Benefits**

Crop	Pest	Yield per acre		Cost of treatment ($ per acre)	Net gain ($ per acre)
		Untreated	Treated		
Seed cotton	Bollworm	7,203 lb	7,860 lb	16	126.50
Pea seed	Fungi	456 lb	610 lb	0.70	21.25
Tomatoes	Diseases	5.4 tons	11.8 tons	40	100
Sugar beets	Root maggot	11.8 tons	14.5 tons	2.25	32.20

Some examples of gains in terms of yield and net dollar returns from pesticide use on crops in the United States.
Source: Reference 1.

357

adopted. But when the growers turned to modern organic pesticides to solve their pest problems, a situation developed that produced near-disaster.

In the years before the new pesticides were introduced, there were three important pests of cotton in the Canete Valley: *Anthonomus vestitus*, which is related to the American boll weevil; a cotton leafworm; and a caterpillar, *Mescinia*. Before and during World War II, control of these pests was based on inorganic pesticides such as nicotine sulfate and various kinds of arsenicals. In the late 1940s some use began to be made of the organic insecticides. At the same time the practice of ratooning (letting the cotton remain in the ground for 2 or 3 years) was adopted. Two more pests, the cotton aphid and a bollworm, *Heliothis virescens*, then became important. During the period when mostly inorganic pesticides had been used, average yields ranged from 466 to 591 kilograms per hectare. In 1949, however, with the heavy outbreak of bollworm and aphids, the yield dropped to 366 kilograms per hectare.

In the face of this drop of yield the growers decided to use more organic insecticides, principally DDT, BHC, and toxaphene. Some new strains of cotton were also introduced; the irrigation was improved, and there were changes in cultural practices. What followed is described by Ray Smith of the University of California at Berkeley[4]:

At first these procedures were very successful. Cotton yields nearly doubled. The average yield in the Canete went from 494 kilograms per hectare in 1950 to 728 kilograms per hectare in 1954. The cotton farmers were enthusiastic and developed the idea that there was a direct relationship between the amount of pesticides used and the cotton crop, i.e. the more the better. The insecticides were applied like a blanket over the entire valley. Trees were cut down to make it easier for the airplanes to treat the fields. The birds that nested in these trees disappeared. Other beneficial forms such as insect parasites and predators disappeared. As the years went by the number of treatments was increased: Also, each year the treatments were started earlier because of the earlier attacks of the pests. In late 1952, BHC was no longer effective against aphids. In the summer of 1954, toxaphene failed to control the leafworm. In the 1955–6 season, *Anthonomus* reached high levels early in the growing season; then *Argyrotaenia sphaleropa* (a leaf-rolling caterpillar) appeared as a

new pest. Next *Heliothis virescens* developed a very heavy infestation and showed a high degree of resistance to DDT. Substitution of organophosphorus compounds for the chlorinated hydrocarbons became necessary. The interval between treatments was progressively shortened from a range of 8–15 days down to 3 days. Meanwhile, a whole complex of previously innocuous insects rose to serious pest status.

In the 1955–56 season the situation came to a head. Yields plummeted to 332 kilograms per hectare and, with the high cost of pesticides, the growers experienced heavy losses. In nearby valleys which had either not used organic pesticides or used them only to a limited extent, the problems did not arise and the yields remained high. The situation was retrieved by the adoption of integrated control techniques; we shall discuss this later in the chapter.

Drawing on these experiences in the United States and Peru and on close observation of cotton growing elsewhere in the world, Smith shows that there is a discernible pattern in the history of cotton pest control which applies almost ubiquitously. He identifies five phases, progressing from subsistence farming through exploitative agriculture, to crisis, then to disaster, which, in turn, may be overcome by resorting to integrated control (see Table II).

OTHER CROPS

The story of cotton as an illustration of the failings of modern pesticides is by no means unique. Similar dramatic case histories could be described for alfalfa, apples, and citrus in the United States; for cocoa and oil palms in Malaysia; for tea in Ceylon; and for greenhouse crops in England.[6,7] There is also growing evidence that the same pattern of pesticide failure is being repeated on the new cereal varieties of wheat, corn, and rice, which are being grown as part of the Green Revolution throughout the developing countries (Chapter 3). The problem then is not a minor one, restricted to special situations. It is a general and growing problem that now threatens the ability of the world to feed and clothe its rapidly increasing population.

Why has the problem come about? In analyzing the experience with cotton and other crops it becomes clear that there are three major reasons why pesticide control is breaking down: (1) pesticides are nonregulatory; (2) they are countered by pest resistance; and (3) they interfere with natural regulatory mechanisms. We shall examine each of these in turn.

Causes of the Problem

PESTICIDES AS NONREGULATORS

In the first chapter we saw that many or most organisms are kept relatively stable in numbers by natural regulating mechanisms. Processes within populations or agents acting from outside govern population size by producing changes in numbers which are density-dependent. The proportionate mortality increases with rising density and vice versa (Figure 1a). Moreover, the stability is more or less permanent. The agents that bring it about are continuously present in the ecosystem and come into play whenever there is a shift in population size.

However, pesticides do not act in a density-dependent manner. When a pesticide is applied, it produces a certain percentage kill of the pest population, a percentage that is unrelated to the numbers of the pest present. The pesticide may persist for a time, but as its effect wears off, the pest population rebounds to its previous level (Figure 1b). The population is reduced, but only for a relatively brief period because no mechanism responsive to population change has been introduced. The only way in which a low, stable pest population can be achieved is by repeated applications of pesticide timed to hit the population whenever it begins to recover or by using a compound that persists for a very long period (Figure 1c).

One consequence of this characteristic of pesticide use is that environmental contamination is *persistent*. Wildlife and man in affected ecosystems are being constantly exposed to pesticides. If the exposure was less prolonged, there would be fewer environmental problems. For pest control the direct consequence is that pesticide use is costly in materials, time, and manpower. Often the need for repeated pesticide applications results in farmers adopting spray schedules whereby the pesticides are applied on a regular basis determined solely

II   Cotton Pest Control Phases

I. Subsistence Phase
   1. Cotton is not irrigated, part of subsistence agriculture.
   2. Yields below 200 kilograms lint per hectare.
   3. Crop protection dependent on natural control, inherent resistance, hand picking, cultural practices, rare insecticide treatments, and luck.

II. Exploitation Phase
   1. Cotton grown on newly irrigated land.
   2. Crop protection schemes based on chemical pesticides.
   3. Intensive pesticide use, often on fixed schedules.
   4. Initially high yields.

III. Crisis Phase
   1. More frequent pesticide applications needed.
   2. Treatments start earlier in growing season and extend later into harvest period.
   3. Pest populations resurge rapidly to new higher levels after treatment.
   4. New pests arise.

IV. Disaster Phase
   1. Pesticide usage increases costs.
   2. Marginal land removed from production.
   3. Eventually cotton can no longer be grown profitably in the area.

V. Integrated-Control Phase
   1. Crop protection system devised which relies on a variety of control methods.
   2. Environmental factors are modified and fullest use is made of natural mortality and biological control.

The five phases that have occurred worldwide in cotton pest control.
Source: Reference 5.

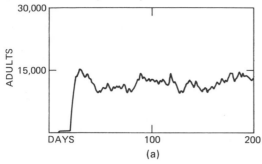

(a)

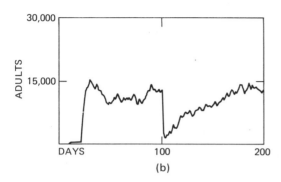

(b)

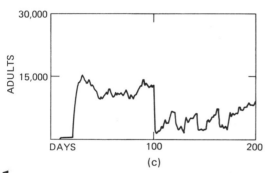

(c)

1 Simulated Insect Control. A hypothetical insect population derived from a computer model. The model is built so that the population is highly regulated by density-dependent reproduction and mortality. (a) **Insect Population in Equilibrium.** No insecticide applied. (b) **Population Treated Once.** On day 100, treatment with an insecticide is simulated. The treatment is designed to kill 90 percent of the adult insects in the first day and then decreases the kill by 5 percent each subsequent day. (c) **Population Treated Repeatedly.** Four insecticide treatments applied at 3-week intervals beginning on day 100. *Source:* Reference 8.

by the calendar; the farmer simply "sprays every Monday." In crops such as cotton, the costs of such spray schedules have threatened to make the crop uneconomic. (In 1973–1974, the price of cotton increased sharply, making

such behavior economically possible; but such high prices probably are only temporary.) However, perhaps more important than this effect, the need for persistent or repeated pesticide applications contributes to the other two components of pesticide failure.

RESISTANCE TO PESTICIDES

This is not a new problem; pest populations developed resistance to the older inorganic compounds. As early as 1908, for example, the San Jose scale, a sap-sucking insect that is a pest of apples, showed resistance to lime sulfur. By the 1940s there were some 10 pest species exhibiting resistance, principally to hydrogen cyanide and arsenic compounds. But with the introduction of organic pesticides, the number of resistant species began to increase rapidly. By 1958 there were 60 such species, and by 1968 over 200[9] (see Figure 2). If the present rate of increase continues, all of the estimated 5,000 species of insect and spider pests in the world will be showing resistance to one or more pesticides by the year 1990! The increase is unlikely to continue at the same rate, but this calculation gives some idea of the magnitude of the problem.

The term "resistance" in this context does not imply that individual insects develop immunity to pesticides in the way that human beings develop immunity to disease organisms. Pesticide resistance is a genetic phenomenon.[10] It occurs because a few individuals in a population already possess a genetic structure that happens to confer on them the ability to survive exposure to a pesticide. When the pesticide is applied, all, or nearly all, of the susceptible individuals in the population are destroyed and the survivors are predominantly those with the genes for resistance. Since the next generation arises from these survivors, it will contain a much higher proportion of individuals with resistance. As applications of the pesticide are repeated, this proportion will continue to increase until finally the whole population is resistant.

The genes actually confer resistance by setting up mechanisms in the individual which detoxify the pesticide, or sometimes by producing behavior that leads the pest to avoid pesticide. In natural populations untouched by pesticides, the genes that produce these mechanisms are usually rare. Presumably this is because they confer other, disadvantageous characteristics. It is only in the presence of pesticides that the resistant genes have a clear advantage and so spread, in time and space, through the population. This process of acquiring pesticide resistance is thus one of simple natural selection in the Darwinian sense. It is a classic example of microevolution.

Since acquiring resistance depends on selection in this way, it usually follows that the stronger the selection pressure, the more rapidly does the population become resistant. With very complete coverage of the population, very few susceptible individuals escape the pesticide. Most of the survivors are resistant individuals, and hence a very high proportion of the next generation will be resistant. We have already pointed out that because pesticides are nonregulatory, they have either to be very persistent or to be applied repeatedly. Now we can clearly see that this failing, important in itself, leads to far worse resistance problems than would occur otherwise.

In addition to selection pressure, the rapidity with which resistance is acquired depends on the generation time of the pest. Species with short generations which breed continuously throughout the year develop resistant populations that much faster. The two factors combined set up a vicious circle. Rapidly breeding pests elicit frequent pesticide applications, which, in turn, create a strong selection pressure for resistance. Then as populations recover earlier, applications become even more frequent and resistance is developed progressively faster. Some idea of the speed and spread of population resistance can be gained from Figure 3.

It is not surprising, then, that the first species to show resistance were those which breed rapidly and live in continuously favorable environments, such as occur in tropical regions or in human dwelling places. Nor is it surprising that half of the resistant species so far recorded are public health or veterinary pests; these species have been subject to the most intensive pesticide pressures. There is evidence of widespread resistance among houseflies, biting flies, flies of cattle, mosquitoes and midges, and bedbugs and cockroaches. Figure 4 shows the extent of resistance in *Aedes aegypti*, a mosquito that carries yellow fever. Also, resistance has developed in

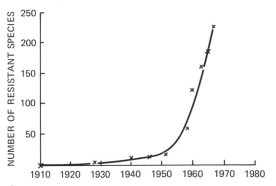

2 **Resistance to Insecticides.** The growth of resistance to insecticides. The number of resistant species refers to the number of different insects and mites that have resistance to one or more pesticides.

several mammal pests. The first case was of resistance in the Norway rat, *Rattus norvegicus*, to anticoagulant rodenticides such as warfarin. This was observed in 1958 in wild Norway rats near Glasgow in Scotland. In 1960 it was found in England and also in Jutland in Denmark; and in the last few years the first cases in the United States have been recorded. There are also now reports of warfarin resistance in the house mouse and the black rat *Rattus rattus*, potentially an even more serious situation.

Resistance, in fact, will develop in most organisms subjected to the pressure of chemical toxicants. Bacteria that cause human, animal, or plant disease readily evolve resistance to various antibiotics. Fungi are also capable of developing resistance to fungicides; for example, *Helminthosporium avenae*, a fungus that attacks oats, has developed resistance to organomercury compounds. Most recently there have been reports of weed resistance to herbicides.

Another aspect of the problem is the occurrence of cross-resistance. When a pest population shows resistance to DDT, for example, it may also show resistance to several related compounds. There are, in fact, four groups of insecticides within each of which resistance to any one member may also confer resistance against any other. Two of the groups are organochlorine insecticides: the DDT group and the cyclodiene derivative group, to which dieldrin belongs. The other two groups comprise the organophosphorus compounds and the carbamates. Resistance may also be present between groups but only in certain directions. Resistance to organophosphorus and

361

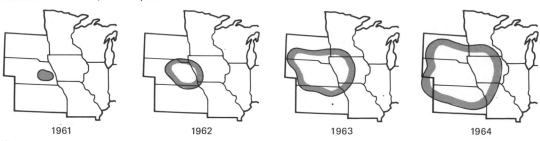

1961	1962	1963	1964

3   Spread of Insecticide-Resistant Population.
Annual spread of resistance to dieldrin-related compounds in the western corn rootworm in
South Dakota, Nebraska, Kansas, Minnesota, Iowa, and Missouri. *Source:* Reference 11.

carbamate insecticides may confer resistance to organochlorines, but the reverse has not been reported.

Taken together, these factors add up to a very serious situation. But the point that needs stressing most is that the resistance genes and the detoxifying mechanisms occur *naturally* in pest populations. They are present in the populations long before any exposure to pesticides. Pests have been, in this sense, preadapted to resist the pesticides that have been devised so far. It is possible, of course, that new and radically different compounds will be developed which are not vulnerable in this way. But the probability is that genes are already in existence which will confer resistance against the pesticides of the future.

INTERFERENCE WITH NATURAL ENEMIES

As we saw in Chapter 1, one way in which animal numbers can be regulated is through the mortality caused by natural enemies, such as parasites or predators. The third failing common to the majority of modern organic pesticides is their ability to interfere with this natural regulation.[13]

Provided that natural enemies are able to respond rapidly in their attack rate to changes in the numbers of their hosts or prey, they can bring about very tight control. In many cases, though, this is at a level that allows the pest to cause economic damage, and additional pest control measures have to be taken. If pesticides are applied and they kill off relatively more of the natural enemies than the pest or interfere with the regulation in some other way, the pest is liable to rebound following the application and may well reach a level much higher and more damaging than before. This phenomenon is termed *resurgence*.

In recent years resurgences have become increasingly common in a wide variety of pests. To quote one example, populations of cyclamen mite on strawberries can increase from 15 to 35 times following parathion spraying, because this insecticide is less lethal to the cyclamen mite than to another mite which preys upon it. In another instance, resurgences of bagworms on oil palms in Malaya occurred apparently because pesticide applications broke the synchrony between the life cycles of the pest and its natural enemy. Before the pesticide was applied, the pest had continuous, overlapping generations. At any one time, all stages of the pest were present on the crop and the natural enemy, which had a shorter life cycle, was able to keep up. But the pesticide killed all but one stage of the pest. Subsequent generations became distinct and the enemy species died out because its attack stage did not coincide with the susceptible stage of the pest. Figure 5 gives the dramatic results of an experiment designed to illustrate the role of pesticides in causing resurgences of bagworms.

Natural enemies also keep many potential pests more or less permanently at harmless levels. A pesticide may have little or no effect on the natural enemies of the pest against which it is applied but instead may kill off the enemies of hitherto harmless species which will then build up to pest status. New pests caused in this way are usually referred to as *upset pests*.

The best-known group of upset pests are the spider mites, which have become important pests in fruit orchards following intensive pesticide use.[15] In England a red spider mite arose on apples in the 1920s following the first use of tar oils as winter washes. These oils did not

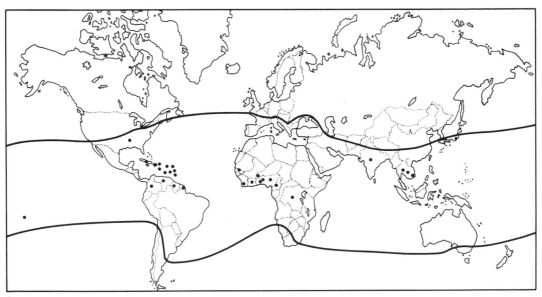

4 **Insecticide Resistance in a Mosquito.** Worldwide distribution of the mosquito *Aedes aegypti,* showing countries from which it is reported to be resistant to one or more organochlorine insecticides. *Source:* References 9 and 12.

affect the winter eggs of the mite, but they did kill many of its hibernating enemies, such as ladybird beetles and predacious mites, both directly and by destroying the mosses and lichens which gave them protection. The outbreaks of red spider mite which followed were, nevertheless, small compared with those triggered off by the use of DDT and BHC after World War II. At that time, over 40 species of beneficial insect were known to occur in apple orchards, and these insecticides were lethal to nearly all of them. In California, the citrus red mite, which was of little importance 25 years ago, now costs the citrus industry $10 million per year.

Sometimes the response to one upset pest only serves to create another. In this way a whole chain of pests is created, as has occurred on tea in Ceylon.[16] Dieldrin was used there against a shot-hole borer which attacks the twigs and branches of the tree. It controlled the borer but eradicated a parasite that had been introduced to regulate a leaf-feeding caterpillar. DDT was then used against the caterpillar, but it in turn produced outbreaks of various mites and other caterpillars. In time, a chain of events such as this may lead to a pest situation in a crop totally different from, although just as serious as, that at the beginning.

There are several reasons why many pesticides appear to be so lethal to natural enemies. The way in which predators concentrate pesticides has been described in Chapter 13; a predator can be killed by feeding on a number of pest individuals, each of which has received only sublethal doses of a pesticide. Parasites, on the other hand, are probably more vulnerable, because they are usually smaller and more delicate in structure than pest species. Both the predators and parasites experience greater risk, because they tend to be more mobile. They have to search out their respective prey or hosts, and in so doing are more liable to encounter insecticide deposits. Some insecticides act only after they have been ingested, but the majority can be picked up by insects through the "skin" or cuticle and so are particularly lethal to natural enemy species.

In this third component of pesticide failure we can once again see the link with the wider environmental problems that have occurred. A great many of the wild animals whose populations have been diminished in recent years are predators or other animals which range widely through diverse habitats and so pick up and concentrate pesticide deposits. The common cause of both pest control failure and environmental damage is that most pesticides are insufficiently specific in their toxicity.

363

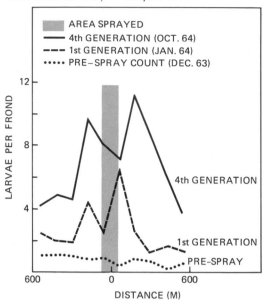

5 **Pest Resurgence Caused by Pesticide.** Pre- and postspray counts of bagworms around a plot of 2 acres sprayed with dieldrin. *Source:* Reference 14.

The Alternatives

Our analysis of pesticide failure provides us with three characteristics to look for in alternative methods: (1) they should be capable of more or less permanently regulating pests at harmless densities, (2) they should not be vulnerable to the development of resistance, and (3) they should work with and not against the natural controlling mechanisms provided by parasites and predators. If we can combine these features we will, moreover, have gone a considerable way toward minimizing the environmental problems that pest control has hitherto left in its wake.

First we shall discuss how far it is possible to satisfy these objectives simply by using pesticides in better ways. Then we shall turn to the several categories of nonpesticidal techniques which bring about control through regulatory mechanisms. Finally, we shall look at a number of new nonpesticidal techniques which hold out the promise of pest eradication.

BETTER USE OF PESTICIDES

Pesticides can often be a useful tool, producing little environmental hazard, provided that the right compounds are chosen and are used efficiently and with care. A basic prerequisite for better pesticide use is accurate *pest monitoring*. By keeping a regular account of the numbers and composition of pest populations, pesticide applications can be planned more rationally. More efficient control can be obtained but with fewer, well-timed applications. On cotton, for example, the young bollworm caterpillar can only be killed in the brief interval after hatching from the egg and before it enters the cotton boll. Careful monitoring can detect when the main egg hatch is to occur and so find the best time to spray (Figure 6).

In many cases the numbers of a pest may be largely determined by the climate, so that the pest assumes an importance only in some years. Where pesticides are being applied according to a fixed schedule, the result is that many unnecessary applications will be made. But with a well-designed forecasting system, sufficient warning of pest buildup can be given and applications made only as required. Forecasting systems are already in operation for many pests in the United States and elsewhere. In the Corn Belt states there are schemes which warn of buildup of the European corn borer, for example. Also, individual farmers can sometimes contract on a private basis with independent agencies to obtain forecasts and assessments for their farms.

The extent to which pesticide applications can be reduced through pest monitoring is very considerable. A recent report on cotton growing in Columbia states that, whereas a few years ago the crop there was sprayed 18 to 20 times per season, now, through careful observation of pest incidence, only 9 or 10 applications are needed. And in the United States it is estimated that pesticide application could be reduced by up to 50 percent if this approach were fully adopted. Such a reduction would be the surest way of postponing the development of resistance, and it would also be a major step in minimizing environmental pollution.

How and where the pesticide is applied is important also. Often crop plants are literally swamped with insecticide in an attempt to kill pests that feed in inaccessible places, for example in the buds or leaf axils. But use of specially designed application machinery and ultra-low-volume spraying can give better kills and far less environmental contamination.

The timing and placing of pesticides can be specifically designed to give control without upsetting natural enemy regulation. Some parasites and predators live in very close association with their hosts or prey. But, in many cases, pest species and their natural enemies only meet at certain times or in well-defined localities. In Malaya the grubs of cockchafer beetles, which feed on the roots of rubber trees, are parasitized by scoliid wasps and tachinid flies. However, it was found that the parasites complete their life cycle and emerge from the soil before the cockchafer grubs pupate. A soil insecticide could thus be safely applied in the period between the emergence of the parasites and the emergence of the cockchafer beetles.

One method which ensures that only the pest species will be affected is to combine the pesticide with a specific bait or sex-attractant chemical. Fruit flies have been very successfully controlled in this way and baits are particularly suited to control of mammal or bird pests. Provided that the bait is attractive only to the pest species or the bait container allows only the pest to gain access, the number of animals affected can be greatly minimized.

SELECTIVE PESTICIDES

Ideally we should use selective pesticides only, that is, compounds which kill the target pests but are harmless to natural enemies or to other animals that may become exposed to them. By their nature, contact-acting pesticides rarely have this ability; they kill most organisms they contact. But varying degrees of selectivity are to be found among the stomach-poison and systemic pesticides. The stomach poisons, as their name suggests, act only after ingestion. When applied to crop plants they only affect those insects which actually eat the plant material. The systemics act in a similar way. They are taken up by plants and translocated through the sap and in consequence tend to affect only sap-sucking insects, such as aphids or leafhoppers. In general, stomach poisons and systemics can be relied on to spare parasites, but predators can still be affected through the prey they feed on.

When the many problems associated with

6 **Monitoring Cotton Pests in Africa.** Farmer counting bollworm eggs on cotton in Malawi, with aid of pegboard. Photo by G. A. Matthews.

broad-spectrum contact-acting pesticides were first realized, much hope was placed on their being replaced by selective compounds. However, the number developed so far is very small and few are really truly selective. Part of the problem is due to the continued lack of detailed knowledge of the mode of action of pesticides. It is still not possible to produce a "blueprint" for a selective pesticide. Also, the sheer cost of discovering, testing, registering, and developing them (about an average $11 million per compound) discourages production of highly selective forms. Broad-spectrum compounds with a variety of applications can be sold over a wide market and bring a better economic return on the high development cost.

REGULATORY TECHNIQUES

Even if enough selective pesticides could be developed, they would still have the drawback of only producing a simple density-independent mortality. As we saw in Figure 1, when a pesticide is applied to a hypothetical insect population, the numbers drop and then recover to their previous level as the effect of the pesticide wears off. If we now look at Figure 7, we see that something has been done to the model population that permanently changes the equilibrium to a new and lower level. In this case the change has been brought

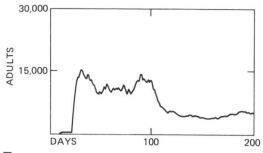

7 **Example of Cultural Control.** A hypothetical insect population similar to that in Figure 1a except that on day 100, partial destruction of the insects' breeding sites is simulated. This brings the population down to a new equilibrium level. *Source:* Reference 8.

about by simulating the partial destruction of the breeding sites of the population. The "carrying capacity" of the pests' environment is reduced and the population size becomes permanently limited through competition for space, food, and other resources.

Environmental manipulation which brings about population regulation in this way is referred to as *cultural control*. Also under this heading come those measures which allow existing regulatory agents such as natural enemies to act in a more efficient manner. A second approach, biological control, brings about the same effect by introducing new natural enemies into an environment. Finally, populations of pests can be regulated, at least with respect to the crops themselves, by developing pest-resistant plants or by creating physical or chemical barriers which limit the number of attacking organisms.

CULTURAL CONTROL

Mosquito control furnishes perhaps the best-known example of the value of destruction of breeding sites. Many fly species, including mosquitoes, biting midges, houseflies, and deerflies, have larval stages that live in or near to water. The drainage and filling of ponds and marshes and the removal of shoreline vegetation in lakes and rivers can greatly reduce populations of pests such as these. Over much of Europe and North America malaria had disappeared long before the appearance of modern pesticides, because industrialization and urban growth had brought about the destruction of mosquito breeding sites.

With some pests, plants other than the crop may serve as breeding sites or as hosts for alternate generations, and removing these can bring about control. For example, the sorghum midge overwinters on johnson grass which grows along the borders of the fields, and the midge is able to build up for two or three generations on this plant before moving onto the sorghum blooms.

Pests can also be limited environmentally by crop rotation. The principle involved is a simple one. By alternating susceptible with non-susceptible crops the growth of pest populations is periodically checked. It is a method particularly suited to plant diseases, and in the United States approximately 24 diseases on 17 different important crops can be controlled in this way. But it is also effective against a number of insect pests and against nematodes. In the southeastern United States rotation has virtually eliminated the wheat gall nematode as a pest. Much the same kind of effect can be produced by destroying the residue of a crop after harvesting.

BIOLOGICAL CONTROL

Several techniques come under the general heading of *biological control*. These include: (1) augmentation of "natural enemies" already present, either by cultural methods or by breeding and releasing more of them; (2) introduction of an enemy species from a foreign country, usually to control a pest which itself had come from that country; and (3) the use of microorganisms — viruses, bacteria, or fungi.

Cultural measures such as rotation and destruction of residues aim to limit population size by providing a break in continuity. However, when cultural measures are used to try to enhance natural enemy regulation, they are designed to work in exactly the opposite way. The principle is again a simple one. By producing continuity both for the pests and for the parasites and predators that are their "natural enemies", the efficiency and permanence of regulation can be increased. When crops are grown in rotation, each crop in turn goes through a similar sequence of plant growth, invasion, and buildup of pests, which is in turn followed by an increase in natural enemies. Each phase lags behind the other, and frequently it is only after much of the damage by the pest species has been done that

8 Strip Harvesting of Alfalfa. Field in California showing alternate strips of half-grown and newly cut alfalfa. Natural enemies of the alfalfa pests are maintained on the half-grown plants but would be destroyed if the whole field were cut. Photo by A. Stern.

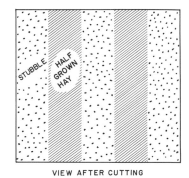

SCHEMATIC FIELD VIEW BEFORE CUTTING

VIEW AFTER CUTTING

← MATURE

← ½ GROWN

PROFILE BEFORE CUTTING

PROFILE AFTER CUTTING

the natural enemies build up sufficiently to exercise control. However, in perennial crops and when the same annual crops are grown in succession, the buildup phase tends only to occur once, when the crop is first planted; thereafter, regulation may remain more or less uninterrupted. For example, in South Africa a number of pests of cabbages and related crops can be kept under control by growing several varieties of these crops in close succession. In California the principle is recognized in the practice of harvesting alfalfa in alternate strips (Figure 8). This approach tends to be most useful where climate is equable, because the continuity is unlikely to be upset by changes in temperature or rainfall.

In some cases, though, continuous cropping may not be sufficient to provide all the continuity that is required. Some natural enemies require plants other than the crop to complete their life cycle. For example, it was recently found that a leafhopper on vines in California could be controlled effectively by providing blackberry plants around the vine fields. This was because the parasite of the leafhopper could only survive the winter months by parasitizing another noneconomic leafhopper, which lives on blackberry.[5]

Hedgerows, and uncultivated land in general, have an important influence on the presence of natural enemies in crops. As a rule, the diversity of animal species increases in the vicinity of hedgerows (Figure 9), and this implies that there are many more natural enemy species present. It has been found in work on cabbage aphids that there are usually fewer aphids near hedgerows because of the higher incidence of predation by ladybug beetles and hoverfly larvae. In the latter case this is because the adult hoverflies feed on the pollen in the wild hedgerow flowers.

Natural enemies may also be favored by changes in shade or in cover crops or by different spacing and pruning practices. The possibilities of environmental manipulation are endless. The advantages of the method lie in

367

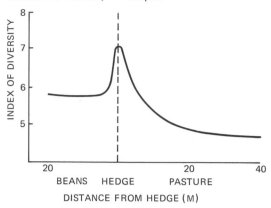

INDEX OF DIVERSITY

DISTANCE FROM HEDGE (M)

BEANS   HEDGE   PASTURE

9   Fauna Near a Hedgerow. Changes in diversity of the fauna in the air above a bean field, a pasture, and the intervening hedge. The hedge is approximately 2 meters high. *Source:* Reference 17; used by permission of the British Ecological Society.

the permanence of control that can be attained and the low cost usually involved. However, there may be serious disadvantages. Changes in cropping practice may bring about good pest control but may reduce yield or farming efficiency in other ways. Hedges may be costly to keep up and may interfere with the efficient use of farm machinery. Furthermore, environmental changes favorable for pest control may be deleterious to wild animals and plants. For example, draining ponds and marshes as part of mosquito control has severely reduced the populations of many fish and birds.

Introducing *natural enemies* (parasites and predators) to control pests is a practice of great antiquity. The cat was domesticated by man to control rats and mice in his granaries and home, and in many parts of the world its primary role is still that of a predator rather than a pet. The modern interest in biological control, however, stems from two highly successful introductions against, respectively, an insect pest at the end of the last century and a weed in the 1920s. Both these stories are by now classics of biological control. The full details make colorful reading, but the bare outline is as follows.

In 1887 the citrus industry in California had become very seriously threatened by infestations of the cottony-cushion scale insect *Icerya purchasi*, which had been introduced accidentally a few years earlier. A prominent government entomologist, C. V. Riley, thought that

the scale had probably been introduced from Australia or New Zealand and suggested that a search should be made there for the natural enemies of the scale, which could then be imported to California. The next year, after some opposition, a search was made and a number of parasites and also some predators were found, including a ladybug beetle, *Rodolia cardinalis*. It was first thought that the parasites were likely to be the most valuable, but when the various species were tried out in the field, the ladybug beetle showed the greatest potential. It was a voracious feeder on the scale, and farmers who heard of its powers were soon eager to obtain colonies for their own orchards. Within a year Los Angeles County had tripled the citrus fruit harvest. Soon the scale was reduced to insignificant numbers throughout the state and it has remained so for the last 80 years, except for a brief period in the 1950s when in several counties DDT and malathion sprays eradicated the ladybugs.[18]

The second story also involves the Americas and Australasia, but the roles are reversed. At the end of the last century, several species of a cactus plant, the prickly pear, *Opuntia*, were introduced as garden plants into Australia from their home in Mexico. But they spread from the gardens and became very serious weeds of pasture land. By the 1920s, 60 million acres had become infested, and over half of this had so dense a cover of the cactus that it was impenetrable to man or large animals. A search was then mounted for suitable plant-eating insects in the Americas. Some 50 species were found and several of these showed good potential. But in 1925 the ideal insect was found in Argentina. It was *Cactoblastis cactorum*, the larvae of which tunnel in the cactus pads and bring about complete destruction of the plants. A single importation of nearly 3,000 eggs was made. From these a breeding stock was established and in the next 3 years eggs were released in the field. Massive destruction of the prickly pear occurred and by the early 1930s large areas were under control. Today it only occurs as individual plants or in small patches regulated at this very low level by *Cactoblastis*.[19]

In the last 40 years there have been many other examples of classical biological control (Figure 10). A recent estimate puts the number

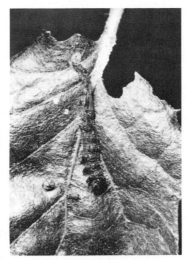

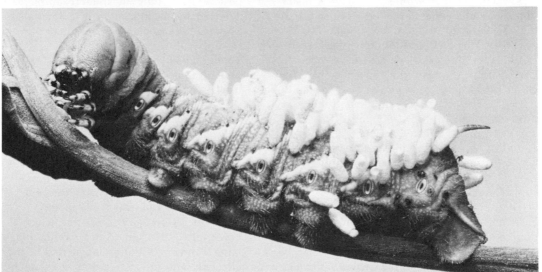

10   **Biological Control Agents.** Cocoons of the moth *Cactoblastis cactorum* (Berg) on pad of tree cactus (left); larva of California oak moth *Phryganidia californica* (Packard) killed by virus disease (center); the vedalia beetle *Rodolia cardinalis* (Mulsant) feeding on cottony-cushion scale (right); cocoons of the parasitic insect *Apanteles congregatus* (Say) on a hornworm (bottom). Top left photo by J. K. Holloway, others by F. E. Skinner.

at over 100, spread over 60 countries. The principal features of most of the successes are well illustrated by the examples already given.

1. Classical biological control has been most readily achieved when the pest species itself is exotic, that is, has been introduced from another region of the world. Endemic pests usually have a full complement of natural enemies or diseases and if these are not already producing a sufficient degree of control, it is not easy to find and introduce new agents of control that will improve the situation. On the other hand, introduced pests will have brought few of their natural enemies or diseases with them. The natural enemy or disease niche is wide open and there is a good chance that it can be filled by the original enemy species or disease agents of the pest, if they can be located.

2. If the introduced parasites, predators, or microorganisms do establish themselves,

369

they can frequently bring about a very high degree of regulation, which will persist almost indefinitely if there are no major environmental changes or there is no interference from pesticides.

3. Biological control can be achieved very inexpensively. The cost of introducing the coccinellid against cottony-cushion scale in California was only $2,000, which was a small price to pay, even in 1889.

Insect pests are not only killed by other insects, they also die from disease caused by viruses, bacteria, fungi, and many other microorganisms.[20] Disease epidemics in insect populations have often been observed, and sometimes a severe pest outbreak will collapse, without man's intervention, as a result of such an epidemic. However, it is also possible deliberately to use microorganisms as agents of biological control. A very successful example of this was the use of a virus to control the European sawfly, which had become established as a pest of pine forests in Canada. The virus was imported from Sweden, cultured in the laboratory, and then sprayed on the infested forests from the air. The virus spreads from infected to noninfected larvae by contact and will survive through the winter, so providing a persistent degree of control.

Alternatively, microorganisms can be used as living pesticides, their value lying in the direct kill they produce rather than the spread of disease. A microorganism that is commonly employed in this way is the bacterium *Bacillus thuringiensis*. It is particularly effective against the larvae of butterflies and moths but does not harm their natural enemies.

PLANT AND ANIMAL RESISTANCE

When we considered the drawbacks to pesticide use we saw how pests can evolve resistance to pesticides. In a similar way, plants and animals can evolve resistance to the attack of pests, and this can be utilized in pest control. In nature the resistance of an organism is essentially its ability to avoid being wiped out by organisms that feed on it. This resistance evolves through the process of natural selection and tends to result in a fairly stable balance between plant and herbivore and between herbivore and parasite or predator. But

in a crop plant (or domestic animal) the resistance may be insufficient or absent. The crop plant may be able to survive the attacks of pests but at the cost of a loss in yield. Alternatively, the resistant factors of the natural plant may have been lost during the process of breeding a cultivable variety. Finally, it is possible for the plant to be resistant to the organisms that attack it in its natural environment, but in the crop environment to be exposed to quite different organisms, to which it has no resistance. There are three principal mechanisms of resistance:

1. Resistance may be due to a form of deterrence. The plant or animal may have a physical or chemical characteristic that prevents pest attack. Biting flies, for example, are deterred by the thickness of animal skins. The structure of plant surfaces may prevent the entry of fungal spores.
2. There may be a positive antibiosis. The host may be able to kill the pest or at least affect its reproductive capacity. Cotton, for example, produces the compound gossypol, which retards growth in the bollworm *Heliothis zea*.
3. The resistance may be simply that the host is able to contain the activities of the pest so that no vital damage is caused. In Borneo it was found that a single boring caterpillar of the moth *Endoclita* was liable to kill a cocoa tree. Yet a secondary forest tree which the borer also attacked could accommodate large numbers of boring larvae and be little affected. The forest tree was resistant but the cocoa was not.

One of the earliest ways of improving the resistance of a crop was by grafting the susceptible plant onto a related resistant variety or species. This works well providing the pest only attacks a certain part of the plant. Species of the grape *Vitis* occur both in Europe and America. Over a hundred years ago an aphid, *Phylloxera*, was imported to Europe on American vines. The aphid feeds only on the leaves of American vines and does little harm to them, but on the European vines it also attacks the roots, causing galls. Attacked vines usually died, and soon large areas of the French wine-growing regions were affected. In 13 years, over 1.5 million hectares of vines were destroyed and wine production was cut by two-thirds. Much effort was wasted on insecticidal and other forms of control, but then the simple suggestion of grafting the French vines onto American roots was made. This immediately solved the problem. The practice was widely adopted and continues to be used to the present day.[21]

However, improved resistance has been mostly brought about by crossing and selection. In general, the degree of resistance in a crop will vary from plant to plant. By careful screening it is possible to identify those individuals in which the resistance is exceptionally high. These can then be selected and used as parents to produce more resistant lines. Alternatively, the resistance may have to come from outside, and it then becomes necessary to screen related varieties and species. In some cases the process involves widespread searching throughout the countries where the crop and its relatives occur. The resistant species and the crop plant are then crossed to produce a hybrid, which will combine the characters of both parents. The hybrid is then repeatedly back-crossed with the crop plant parent until finally a plant is obtained which is identical with this parent except that it retains the resistant factors of the other.

Breeding for resistance has been perhaps the most successful of the nonpesticidal approaches to pest control. Today most crop varieties have a built-in resistance to one or more insect pests or diseases. Resistance makes a particularly high contribution to disease control. For example, over 95 percent of the acreage of small grains and a similar proportion of the alfalfa acreage in the United States is now planted to disease-resistant varieties.

The advantage of the resistance approach is that it can bring about more-or-less permanent control with little recurring expense. Most of the cost is incurred during the breeding program. In some instances this can be fairly short; for example, new varieties of alfalfa resistant to spotted aphid were developed in 3 to 5 years. Nevertheless, it usually takes up to 15 or 20 years to perfect a new resistant variety, and, where disease is involved, the control may *not* be very long-lived. Fungi and bacterial pathogens all have short generation times and there is thus a great likelihood of their evolving characteristics that will overcome the resistance. For instance, new races of rust disease are thrown up every 3 or 4 years in the United States, and each time this happens there has to be a new countereffort in breeding wheat. In such situations pest control involves a neck-and-neck race between the pathogen and the crop breeder.

THE POSSIBILITY OF ERADICATION

The alternatives considered so far have been aimed at regulating pest numbers. Often in the past, pesticides have been applied in the hope that pest populations would be totally wiped out, but this has rarely occurred. However, in recent years a number of techniques have been developed which hold the promise of eradicating pests on a much wider scale.

These techniques come under the general heading of genetic control, because in one way or another they involve manipulating or changing the hereditary characteristics of the pests themselves. There are two kinds of technique: (1) the pest population is eradicated by increasing the proportion of matings that are sterile, and (2) eradication by introducing a lethal hereditary factor into the population.

The most successful form of the *sterile mating technique* involves breeding large numbers of the pest species in the laboratory. These are then sterilized by irradiation and released in the field in sufficient numbers to swamp the natural population. Provided that the sterilized and normal individuals mate readily with one another, a high proportion of the ensuing matings will produce no offspring, and this will result in much smaller populations in the next generation. If the releases are continued in each subsequent generation, a higher and higher proportion of matings will be sterile until finally the population becomes extinct. The principle can be clearly shown in terms of simple arithmetic, as in Table III.

The sterile mating method first received worldwide attention 10 years ago, when it was used to wipe out the screwworm in the southeastern United States.[23] The screwworm, *Cochliomya hominivorae*, is a serious pest of cattle. The adult fly lays its eggs in the skin of the animals and the larvae that hatch out feed on the flesh, causing large wounds. Before 1958 the screwworm produced considerable losses in the livestock industry in Florida and in the southwestern states along the Mexican border. It was known, however, that the adult female fly mated only once, and this gave E. F. Knipling the idea that the pest could be eradicated by releasing large numbers of sterile male flies. A massive operation to breed sterile males was begun in 1957, and from January 1958 over 50 million such flies were produced for release each week in Florida and 100 to 150 million in the Southwest. The operation was spectacularly successful, and within 1 year the screwworm was eradicated from Florida and a continuing release program in the Southwest keeps the populations at low levels.

There have been one or two other successes since the screwworm program. In 1962 on the island of Rota in the Pacific, the melon fly was eradicated by releases of sterile flies of both sexes. The melon fly, unlike the screwworm, can mate several times, so this success demonstrated that monogamy was not an essential prerequisite for the technique. However, as in the case of the screwworm, the pest population was small and isolated. Prior to the releases the melon flies were, in fact, deliberately reduced by insecticide applications.

It seems unlikely that the release of sterile insects will succeed against large populations which are continuous over wide areas. The problem lies in the logistics of rearing and releasing the immense numbers of insects that would be required. An alternative approach that has been suggested is to directly sterilize populations in the field by means of chemosterilants mixed with bait. A great deal of attention is being paid to finding suitable compounds, but so far the method has not been tried successfully on a large scale. One prob-

lem, of course, lies in devising very specific baits or in finding sterilizing compounds that will be harmless to other animals.

The *lethal-factor technique* is even more experimental. Quite a number of lethal mutant genes have been identified in pest species. In mosquitoes of the *Aedes* genus, for example, two mutants appear to have considerable potential. The first is called bronze. Females with this mutant lay eggs with poor shells which die within a few hours of oviposition. The second mutant, proboscipedia, produces female mosquitoes with tarsi or leg joints in the place of a proboscis. The females cannot feed and hence do not produce eggs. By introducing sufficient numbers of individuals with these mutants into natural populations it is theoretically possible to cause eradication.[24]

A particularly intriguing possibility for genetic control was suggested by the discovery that mutants can occur on the sex chromosomes of insects which result in a biased sex ratio.[25] Sex is usually determined by the X and Y chromosomes, which are the sex chromosomes. Commonly if a Y chromosome is present, the insect is male; and if it is absent, the insect is female. It has been found, though, that a mutant can occur on the Y chromosome which enables the Y-bearing sperm to fertilize more readily than the X-bearing sperm. Males carrying the Y mutant thus produce nothing but male offspring. It is therefore theoretically possible that if a few individuals with the mutant are introduced into a natural population, the mutant will spread, the sex ratio will become more and more biased, and eventually the population will become extinct.

The potential power of genetic control lies in

III   Effect of Sterile Mating

Generation	No. fertile insects	No. sterile insects	Ratio of sterile to fertile insects	No. insects reproducing
Parent	1,000,000	9,000,000	9:1	100,000
F₁	500,000	9,000,000	18:1	26,316
F₂	131,560	9,000,000	68:1	1,907
F₃	9,535	9,000,000	944:1	10
F₄	50	9,000,000	180,000:1	0

A simple arithmetic model showing how a population is eradicated by repeated releases of sterile male insects. It is assumed that there is a fivefold rate of increase in the normal population.
Source: Reference 22.

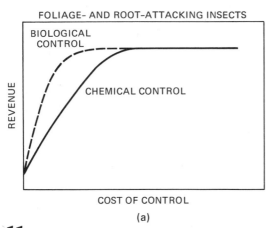

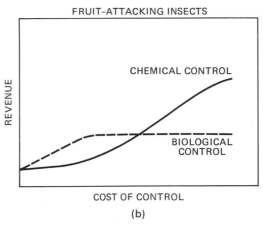

11 **Economics of Pest Control.** Relative economic advantages of biological and chemical control on fruit trees where target pests attack (a) foliage and roots or (b) fruit. *Source:* Reference 26.

the fact that these techniques are autocidal; that is, they cause the insect pests to turn on themselves. The high genetic variability shown by most pests enables them to evolve resistance to most pest control techniques. But in genetic control it is this variability itself that is used. Coupled with the speed with which eradications can occur, it ensures that there will be little chance of resistance occurring. Also, it follows that the techniques are highly specific, and as a consequence there is less likelihood of environmental hazards. The major problem, though, is that there are very few situations where the pest population is sufficiently small and isolated for the technique to work.

The Systems Approach to Pest Control

So far we have discussed a number of major alternatives to using broad-spectrum, contact-acting pesticides in pest control. Many of these methods have been shown in the field to be really viable alternatives, producing effective control, often at less cost, and with less environmental contamination and upset. But the position is not a simple one of contrast between good and bad methods. Some of the alternatives can bring about serious and deleterious environmental effects. For example, the older but more selective compounds, such as the arsenicals, can be as toxic to wildlife as the modern organic compounds. And we have

seen that the destruction of habitats which occurs when breeding sites or the alternative hosts of pests are removed can produce even more far-reaching effects than pesticide use. Under certain conditions biological control, too, can get out of hand. For example, herbivores brought in to control weeds may turn to crop plants, and it is conceivable that microorganisms used in biological control could mutate and affect man.

Furthermore, these alternatives vary in their effectiveness from pest to pest and from one situation to another. Often a method that works against a species on one crop will not work against the same species on a different crop. Similarly, a method of control (e.g., crop rotation) that is effective against one member of the pest complex of a crop may well reduce control of another.

Economic factors are important, also. Figure 11 illustrates the point that the economic advantages of biological versus chemical control depend on the nature of the damage that the pests are causing. If we are concerned with pests attacking, say, the roots or foliage of fruit trees, biological control will give a return that is bigger than or equal to chemical control for a similar level of expenditure. But where the final product (e.g., the fruit) is being attacked, there will soon come a level of expenditure at which chemical control has the advantage. Biological control implies low persistent pest populations. These may be tolerated on roots and foliage, but any damage to fruit is a loss of potential income.

373

INTEGRATED CONTROL

What has been needed for some time is an approach that looks at each pest situation *as a whole* and then devises a program that integrates the various control methods in the light of all the factors present. The origins of such an approach lie in early attempts to apply chemicals selectively so that they would not interfere with the effectiveness of natural or imported enemies. As early as the late 1940s, Pickett and his colleagues in Nova Scotia developed an integrated program for control of apple and pear pests. It was very successful and at one time was in use in over 80 percent of the orchards in Nova Scotia. The term "integrated control" was, however, invented by a group of Californian entomologists to describe the program they had developed to deal with the spotted alfalfa aphid. This had been introduced accidentally to California in 1954, where it had soon increased to the extent that it threatened the whole crop. Initially, broad-spectrum organophosphorus insecticides were used, but these interfered with the buildup of a number of native predators and of parasites that were specially imported. The problem was solved very successfully (costs and losses attributable to the aphid dropped to one-sixth) by replacing the pesticides then being used by different and more selective compounds, which allowed the natural control to be retained.

At about the same time, in Peru, an integrated approach much broader in concept was being developed in an attempt to solve the problems of the Canete Valley discussed at the beginning of this chapter. The cotton crop, it will be remembered, had yielded less than 350 kilograms per hectare in the 1955–1956 season. A new plan was evolved that consisted of the following key measures:

1. Cotton was no longer to be grown on the marginal land in the valley.
2. There was to be no more ratooning, that is, allowing the crop to persist for 2 or 3 years.
3. The soil was to be "dry-cultivated" to kill the bollworm pupae.
4. Beneficial insects were to be reintroduced from other valleys.
5. Dates were to be fixed for planting and irrigation and for destroying crop residues.
6. Synthetic organic pesticides were to be prohibited except by dispensation of a special commission. Reliance would be placed instead on arsenical and botanical insecticides. If organic compounds were used, they should be of the selective kind.

The plan was approved by the Peruvian Ministry of Agriculture and put into action. The result was dramatic. The upset pests returned to their former unimportant role and the major pests diminished. In the years 1957 through 1963 average yields were 790 kilograms per hectare.

In the last 10 years many similarly successful integrated-control programs have been developed in widely different crop situations throughout the world (examples are reviewed in the book edited by Huffaker, listed in the Further Reading). The integrated approach has itself evolved, particularly through the recognition of two important concepts: key pests and economic threshold.

A *key pest* in a crop is one whose control is critical to the control of the other pests in the crop. Often we find that an entire complex of pest problems can be attributed to adopting the wrong control measures against just one or two pests. An example of this situation was described earlier for tea in Ceylon. Correct identification of the key pest and an appropriate change in control strategy can bring about great improvements for little cost. In California, for example, it has been found that spraying *Lygus* bugs which appear on cotton early in the season is the cause of the later pest problems. By avoiding spraying *Lygus*, all later problems are minor.

The *Lygus* bug problem also illustrates the use of the concept of an *economic threshold*. Clearly, one way of reducing the spraying of this key pest is to replace spray schedules based on the calendar with a system of population monitoring and to then spray only when the population exceeds a level at which profitable return from spraying can be expected. This population level is known as the economic threshold. It simply measures the point at which the potential loss of yield is worth more than the cost of successful control. In practice, of course, this may be difficult to measure accurately. Crops such as cotton, for example, are able to compensate for a certain amount of damage by faster growth rates, and this must be taken into account. But, in general, this is a useful and practical concept.

AGRICULTURAL SYSTEMS

Today, integrated pest control is itself being seen in a wider context, as a component of the integrated management of whole agricultural systems. In California, attempts are being made to manage the pest problems of alfalfa and cotton as a single problem in regions where the crops are grown together. There is a parallel development in the control of insect-borne disease, such as malaria, through the concerted use of insecticides, chemotherapy, environmental sanitation, and immunization.

Perhaps the best illustration of this trend comes from the recent history of the Green Revolution in southeastern Asia.[27] In Chapter 3, Brown and Eckholm described the development of the new rice varieties at the International Rice Research Institute in The Philippines. The early varieties, bred for their high yield response to heavy fertilizer treatment, originated in narrow genetic stock which possessed only a small degree of resistance to pests and diseases. They were soon grown on an extensive scale, replacing numerous local varieties selected over centuries for their suitability to local environments. As had been predicted, they began to suffer from pest and disease outbreaks. The most serious recent problem has been caused by increasing populations of the rice leafhopper, an insect that carries a virus called Tungro. In 1971 a major outbreak of the disease in The Philippines was a principal cause of a large drop in rice production. When the new rice pest problems first appeared, chemicals were widely used. But insecticide resistance developed and upset pests appeared. Varieties with broader resistance to pests were developed in response and used as a basis for integrated control.

However, pest and disease outbreaks are only one of a set of interrelated problems that have appeared in the wake of the Green Revolution. These problems seem to have a common cause in the reliance on extensive single-variety monoculture. Recently, agronomists have begun to develop alternative agricultural systems based on a mixture of crops, which attempt through their diversity to recreate some of the stability of natural climax ecosystems (see Chapter 1). Cropping systems of this kind are, of course, not new; they have been the basis of peasant agriculture for centuries but have been neglected by modern agricultural science. The new research involves ways of improving the productivity of these systems by new cultivation techniques and by the judicious use of the new grain varieties.

There are innumerable possible cropping combinations from which to choose. In one system developed in The Philippines a sequence of five crops in a year—rice, sweet potato, soya, sweet corn, soya — are grown in overlapping rotation. The crops are planted in alternate rows, a new crop being planted between the rows of the old before it is harvested. First results show that intercropping of this kind does indeed reduce pest problems. Table IV shows the reduction in corn borer infestation where corn and peanuts are interplanted. But research is showing that these systems, when well designed, will also lessen the need for weed control, reduce fertilizer requirements, promote good soil structure, and provide insurance against bad weather conditions. Even in the worst years the farmer has a harvest to look forward to.

SYSTEMS ANALYSIS

The problem of pest control can now be seen to be very much more than a simple question of killing bugs. Pest control, like any other sphere of management, is a process of decision making in the face of complexity and uncertainty. To do his job well the pest control worker needs adequate analytical tools. At any time he must have

1. An understanding of the current state of the system in terms of levels of the pest population and the damage or illness it is causing.
2. An expectation of the potential state of the system at some time in the future if no control action is taken.
3. A knowledge of the modes of action and costs (both social and economic) of a wide range of possible control measures.

IV **Pest Control Through Intercropping**

Crop	Plants infested (%)	Borers per plant
Corn alone	86	2.6
Corn and peanuts	30	0.4

Effect of growing corn alone or interplanted with peanuts on the incidence of attack by corn borer.
Source: Reference 28.

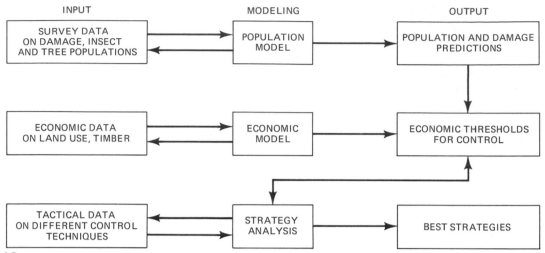

12 **Pest Control System.** Scheme for analysis of control of the pine bark beetle in the United States. *Source:* Reference 29.

4. A prediction of the state of the system that would arise following each or any combination of these control measures.

Armed with this knowledge he can decide whether or not to act and, if he acts, what measures to take.

Inevitably, if this procedure is followed, the pest control worker must simplify the complexity of the situation he faces. He builds a model of the real world. Traditionally, he has used mental models, abstracting and ordering .information gained from experience and from experiments. Often, though, faced with a crisis, he has acted without sufficient thought or analysis, turning to the simplest, most readily available control measure — the broad-spectrum pesticide.

The problem is the sheer complexity of the systems that he deals with, a problem that becomes even more difficult as the integrated-control approach is in turn integrated into wider agroecosystem management. However, some 30 years ago, operations research and systems analysis were developed specifically to enable complex systems to be analyzed and managed. They rely on mathematical models as a more efficient means of ordering facts and testing the consequences of different hypotheses.

Figures 1 and 7 are the outcomes of very simple mathematical models of pest populations programmed for a computer.

Currently, attempts are being made to develop more realistic and sophisticated models.

Unfortunately, however, our knowledge is often meager or the information has been gathered in such a way that it is difficult to put it together to produce a meaningful and useful model. The modeling component has to be present at the outset of the research. A large program has recently been initiated under the auspices of the U.S. contribution to the International Biological Program and the National Science Foundation which aims to use systems analysis as a framework for investigation of the pest problems of five major crop systems. One of these projects concerns the bark beetle that attacks U.S. pine forests. Figure 12 is a simplified version of the type of model that is being built.

FUTURE PROSPECTS

In the early years of pesticide use it was widely thought that the great majority of pest problems would soon disappear. But there is little sign of this today. Many problems have resulted from bad pest control and many more are being created by the pressure for greater agricultural production. Some of the biggest problems are yet to come and will pose a tremendous challenge to future generations of pest control workers. Pest control still has a long way to go before it is an applied science as sophisticated as, say, civil engineering. As we have seen, it is still predominantly empirical and narrowly reliant on the use of unselective pesticides. Nevertheless, developments of recent years suggest that it is quite firmly moving away from this. The mainstream is in integrated control and system management, and as the new techniques of analysis become developed, its position will become more and more assured. In all of this, concern with wider environmental effects must play an increasingly important role. As we have seen, the environmental problems are commonly linked with the inability to get good pest control. As one improves, so, we hope, will the other.

References

1. Rudd, R. L. 1966. *Pesticides and the Living Landscape*. University of Wisconsin Press, Madison, Wis.

2. Watson, J., et al. 1965. *Cotton Insects*. A report of a panel of the President's Science Advisory Committee. The White House, Washington, D.C.

3. Smith, R. F., and van den Bosch, R. 1967. Integrated control. In *Pest Control — Biological, Physical, and Selected Chemical Methods* (W. W. Kilgore and R. C. Doutt, eds.). Academic Press, Inc., New York. pp. 295–340.

4. Smith, R. F. 1969. The new and the old in pest control. *Proc. Accad. Nazion. dei Lincei* 366(128): 21–30.

5. Doutt, R. L., and Smith, R. F. 1971. The pesticide syndrome — diagnosis and suggested prophylaxis. In *Biological Control* (C. B. Huffaker, ed.). Plenum Publishing Corporation, New York. pp. 3–15.

6. Wood, B. J. 1973. Integrated control: critical assessment of case histories in developing economies. In *Insects: Studies in Population Management* (Mem. 1) (P. W. Geier, L. R. Clark, D. J. Anderson, and H. A. Nix, eds.). pp. 196–220. Ecological Society of Australia, Canberra.

7. Smith, R. F. 1972. The impact of the green revolution on plant protection in tropical and sub-tropical areas. *Bull. Entomol. Soc. Amer. 18*: 7–14.

8. Conway, G. R. 1970. Computer simulation as an aid to developing strategies for anopheline control. *Misc. Publ. Entomol. Soc. Amer. 7*: 181–193.

9. Brown, A. W. A. 1971. Pest resistance to pesticides. In *Pesticides in the Environment* (R. White-Stevens, ed.) Marcel Dekker, New York. Vol. 1, Pt. 2.

10. Georghiou, G. P. 1972. The evolution of resistance to pesticides. *Ann. Rev. Ecol. Syst. 3*: 163–168.

11. Geigy Chemical Corp. 1964. *Geigy Agricultural Chemicals* (Circ. F.A.C. 100-203). Geigy, Ardsley, N.Y.

12. Christophers, S. R. 1960. *Aedes aegypti* (L.), *The Yellow Fever Mosquito: Its Life History, Bionomics and Structure*. Cambridge University Press, London.

13. Reynolds, H. T. 1971. A world review of the problem of insect population upsets and resurgences caused by pesticide chemicals. *Agricultural Chemicals —Harmony or Discord for Food–People–Environment* (J. E. Swift, ed.). University of California, Division of Agricultural Science, Berkeley, Calif. pp. 108–112.

14. Wood, B. J. 1971. Development of integrated control programs for pests of tropical perennial crops in Malaysia. In *Biological Control* (C. B. Huffaker, ed.). Plenum Publishing Corporation, New York. pp. 422–457.

15. McMurtry, J. A., Huffaker, C. B., and van der Urie, M. 1970. Ecology of tetranychid mites and their natural enemies: a review. I. Tetranychid enemies. Their biological characters and the impact of spray practices. *Hilgardia 40*: 331–390.

16. Dan Thanarayana, W. 1967. Tea entomology in perspective. *Tea Quart. 38*: 153–177.

17. Lewis, T. 1969. The diversity of the insect fauna in a hedgerow and neighboring field. *J. Appl. Ecol. 6*: 453–458.

18. Doutt, R. L. 1958. Vice, virtue and the vedalia. *Bull. Entomol. Soc. Amer. 4*: 119–123.

19. Holloway, J. K. 1964. Projects in biological control of weeds. In *Biological Control of Insect Pests and Weeds* (P. de Bach, ed.). Van Nostrand Reinhold, New York. pp. 650–670.

20. Steinhaus, E. A. 1964. Microbial diseases of insects. In *Biological Control of Insect Pests and Weeds* (P. de Bach, ed.), Van Nostrand Reinhold, New York. pp. 515–547.

21. Ordish, G. 1967. *Biological Methods in Crop Pest Control*. Constable & Co., Inc., London.

22. Knipling, E. F. 1964. *The Potential Role of the Sterility Method for Insect Population Control*

with Special Reference to Combining This Method with Conventional Methods (Agricultural Research Service 33-98). U.S. Department of Agriculture, Washington, D.C.

23. Knipling, E. F. 1960. The eradication of the screw-worm fly. *Sci. Amer. 203:* 54–61.

24. Craig, G. C. 1967. Genetic control of *Aedes aegypti* (L.). *Bull. World Health Organization 36:* 628–632.

25. Hamilton, W. D. 1967. Extraordinary sex ratios. *Science 156:* 477–488.

26. Southwood, T. R. E., and Norton, G. A. 1973. Economic aspects of pest management strategies and decisions. In *Insects: Studies in Population Management* (Mem. 1). (P. W. Geier, L. R. Clark, D. J. Anderson, and H. A. Nix, eds.). Ecological Society of Australia, Canberra, pp. 168–184.

27. Conway, G. R. 1973. Aftermath of the Green Revolution. In *Nature in the Round* (N. Calder, ed.). George Weidenfeld & Nicholson Ltd., London. pp. 226–235.

28. International Rice Research Institute. 1973. *Annual Report for 1972*. Los Baños, The Philippines.

29. Stark, R. W. 1973. The systems approach to insect pest management — a developing program in the United States of America: the pine bark beetles. In *Insects: Studies in Population Management* (Mem. 1). (P. W. Geier, L. R. Clark. D. J. Anderson, and H. A. Nix, eds.). Ecological Society of Australia, Canberra, pp. 265–273.

Further Reading

Van den Bosch, R., and Messenger, P. S. 1973. *Biological Control*. Intext Educational Publishers, New York.

Farvar, T. G., and Milton, J. (eds.). 1971. *The Careless Technology: The Ecology of International Development*. Doubleday & Company, Inc. (Natural History Press), Garden City, N.Y.

Geier, P. W., Clark, L. R., Anderson, D. J., and Nix, H. A. (eds.). 1973. *Insects: Studies in Population Management* (Mem. 1). Ecological Society of Australia, Canberra.

Huffaker, C. B. (ed.). 1971. *Biological Control*. Plenum Publishing Corporation, New York.

Kilgore, W. W., and Doutt, R. L. 1967. *Pest Control: Biological, Physical and Selected Chemical Methods*. Academic Press, Inc., New York. pp. 293–342.

McGovran, E. R., et al. 1969. *Insect-Pest Management and Control*. Principles of Plant and Animal Pest Control, Vol. 3 (National Academy of Sciences Publ. 1695). The Academy, Washington, D.C.

Rabb, R. L., and Guthrie, F. E. (eds.) 1970. *Concepts of Pest Management*. North Carolina State University Press, Raleigh, N.C.

Editor's Commentary—
Chapter 15

HE FINAL CHAPTER in this section, on weather and climate, takes a look at Man's effects on his environment on a much larger, global, scale than the earlier chapters, and looks at his likely effects over a much larger time scale. In a sense this is a much more disturbing problem than the more immediate, short-range issues such as air and water pollution. We do not have a clear understanding of the workings of global climate and, just as in analyzing nuclear safety, we have to rely on computer models of the atmosphere and oceans to evaluate the probable consequences of our actions. The conclusions are no more reliable than the assumptions that go into the model, and these are, unfortunately, surrounded with uncertainty.

One difficulty with understanding and predicting climate is that such features as wind and oceanic current patterns, sea ice and snow cover all interact in poorly understood ways. An additional danger, too, is that interaction among these various factors can lead to *positive* feedback, so that an initial destabilizing change may become gradually worse. In the brief period since Dr. MacDonald completed this chapter several new models have explored these interactions. Models developed in Leningrad and at the University of Arizona agree that a decrease in temperature might increase the ice cover, which would reflect more sunlight into space, thus driving the temperature even lower. Another model (developed at M.I.T.) shows that loss of vegetation in an arid region, such as is caused by over-grazing, might increase the sun's reflection (the albedo) which can result in a much drier atmosphere in the area and hence lead to lower rainfall and the development of a desert. Finally, a set of model results from M.I.T. lead to the conclusion that nitrogen oxides from a fleet of 500 SSTs, with each plane flying eight hours per day, would deplete the ozone shield around the earth by 10-15 percent, causing a significant increase in ultraviolet radiation (which destroys biological tissue).

In many popular discussions of Man's effect on climate, attention is paid solely to questions such as whether there will be an ice age, or if the polar ice caps will melt. Clearly these are not insignificant problems! But it is important to remember that, long before such massive changes occur, even small increases or decreases in average world temperature can cause rapid alteration of the world's climatic *patterns*, causing anomalies such as failure of seasonal rains, or rain out of season. There is some scattered evidence that this is occurring now and is partly to blame for the poor harvests in various parts of the world.

It is clear from Dr. MacDonald's chapter that Man in fact is affecting global climate. It is also clear that the size of our effects will increase according to the scope of our activities. Since we understand so poorly the mechanisms controlling climate we need to be more cautious in the future than we have been in the past.

The skyline of lower Manhattan. Photo by Minoru Aoki.

15

Man, Weather and Climate

Gordon J. F. MacDonald   *is Professor of Environmental Science at Dartmouth. He was previously Vice Chancellor for Research and Graduate Affairs at the University of California, Santa Barbara, and an original member of the President's Council on Environmental Quality. He did both his undergraduate and graduate work at Harvard, and was a junior fellow there from 1952 to 1954. After teaching at MIT, he went to UCLA, where he became chairman of a new department of Planetary and Space Science. Professor MacDonald has served on the President's Science Advisory Committee and has been a consultant to NASA and the U.S. Department of State. He has made important contributions to the study of the interior of the Earth, the upper atmosphere, the modification of weather, and the origin of the moon and the planets.*

The evidence continues to mount that human activities are changing the state of the atmosphere locally and possibly globally. It is important to examine this subject at this time for at least three reasons. First, the growing population and the more complex activities of man may be altering climate in a way that could, in the long term, be undesirable. Although, as I shall emphasize, we do not know in detail what man is doing to his climate, the delicate balances within the atmosphere and the history of climate in the past suggest that man, through his inadvertent actions, may cause a disastrous ice age or an equally disastrous melting of the ice caps. The increasingly complex technologies developing today may bring about other undesirable changes. In the 1960s there was little, if any, concern about what might happen to life on earth as a result of airplanes flying in the upper regions of the atmosphere. In the 1970s we see a race between those trying to understand the effects of the supersonic transport (SST) on climate and on man and those who wish to develop and fly these airplanes.

Second, the study of inadvertent modification, both on the planetary and local scale, may identify means by which weather and climate can be altered to produce desirable effects. The science and engineering of weather modification are still in their infancies, although they are rapidly developing as more becomes known about how to remove fog and change rainfall and snowfall. These technologies may grow more rapidly if we understand what is happening in the atmosphere as a result of what man is doing. Further, we may achieve a better understanding of what changes, if any, will be of overall benefit to man.

Finally, it has long been known that the climate of cities differs from that of the surrounding countryside. Only recently has it been recognized that these climatic differences may affect the behavioral and social attitudes of the urban population. The feelings of alienation and discouragement that are so prevalent today in the cores of cities are not independent of the problems of suffocating smog, agitating noise, and city heat. Because of these barely perceived but important connections, the problems of inadvertent weather modification are more than important, they are fundamental.

Changing Climate

Man in his everyday experience and observation of hot or cold, sunny or cloudy, days recognizes what is called *weather*. *Climate* is deduced from weather and in a sense is a fiction created by man's mind. Climate is basically what the weather is like on the average. For example, the temperature on a given day at a particular time may fluctuate wildly from year to year, but the average temperature over any 10-year period will be much like any other 10-year period. This average temperature determines in part what is meant by climate. When the averages of temperature or cloudiness or precipitation over a period such as 10 or 20 years change, it can be said that the climate is changing.

The study of geologic and historical records clearly shows that major climatic changes have taken place in the past. Since the last advance of the ice sheet in Eurasia, about 10,000 years ago, the permanent ice cover in the Northern Hemisphere has been limited largely to the Arctic Ocean, to some islands in the higher latitudes, and to elevated mountain regions. Even during the last 10,000 years, periods of marked warming and cooling with time scales of centuries have been noted. Over the last two centuries, for which instrumental data are available, climatic fluctuations have continued.

Climate is much too complicated to be described by a single parameter. The number of days of snow cover or frost are important indicators of climate, particularly to the farmer. The amount and seasonal variation of precipitation are other critical factors that help to determine climate. A useful but imperfect guide is the temperature of the atmosphere measured at the earth's surface and averaged for 1 year over the entire earth. This number is also used because measurements of temperature are more easily made and recorded than many of the other overall determinants of climate. Records kept by meteorological stations show that from 1880 to 1940 this average temperature increased by about 0.6°C, whereas in the last 25 years the average temperature has decreased by about 0.3°C.[1,2] Thus, during the last three decades, one-half of the warming that had occurred during the preceding six decades has been erased.

Associated with increasing temperature were northward movements of the frost and ice boundaries, pronounced aridity in the south-central parts of Eurasia and North America leading to dust bowl conditions, strong mean motion (wind) parallel to the lines of latitude in the Northern Hemisphere, and a northward displacement of cold air masses. In more recent times the lowering of temperatures has been associated with the shifting of the frost and ice boundaries to the south, a weakening mean air motion along lines of latitude, and marked increases in rainfall in parts of the previously arid continental areas. For example, sea-ice coverage in the North Atlantic in 1968 was the most extensive in over 60 years. As a result, Icelandic fishermen suffered losses, and the accompanying colder weather shortened the growing season. In contrast, the rains in the central continental regions, particularly in India and in east Africa, contributed to very high wheat yields. These experiences emphasize two additional points about climate. The complex pattern of human activity is sensitive to relatively small changes in climate. This is particularly true in those regions in the world where agriculture and other activities are marginal. In northern countries such as the USSR and Canada, a small decrease in the growing season due

to lowered temperatures can bring about crop disasters. The changes in temperature need be only a few tenths of a degree or less to bring about such major alterations in food productivity, if the experience of Iceland is any guide. Similarly, small percentage decreases in precipitation in simiarid regions, such as in sub-Sahara Africa, result in widespread famine. These potential dangers are accentuated by the fact that our ability to predict changes in climate is very limited.

The fluctuations observed in recent times, with a time scale of decades, are still small compared with the climatic variations obtained during the "Little Ice Age" of 1550–1700, the warmer period between A.D. 1150 and 1300, and the still larger variation that was associated with the Ice Age and its retreat. In the middle latitudes the temperature during the Little Ice Age was about 0.6°C cooler on the average than today; in the preceding warmer period, temperatures were about 0.8°C higher. The minimum average temperature during the last glaciation, 20,000 years ago, was 5 to 6°C cooler than today. It is important to note that, while fluctuations of only a few tenths of a degree are involved in recent changes, fluctuations of about 4 to 6°C either way can lead to melting of the ice caps or to a new ice age.

Causes of Climatic Fluctuation

Over the years numerous theories have been advanced to explain the onset of the glacial age and the subsequent retreats of the ice. Among the suggested causes of glaciation is extensive volcanism, which would result in high dust levels blocking out the incoming solar radiation. The variability of the sun's radiation and the variation of the orbit of the earth about the sun have also been advanced as possible reasons for setting off a glacial period.[3] An interesting hypothesis has recently been advanced by Wilson.[4] Wilson assumes that the Antarctic ice sheet is unstable and that it can slide seaward as a result of melting along the bottom of the glacier. The melting is induced by the heat flowing from the earth's interior with the melting water lubricating the outward flow of ice. The increase in sea-ice coverage in the southern oceans would result in an outward reflection of the sun's radiation, the earth's albedo (the amount of light reflected) would increase, and the earth would cool.

None of the theories advanced so far has gained wide acceptance. Almost all theories, however, invoke a strong positive feedback mechanism. Once a sheet begins to spread, the albedo of the earth increases and in theory the temperature of the earth should continue to drop. In fact, a major unsolved puzzle is: Once an ice age begins, how does it stop?

The understanding of the cause of much smaller climatic changes observed in recent years is in a similarly unsatisfactory state. Very large computer models of the atmosphere have been created, and these have led to distinct improvements in forecasting weather, but as yet they have provided little insight into the causes of natural climatic fluctuations.

Given the imperfect state of knowledge of why climatic changes take place, it is exceedingly difficult to determine man's contribution to these changes. There is no simple cause-and-effect relationship and the isolation of man's contribution to climatic fluctuation depends on the ability to correctly predict what would have happened in the absence of man's intervention. In view of these uncertainties, one method of attempting to isolate human intervention is to examine the overall energy budget of the atmosphere and determine if man can influence the energy budget or parts of it. Weather and climate are determined, in the end, by how much energy reaches the earth, where that energy goes in time and space, and what processes alter its flow through the atmosphere.

The overall energy budget for the atmosphere is shown in Figure 1. By far the greatest source of energy (over 99.9 percent) is the incoming shortwave solar radiation, part of which is trapped within the atmosphere and causes the changes in the atmosphere which we call weather. Part of this energy does not affect the earth; it is reflected back into space as shortwave radiation. Some is absorbed by the atmosphere and is radiated into space as longwave (infrared) radiation (i.e., as heat). The remainder reaches the earth's surface and eventually is radiated back into space as infrared radiation. As can be seen from Figure 1, it is this longwave radiation that provides most of the heat to the atmosphere, which is mainly heated from below. A very small fraction of energy reaches the earth through the action of tides (3×10^{12} watts) and

from the interior by conduction, volcanoes, and hot springs (33×10^{12} watts). If man is to influence climate over the long term, he will have to alter in some way either the amount of energy flowing into or out of the atmosphere or the ways in which energy is transformed within the atmosphere. At present we can identify at least six ways in which man could perturb the atmospheric heat balance and thus affect the climate in a significant way. These are as follows:

1. Increasing the carbon dioxide content of the atmosphere by burning fossil fuels (carbon dioxide pollution) and thus increasing the heat trapped in the atmosphere.
2. Decreasing the atmospheric transparency by small particles (aerosols) resulting from industry, surface transportation, jet planes, and home heating units (particle pollution).
3. Decreasing atmospheric transparency by dust put into the atmosphere as a result of improper agricultural practices (dust pollution).
4. Direct heating of the atmosphere by burning fossil and nuclear fuels (thermal pollution).
5. Changing the albedo (fraction of the incoming solar radiation that is diffusely reflected out into space by land, sea, and cloud cover) of the earth's land surface through urbanization, agriculture, deforestation, and construction of large water reservoirs (surface pollution).
6. Altering the rate of transfer of thermal energy and momentum between the oceans and atmosphere by an oil film resulting from the incomplete combustion and the oil spilled from ocean-going vessels (ocean oil pollution).

These activities directly affect the energy budget of the atmosphere. Other man-made activities may have indirect effects. For example, certain kinds of industrial processes emit small particles that cause condensation and thus affect the frequency of fog and low stratus, which, in turn, affect the radiation budget. Also, forest fires produce cloud condensation and nuclei that induce the formation of ice as well as large quantities of heat and water vapor. In this connection the large-scale burning of trash forest products and of accidentally set forest fires, which are particularly prevalent in the

384

Western states, might result in serious modification of climate as well as of local weather.

Carbon dioxide pollution has long been recognized as potentially capable of affecting worldwide climate. The possible effects of urban, industrial, and agricultural activities on climate, as opposed to local weather, however, have only recently been noted.[5,6]

Atmospheric Carbon Dioxide

Carbon dioxide makes up only one three-thousandth of the atmosphere, yet it influences how much of the sun's heat is retained in the atmosphere and in so doing is important in determining climate. Further, carbon dioxide in both the atmosphere and oceans (one ten-thousandth of the ocean is dissolved carbon dioxide) is of vital importance to all biologic activity. Carbon is the key element of all living matter and carbon is derived either directly or indirectly from atmospheric carbon dioxide (Chapter 1).

Most of the energy used by man to drive modern technology is derived from fossil fuels. The carbon contained in these fuels was once present as carbon dioxide in the atmosphere, it was then incorporated by organisms, and the process of oxidation during burning of the fuels returns carbon dioxide to the atmosphere. The interchange is a speeded-up version of part of a cyclic process in which carbon participates over geologic time.

Natural carbon dioxide enters the atmosphere through the action of volcanoes and hot springs. Some of this gas may represent material buried since the origin of the earth, and other parts result from the melting and degasifying of limestone and other carbon-bearing rocks. Indeed, carbon dioxide goes through a complicated life cycle which is illustrated in Figure 2. During the 4.5 billion years of earth history, volcanic and related activity have poured carbon dioxide into the atmosphere. The total amount introduced is uncertain, but it is at least 40,000 times the amount now present in the atmosphere. Part of the carbon dioxide released from the earth's interior has dissolved in the oceans. Part of the carbon dioxide in the oceans (almost 8 percent) is in the upper few tens of meters of the ocean. This carbon dioxide can interchange with the lower atmosphere with a time scale of about 2 to 10 years (see Figure 2). Most of the carbon dioxide in the oceans is at great depths and the primary mechanism for interchange with the

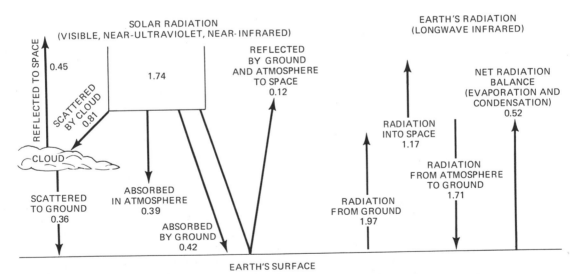

1 **Radiation Balance in Atmosphere.** The higher values for radiation flux in the infrared are due to the greenhouse effect of H_2O and CO_2. Units: 10^{17} watts. Note that total radiation absorbed by the ground equals energy per second entering the atmosphere from below (0.36 + 0.42 + 1.71 = 1.97 + 0.52). Similarly, the total energy per second entering the atmosphere equals energy leaving the atmosphere (0.39 + 1.97 + 0.52 = 1.17 + 1.71). The reflected solar energy does not interact with the ground or atmosphere.

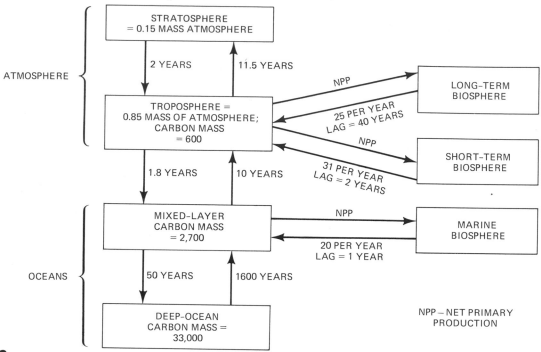

2 **Sources, Sinks, and Exchange Rates for CO_2.** Mass units in 10^9 metric tons; exchange times in years. *Source:* Reference 7.

surface layers is slow diffusion rather than the more rapid, turbulent mixing characteristic of the upper layers. In the ocean, carbon dioxide can combine with calcium and magnesium, supplied by the weathering of rocks, to form limestone and dolomite. A portion of the carbon dioxide dissolved in the ocean and of the atmospheric carbon dioxide enters the biosphere and is taken up by plants. Some of the carbon taken up by plants returns to the atmosphere relatively quickly and some remains fixed in living matter (trees, etc.) for several decades. About one-fourth of the total quantity of carbon dioxide taken up by plants ends up as carbon compounds, which are eventually buried in sediments. A fraction of this material has been transformed by a variety of geologic processes into gas, coal, lignite, oil shale, tar sand, and petroleum. The total amount of carbon buried in a form in which it can be recovered for use as a fuel is most uncertain, but it is probably at least a quadrillion (10^{15}) tons.

The burning of fossil fuels (primarily lignite, coal, petroleum, and natural gas) now releases annually about 15 billion tons of carbon dioxide. This figure should be compared with the amount of carbon dioxide annually consumed in photosynthesis, about 110 billion tons. Thus, today the combustion of fossil fuels produces an amount of carbon dioxide that is an appreciable fraction (approximately one-seventh) of the carbon dioxide that enters the photosynthesis cycle each year. Indeed, a major question today is: What are the short- and long-term effects of increased carbon dioxide on biological activity? It is, furthermore, about four orders of magnitude (10,000 times) greater than the return of carbon to the fossil reservoir; the rate of release of carbon dioxide into the atmosphere by natural oxidation during respiration of recently grown organic materials (i.e., by plants and decomposers, see Chapter 1) replaces all but one ten-thousandth of the amount consumed in photosynthesis. The significance of these numbers is that man in his use of fossil fuels is rapidly approaching the rate at which all plants, all over the world, are putting carbon dioxide into the atmosphere. A balance between production and take-up by natural processes developed over geologic time is being upset in a time span of a few decades.

Until recently, it was not clear how much of

the carbon dioxide being released by combustion accumulated in the atmosphere and how much entered the oceans and the terrestrial biomass. Callendar[8] calculated that the carbon dioxide had increased at approximately a constant rate from the nineteenth-century level of about 290 parts per million (ppm) to about 330 ppm in 1960 (0.03 percent of the atmosphere). To sustain such an increase, about three-fourths of the carbon dioxide released through combustion would have to remain in the atmosphere. More recently, Keeling has undertaken a detailed monitoring program at Mauna Loa Observatory in Hawaii and at the South Pole station in Antarctica. He has found that the annual average CO_2 levels measured at Mauna Loa and at the South Pole station agree to within 1 ppm. Furthermore, both stations show a consistent increase over the past few years. Keeling[9] finds a concentration of CO_2 of about 314 ppm and that the rate of increase averages about 0.2 ppm per year. This implies that each year the mass of carbon dioxide in the atmosphere is increased by 5 billion tons. Thus, of all the CO_2 produced by combustion, one-third remains in the atmosphere and two-thirds are taken up by the oceans and by the biomass. Machta's calculations[7] (discussed below) suggest that one-half remains in the atmosphere. At the current rate of deposition in the atmosphere of carbon dioxide, the amount of man-made CO_2 doubles every 23 years. Assuming a mass of atmospheric CO_2 of 2.2 trillion tons, this is an increase in atmospheric CO_2 of over 0.2 percent per year or 2 percent per decade. Machta's model, on which Figure 2 is based, provides an estimate of the past and future variation of the CO_2 content of the atmosphere. If present practices of the use of fossil fuels are continued, we can expect a variation of carbon dioxide in the atmosphere, as shown in Figure 3.

SOURCES OF INCREASING CARBON DIOXIDE

At present, the world's use of energy is increasing annually at about 5 percent, which corresponds to a doubling time of about 14 years. Thus, in 1985 the world will be using twice the energy that it used in 1972. In 1972 the world rate of energy usage was 8 trillion watts (Chapter 5). Of all the energy produced, about 98 percent comes from oil, coal, and natural gas, while water power contributes only 2 percent

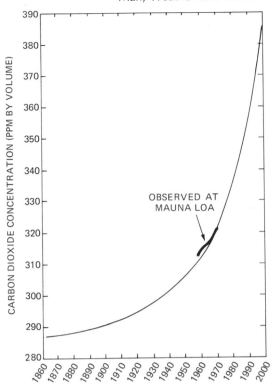

3 **Increase in Atmospheric CO_2 Calculated by a Mathematical Model.** The amount increases because of the combustion of fossil fuels. Note that the observations at Mauna Loa fit the predicted values closely. *Source:* Reference 7.

and on a global scale nuclear energy is still negligible. The total thermal energy produced from crude oil, coal, and lignite in 1973 corresponds to 0.17 watt per square meter (watts/m²) averaged over the total surface of the earth. This is five times the 1940 energy production of 0.003 watt/m² averaged over the earth's surface.

Not only has the total amount of energy used by modern industrial society increased very greatly but, also, the sources of the energy have changed. Since 1800, the principal sources of the world's industrial energy have been the fossil fuels and water power. Before 1900, the energy derived from oil, as compared with that obtained from coal, was almost negligible. Since 1900, the contribution of oil to the total energy supply has steadily increased and is now approximately equal to that of coal, and increasing more rapidly (Chapter 5). If natural gas and natural gas liquids are added to crude oil, the petroleum group of fuels represents about 60 percent of all the energy derived from coal, petroleum, and water power. Since World War II the production of coal has grown at a constant rate of about 3.6 percent with a doubling period of 20 years. World production of crude oil, except for a slight retardation during the Depression of the 1930s and during World War II, has increased from 1890 to the present at a nearly constant exponential rate of 6.9 percent per year, with a doubling period of 10 years.

The rates of growth of a highly industrialized nation differ substantially from the world average. The U.S. production of total energy from coal, oil, natural gas, and water power divides into two distinct growth periods. From 1850 to 1907 energy in the United States was produced at a growth rate of 7 percent per year, with a doubling period of 10 years. From 1907 to 1970 the average growth rate dropped to 1.8 percent per year, with a doubling period of 39 years. However, since 1965 the rate of energy use has increased, so that in 1972 about 4 percent more energy was used than in 1971.

Machta's[7] model of the current increase of CO_2 in the atmosphere indicates that about one-half of the CO_2 released through combustion enters the atmosphere, implying that man's activities have added about 2.8×10^{17} grams of carbon dioxide to the atmosphere. This is 13 percent of the 2.2×10^{18} grams now present. In consideration of climate, it is not only the total amount of CO_2 introduced into the atmosphere that is important, but also the rate at which it is introduced. Although coal has been mined for about 800 years, one-half of the coal produced during that period has been mined during the last 31 years. Half of the world's cumulative production of petroleum has occurred during the 13-year period since 1956. Thus, 9×10^{16} grams of CO_2 have been introduced into the atmosphere since 1950, and 1.3×10^{17} grams have been introduced since the mid-1930s.

An estimate of the maximum amount of CO_2 that might be introduced in the future by man into the atmosphere can be made from estimates of the total fossil fuels that are available. Averitt[10] estimates that 7.6×10^{12} metric tons of coal represents the maximum available. This is about twice the coal resources established by actual geologic mapping. Estimates of petroleum resources vary considerably. Weeks[11] and Ryman[12] estimate that approximately 2×10^{12} barrels of oil are ultimately recoverable.

Hubbert[13] appears to favor the somewhat lower figure of 1.35×10^{12} barrels. If these fossil fuels were burned, they would add to the atmosphere 3.3×10^{18} grams of CO_2. This figure should be compared with the 2.2×10^{18} grams now in the atmosphere. Thus, man is capable of increasing the current CO_2 content by about 150 percent.

Hubbert[13] estimates that 80 percent of the ultimate resources of the petroleum family — crude oil, natural gas, tar sand, oil, and shale oil — will be exhausted in less than a century. The time required to exhaust 80 percent of the world's coal resources would be 300 to 400 years, but only 100 to 200 years if coal rather than nuclear power is used as the main energy source. These rates of consumption imply that, in the next century, the content of the CO_2 in the atmosphere could be doubled. We can also estimate that over the next 30 years it could be increased by about one-third.

Projected possible increases of carbon dioxide in the atmosphere must be viewed with caution. The oceans, for example, contain 60 times more carbon dioxide than does the atmosphere. If the atmosphere's average temperature increased, either as a result of the increasing carbon dioxide content or through some other means, the heating would tend to drive some of the carbon dioxide now dissolved in the oceans back into the atmosphere. Alternatively, the increasing flow of nutrients into the ocean resulting from improper agricultural practices and inadequate waste treatment would stimulate the growth of phytoplankton and increase, through photosynthesis, removal of carbon dioxide from the atmosphere. Looking to the future, we cannot say with any precision what the carbon dioxide content of the atmosphere actually will be. If present trends continue, Figure 3 is as good a projection as currently can be made. But changes in fuels, alterations in fuel use, and climatic changes, either natural or man-made, can alter these projections. What we can say is that man has changed the total amount of carbon dioxide by several percent and, further, that he is capable of more than doubling the carbon dioxide content over the next hundred years.

THERMAL EFFECTS OF CARBON DIOXIDE

Weather and climate are determined by how much of the solar energy is converted by the atmosphere–ocean–land system into the motion of the atmosphere. The heat-energy input into the atmosphere depends on the albedo of the earth, including clouds, the interchange of heat by the atmosphere with the land and water and by the absorption of radiation by gaseous constituents in the atmosphere (Figure 1). Carbon dioxide is nearly transparent to visible light, so it has only minor effect on the incoming high-temperature solar radiation. However, it is a strong absorber of infrared radiation given off by the land, sea, and clouds, at temperatures much lower than the solar radiation. Carbon dioxide then radiates back to the surface a portion of the absorbed thermal energy so that a fraction of the heat that would be lost in space remains to warm the atmosphere and the land. This effect is universally but misleadingly termed a *greenhouse effect*. A greenhouse, however, operates not only by trapping radiation by its glass structure but also by reducing heat lost to the air that otherwise would move through the greenhouse.

The most complete calculations of the net effect of altering the carbon dioxide content of the atmosphere are those of Manabe and Wetherald.[14] These numerical calculations show that increasing CO_2 content results in warming of the entire lower atmosphere, the amount of warming being dependent in part on the water-vapor concentration of the atmosphere. With the assumption of fixed concentration of water vapor, together with conditions of albedo, cloudiness, radiation, and other parameters chosen as typical of the mid-latitudes, an increase of 10 percent in the CO_2 concentration leads to a warming of 0.2°C. For the assumption of a fixed relative humidity (actual water-vapor pressure as a percentage of vapor pressure required for condensation), with all other conditions remaining the same, a 10 percent increase in content of CO_2 raises the temperature by 0.3°C. Manabe and Wetherald consider the latter assumption of fixed relative humidity conditions as being more realistic. If the amount of carbon dioxide in the atmosphere is doubled at constant relative humidity, the temperature is increased by 2.4°C. These calculations are significant, in that they indicate the magnitude of the primary change. Increasing carbon dioxide, maintaining other variables constant, will lead to a rise in temperature. There are, however, a number of possible important feedback loops. An increase in tempera-

ture could result in a further increase in the carbon dioxide content of the atmosphere. For example, Revelle[15] calculates that a 1°C warming of the surface waters "would cause about a 6 percent increase in atmospheric CO_2." This, in turn, would lead to further heating of the atmosphere, which would increase the humidity, which might raise the level of cloudiness, thus affecting the albedo; and the totality of effects remains uncertain.

In addition to carbon dioxide, water vapor also absorbs infrared radiation. In a humid atmosphere the net effect of carbon dioxide is less than in a dry atmosphere. The third component of importance in the atmosphere is ozone. In the upper region of the atmosphere, above 40,000 to 50,000 feet, ozone absorbs some infrared, but its principal effect on air temperature is due to its absorption of ultraviolet and visible sunlight. Of the three components that affect the radiation balance, so far man has only changed in a measurable way the concentration of carbon dioxide. However, the possible introduction of a fleet of commercial supersonic transports presents a potential problem for the future. High-flying aircraft can introduce water, nitrogen oxides, particulates, and other pollutants into the rarefied atmosphere.

Aircraft add water as a result of combustion of jet fuel. In the lower atmosphere this water mixes rapidly and does not change in any significant way the radiative properties of the lower atmosphere. However, the higher atmosphere is relatively dry, with a water content of only about 2 parts per million. The water introduced in the stratosphere (about 10 to 30 kilometers altitude) remains there for 18 months to 2 years. This water can lead to an increase in temperature at the surface by trapping outgoing radiation; a fivefold increase in concentration would raise the surface temperature by about 2°C. A fleet of 400 SSTs flying in this region would change the concentration by about 25 percent if the water were distributed uniformly through the stratosphere.

Another potential effect of the water introduced into the stratosphere by SSTs is on the concentration of ozone. Water can react with ozone, converting it to molecular oxygen in photochemical reactions. These reactions could lower the ozone concentration. Ozone plays a key role in determining the temperature structure in the atmosphere. While carbon dioxide

389

and water vapor trap the outgoing infrared radiation generated at low temperatures at the earth's surface, ozone acts as a shield against the ultraviolet radiation given off at high temperatures by the sun. If ozone were removed to any appreciable extent, greater quantities of ultraviolet radiation would reach the surface. Small increases in the intensity of ultraviolet radiation could result in increased incidence of skin cancers and related diseases; large increases in ultraviolet could bring about massive changes in all terrestrial biological systems. However, uncertainties in the nature of the reactions, their rates, and the natural concentration of ozone and water vapor make any estimate of the overall change in ultraviolet radiation and atmospheric temperature most uncertain.

An additional possible effect of SSTs on the upper atmosphere is associated with sulfur in the fuel burned by the aircraft. The sulfur is partially oxidized during combustion to form sulfate particles. These fine particles can remain suspended in the high atmosphere for many months or years and could contribute to shielding the incoming solar radiation. These particles are a particular example of a more general problem: our advanced technology can and does obscure the atmosphere by placing into it a variety of tiny specks of waste.

Particles and Turbidity of the Atmosphere

Is the observed decrease in the temperature since about 1940 due to man, or is it the result of whimsical nature that is now overwhelming the effects of carbon dioxide which should lead to a temperature increase? The answer is by no means clear, but Bryson has argued that there is increasing evidence that urban and industrial pollution, perhaps aided by agricultural pollution, may be in part responsible for decreasing the surface temperature at a rate that is large compared with the effects of carbon dioxide in increasing the temperature. What is not clear is whether agricultural and industrial pollution have effects that are comparable to those due to natural volcanic activity, which produces large quantities of natural fine particles that can remain suspended in the atmosphere for long periods.

Particle pollution, whether urban and industrial aerosols or whether agricultural dust, can affect the thermal balance in at least two ways, one direct and one indirect. The presence of small particles in the atmosphere decreases the transparency of the atmosphere to the incoming solar radiation. This partial shielding of the surface is usually described in terms of the turbidity of the atmosphere. The decreased solar radiation reaching the surface will lead to lower temperature, but the small particles also affect the outgoing longwave radiation. The net effect on the total radiation balance depends on the abundance and size distribution, and the altitude range of the small particles. Calculation of this net effect is still primitive; however, Bryson[5] estimates that a decrease of atmospheric transparency of only 3 percent or 4 percent could lead to a reduction in surface temperature of 0.4°C. What is important is the net amount of radiation reaching the surface, and this is composed both of the direct solar radiation and the infrared radiation, which is both reflected and reradiated back by clouds, carbon dioxide, and water vapor.

The total amount of particulate matter injected into the atmosphere is in doubt and there are few reliable data on how much is due to man. Volcanic eruptions are estimated to introduce about 800 million to 2 billion tons per year. Man-made particles, mostly sulfates resulting from combustion in the lower atmosphere, contribute from 200 to 400 million tons per year. The particles introduced in the lower atmosphere (troposphere) have a relatively short lifetime and rain out after a period of a few days or weeks. If the steady-state concentration of particles, either man-made or natural, increases, the energy balance of the atmosphere is affected. The particles interact both with the incoming short-wavelength solar radiation and the outgoing long-wavelength infrared radiation. Which interaction dominates depends on the particle sizes and vertical distribution. However, it would appear that the suspended dust tends to increase the albedo of the earth, increasing the fraction of radiation reflected into space. Dust particles can also act as condensation nuclei and thus assist in forming clouds made up of small water droplets. Thus, increased steady-state concentration of dust particles could lead to increased cloudiness. This can be significant, because only a 2.4 percent increase in cloud cover could lead to a 2°C drop in temperature by increasing the earth's albedo.[14] Those particles that reach the high atmosphere (stratosphere) and remain there for 1 to 2 years can also significantly affect the energy budget of the atmosphere.

Second, the indirect effect of particles introduced by man on the thermal budget of the atmosphere arises because air, particularly the air that has been over the oceans for some time, is often deficient in cloud condensation nuclei and ice nuclei. The introduction of man-made condensation nuclei into the atmosphere aids the formation of fog and low cloud layer from water-vapor-laden air. These artificially stimulated clouds will reflect some fraction of solar radiation back out into space and block the infrared radiation from reaching the surface.

The effect of particles in encouraging the formation of low clouds and fog further enhances the modification of the thermal balance. Low clouds have a large effect on the net energy reaching the surface; Manabe and Wetherald[14] show that a 1 percent change in the average low cloud cover the world over will bring about a decrease in temperature of 0.8°C, four times the observed drop over the past two decades. At present, on the average, about 31 percent of the earth's surface is covered by low cloud; increasing this percentage to only 36 percent would drop the temperature about 4°C, a decrease very close to that required for the return of an ice age!

There are few reliable observations of the transparency or turbidity of the atmosphere extending over long periods of time. McCormick and Ludwig[6] discuss data from Washington, D.C., and Davos, Switzerland, that indicate increases of turbidity of approximately 10 to 20 percent per decade, respectively, for the period 1910 to 1960. Somewhat larger changes have been reported from the results obtained at the observatory in Mauna Loa, but these observations have been questioned on instrumental grounds.

Davitaya[16] presented data on dustfall on snow in the high Caucasus Mountains. The quantities increased by an order of magnitude (tenfold) between 1930 and 1960, with most of the changes occurring since 1950; however, there are uncertainties as to how much of the dust is locally derived and how much is representative of the overall atmospheric content.

Although the local increases of fog and low cloud cover over urban areas are well documented, the overall increase of cloud cover, if indeed there has been one, is not known. Only with the availability of satellites have data been obtained on a fairly routine basis over ocean areas, but these data are often incomplete; high clouds may hide clouds at lower levels, and the perturbation in the thermal balance depends critically on cloud-height distribution.

Mitchell[2] has critically examined the hypothesis that man-made aerosols play a significant role in controlling the radiation balance. He concludes that a major part of the turbidity variation is due to the introduction into the stratosphere of fine-grained particles as a result of volcanic activity. Mitchell's calculations are based on estimates of the total debris associated with volcanic explosion, an assignment of 1 percent of this total to stratospheric dust, and an assumption of a 14-month lifetime for the stratospheric dust. All these numbers can be questioned but, if they are accepted, it would appear clear that volcanic dust dominates urban and agricultural pollution.

The above remarks are not intended to imply that there is a proved case that urban, industrial and agricultural pollution, or volcanic dust is the principal cause of the recent cooling trend. What is significant is that the apparent changes in atmospheric transparency may be sufficient to bring about the observed cooling of the earth's surface. Further, the direct effects of transparency can be amplified by increased formation of low clouds and fogs, which greatly affect the thermal balance of the earth's surface.

If the pollution interpretation is correct, we face an urgent problem of global climate modification, because atmospheric pollution is increasing at an exponential rate as energy use increases worldwide at a rate of about 5 percent per year; and there are, as yet, no effective or widely acceptable means of impeding this growth.

Thermal Pollution

The earth's cross section, a disk in the sky of area πr^2 (where r is the earth's radius), intercepts on the average solar energy at a rate of 1.74 $\times$ 10^{17} watts. This energy, averaged over the total area of $4\pi r^2$, is 1,395 watts/m^2, which is roughly the rate at which a kitchen oven generates heat. Since the temperature of the earth has not undergone large fluctuations in recent times, the earth radiates out into space energy at the same rate. Part of the incoming energy (about one-third) is directly reflected back out into space, mostly by the cloud cover but also by dust in the atmosphere and the surface itself (this reflected amount is the albedo). Part of the energy (over 40 percent) is reradiated out by the atmosphere as low-temperature infrared radiation, after heating the atmosphere, but an important part (almost all of the remaining energy) does evaporative and convective work in the atmosphere. This energy that is neither immediately reflected nor reradiated is termed the *radiation balance* and is about 100 watts/m², although it varies with latitude.

Most of this radiation balance is used in the evaporation of water, heating of the atmosphere, and driving various meteorological processes. A tiny part, less than 1 percent, is used in the photosynthesis of green plants and is turned into a relatively stable form of chemical energy. Man's industrial activities directly affect this thermal budget because industrial heat represents a new source of heat additional to that from solar radiation.

The total rate of energy use worldwide in 1973 was about 8.5×10^{12} watts, or about 2,200 watts per person. This corresponds to a contribution of 0.006 watt/m², and over the continents of 0.017 watt/m². These numbers are small compared with the net radiation, that part of the radiation that goes into evaporating water and causing convective motion. Since the industrial heat is only about 1.7×10^{-4} (roughly 1/5,000) of the net radiation of 100 watts/m², it cannot affect climate on a large scale, but it can certainly alter microclimate. For example, in Manhattan over an area of about 60 km², the artificial input of energy on an annual average is 710 watts/m², while the net radiation at this latitude is 93 watts/m². Thus, industry, heating and air conditioning, transportation, and other means of energy use completely dominate the natural flux over this small area.

In a primitive society, the utilization of energy is pretty well limited to the food used by the individual. This corresponds to about 150 thermal watts per capita. The present world average is 2 kilowatts per person, while a highly industrialized society such as the United States used in 1973 about 13,000 thermal watts (13 kilowatts)

per capita. Projecting into the future, we see that if the world population goes to 5 billion and if the worldwide average of energy use is 10 kilowatts per person, the total rate of energy use would be 50×10^{12} watts, about six times present usage. The direct energy input into the atmosphere would be 0.1 watt/m² averaged over the globe or about 1/1,000 that of the natural radiation balance. Indeed, if the present rate of energy increase of 5.6 percent is maintained (a doubling time of about 12 years), in 100 years artificial energy input into the atmosphere would equal 5 percent of the radiation balance. As Budyko[17] has argued, an increase of only a few tenths of 1 percent in the radiation balance (0.2 to 0.6 watt/m²) would be sufficient to cause the melting of polar ice. With a doubling time of 12 years, we would increase energy production 25-fold in about 60 years, to reach an artificial energy input of 0.4 watt/m².

The combined effect of carbon dioxide pollution and direct heat pollution works toward warming the earth's atmosphere. On the other hand, urban and agricultural pollution tend to lower the earth's temperature by increasing the earth's albedo both directly and indirectly through enhanced cloud formation. Which pollution will win in the end? Will we drown or freeze? Can we, as we obscure the sky, neatly balance the lost solar radiation through energy generated by fission and fusion? These will become critical questions in the not-too-distant future.

Albedo

The albedo of the earth averages about 0.33; that is, about one-third of the incoming radiation is reflected back out into space. The average cloud cover, of about one-half the earth's surface, is responsible for about 80 percent of the albedo. Increasing the albedo will result in a lower average surface temperature because less heat is available to the atmosphere. Variations in albedo can be of great significance because of the feedback mechanisms referred to earlier in connection with the ice cap. A forested region transformed into a desert absorbs less radiation and favors drier and more infrared-transmitting air above it. These conditions can lead to a further increase in desert climate.

Manabe and Wetherald[14] calculate that a unit increase in albedo of the earth's surface will produce a decrease in average surface temperature of 1°C. Thus, man-made alteration of the albedo at the earth's surface, if large enough, can bring about substantial changes in climatic conditions. Densely builtup regions have a higher albedo than do forests and cultivated soils. Deserts, some of which may have resulted from man's activities, have a much higher albedo than do grass-covered fields. While local changes of albedo have been measured, the long-term global variation is unknown even as to sign. The vast proliferation of urban areas and highway systems suggests that man may, at present, be increasing the surface albedo.

Man's construction, lumbering, and farming change not only the radiative properties, but other surface features that can influence the thermal state of the atmosphere. The ground exchanges heat with the air primarily through the irregular motion in the atmosphere known as *turbulence*. The degree to which the flow of air is turbulent depends on the roughness of the surface. For example, on a small scale, the air temperature at a height of a few feet over a hot airfield runway can be 1 to 2°C cooler than over the cooler grassy surroundings. The flow of the air over the smooth surface is less turbulent than over grass, and less heat is transferred from the hot surface upward into the air. Further, the roughness also affects the actual flow of the air, particularly in the vertical direction.

The calculation of the net effect of changing surface characteristics is uncertain because of incomplete information on the overall thermal effects of changing albedo and surface roughness. However, both the magnitude of man-made surface changes (direct, as in construction, and indirect, as in the formation of deserts) and the sensitivity of the atmosphere to such alteration suggest that these should be carefully studied.

Ocean Pollution

Oil released by discharges from tankers or by drilling operations can form a thin film, on the order of 1 micrometer (a thousandth of a millimeter) thick, on the surface of the ocean. The film can spread out over wide areas, eventually to be destroyed by chemical degradation or bacterial action. The surface currents can concentrate the film in specific areas, particularly near lanes of heavy tanker traffic.

The possible effects of spreading oil in a very thin film on the ocean surface are poorly understood. It is generally assumed that the oceans, with their vast store of thermal energy, act as a balance wheel to climate. The atmosphere exchanges energy with the ocean not only through the radiative processes but through the mechanical processes associated with air moving over a wave-roughened surface. The strength of the mechanical interaction depends on the roughness of the surface at various length scales, and the roughness is determined by the surface properties of the water as well as by the velocity and the irregularity of the wind blowing over the surface. Very thin oil films can perturb the interchange by reducing the turbulent flux of the heat and of the momentum, reducing the evaporation, and lowering the amount of energy radiated by the surface.

We do not know whether oil pollution is a significant factor in climatic change. Data on the extent of oil pollution, the lifetime of an oil film on the sea's surface, and the detailed thermal effects of such a film are not available, so we do not even know whether ocean oil pollution lowers or raises the surface temperature.

Urban and Rural Climates

Man-made changes in global climate are at the best speculative. However, there is no doubt that the activities of man are changing both weather and climate on a local and sometimes regional basis. For example, Manhattan Island houses some 1.7 million people in an area of about 6×10^7 m² (60 square kilometers). The total energy output from all activities is about 700 watts/m²! If winds did not carry off this excess heat, the people living in Manhattan would literally fry.

It has long been recognized that the climate of cities, which represent the most concentrated form of environmental modification by man, differs appreciably from the climate of adjacent rural areas. In addition, there is a growing body of literature showing that industrial activities are modifying atmospheric properties and in some cases may be profoundly changing the weather.

Cities on the average have a temperature 0.5° to 1°C higher than the surrounding countryside. The effect is even more marked in the winter, when the minimum temperature reached may be 1° to 2°C higher than that of the surrounding areas. In addition to the difference in temperature, cities are more frequently covered with clouds and the frequency of fog in wintertime can be twice that of the suburbs. Accompanying the increased frequency of cloud cover is a higher total precipitation. Table I summarizes the relative climates of cities as compared with the adjacent countryside. These are general averages; in very large metropolitan areas the differences can be greater.

Several factors contribute to the higher temperatures in urban centers. The direct energy input from home heating units, industry, and air conditioning all lead to the higher annual mean temperature and to raising the minimum temperature. In addition, buildings naturally shelter the city so that wind speeds are less and the turbulent transfer of energy from the city to moving weather systems may be decreased. The increased cloudiness over cities, together with the effect of buildings and pavements on the amount of solar heat retained, can further accentuate the direct input of heat into the atmosphere.

I   Climatic Changes Produced by Cities

Parameter	City compared with rural surroundings
Temperature	
Annual mean	0.5–1.0°C higher
Winter minima	1.0–2.0°C higher
Cloudiness	
Clouds	5–10 percent more
Fog, winter	100 percent more
Fog, summer	30 percent more
Dust particles	10 times more
Wind speed	
Annual mean	20–30 percent lower
Extreme gusts	10–20 percent lower
Precipitation	5–10 percent more

Source: References 1 and 2.

The increased precipitation in cities is probably also due to a number of factors. The heat of the city in effect causes a "thermal mountain" in the air flow. The heat of the sun, together with the heat produced by furnaces, automobiles, factories, and homes, warms the air in the city. The warmer, less dense air rises in comparison with the cooler air of the suburbs and surrounding regions. As the air rises, it expands and therefore cools. When the air contains water vapor, the cooling can cause the water vapor to change to liquid drops or even ice crystals. When this happens, a cloud forms. If the droplets or ice crystals coagulate to a large-enough size, they drop from the clouds as rain or snow. The actual formation of precipitation is much more complex than is indicated in this simplified account, but over cities the basic mechanisms involve the heating, rising, and expanding of the air.

Laboratory experiments have shown that if one cleans the air of all solid particles, the air can hold much more water at a given temperature before condensing or crystallizing than it can when small particles are present. In nature, the small particles act as nuclei to which the water molecules can become attached. The particles can be dust or salt particles from the ocean, and over cities, the bits of matter produced by the burning of oil, coal, and wood, and to a large extent the exhaust particles of internal combustion engines. These artificial nuclei can aid the formation of precipitation when there is a deficiency of natural nuclei. In cases where the artificial nuclei are abundant, many droplets form and a fine drizzle or fog is produced.

Support for the theory that particle pollution plays a role in precipitation has recently been offered by R. H. Frederick of the Environmental Science Services Administration.[18] He shows the existence of a rather systematic tendency for cold-season precipitation at 22 urban Weather Bureau stations in the eastern United States to average several percent greater on weekdays than on weekends, when there is less dust around. This is an important finding which, if substantiated in further studies on other urban areas, would go far in explaining the difference in precipitation between cities and the country.

The most dramatic cases of the relation of industrial pollution and weather are found in regions where the pollutants are concentrated in valleys. Hosler[19] has illustrated how the emission of water vapor and condensation nuclei from the stack of a single wood pulp mill in Pennsylvania can cause fog formation. Fog sometimes fills a valley several miles wide and 20 miles long and spills into adjacent valleys. Hobbs and co-workers[20,21] have shown that Washington State pulp and paper mills are prolific sources of cloud condensation nuclei and that clouds often form downwind of these activities. Hobbs argues that regions of abundant artificial condensation nuclei show an annual precipitation in the last 20 years which is 30 percent greater than in the previous 30 years. Furthermore, these studies have established a higher ice nuclei abundance by as much as an order of magnitude over urban Seattle, compared with stations removed from industrial activity. These observations, taken along with those of Frederick, clearly show the potential importance of man-introduced particulate matter in influencing the weather in regions of high industrial activity. The effects of individual urban centers produce local and regional changes in weather. The sum of the contributions of many metropolitan areas can bring about changes in climate over large regions.

Conclusions

I have briefly reviewed six major ways in which man could be altering the planet's climate; there may be others of comparable importance, such as contrails from jets and supersonic aircraft. Of these six ways, we know most about the effect of carbon dioxide, but, even here, the uncertainties are large.

Examination of the possible ways in which man affects climate on a large scale leads to certain generalizations:

1. Large-scale man-made changes may be taking place in the environment. These are, for the most part, inadvertent, and some have been only recently recognized.
2. The magnitude of the changes produced by man is of the same order as that caused by nature. For example, the carbon dioxide added to the atmosphere can bring about a change of several tenths of a degree in the average temperature; and changes of this magnitude have been observed to occur naturally.

3. The alterations produced by man can no longer be regarded as local. Direct heat input by a city changes the microclimate of that city. The combined effect of many cities can change the climate of a region, such as southern California, and all these changes can influence the global climate.
4. Our understanding of the physical environment is sufficient to identify inadvertent modification, but it is far too primitive to predict confidently all the consequences of man's unwise use of his resources.
5. Despite the very long-term importance of understanding changes in our environment if we are to maintain a habitable planet, inadvertent modification is a neglected area of research — neither fashionable to the scientists and engineers nor, until recently, of high priority to the money-distributing government agencies. For example, there are at most a handful of small research groups throughout the United States studying, in a professional way, the influence of man's activity on climate. Of the monies provided by the U.S. government for research in weather modification, only about 1 percent is for the support of programs studying inadvertent modification.

In looking toward the future, clearly two effects on the atmosphere, both closely related, need careful monitoring. The direct input of heat into the atmosphere by the burning of fossil fuels and other uses of energy, and the production of carbon dioxide, reinforce each other in tending toward increasing the temperature of the atmosphere. The particulate production, working in the opposite sense of tending to lower the temperature, does not seem to pose as serious a long-term problem. Most of the small particles produced by man stay in the atmosphere for a relatively short period of time (about 1 week). Only if the particles are injected into the high atmosphere such as by the SST can these become significant in influencing overall climatic patterns.

In the near term (50 to 100 years) both direct energy input and carbon dioxide could swamp any natural lowering of overall temperature. If energy use on a worldwide scale continues to grow at 5 percent per year, the world average temperature might rise by as much as 1 to 2°C, provided that other feedback mechanisms do not come into play. For example, a higher temperature would lead to a greater water content of the atmosphere, but this could either lead to cooling by increased cloud cover or heating as water molecules trap the outgoing infrared radiation. Clearly, an understanding of these processes is essential if life is to be maintained on the earth.

As to the longer term, the basic limitation to industrial growth is that derived from the second law of thermodynamics. No matter how efficient we become, energy will be wasted, and even energy converted into useful work will, at least in part, end up in the atmosphere. Only a small fraction will radiate out into space and be of no further concern. Part, however, will remain trapped and perturb the atmosphere, its motion, the weather, and thus the climate. Today we are using energy to provide power to society on a worldwide basis at a rate that is 1/100 of 1 percent of the energy that actually moves the winds and waves. Exponential growth at 5 percent per year implies that in 150 years man's activities would release into the atmosphere about 2 percent of the energy now driving all weather and climate. Even though the flywheels of stability—the atmosphere with a mass of 5.1×10^{15} tons of air and the upper layers of the oceans with a mass of 1.4×10^{18} tons of water — are great, several decades of massive energy input will disrupt the present climatic regime.

Other limits to growth may come into play long before massive climatic changes take place. However, if these other limits can be overcome, there is one last barrier that thermodynamics places on what man can do. Work means heat and heat placed into the atmosphere will bring about changes that in 1975 we cannot predict. But we do know how much energy the earth and its atmosphere can absorb. That limit is finite, and at present growth rates we will bring about major changes in climate in the next hundred years.

References

1. Mitchell, J. 1963. *Changes of Climate* (UNESCO Symposium). U.N., New York.
2. Mitchell, J. 1969. In *Global Effects of Environmental Pollution* (S. F. Singer, ed.) (AAAS Symposium). Boston.
3. Matthews, W. H., Kellogg, W. H., and Robinson, G. D. (eds.). 1971. *Man's Impact on the Climate*. The MIT Press, Cambridge, Mass. pp. 35–36.
4. Wilson, A. 1964. *Nature 201:* 477–478.
5. Bryson, R. 1968. *Weatherwise 21:* 56–61.
6. McCormick, R., and Ludwig, J. 1967. *Science 156:* 1358–1359.
7. Machta, L. 1972. The role of oceans and biosphere in the carbon dioxide cycle. In *The Changing Chemistry of the Oceans* (D. Dyrssen and D. Jagner, eds.). Almqvist & Wiksell Förlag AB, Stockholm, pp. 121–145.
8. Callendar, G. 1958. On the amount of carbon dioxide in the atmosphere. *Tellus 10:* 243–248.
9. Keeling, C. D. 1970. As reported by J. Murray Mitchell Jr. in "A preliminary evaluation of atmospheric pollution as a cause of the global temperature fluctuations of the past century." In *Global Effects of Environmental Pollution* (S. F. Singer, ed.). D. Reidel Publ. Co., Dordrecht, Holland.
10. Averitt, P. 1969. *Coal Resources of the United States, Jan. 1, 1967 (U.S. Geol. Survey Bull. 1275)*. Government Printing Office, Washington, D.C.
11. Weeks, L. H. 1958. Fuel reserves of the future. *Amer. Assoc. Petrol. Geol. Bull. 42:* 431–438; see also Reference 13.
12. Ryman, W. P. 1969. Quoted by Hubbert, Reference 13.
13. Hubbert, M. 1969. Energy resources. In *Resources and Man* (P. Cloud, ed.). W. H. Freeman and Company, San Francisco.
14. Manabe, S., and Wetherald, R. T. 1967. Thermal equilibrium of the atmosphere with a given distribution of relative humidity. *J. Atmos. Sci. 24:* 241–259.
15. Matthews, W. H. (ed.). 1970. *Man's Impact on the Global Environment*. SCEP. 1970. The MIT Press,
16. Davitaya, F. 1965. *Trans Soviet Acad. Sci. Geogr. Sov. 2.*
17. Budyko, M. 1967. Changes of climate. *Meteorologiya i gidrologiya 11.*
18. Frederick, R. H. Personal communications.
19. Hosler, C. L. 1968. *Weatherwise 21:* 110.
20. Hobbs, P. V., and Locatelli, J. D. 1970. *J. Atmos. Sci. 27:* 90–100.
21. Hobbs, P. V., Radke, L. F., and Shumway, S. E. 1970. *J. Atmos. Sci. 27:* 81–89.

Environment and Society

PHOTO BY GEORG GERSTER/RAPHO GUILLUMETTE

Editor's Commentary—
Chapters 16-19

THE FOLLOWING COMMENTS are introductory to the last four chapters, which deal with the broadly social aspects (economics, law, politics, institutions) of environmental problems. These chapters present quite openly biased views, and this is an important *positive* feature, because there is no single, totally objective, "best" solution to environmental problems. Each of us will judge a solution to be better or worse against a set of criteria that reflects not only the physical reality of the problem, but also the social consequences of the solution, and, in turn, each person's view of good or bad social consequences is a result of her or his values and experiences. Scientists are inclined to claim that we must view problems "objectively", but with regard to social issues, of which environmental problems are an example, this is a dangerous smokescreen; our views on problems that have social implications are inextricably bound up with our value system, *and quite rightly so;* to claim objectivity is to obscure this fact. (In fact, scientists are not even truly objective about science, but see it through lenses shaped by existing theory and expectations.) However, once we have admitted that we are not wholly objective in analyzing environmental issues, it is important to make clear just what our bias is, so that others can evaluate it separately from other aspects of our argument.

The first two chapters in this section are written with a bias towards free enterprise, capitalist values. The basic hypothesis is that, given the opportunity, the "market" processes of capitalism, by operating through the profit motive and pricing mechanisms, can produce the desired environmental quality with the highest efficiency. Of course regulation, administration, legislation and the courts are needed to set a proper framework for these mechanisms and to ensure their proper functioning. While this may seem the standard, or establishment point of view, the reader should realize that Dr. Krier's fusing of economic and legal theory in this way is in fact a new development, for which he has been largely responsible. Notice, too, that neither Dr. Ruff nor Dr. Krier is suggesting that the pricing mechanism will produce perfect results, but that, given the alternatives, it is likely to maximize efficiency, while being constrained by considerations of fairness.

The last two chapters are more clearly "anti-establishment". While efficiency is not rejected as a goal, the goals of fairness, or economic equality, and of the ability of citizens to control the decisions that shape their environment, are held to be of prime importance. Also implicit is the view that under corporate capitalism the pricing mechanism does not operate either efficiently or fairly, it doesn't lead to a healthy environment, and that in our huge society capitalism does not result in the citizen's control over the quality of her or his environment. While the last chapter deals with these and other environmental questions, in general and on a large scale, Dr. Molotch's chapter specifically attacks the question of growth where it most immediately affects us all, namely at the local level.

In addition to gaining information and concepts from these four chapters, I hope the reader will learn from them how to see environmental problems in a broad social perspective, and how to recognize and evaluate the social bias in proposed solutions.

398

Cars stacked in junkyard near Escondido, California. EPA—Documerica
photo by Gene Daniels.

16

Environment and Economics

Larry E. Ruff *is Program Officer in the Office of Resources and the Environment of the
Ford Foundation, New York, N.Y. He received his B.S. in physics at the California Institute of
Technology in 1963 and his Ph.D. in economics at Stanford in 1968. After teaching economics at
the University of California, San Diego, for three years, he went to Washington to work on
environmental economics and policy in several Federal agencies. In 1974, he left the position of
Director of Social Science Research, U. S. Environmental Protection Agency, to join the Ford
Foundation.*

We are going to make very little real progress in solving the problem of pollution until we recognize it for what, primarily, it is: an economic problem, which must be understood in economic terms. Of course, there are *noneconomic* aspects of pollution, as there are with all economic problems, but all too often, such secondary matters dominate discussion. Engineers, for example, are certain that pollution will vanish once they find the magic gadget or power source. Politicians keep trying to find the right kind of bureaucracy; and bureaucrats maintain an unending search for the correct set of rules and regulations. Those who are above such vulgar pursuits pin their hopes on a moral regeneration or social revolution, apparently in the belief that saints and socialists have no garbage to dispose of. But as important as technology, politics, law, and ethics are to the pollution question, all these approaches are bound to have disappointing results, for they ignore the primary fact that pollution is an economic problem.

Before developing an economic analysis of pollution, however, it is necessary to dispose of some popular myths.

First, pollution is not new. Spanish explorers landing in the sixteenth century noted that smoke from Indian campfires hung in the air of the Los Angeles Basin, trapped by what is now called the inversion layer. Before the first century B.C., the drinking waters of Rome were becoming polluted.

Second, most pollution is not due to afflu-

This chapter previously appeared in The Public Interest, No. 19 (Spring 1970), 69–85, under the title "The Economic Commonsense of Pollution." Reprinted by permission of The Public Interest. Copyright © by National Affairs, Inc., 1970.

ence, despite the current popularity of this notion. In India, the pollution runs in the streets, and advice against drinking the water in exotic lands is often well taken. Nor can pollution be blamed on the self-seeking activities of greedy capitalists. Once-beautiful rivers and lakes that are now open sewers and cesspools can be found in the USSR as well as in the United States, and some of the world's dirtiest air hangs over cities in eastern Europe, which are neither capitalist nor affluent. In many ways, indeed, it is much more difficult to do anything about pollution in noncapitalist societies. In the USSR, there is no way for the public to become outraged or to exert any pressure, and the polluters and the courts there work for the same people, who often decide that clean air and water, like good clothing, are low on their list of social priorities.

In fact, it seems probable that affluence, technology, and slow-moving, inefficient democracy will turn out to be the cure more than the cause of pollution. After all, only an affluent, technological society can afford such luxuries as moon trips, 3-day weekends, and clean water, although even our society may not be able to afford them all; and only in a democracy can the people hope to have any real influence on the choice among such alternatives.

What *is* new about pollution is what might be called the *problem* of pollution. Many unpleasant phenomena — poverty, genetic defects, hurricanes — have existed forever without being considered problems; they are, or were, considered to be facts of life, like gravity and death, and a mature person simply adjusted to them. Such phenomena become problems only when it begins to appear that something can and should be done about them. It is evident that pollution has advanced to the problem stage.

400

Now the question is what can and should be done?

Most discussions of the pollution problem begin with some startling facts: Did you know that 15,000 tons of filth are dumped into the air of Los Angeles County every day? But by themselves, such facts are meaningless, if only because there is no way to know whether 15,000 tons is a lot or a little. It is much more important for clear thinking about the pollution problem to understand a few economic concepts than to learn a lot of sensational-sounding numbers.

Marginalism

One of the most fundamental economic ideas is that of *marginalism*, which entered economic theory when economists became aware of the differential calculus in the nineteenth century and used it to formulate economic problems as problems of "maximization." The standard economic problem came to be viewed as that of finding a level of operation of some activity which would maximize the net gain from that activity, where the net gain is the difference between the benefits and the costs of the activity. As the level of activity increases, both benefits and costs will increase; but because of diminishing returns, costs will increase faster than benefits. When a certain level of activity is reached, any further expansion increases costs more than benefits. At this "optimal" level, "marginal cost" — or the cost of expanding the activity — equals "marginal benefit," or the benefit from expanding the activity. Further expansion would cost more than it is worth, and reduction in the activity would reduce benefits more than it would save costs. The net gain from the activity is said to be maximized at this point.

This principle is so simple that it is almost embarrassing to admit that it is the cornerstone of economics. Yet intelligent men often ignore it in discussion of public issues. Educators, for example, often suggest that, if it is better to be literate than illiterate, there is no logical stopping point in supporting education. Or scientists have pointed out that the benefits derived from "science" obviously exceed the costs and then have proceeded to infer that their particular project should be supported. The correct comparison, of course, is between *additional* benefits created by the proposed activity and the *additional* costs incurred.

The application of marginalism to questions of pollution is simple enough conceptually. The difficult part lies in estimating the cost and benefits functions, a question to which I shall return. But several important qualitative points can be made immediately. The first is that the choice facing a rational society is *not* between clean air and dirty air, or between clear water and polluted water, but rather between various *levels* of dirt and pollution. The aim must be to find that level of pollution abatement where the costs of further abatement begin to exceed the benefits.

The second point is that the optimal combination of pollution control methods is going to be a very complex affair. Such steps as demanding a 10 percent reduction in pollution from all sources, without considering the relative difficulties and costs of the reduction, will certainly be an inefficient approach. Where it is less costly to reduce pollution, we want a greater reduction, to a point where an additional dollar spent on control anywhere yields the same reduction in pollution levels.

Markets, Efficiency, and Equity

A second basic economic concept is the idea — or the ideal — of the self-regulating economic system. Adam Smith illustrated this ideal with the example of bread in London: the uncoordinated, selfish actions of many people — farmer, miller, shipper, baker, grocer — provide bread for the city dweller, without any central control and at the lowest possible cost. Pure self-interest, guided only by the famous "invisible hand" of competition, organizes the economy efficiently.

The logical basis of this rather startling result is that, under certain conditions, competitive prices convey all the information necessary for making the optimal decision. A builder trying to decide whether to use brick or concrete will weigh his requirements and tastes against the prices of the materials. Other users will do the same, with the result that those whose needs and preferences for brick are relatively the strongest will get brick. Further, profit-maximizing producers will weigh relative production costs, reflecting society's productive capabilities, against relative prices, reflecting society's tastes and desires, when deciding how much of each good to produce. The end result is

401

that users get brick and cement in quantities and proportions that reflect their individual tastes and society's production opportunities. No other solution would be better from the standpoint of all the individuals concerned.

This suggests what it is that makes pollution different. The efficiency of competitive markets depends on *private* costs and *social* costs being exactly the same. As long as the brick–cement producer must compensate somebody for every cost imposed by his production, his profit-maximizing decisions about how much to produce, and how, will also be socially efficient decisions. Thus, if a producer dumps wastes into the air, river or ocean; if he pays nothing for such dumping; and if the disposed wastes have no noticeable effect on anyone else, living or still unborn; then the private and social costs of disposal are identical and nil, and the producer's private decisions are socially efficient. *But if these wastes do affect others, the social costs of waste disposal are not zero. Private and social costs diverge, and private profit-maximizing decisions are not socially efficient.* Suppose, for example, that cement production dumps large quantities of dust into the air, which damages neighbors, and that the brick–cement producer pays these neighbors nothing. In the social sense, cement will be overproduced relative to brick and other products because users of the products will make decisions based on market prices which do not reflect true social costs. They will use cement when they should use brick, or when they should not build at all.

This divergence between private and social costs is the fundamental cause of pollution of all types, and it arises in any society where decisions are at all decentralized—which is to say, in any economy of any size which hopes to function at all. Even the socialist manager of the brick–cement plant, told to maximize output given the resources at his disposal, will use the People's Air to dispose of the People's Wastes; to do otherwise would be to violate his instructions. And if instructed to avoid pollution "when possible," he does not know what to do: how can he decide whether more brick or cleaner air is more important for building socialism? The capitalist manager is in exactly the same situation. Without prices to convey the needed information, he does not know what action is in the public interest, and certainly would have no incentive to act correctly even if he did know.

Although markets fail to perform efficiently when private and social costs diverge, this does not imply that there is some inherent flaw in the idea of acting on self-interest in response to market prices. Decisions based on private cost calculations are typically correct from a social point of view; and even when they are not quite correct, it often is better to accept this inefficiency than to turn to an alternative decision mechanism, which may be worse. Even the modern economic theory of socialism is based on the high correlation between managerial self-interest and public good. There is no point in trying to find something — some omniscient and omnipotent *deus ex machina* — to replace markets and self-interest. Usually it is preferable to modify existing institutions, where necessary, to make private and social interest coincide.

And there is a third relevant economic concept: the fundamental distinction between questions of efficiency and questions of equity or fairness. A situation is said to be efficient if it is not possible to rearrange things so as to benefit one person without harming any others. That is the *economic* equation for efficiency. *Politically*, this equation can be solved in various ways; though most reasonable men will agree that efficiency is a good thing, they will rarely agree about which of the many possible efficient states, each with a different distribution of "welfare" among individuals, is the best one. Economics itself has nothing to say about which efficient state is the best. That decision is a matter of personal and philosophical values, and ultimately must be decided by some political process. Economics can suggest ways of achieving efficient states, and can try to describe the equity considerations involved in any suggested social policy; but the final decisions about matters of "fairness" or "justice" cannot be decided on economic grounds.

Estimating the Costs of Pollution

Both in theory and practice, the most difficult part of an economic approach to pollution is the measurement of the cost and benefits of its abatement. Only a small fraction of the costs of pollution can be estimated straightforwardly. If, for example, smog reduces the life of automobile tires by 10 percent, one component of the cost of smog is 10 percent of tire expenditures. It has been estimated that, in a moderately polluted area of New York City, filthy air imposes extra costs for painting, washing, laundry, etc., of $200 per person per year. Such costs must be included in any calculation of the benefits of pollution abatement, and yet they are only a part of the relevant costs — and often a small part. Accordingly, it rarely is possible to justify a measure like river pollution control solely on the basis of costs to individuals or firms of treating water because it usually is cheaper to process only the water that is actually used for industrial or municipal purposes, and to ignore the river itself.

The costs of pollution that cannot be measured so easily are often called "intangible" or "noneconomic," although neither term is particularly appropriate. Many of these costs are as tangible as burning eyes or a dead fish, and all such costs are relevant to a valid economic analysis. Let us therefore call these costs "nonpecuniary."

The only real difference between nonpecuniary costs and the other kind lies in the difficulty of estimating them. If pollution in Los Angeles harbor is reducing marine life, this imposes costs on society. The cost of reducing commercial fishing could be estimated directly: it would be the fixed cost of converting men and equipment from fishing to an alternative occupation, plus the difference between what they earned in fishing and what they earn in the new occupation, plus the loss to consumers who must eat chicken instead of fish. But there are other, less straightforward costs: the loss of recreation opportunities for children and sportsfishermen and of research facilities for marine biologists, etc. Such costs are obviously difficult to measure and may be very large indeed; but just as surely as they are not zero, so too are they not infinite. Those who call for immediate action and damn the cost, merely because the spiney starfish and furry crab populations are shrinking, are putting an infinite marginal value on these creatures. This strikes a disinterested observer as an overestimate.

The above comments may seem crass and insensitive to those who, like one angry letter-writer to the Los Angeles *Times*, want to ask: "If conservation is not for its own sake, then what in the world *is* it for?" Well, what *is* the purpose of pollution control? Is it for its own sake? Of course not. If we answer that it is to make the air and water clean and quiet, the question arises: What is the purpose of clean air and water? If the answer is, to please the nature gods, it must be conceded that all pollution must cease immediately because the cost of angering the gods is presumably infinite. But if the answer is that the purpose of clean air and water is to further human enjoyment of life on this planet, we are faced with the economists' basic question: Given the limited alternatives that a niggardly nature allows, how can we best further human enjoyment of life? And the answer is, by making intelligent marginal decisions on the basis of costs and benefits. Pollution control is for lots of things: breathing comfortably, enjoying mountains, swimming in water, for health, beauty, and the general delectation. But so are many other things, like good food and wine, comfortable housing and fast transportation. The question is not which of these desirable things we should have, but rather what combination is most desirable. To determine such a combination, we must know the rate at which individuals are willing to substitute more of one desirable thing for less of another desirable thing. Prices are one way of determining those rates.

But if we cannot directly observe market prices for many of the costs of pollution, we must find another way to proceed. One possibility is to infer the costs from other prices, just as we infer the value of an ocean view from real estate prices. In principle, one could estimate the value people put on clean air and beaches by observing how much more they are willing to pay for property in nonpolluted areas. Such information could be obtained; but there is little of it available at present.

Another possible way of estimating the costs of pollution is to ask people how much they would be willing to pay to have pollution reduced. A resident of Pasadena might be willing to pay $100 per year to have smog reduced 10 or 20 percent. In Barstow, where the marginal cost of smog is much less, a resident might not pay $10 per year to have smog reduced 10 percent. If we knew how much it was worth to everybody, we could add up these amounts and obtain an estimate of the cost of a marginal amount of pollution. The difficulty, of course, is that there is no way of guaranteeing truthful responses. Your response to the question, how much is pollution costing *you*, obviously will depend on what you think will be done with this information. If you think you will be compensated for these costs, you will make a generous estimate; if you think that you will be charged for the control in proportion to these costs, you will make a small estimate.

In such cases it becomes very important how the questions are asked. For example, the voters could be asked a question of the form: Would you like to see pollution reduced x percent if the result is a y percent increase in the cost of living? Presumably a set of questions of this form could be used to estimate the costs of pollution, including the "unmeasurable" costs. But great care must be taken in formulating the questions. For one thing, if the voters will benefit differentially from the activity, the questions should be asked in a way that reflects this fact. If, for example, the issue is cleaning up a river, residents near the river will be willing to pay more for the cleanup and should have a means of expressing this. Ultimately, some such political procedure probably will be necessary, at least until our more direct measurement techniques are greatly improved.

Let us assume that, somehow, we have made an estimate of the social cost function for pollution, including the marginal cost associated with various pollution levels. We now need an estimate of the benefits of pollution — or, if you prefer, of the costs of pollution abatement. So we set the Pollution Control Board (PCB) to work on this task.

The PCB has a staff of engineers and technicians, and they begin working on the obvious question: for each pollution source, how much would it cost to reduce pollution by 10 percent, 20 percent, and so on. If the PCB has some economists, they will know that the cost of reducing total pollution by 10 percent is *not* the total cost of reducing each pollution source by 10 percent. Rather, they will use the equimarginal principle and find the pattern of control such that an additional dollar spent on control of any pollution source yields the same reduction. This will minimize the cost of achieving any given level of abatement. In this way the PCB can generate a "cost of abatement" function and the corresponding marginal cost function.

Although this procedure seems straightforward enough, the practical difficulties are tremendous. The amount of information needed by the PCB is staggering; to do this job right, the PCB would have to know as much about each plant as the operators of the plant themselves. The cost of gathering these data is obviously prohibitive, and, since marginal principles apply to data collection, too, the PCB would have to stop short of complete information, trading off the resulting loss in efficient control against the cost of better information. Of course, just as fast as the PCB obtained the data, a technological change would make it obsolete.

The PCB would have to face an additional complication. It would not be correct simply to determine how to control existing pollution sources given their existing locations and production methods. Although this is almost certainly what the PCB would do, the resulting cost functions will overstate the true social cost of control. Muzzling existing plants is only one method of control. Plants can move, or switch to a new process, or even to a new product. Consumers can switch to a less-polluting substitute. There are any number of alternatives, and the poor PCB engineers can never know them all. This could lead to some costly mistakes. For example, the PCB may correctly conclude that the cost of installing effective dust control at the cement plant is very high and hence may allow the pollution to continue, when the best solution is for the cement plant to switch to brick production while a plant in the desert switches from brick to cement. The PCB can never have all this information and therefore is doomed to inefficiency, sometimes an inefficiency of large proportions.

Once cost and benefit functions are known, the PCB should choose a level of abatement that maximizes net gain. This occurs where the marginal cost of further abatement just equals the marginal benefit. If, for example, we could reduce pollution damages by $2 million at a cost of $1 million, we should obviously impose that $1 million cost. But if the damage reduction is only $500,000, we should not and in fact should reduce control efforts.

This principle is obvious enough but is often overlooked. One author, for example, has written that the national cost of air pollution is $11 billion per year but that we are spending less than $50 million per year on control; he infers from this that "we could justify a tremendous strengthening of control efforts on purely economic grounds." That *sounds* reasonable, if all you care about are sounds. But what is the logical content of the statement? Does it imply that we should spend $11 billion on control just to make things even? Suppose that we were spending $11 billion on control and thereby succeeded in reducing pollution costs to $50 million. Would this imply that we were spending too *much* on control? Of course not. We must compare the *marginal* decrease in pollution costs to the *marginal* increase in abatement costs.

Difficult Decisions

Once the optimal pollution level is determined, all that is necessary is for the PCB to enforce the pattern of controls that it has determined to be optimal. (Of course, this pattern will not really be the best one, because the PCB will not have all the information it should have.) But now a new problem arises: How should the controls be enforced?

The most direct and widely used method is in many ways the least efficient: direct regulation. The PCB can decide what each polluter must do to reduce pollution and then simply require that action under penalty of law. But this approach has many shortcomings. The polluters have little incentive to install the required devices or to keep them operating properly. Constant inspection is therefore necessary. Once the polluter has complied with the letter of the law, he has no incentive to find better methods of pollution reduction. Direct control of this sort has a long history of inadequacy; the necessary bureaucracies rarely manifest much vigor, imagination, or devotion to the public interest. Still, in some situations there may be no alternative.

A slightly better method of control is for the PCB to set an acceptable level of pollution for each source and let the polluters find the cheapest means of achieving this level. This reduces the amount of information the PCB needs, but not by much. The setting of the acceptable levels becomes a matter for negotiation, political pull, or even graft. As new plants are built and new control methods invented, the limits should be changed; but if they are, the incentive to find new designs and new techniques is reduced.

A third possibility is to subsidize the reduction of pollution, either by subsidizing control equipment or by paying for the reduction of pollution below standard levels. This alternative has all the problems of the above methods, plus the classic shortcoming that plagues agricultural subsidies: the old joke about getting into the not-growing-cotton business is not always so funny.

The PCB will also have to face the related problem of deciding *who* is going to pay the costs of abatement. Ultimately, this is a question of equity or fairness which economics cannot answer; but economics can suggest ways of achieving equity without causing inefficiency. In general, the economist will say: if you think polluter A is deserving of more income at polluter B's expense, by all means give A some of B's income; but do *not* try to help A by allowing him to pollute freely. For example, suppose that A and B each operate plants which produce identical amounts of pollution. Because of different technologies, however, A can reduce his pollution 10 percent for $100, while B can reduce his pollution 10 percent for $1,000. Suppose that your goal is to reduce total pollution 5 percent. Surely it is obvious that the best (most efficient) way to do this is for A to reduce his pollution 10 percent while B does nothing. But suppose that B is rich and A is poor. Then many would demand that B reduce his pollution 10 percent while A does nothing because B has a greater "ability to pay." Well, perhaps B does have greater ability to pay, and perhaps it is "fairer" that he pay the costs of pollution control; but if so, B should pay the $100 necessary to reduce A's pollution. To force B to reduce his own pollution 10 percent is equivalent to taxing B $1,000 and then blowing the $1,000 on an extremely inefficient pollution control method. Put this way, it is obviously a stupid thing to do; but put in terms of B's greater ability to pay, it will get considerable support, although it is no less stupid. The more efficient alternative is not always available, in which case it may be acceptable to use the inefficient method. Still, it should not be the responsibility of the pollution authorities to change the distribution of welfare in society; this is the responsibility of higher authorities. The PCB should concentrate on achieving economic efficiency without being grossly unfair in its allocation of costs.

Clearly, the PCB has a big job that it will never be able to handle with any degree of efficiency. Needed is some sort of self-regulating system, such as a market, which will automatically adapt to changes in conditions, provide incentives for development and adoption of improved control methods, reduce the amount of information the PCB must gather and the amount of detailed control it must exercise, and so on. This, by any standard, is a tall order.

Putting a Price on Pollution

And yet there is a very simple way to accomplish all this. *Put a price on pollution.* A price-based control mechanism would differ from an ordinary market transaction system only in that the PCB would set the prices, instead of their being set by demand–supply forces, and that the state would force payment. Under such a system, anyone could emit any amount of pollution as long as he pays the price which the PCB sets to approximate the marginal social cost of pollution. Under this circumstance, private decisions based on self-interest are efficient. If pollution consists of many components, each with its own social cost, there should be different prices for each component. Thus, extremely dangerous materials must have an extremely high price, perhaps stated in terms of "years in jail" rather than "dollars," although a sufficiently high dollar price is essentially the same thing. In principle, the prices should vary with geographical location, season of the year, direction of the wind, and even day of the week, although the cost of too many variations may preclude such fine distinctions.

Once the prices are set, polluters can adjust to them any way they choose. Because they act on self-interest they will reduce their pollution by every means possible up to the point where further reduction would cost more than the price. Because all face the same price for the same type of pollution, the marginal cost of abatement is the same everywhere. If there are economies of scale in pollution control, as in some types of liquid waste treatment, plants can cooperate in establishing joint treatment facilities. In fact, some enterprising individual could buy these wastes from various plants (at negative prices — i.e., they would get paid for carting them off), treat them, and then sell them at a higher price, making a profit in the process. (After all, this is what rubbish removal firms do now.) If economies of scale are so substantial

that the provider of such a service becomes a monopolist, the PCB can operate the facilities itself.

Obviously, such a scheme does not eliminate the need for the PCB. The board must measure the output of pollution from all sources, collect the fees, and so on. But it does not need to know anything about any plant except its total emission of pollution. It does not control, negotiate, threaten, or grant favors. It does not destroy incentive, because development of new control methods will reduce pollution payments.

As a test of this price system of control, let us consider how well it would work when applied to automobile pollution, a problem for which direct control is usually considered the only feasible approach. If the price system can work here, it can work anywhere. Suppose that a price is put on the emissions of automobiles. Obviously, continuous metering of such emissions is impossible. But it should be easy to determine the average output of pollution for cars of various makes, models, and years, having different types of control devices and using various types of fuel. Through graduated registration fees and fuel taxes, each car owner would be assessed roughly the social cost of his car's pollution, adjusted for whatever control devices he has chosen to install and for his driving habits. If the cost of installing a device, driving a different car, or finding alternative means of transportation is less than the price he must pay to continue his pollution, he will presumably take the necessary steps. But each individual remains free to find the best adjustment to his particular situation. It would be remarkable if everyone decided to install the same devices which some states currently require; and yet that is the effective assumption of such requirements.

Even in the difficult case of auto pollution, the price system has a number of advantages. Why should a person living in the Mojave desert, where pollution has little social cost, take the same pains to reduce air pollution as a person living in Pasadena? Present California law, for example, makes no distinction between such areas; the price system would. And what incentive is there for auto manufacturers to design a less polluting engine? The law says only that they must install a certain device in every car. If GM develops a more efficient engine, the law will eventually be changed to require this engine on all cars, raising costs and reducing sales. But will such development take place? No collusion is needed for manufacturers to decide unanimously that it would be foolish to devote funds to such development. But with a pollution fee paid by the consumer, there is a real advantage for any firm to be first with a better engine, and even a collusive agreement would not last long in the face of such an incentive. The same is true of fuel manufacturers, who now have no real incentive to look for better fuels. Perhaps most important of all, the present situation provides no real way of determining whether it is cheaper to reduce pollution by muzzling cars or industrial plants. The experts say that most smog comes from cars; but *even if true, this does not imply that it is more efficient to control autos rather than other pollution sources*. How can we decide which is more efficient without mountains of information? The answer is, by making drivers and plants pay the same price for the same pollution, and letting self-interest do the job.

In situations where pollution outputs can be measured more or less directly (unlike the automobile pollution case), the price system is clearly superior to direct control. A study of possible control methods in the Delaware estuary, for example, estimated that, compared to a direct control scheme requiring each polluter to reduce his pollution by a fixed percentage, an effluent charge that would achieve the same level of pollution abatement would be only half as costly—a saving of about $150 million. Such a price system would also provide incentive for further improvements, a simple method of handling new plants, and revenue for the control authority.

In general, the price system allocates costs in a manner that is at least superficially fair: those who produce and consume goods that cause pollution pay the costs. But the superior efficiency in control and apparent fairness are not the only advantages of the price mechanism. Equally important is the ease with which it can be put into operation. It is not necessary to have detailed information about all the techniques of pollution reduction, or estimates of all costs and benefits. Nor is it necessary to determine whom to blame or who should pay. All that is needed is a mechanism for estimating, if only roughly at first, the pollution output of all polluters, together with a means of collecting fees. Then we can simply pick a price — any price — for each

category of pollution, and we are in busir The initial price should be chosen on the ba; some estimate of its effects but need not be the optimal one. If the resulting reduction in pollution is not "enough," the price can be raised until there is sufficient reduction. A change in technology, number of plants, or whatever can be accommodated by a change in the price, even without detailed knowledge of all the technological and economic data. Further, once the idea is explained, the price system is much more likely to be politically acceptable than some method of direct control. Paying for a service, such as garbage disposal, is a well-established tradition and is much less objectionable than having a bureaucrat nosing around and giving arbitrary orders. When businessmen, consumers, and politicians understand the alternatives, the price system will seem very attractive indeed.

Who Sets the Prices?

An important part of this method of control obviously is the mechanism that sets and changes the pollution price. Ideally, the PCB could choose this price on the basis of an estimate of the benefits and costs involved, in effect imitating the impersonal workings of ordinary market forces. But because many of the costs and benefits cannot be measured, a less "objective," more political procedure is needed. This political procedure could take the form of a referendum, in which the PCB would present to the voters alternative schedules of pollution prices, together with the estimated effects of each. There would be a massive propaganda campaign waged by the interested parties, of course. Slogans such as "Vote NO on 12 and Save Your Job," or "Proposition 12 Means Higher Prices," might be overstatements but would contain some truth, as the individual voter would realize when he considered the suggested increase in gasoline taxes and auto registration fees. But the other side, in true American fashion, would respond by overstating *their* case: "Smog Kills, Yes on 12," or "Stop *Them* From Ruining *Your* Water." It would be up to the PCB to inform the public about the true effects of the alternatives; but, ultimately, the voters would make the decision.

It is fashionable in intellectual circles to object to such democratic procedures on the ground that the uncultured masses will not make correct decisions. If this view is based on the fact that the technical and economic arguments are likely to be too complex to be decided by direct referendum, it is certainly a reasonable position; one obvious solution is to set up an elective or appointive board to make the detailed decisions, with the expert board members being ultimately responsible to the voters. But often there is another aspect to the antidemocratic position — a feeling that it is impossible to convince the people of the desirability of some social policy, not because the issues are too complex but purely because their values are "different" and inferior. To put it bluntly: many ardent foes of pollution are not so certain that popular opinion is really behind them, and they therefore prefer a more bureaucratic and less political solution.

The question of who should make decisions for whom, or whose desires should count in a society, is essentially a noneconomic question that an economist cannot answer with authority, whatever his personal views on the matter. The political structures outlined here, when combined with the economic suggestions, can lead to a reasonably efficient solution of the pollution problem in a society where the tastes and values of all men are given some consideration. In such a society, when any nonrepresentative group is in a position to impose its particular evaluation of the costs and benefits, an inefficient situation will result. The swimmer or tidepool enthusiast who wants Los Angeles harbor converted into a crystal-clear swimming pool, at the expense of all the workers, consumers, and businessmen who use the harbor for commerce and industry, is indistinguishable from the stockholder in Union Oil who wants maximum output from offshore wells, at the expense of everyone in the Santa Barbara area. Both are urging an inefficient use of society's resources; both are trying to get others to subsidize their particular thing — a perfectly normal, if not especially noble, endeavor.

If the democratic principle upon which the above political suggestions are based is rejected, the economist cannot object. He will still suggest the price system as a tool for controlling pollution. With any method of decision — whether popular vote, representative democracy, consultation with the nature gods, or a dictate of the intellectual elite — the price system can simplify

control and reduce the amount of information needed for decisions. It provides an efficient comprehensive, easily understood, adaptable, and reasonably fair way of handling the problem. It is ultimately the only way the problem will be solved. Arbitrary, piecemeal, stop-and-go programs of direct control have not and will not accomplish the job.

Some Objections Are Not an Answer

There are some objections that can be raised against the price system as a tool of pollution policy. Most are either illogical or apply with much greater force to any other method of control.

For example, one could object that what has been suggested here ignores the difficulties caused by fragmented political jurisdictions; but this is true for any method of control. The relevant question is: What method of control makes interjurisdictional cooperation easier and more likely? And the answer is: a price system, for several reasons. First, it is probably easier to get agreement on a simple schedule of pollution prices than on a complex set of detailed regulations. Second, a uniform price schedule would make it more difficult for any member of the "cooperative" group to attract industry from the other areas by promising a more lenient attitude toward pollution. Third, and most important, a price system generates revenues for the control board, which can be distributed to the various political entities. While the allocation of these revenues would involve some vigorous discussion, any alternative methods of control would require the various governments to raise taxes to pay the costs, a much less appealing prospect; in fact, there would be a danger that the pollution prices might be considered a device to generate revenue rather than to reduce pollution, which could lead to an overly clean, inefficient situation.

Another objection is that the Pollution Control Board might be captured by those it is supposed to control. This danger can be countered by having the board members subject to election or by having the pollution prices set by referendum. With any other control method, the danger of the captive regulator is much greater. A uniform price is easy for the public to understand, unlike obscure technical arguments about boiler temperatures and the costs of electrostatic collectors versus low-sulfur oil from Indonesia; if pollution is too high, the public can demand higher prices, pure and simple. And the price is the same for all plants, with no excuses. With direct control, acceptable pollution levels are negotiated with each plant separately and in private, with approved delays and special permits and other nonsense. The opportunities for using political influence and simple graft are clearly much larger with direct control.

A different type of objection occasionally has been raised against the price system, based essentially on the fear that it will solve the problem. Pollution, after all, is a hot issue with which to assault The Establishment, Capitalism, Human Nature, and Them; any attempt to remove the issue by some minor change in institutions, well within The System, must be resisted by The Movement. From some points of view, of course, this is a perfectly valid objection. But one is hopeful that there still exists a majority more concerned with finding solutions than with creating issues.

There are other objections which could be raised and answered in a similar way. But the strongest argument for the price system is not found in idle speculation but in the real world, and in particular, in Germany. The Rhine River in Germany is a dirty stream, recently made notorious when an insecticide spilled into the river and killed millions of fish. One tributary of the Rhine, a river called the Ruhr, is the sewer for one of the world's most concentrated industrial areas. The Ruhr River valley contains 40 percent of German industry, including 80 percent of coal, iron, steel and heavy chemical capacity. The Ruhr is a small river, with a low flow of less than half the flow on the Potomac near Washington. The volume of wastes is extremely large—actually exceeding the flow of the river itself in the dry season! *Yet people and fish swim in the Ruhr River.*

This amazing situation is the result of over 40 years of control of the Ruhr and its tributaries by a hierarchy of regional authorities. These authorities have as their goal the maintenance of the quality of the water in the area at minimum cost, and they have explicitly applied the equimarginal principle to accomplish this. Water quality is formally defined in a technolog-

ical rather than an economic way; the objective is to "not kill the fish." Laboratory tests are conducted to determine what levels of various types of pollution are lethal to fish, and from these figures an index is constructed which measures the "amount of pollution" from each source in terms of its fish-killing capacity. This index is different for each source, because of differences in amount and composition of the waste, and geographical locale. Although this physical index is not really a very precise measure of the real economic *cost* of the waste, it has the advantage of being easily measured and widely understood. Attempts are made on an *ad hoc* basis to correct the index if necessary—if, for example, a nonlethal pollutant gives fish an unpleasant taste.

Once the index of pollution is constructed, a price is put on the pollution, and each source is free to adjust its operation any way it chooses. Geographical variation in prices, together with some direct advice from the authorities, encourage new plants to locate where pollution is less damaging. For example, one tributary of the Ruhr has been converted to an open sewer; it has been lined with concrete and landscaped, but otherwise no attempt is made to reduce pollution in the river itself. A treatment plant at the mouth of the river processes all these wastes at low cost. Therefore, the price of pollution on this river is set low. This arrangement, by the way, is a rational, if perhaps unconscious, recognition of marginal principles. The loss caused by destruction of *one* tributary is rather small, if the nearby rivers are maintained, while the benefit from having this inexpensive means of waste disposal is very large. However, if *another* river were lost, the cost would be higher and the benefits lower; one open sewer may be the optimal number.

The revenues from the pollution charges are used by the authorities to measure pollution, conduct tests and research, operate dams and regulate stream flow, and operate waste treatment facilities where economies of scale make this desirable. These facilities are located at the mouths of some tributaries, and at several dams in the Ruhr. If the authorities find that pollution levels are getting too high, they simply raise the price, which causes polluters to try to reduce their wastes, and provides increased revenues to use on further treatment. Local governments influence the authorities, which helps to maintain recreation values, at least in certain stretches of the river.

This classic example of water management is obviously not exactly the price system method discussed earlier. There is considerable direct control, and the pollution authorities take a very active role. Price regulation is not used as much as it could be; for example, no attempt is made to vary the price over the season, even though high flow on the Ruhr is more than 10 times larger than low flow. If the price of pollution were reduced during high flow periods, plants would have an incentive to regulate their production and/or store their wastes for release during periods when the river can more easily handle them. The difficulty of continuously monitoring wastes means this is not done; as automatic, continuous measurement techniques improve and are made less expensive, the use of variable prices will increase. Although this system is not entirely regulated by the price mechanism, prices are used more here than anywhere else, and the system is much more successful than any other.[1] So, both in theory and in practice, the price system is attractive, and ultimately must be the solution to pollution problems.

"If We Can Go to the Moon, Why . . ." etc.?

"If we can go to the moon, why can't we eliminate pollution?" This new, and already trite, rhetorical question invites a rhetorical response: "If physical scientists and engineers approached their tasks with the same kind of wishful thinking and fuzzy moralizing which characterizes much of the pollution discussion, we would never have gotten off the ground." Solving the pollution problem is no easier than going to the moon, and therefore requires a comparable effort in terms of men and resources and the same sort of logical hard-headedness that made Apollo a success. Social scientists, politicians, and journalists who spend their time trying to find someone to blame, searching for a magic device or regulation, or complaining about human nature, will be as helpful in solving the pollution problem as they were in getting us to the moon. The price system outlined here is no magic formula, but it attacks the problem at its roots and has a real chance of providing a long-term solution.

Reference

1. Kneese, A. V. 1964. *The Economics of Regional Water Quality Management*. The Johns Hopkins Press, Baltimore, Md.

Further Reading

Anthologies: There are a number of recent collections of readings on environmental economics which cover the field quite well. Among them are:

Dorfman, R. and Dorfman, N., 1972. *Economics of the Environment*. W. W. Norton & Company, Inc., New York.
Kneese, A. V., and Bower, B., 1972. *Environmental Quality Analysis*. The Johns Hopkins Press (for Resources for the Future), Baltimore, Md.
Enthoven, A., and Freeman, A. M., III, 1973. *Pollution, Resources, and the Environment*. W. W. Norton & Company, Inc., New York.

Economic Theory: The pure economic theory of environmental problems is developed in a number of books. Beginning with a light, non-technical presentation, through a serious but brief and self-contained exposition, and concluding with a highly abstract, mathematical treatment, the interested reader might refer to:

Crocker, T., and Rogers, A., III, 1971. *Environmental Economics*, Dryden Press, Hinsdale, Ill.
Kneese, A. V., and Bower, B., 1968. *Managing Water Quality: Economics, Technology, Institutions*. The John Hopkins Press (for Resources for the Future), Baltimore, Md. Especially Chapter 5.
Maler, K-G., 1974. *Environmental Economics: A Theoretical Inquiry*. The John Hopkins Press (for Resources for the Future). Baltimore, Md.

Empirical Estimates of Costs and Benefits: Aggregate estimates of the costs, benefits, and economic impacts of pollution control efforts are prepared periodically by the U. S. Environmental Protection Agency and the Council on Environmental Quality. *The Economic Impact of Pollution Control* (March 1972), *The Economics of Clean Water* (more or less annually, with the latest published in 1974), *The Cost of*

Clean Air (annually, most recently in 1974), and *Economic Damages of Air Pollution* (T. Waddell, May 1974) are all EPA publications available through the Government Printing Office, Washington, D. C. The annual reports of CEQ, most notably *Environmental Quality—1973*, usually contain good summaries of this aggregate economic information. See also:

Lave, L., 1972. Air pollution damages: some difficulties in estimating the value of abatement. In Kneese and Bower, *Environmental Quality Analysis, op. cit.*

Lave, L., and Seskin, E., 1970. Air pollution and human health. *Science:* August. Reprinted in Enthoven and Freeman, *op. cit.*, and Dorfman and Dorfman, *op. cit.*

Ridker, R., 1967. *Economic Costs of Air Pollution*, Praeger Publishers, Inc., New York.

Effluent Charge Systems: For more detailed statements of the relative advantages of effluent charge systems, and more fully worked out proposals, see:

Chapters 7–9 and 12 in Kneese and Bower, *Managing Water Quality, op. cit.*

Selig, E., 1973. *Effluent Charges on Air and Water Pollution: A Conference Report.* Environmental Law Institute, Washington, D. C.

Freeman, A. M., III, and Haveman, R., 1972. Clean rhetoric and dirty water. *The Public Interest:* Summer. Reprinted in Enthoven and Freeman, *op. cit.*

White, L., 1973. The auto pollution muddle. *The Public Interest:* Summer.

Growth, Environment, and Resources: An extensive collection of readings is to be found in *Growth and Its Implications for the Future*, Pts. 2 and 3, Hearing Appendix for the Subcommittee on Fisheries and Wildlife Conservation and the Environment, of the Committee on Merchant Marine and Fisheries, House of Representatives, 93rd Congress, May 1, 1973, Serial No. 93-28, available from the Government Printing Office, Washington, D. C. Particularly good expressions of the view of thoughtful economists are:

Ridker, R., 1973. To grow or not to grow: that's not the relevant question. *Science:* Dec. 28.

Heller, W., 1972. Coming to terms with growth and the environment. In *Energy, Economic Growth, and the Environment* (Sam Schurr, ed.). The Johns Hopkins Press (for Resources for the Future), Baltimore, Md. Reprinted in Enthoven and Freeman, *op. cit.*

Ayres, R., and Kneese, A., 1971. Economic and ecological effects of a stationary economy. *Ann. Rev. Ecol. Systematics: 2.* Reprinted in part in Enthoven and Freeman, *op. cit.*

Kaysen, C., 1972. The computer that printed out W*O*L*F*. *Foreign Affairs:* July.

411

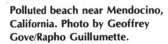

Environmental Law and Its Administration

James E. Krier *is Professor of Law at the University of California, Los Angeles. He did his undergraduate work in economics at the University of Wisconsin and graduated from the Wisconsin Law School in 1966. From 1967 to 1969 he practiced law in Washington, D. C., then joined the faculty at UCLA. His primary interests have been environmental law and the relationship between law and economics. He is the author of a number of articles on both subjects, as well as a book,* Environmental Law and Policy *(Bobbs-Merrill, 1971). He is a consultant to the Environmental Quality Laboratory of the California Institute of Technology and a member of the Executive Committee of the National Academy of Sciences Study on Problems of Pest Control.*

413

Laws and lawyers seem to be everywhere. Observe a problem in our society of almost any character and dimension and you will find it surrounded by, invaded by, indeed, some think created by, laws and lawyers. This is true, you will recognize, of the urban crisis, unsafe cars, international strife, political dirty tricks, campus revolts, and unwanted pregnancies. And it is true of environmental problems. As distasteful and mysterious as some of you may find lawyers and the law, the fact is that you cannot have a rounded understanding of environmental problems without giving some attention to how the legal system deals with them. The purpose of this chapter is to force that attention, to provide an introductory overview of how the legal system does help and might help resolve the "problem" — the conflicting demands placed upon environmental resources.

"Legal system" for our purposes here refers to courts, legislative bodies, and administrative agencies at the federal, state, and local levels. Our aims are to have some concept of the roles played by these components of the legal system, and some familiarity with the tools they employ in carrying out those roles. This will hardly make you an expert in environmental law; indeed, it will not even introduce you to *all* of the substantial body of rules and procedures commonly pulled together under that label. But, by talking in terms of categories and general underlying elements, and by illustrating these with concrete examples, we can get some useful notion of how the legal system bears on environmental problems.

Part of our task, then, is purely descriptive, but we want to do more; we want to assess — to consider not only how the system *does* deal with the problems, but how it *might better* do so as

well. But before undertaking description and assessment, let us face a preliminary question.

The only things new about environmental law are the label itself and the intensity (at times the frenzy) of activity centered about it. The fact is that the legal system has been confronting issues of the use and misuse of environmental resources for centuries, and more than occasionally. In fourteenth-century England it was against the law (a Royal Proclamation) to use coal in furnaces, not because coal was in short supply but because smoke was too abundant. Violation was not treated lightly: in 1307 one offender was executed.[1] In 1611 an English court decided *Aldred's Case*,[2] viewed by many as the first instance of air pollution litigation. You would probably regard the suit as strikingly odd today — it involved a farmer's objections to the odors of his neighbor's pig farm. As to the United States, one can find nineteenth-century cases dealing with air and water pollution, and local ordinances dealing with the same subjects. State legislation came somewhat later, and — as a general rule — federal legislation sometime after that. (We shall focus shortly on this trend, which starts with courts and local legislation, followed by state legislation, followed by federal legislation.) But still we are talking about the fairly distant past; the well-known federal Refuse Act, for example, was passed by Congress in 1899.[3]

What is different about today, then, is not that there is environmental law, but that there is so much of it reaching so far and touching so many concerns: local, state, and federal law on air and water and noise pollution, pesticides, radiation, land use, conservation, strip mining, and on and on almost endlessly — or at least close enough to endlessly that we cannot profitably examine all the law here. If one were forced to pick a date when this body of law began to explode and form its mushroom-shaped cloud of rules and procedures, it would probably be 1969, the year Congress passed the National Environmental Policy Act,[4] which I discuss later. But why the need for this intervention by law at all, and why suddenly so much of it?

We can get at the first part of this question by employing one of the economic concepts discussed in Chapter 16. (We shall draw on those concepts from time to time throughout our treatment of environmental law, so the reader is encouraged to read Chapter 16 in conjunction with the present one.) Ruff, you will recall, pointed out that divergences between private and social costs are an underlying cause of pollution and other environmental problems. Economists often refer to such divergences as *external costs* because they arise when an individual does not bear the full costs of his decisions—when, that is, some or all of the costs are "external" to him. For example, when a factory pollutes the air, large costs are imposed upon society, but the factory owner escapes all or most of these. If, on the other hand, he were to control the pollution, he would bear the full costs of control himself. Quite obviously, it is cheaper to the factory owner to pollute than to control, even though pollution may be the more expensive alternative to society. But the owner does not feel the pollution costs — they are external to him—and so he has no incentive to avoid them through control. The concept of external costs is a powerful explanatory tool in the case of environmental problems. Suppose you drive a car that pollutes because you do not use the best fuel. You could use the best, but it costs more, and you will bear that cost all by yourself. If you decide not to use it, you save money. The air gets dirtier, if at all, imperceptibly. And in any event, the degree to which it gets dirtier is but a tiny cost to you. Of course, it is a tiny cost to everyone that breathes that air, making the aggregate cost high. Moreover, since just about everyone reasons as you do, just about no one uses the good fuel. The pollution problem, accordingly, is aggravated a great deal.

Use this same example to understand a second important concept, that of *collective goods*. What happens if everyone uses the clean fuel?

The air gets cleaner. But it also get' everyone *but you* uses it, and you can ⌐ clean air even though you did nothing to bring about. When a person takes steps to clean up the environment, he produces a good for everyone that uses that environment—a collective good. When people can get a good from the actions of others, without any action of their own, they tend to take a *free ride* on the others' efforts. Unfortunately, most people reason this way, so few do anything. Refer once again to the polluting factory that we talked about earlier. All the people in the community could get together and pay the factory owner to install equipment. If the cost of pollution were greater than the cost of control, the community could make this payment to the owner and still be better off. But would such a payment be made, or would all those friendly neighbors tend to reason as follows? "My contribution will make little difference, and in any event if I don't give but the rest do, the problem will clear up without costing me anything. They cannot withhold the clean air from me." Theory and experience confirm that the answer would be exactly what your intuition suggests.

"External costs" help us understand why people engage in activities that disrupt the environment; "collective goods" suggest why they fail to stop voluntarily, or fail voluntarily to undertake or financially support cleanup efforts. Without the coercive intervention of law—whether it be prohibiting certain acts, or making people pay for the results of those acts, or taxing people to raise revenue to support cleanup—the problem would only get worse.

The analysis above helps explain the need for legal intervention, but how can we explain why all of a sudden there has been so much of it? At least part of the answer would appear to lie in increasing demand for such intervention, in turn brought about by several factors:

1. Exponential (or near-exponential) growth in population and consumption. People and their consumption patterns put heavy strains on the environment, and once sufficient time has passed for the population base to grow large, the environmental problem suddenly reaches "crisis" proportions.
2. A related factor is increasing concentration of population. Concentration usually will mean

quicker depletion of the local environment's capacity to assimilate wastes. Concentration means something else, a rapid increase in the *costs* of environmental degradation. To understand this point, assume as an example that a community's air becomes twice as polluted with a doubling of its population. The *costs* of pollution would increase by far more than a factor of two, because twice as many people would be suffering twice as much pollution! Some observers find in this phenomenon the explanation for the sudden emergence of — and response to — the environmental crisis.[5]

3. The first two factors relate to *absolute* increases in environmental problems. A third, increasing affluence, relates to *relative* increases. As people's real incomes go up, so do their demands for such "luxury goods" as clean air, clean water, open space for recreation, and so forth. This factor suggests that environmental problems can grow even though the environment stays the same in quality, simply because society wants a cleaner environment.

4. Finally, we should not overlook the role that key events and people play in stimulating the demand for environmental protection — the Torrey Canyon oil spill, the Donora air pollution episode, Rachel Carson, Ralph Nader, and so on.

Together these factors suggest some reasons for the sudden emergence of environmental law. Rapidly decreasing environmental quality (or at least the near-term potential for this), rapidly increasing costs, a rising desire for a cleaner environment, serious episodes, and persuasive protagonists generate public outcries and pressures on the legal system for effective response. The result is an increased sense of awareness by lawyers, judges, and legislators and an increased flow of legal output. Much of this output today takes the form of legislation passed at the state and federal levels. At an earlier time this was not the case. Up until 40 years ago, the courts were probably the primary tool of legal intervention into environmental problems. This was sensible during a period when such problems were more or less localized, involved quite simple technical components (e.g., the smelly pigs of William Aldred's neighbor), and in both absolute and

relative terms were small. The courts could deal effectively with the isolated smoke nuisance caused by a discernible source and reaching at most a small handful of people. But with increased population growth and density, and expansion of technological scale and complexity, environmental problems came to be in every sense more far-reaching. The harms they imposed justified the relatively high overhead costs of legislative–administrative response (as compared to judicial intervention); the complexity and scale of the problems, and the need at times for the spending power to deal effectively with them, made such response necessary. Legislative response generally occurred first at the local and state level. Then, with recognition that the problems often reached beyond state boundaries, and called (according to many) for nationwide standards, leadership, and financial support, there began the wave of federal legislation. Today, to name but a few, federal legislation touches and in some instances virtually occupies such areas as air quality, water quality, pesticides, and noise and radiation hazards. This list does not even mention the National Environmental Policy Act[4] nor the fact that the major area of land use is likely to be the next candidate for federal intervention.

In the remainder of this chapter we will look at the various ways in which the legal system can deal with environmental problems. We deal first with the courts, and then with legislative bodies (including both the law-making activity of legislatures and the administration of the laws by administrative agencies). Throughout the discussion, two questions will be important. First, to what extent is the legal system (and, in particular, each of its components that we examine) responsive to fair and efficient resolution of environmental problems? Second, to what extent can the system be improved in these respects?

Environmental Problems and the Courts

In our discussion of the courts, the major concern is with how law which has accumulated over many years, and which has evolved in response to all sorts of important problems, can be accommodated to the particular needs of contemporary environmental issues. We shall look at the two main types of environmental court actions and at the rules courts make and apply to these lawsuits. And we will see that the rules themselves are not rigidly fixed, but rather are fluid concepts, or "theories," whose application in any specific setting may be far from clear. Finally, we shall try to isolate some problems the courts confront in dealing with environmental issues, and we shall consider what might be done about these. Let us begin by looking at the two main types of lawsuits.

TYPES OF LAWSUITS

The first of the two most important classes of environmental litigation involves citizen suits against *private* decision makers.[6] For example, citizen A sues factory B, A claiming that B is injuring him and his property by emitting an unreasonable amount of smoke and fumes. Assume that no statute (a rule made by the legislature) prohibits B's actions; it might still be the case that a common-law rule does so. (*Common-law rules* are those made and from time to time modified by judges as they decide particular cases.) By applying these rules, the court decides whether B can continue its activity, or at least whether it can continue without paying A. The decision, of course, will influence B's actions and hence the quality of the air. Note that the court is deciding whether A's activity is lawful by reference to *its own* (i.e., court-made) rules.

The second important class of suits (probably the most important today) is comprised of those by citizens against *public* decision makers,[6] which are usually regulatory agencies such as a state air pollution control authority or the federal Environmental Protection Agency. Here citizen A sues the control agency, alleging that it is not enforcing a legislative program of air pollution control against violators or that it has passed an emission standard which is more lenient than the legislative program mandates. In this context the *legislature* has made the rules; it has decided what air quality shall be. The court's function is to see that the agency is doing the job given it by the legislature; its task is *not* to decide if the legislature has made the "proper" decision about air quality.

In the citizen suit against public decision makers, the court serves primarily a *policing* function: it is a watchdog of administrative action. Because the policing function is so inti-mately tied to legislative programs, we will postpone its discussion until we consider environmental legislation. For now, we shall take up the citizen suit against private decision makers. Let us first discuss "rights," then "remedies," then some problems and what might be done about them.

RIGHTS

We shall look here at what lawyers call the *law of torts*, that is, civil (as opposed to criminal) wrongs, other than breaches of contract, for which the law will provide a remedy. In particular, we are concerned with several theories of tort liability: negligence, strict liability, trespass, and nuisance. All these legal theories were developed years (and in some cases centuries) ago; all of them were developed not simply for the specific and limited purpose of resolving environmental problems, but for dealing with a broad spectrum of issues of which those involving the environment are but a portion. The law of negligence, for example, applies not only to air pollution and oil spills, but to automobile accidents, medical malpractice, and many other calamities.[7] Finally, all the theories at which we will look prohibit, in essence, the *unreasonable* use of the environment, but the manner in which they do so differs—at times significantly.

The *law of negligence* requires conformity to a standard of "reasonable" conduct — that is, conduct which does not create unreasonable risks for other people. Whether conduct is reasonable turns on balancing its social utility against the probability and degree of risk that it creates. If one engages in conduct that fails to pass this test, and thereby causes injury, he must pay money damages to the person injured. The negligence theory appears from time to time in environmental litigation—air pollution cases, for example, or pesticide or oil spill suits. In such suits, the injured person (plaintiff) claims that the polluter (defendant) conducted himself in unreasonable ways in that, for example, he applied pesticides during a time of high winds, so that they dispersed widely and were inhaled by the plaintiff and injured him. If plaintiff managed to prove all of these claims — the defendant's acts, their unreasonableness, that they caused his injury, and the amount of his damages — to the satisfaction of the judge or jury, he would win a money judgment. (This assumes that defendant had no defense, such as contributory negligence on the part of the plaintiff.)

While the test for negligence seems straightforward enough in the abstract, its application is not always a simple matter. In environmental cases in particular, a common issue concerns whether it is relevant that the defendant did *not* use the best control technology available. Some courts have held that it is; in one air pollution case, for example, the court concluded that if the existence of an improved control system were well known, failure to use that system would imply negligence.[8] But what if that improved system cost a great deal more than the control system in use, yet would reduce the risk of injury only very slightly; is failure to use it wrongful? At least one case has suggested that it is.[9] The holding does not seem sensible, since it calls for the expenditure of a large amount on control measures to diminish the chance of harm from pollution only slightly. As an analogy, would you spend twice as much for an automobile if that automobile would be 1 percent safer in the event of an accident?

We have seen that negligence requires proof of "fault"—of failure to conform to a reasonable standard of conduct. *Strict liability* does not—at least on the surface. To succeed on a theory of strict liability the plaintiff need not establish fault on the part of the defendant. Rather, he need only show that the defendant was engaged in "abnormally dangerous activities" that caused plaintiff injuries. It seems quite plain, however, that the concept of reasonableness works its way into this theory. For, in determining whether an activity is abnormally dangerous, the courts consider: the gravity and probability of the risk created; whether it could be eliminated by the exercise of reasonable care; whether the activity is a matter of common usage; whether it is appropriate to the place it is carried on; and the value of the activity to the community. It seems clear from this listing that some elements of reasonableness are contained in the theory. However, the law of strict liability can still be different from negligence in an important way. An activity might be conducted with all due care (so that there would be no negligence), yet, when the factors above are considered, it still might be abnormally dangerous (so that there would be strict liability). Thus the doctrine can serve a useful purpose in those instances of environmental injury which can

occur without any fault at all, such as sonic booms and offshore oil drilling.

Strict liability has found some application in environmental litigation; the doctrine has been applied in court cases to the following conditions and activities, among others: "Crop dusting with a dangerous chemical likely to drift; drilling oil wells or operating refineries in thickly settled communities; an excavation letting in the sea; factories emitting smoke, dust or noxious gases in the midst of a town. . . ."[7] Some commentators have suggested that strict liability should apply to the sonic boom.[10]

A very recent theory of strict liability can also find application in environmental litigation — *strict products liability*. Under this doctrine, one who manufactures or sells a product that is unreasonably dangerous because of its very design is liable to parties injured as a result, even if the manufacturer or seller was not negligent in making or marketing the product. Several enterprising lawyers have argued, without success thus far, that automobile manufacturers should be strictly liable for air pollution damage under this theory. The argument runs that motor vehicle engines are unreasonably dangerous because their exhaust contributes to health-damaging smog.[11]

The next two theories we take up — *trespass* and *nuisance*—are the subjects of a good deal of confusion on the part of laymen and lawyers alike. In large part, trespass and nuisance are simply special names given to negligent or abnormally dangerous conduct (both of which we have already considered) that interferes with the plaintiff's possession or use *of land*. Generally, trespass denotes interference by a tangible thing such as water, and nuisance refers to interference by an intangible such as noise. So for our purposes we can simply say that if negligent or abnormally dangerous activity interferes with the rights of a landholder (makes him ill, spoils his view, sends him running to avoid a stench), it is called trespass in the case of interference by tangible things, and otherwise nuisance — and it makes very little difference which! There are some exceptions to this statement, but they are far too technical for us to consider profitably here.

Before concluding this section on rights, we should give brief consideration to another nuisance theory common to environmental litigation, *public nuisance*. A public nuisance arises when an act or omission causes damage to members of the public in their exercise of common rights — air pollution, for example — that injures public health. (Recall that private nuisance protects only against interference with the use or enjoyment of *land*.) One would think from this that the public nuisance theory is suited perfectly to environmental problems, but it labors under one severe limitation that has persisted for centuries: to recover damages from a public nuisance the plaintiff must show that he suffered "special damage" *different from that sustained by the general public*! This might mean, for example, that one could not recover if pollution spoiled the suitability of a lake that he and other members of the community used for swimming, unless he could show in addition that the pollution caused him ill health. Some states have done away with this historical artifact by passing statutes which permit any person suffering from a public nuisance to maintain a lawsuit.[12] This seems an enlightened step.

REMEDIES

Suppose that plaintiff in a suit under one of the theories discussed above manages to prove (as plaintiff must) all elements of liability. To what relief, or remedy, is he entitled? The answer is damages, injunctive relief, or both — to compensate plaintiff for those injuries suffered, and to prevent their recurrence in the future.

Damages are the ordinary remedy for the successful plaintiff. The remedy consists of a money judgment which should, at least as a matter of principle, compensate for all injuries suffered and proved to the satisfaction of the judge or jury. The idea is to make plaintiff "whole" — to place him in the position he would have occupied had the unhappy event not happened. Thus, the damage remedy can include payment for injuries to health, medical, and other out-of-pocket expenses, plus other economic losses, such as damage to the plaintiff's property or business.

We will see shortly that the damage remedy should at least help assure that actions harmful to the environment will not be undertaken in the future, by raising the price of those actions. Sometimes, however, plaintiffs want more — an order from the court that the defendant desist from or undertake certain actions in the future.

Such orders are called *injunctions*. For example, the court might order a polluter to stop dumping effluents into a lake, as well as paying the damages of injured parties; it might order the manager of a factory to install air pollution control equipment.

Most environmentalists may want to stop pollution rather than win money. However, damages are the ordinary form of relief, and to obtain an injunction plaintiff must overcome several barriers. The first of these, a showing that damages alone would be an inadequate remedy, is not so serious in the usual context of environmental litigation. Generally speaking, the courts have found damages to be inadequate when the injury is to land, when the extent of injury cannot easily be calculated (as is often the case, for example, with injuries to health), or when the injury is likely to recur if injunctive relief is not granted (thus forcing the injured party to bring suit time and again). Since one or more of these conditions is likely to be met in most environmental cases, the "inadequacy" obstacle is not too important.

The second barrier is more significant. It is imposed by the doctrine of "balancing the equities," under which the courts will consider the relative hardship that will result from granting injunctive relief, the good faith of the parties, and the value to the community of defendant's activities. A recent New York case (and one chastised by most environmental activists) gives some notion of the function of the doctrine. In the case *Boomer v. Atlantic Cement Company*,[13] a group of sixteen landowners brought an action against the Atlantic Cement Company, seeking damages and an injunction to restrain Atlantic from pouring out cement dust that fell on plaintiffs' property. The trial court found that the company's activities constituted a nuisance (despite the fact Atlantic took every available precaution to protect against the dust). It awarded damages but refused to order the plant to stop operations. "Balancing the equities" involved, the court concluded that an injunction would do more harm than good — in light of Atlantic's $50 million investment, its general contribution to the economy (the plant employed 300 people), and its direct aid to the education of children in the area through the payment of taxes.

On appeal this judgment was affirmed. The court noted:

The total damage to plaintiffs' properties is . . . relatively small in comparison with the value of defendant's operation and with the consequences of the injunction which plaintiffs seek.

The ground for the denial of injunction, notwithstanding both that there is a nuisance and that plaintiffs have been damaged substantially, is the large disparity in economic consequences of the nuisance and the injunction.[13]

The court concluded that plaintiffs could be made whole by an award of permanent damages "which would compensate them for the total economic loss to their property present and future caused by defendant's operations." (Such an award, unlike the trial court's judgment of damages *up to the time of trial*, would mean that the landowners would not need to bring more lawsuits in the future.) If defendant paid those damages, the court held, no injunction would be granted. It concluded its opinion with the following:

It seems reasonable to think that the risk of being required to pay permanent damages to injured property owners by cement plant owners would itself be a reasonable effective spur to research for improved techniques to minimize nuisance.[13]

We shall return to this statement shortly. Note for now that in applying the doctrine of balancing the equities, the courts in *Boomer* aligned the interests of the plaintiffs on the one side against the interests of the company and the community on the other. Courts typically apply the doctrine in this manner. Thus they ask whether granting an injunction would benefit the plaintiff only slightly relative to the injury to the defendant and the community (or whether denying injunctive relief would harm the plaintiff to only a small degree relative to the great benefit to the defendant and the public). Commentators who have observed this bias suggest that balancing the equities makes sense only if courts consider the harm to the public of permitting pollution (or the benefit to the public of stopping it).[14]

The Courts: Some Problems and Solutions

What important problems arise when the courts try to resolve conflicts over environmental resources, and what, if anything, might be

done about these? We cannot fasten on problems or suggest means for resolving them without some model in mind of what the courts *should* be doing. One revealing model, and it will underlie the rest of our discussion, is suggested by the earlier reference to externalities. We saw that because the costs of environmental degradation are to a large extent not borne by polluters, they have little or no incentive to undertake control measures that would avoid those costs, and the likely result is that there will be more pollution than is "worthwhile." (A polluter will pollute, for example, even though the total cost of avoiding pollution through control is less than the total cost of the pollution to society.) If, on the other hand, the responsible party might be forced to pay damages to those injured, he has a strong incentive to consider the costs to society of his actions, and to consider further whether he might best avoid them by undertaking control measure. As the court in *Boomer* put it, the "risk" of liability is an "effective spur."[13] The model suggests, then, that litigation should *serve as a means to internalize the costs of environmental degradation* (see Chapter 16).

The discussion thus far should at least have suggested some problems that must be overcome if litigation is to be an effective internalizer of costs. The main problems are: (1) the burden on the plaintiff of proving his case against the polluter, including proof that satisfies the various requirements of the legal theories discussed above; (2) the high costs of bringing lawsuits; and (3) the difficulties of organizing injured persons to bring "group" lawsuits. Our purpose here is to briefly describe these in a more systematic way and suggest techniques for dealing with them.

BURDEN OF PROOF

You will recall that in the discussion of rights we noted the obligation of plaintiff to prove all the elements of his case ("Prove" here means that plaintiff must show his claims more likely than not — any likelihood greater than 50 percent.) This means that plaintiff must show not only negligence or some other brand of unreasonableness, but a causal link between this and his injury, plus the amount of damages. In many instances it is hard to show these, and to the extent that plaintiffs cannot succeed, or find success so unlikely as to discourage even filing suit, internalization is not achieved.

To some degree the law has responded to these problems. Strict liability theories might ease the burden of proving fault. It seems quite likely, for example, that it would often be easier to establish that defendant engaged in an "abnormally dangerous" activity than to prove that he conducted that activity negligently. Even where strict liability would not apply (i.e., where negligence must be proved), the law has created handy "presumptions" that might ease the burden. Thus, if it can be shown that defendant's conduct violated some statute (e.g., a legislatively prescribed emission limitation), often this alone is enough to show that there was negligence. It is then up to defendant to show that he proceeded with all due care. In other instances, plaintiff can prove negligence by showing that the event causing damage is of a sort that usually does not occur in the absence of negligence, and that the event was caused by operations under defendant's control. This approach has been applied in environmental litigation — for example, to an Oregon case involving an unusual scattering of excessive amounts of fluorides from an aluminum plant.[8]

Thus far we have been discussing plaintiff's burden of proving unreasonableness in one guise or another, but this is only a part, often a small part, of the story: causation and damage must also be established. In some cases this will be easy. In the pastoral setting of *Aldred's Case*,[2] for example, clearly the neighbor's pigs were the source of the problem, and surely it was a relatively straightforward task for plaintiff to convince the jury of the annoyance this caused him. But transport yourself to the more typical modern urban context, where there are multiple sources of emissions, effluents, noise, and so forth; where many of these are located miles distant; where complex physical–chemical reactions and interactions contribute to noxiousness; and where the damage involved turns on disputed points of medical and scientific knowledge (and where the injury complained of could well be the result of some cause other than that alleged by plaintiff — cancer caused or aggravated by smoking rather than by air pollution, for example). Take now the general rule that plaintiff must show not only which

among all the possible sources caused his injuries, but what part of his damage was caused by each if he is to recover anything, and you have some general notion of the magnitude that the problem of proof can reach.

Over time, the courts have responded in part to the problem. If the plaintiff can show that each of a group of sources acted wrongfully (negligently, for example) and that at least one of them caused his injuries, the burden of proof shifts to the defendants, so that each, if it is to escape liability, must prove that it has not caused the harm. Or, if the plaintiff can establish that the activities of a group of defendants caused him harm, but cannot show the proportion of contribution of each, the court might allow the jury to apportion the responsibility based on the best evidence available, or it might make each defendant liable for all the damages, leaving it to the defendants to work out the apportionment of damages among themselves.[15]

From the discussion so far we can conclude that proof problems are often substantial for those injured by acts that interfere with environmental quality; that the law has to some degree (but hardly systematically) responded to these problems; and that enough problems remain to ensure that many of the social costs caused by polluting activities are not internalized. The result, of course, is that those polluting activities are conducted more than they would be if the costs were borne by the polluters (internalized). What should be done about this?

One answer would be simply to put the entire burden of proof on *defendants*, to require that they show in all cases that they acted reasonably, or that their actions did not cause the plaintiffs' injuries. However, for a host of historical and other reasons, the burden of proof in our judicial system is generally placed in the first instance on plaintiff. Thus, in all cases of doubt about any element of the case, defendant wins. This might be fair and acceptable if who is plaintiff and who is defendant were determined randomly, but clearly they are not. In fact, it is almost always the case that polluters are defendants. By the nature of things, you have to sue a polluter to stop him; he does not have to sue you to establish the right to pollute.[16]

Since this systematic bias against environmental resource conservation exists, and since our current national policy appears to be to conserve resources in case of doubt, it might seem sensible to put the burden of proof on the polluters. The problem is that this is not easy. Surely most sensible people would not want a regime where *A* could simply come to court and make assertions against *B* which the latter would then have to disprove, without evidence from *A* to support his charges. In at least one case, however, a court has taken a small step in this direction by reducing the "quantum of proof required" of a party who was seeking to protect a wilderness area from commercial exploitation. The court did this, it said, because of a "sympathetic concern" for conservation shown by government.[17] Legislation proposed on the national level (by Senators Hart and McGovern and Representative Udall) would have achieved substantially the same result by requiring plaintiff to show only that the activity of the defendant "resulted in or reasonably may result in unreasonable" impairment of the environment. A recent article described this burden of proof as "negligible."[18] The legislation, however, has not been passed, and it is unlikely that it will be in the near future. In summary, proof problems remain more or less intractable. Changes like those reflected in the wilderness area case[17] discussed above are a step in the right direction, but apparently no other courts have adopted that approach. Even if they did, there would still be proof problems (although smaller ones) for environmentalists.

HIGH COSTS OF LITIGATION

If problems of proof made up the only constraint, matters would be bad enough, but as it happens there are other, perhaps more serious ones. If internalization is to be realized — if suits are to be won — they must first be brought, and it is of course the person aggrieved by the actions injurious to the environment who must bring suit. But bringing a lawsuit is no casual matter; litigation can be an expensive and time-consuming business. And expense is particularly vexing in the context of environmental problems. It is quite typical for the impact of such problems to be *diffuse*. This is to say that a large number of persons might be injured — by air pollution, for example, or drifting pesticides or a sonic boom — but each to only a small degree. Since many people are involved, the aggregate impact (or cost) is high, but damage to any one individual is likely to be too small to justify the possibly high costs of a lawsuit. The result, of course, is that few suits will be brought. Vast amounts of social cost will not be effectively internalized, nor will the possibility of lawsuits be sufficiently threatening, to deter those whose activities interfere with environmental quality. This is a central feature of environmental problems, and it is of crucial importance.

DIFFICULTIES OF ORGANIZING

You might suggest that the obvious answer to this problem is for all the aggrieved individuals to band together, to pool their injuries (and their funds) in order to make the expense of litigation manageable and worth undertaking. Pooling, however, is hardly simple to accomplish. There is, first of all, a straightforward physical problem: environmental impact is not only diffuse but quite typically of broad reach as well. Potential fellow travelers may be spread over a wide area, making it difficult to identify, contact, and unite them. Second, there is what we can term a structural problem. Recall our earlier discussion of collective goods and free riders — people who refuse to contribute to the production of a collective good because they realize they can enjoy that good, once produced, whether they contribute or not. Environmental lawsuits produce collective goods: to the extent they are successful (or even convincingly threatening), they deter activities that cause environmental injury. People who recognize this have little incentive to join in supporting a group lawsuit. If the lawsuit is successful and as a result deters future activity of the sort complained of, everyone in the area of injury will benefit, whether he helped support the suit or not. It is true that people who do not join the suit might not share in any damage award, but that is not a matter of real importance. As pointed out earlier, each individual's damage is likely to be small anyway; what is important to each is deterring the defendant's harmful actions in the future. And the effects of that deterrence cannot be withheld from those who contribute nothing to achieving them.

There are several ways in which this shortcoming of environmental litigation might be overcome: (1) *class actions*, which have existed for a very long time; (2) *public interest lawyers*, phenomena of quite recent origin; and (3) *institutionalized advocates*, which have little more than the status of a proposal. Let us look at these in turn.

The *class action* is a device that, in effect, permits injured parties to pool resources, but in a manner that overcomes the free-rider problem. One injured party may bring a class action on behalf not only of himself but of other persons "similarly situated," persons allegedly suffering the same sort of injury, caused by the same conduct on the part of the defendant. Such suits, if successful, might result in very large damage awards, and the promise of this can be useful in managing the costs of litigation. Attorneys who would otherwise be out of reach might be interested in taking such a class action on a contingent basis (taking their fees only out of any judgment recovered).

Thus, the class action can aid internalization: suits are brought that otherwise would not be; in any successful suit, the costs imposed by defendant upon many (rather than just a few) injured parties are accounted for. This is hardly to say, however, that the class action is a panacea. Some states, for example, have so narrowly confined their class-action rules that they have little or no application to the typical environmental problem.[19] In the federal courts, a particularly disturbing problem arises in this context. As a general matter a federal court lacks jurisdiction to hear environmental lawsuits unless, among other things, the amount in controversy exceeds $10,000. The United States Supreme Court held in *Zahn v. International Paper Company*[20] that individual claims could *not* be pooled to reach this minimum; each member of the class must have individual damages in an amount greater than $10,000. But in most environmental cases, of course, no one person suffers in that amount; it was this very fact of diffused damages (but large aggregate damages), coupled with the problem of collective goods, which we earlier suggested made a device like the class action necessary if cost internalization is to be realized. Thus, *Zahn* drastically limits the usefulness of environmental class actions in the federal courts.

Class actions in state courts, of course, do not run up against the federal jurisdictional-minimum barrier. Some states, moreover, do have quite liberal class-action rules under which a plaintiff could plausibly fit an environmental suit. But problems still remain. Many states, and the federal rules, too, require the plaintiff to give notice of the suit to all members of the alleged class. This can be very expensive; one case has been cited in which the estimated cost of mailing notices to class members would have exceeded $400,000.[21] Needless to say, if expenses such as these exceed the discounted (by the risk of losing) value of the damages sought, no suit will be brought. And, of course, if the class suit seeks only injunctive relief (as opposed to damages), it would appear to have no value whatsoever as a means to attract attorneys with the lure of a large contingent fee.

The comments above touch only on a few, although probably the major, limitations of the class action, but they are sufficient to show that the device is not nearly so effective as it seems at first glance.

The second possible response to the problem is the *public interest lawyer or law firm*, such as the Environmental Defense Fund, the Sierra Club, and the Center for Law and Social Policy in Washington, D.C. — supported in largest part not by legal fees but by grants from foundations and contributions from concerned members of the public. (Are these contributions likely to be large, or will most "concerned" members of the public opt instead for a "free ride"?) Public interest lawyers and law firms began to spring up in noticeable numbers in the mid-1960s, and they handle — in absolute terms — a fairly large quantity of environmental litigation. The supply, however, is much smaller than the demand. And to date, public interest lawyers have allocated most of their resources to suits against government agencies, not to private damage suits. Public interest lawyering is a healthy and helpful trend, but it is unlikely ever to be a sufficient answer to the problems that we are talking about.

There has been some suggestion, in response to these problems for *institutionalized environmental advocacy*, in effect a public interest law firm created and funded by the legislature.[22] The funds, of course, would be taken out of government revenues raised through income and other taxes, so there would be no problems of free riding; people would be forced to contribute whether they wished to or not. The idea is that the public advocate would take those environmental cases which appeared to have some merit and which, but for the public advocate, would go unrepresented. The theory is that this would respond in a quite systematic way to the structural underrepresentation of environmental interests occasioned by diffusion and free riding. A bill proposing something akin to this notion was introduced in Congress in 1970 by Senator Tunney[23] (then a member of the House of Representatives), but it got nowhere; there is no indication that the idea of institutionalized advocacy in the environmental area will be realized in the foreseeable future.

INJUNCTIVE RELIEF

From the discussion thus far, it appears that lawsuits against private decision makers to recover the environmental damages caused by their activities are unlikely to be effective internalizers of costs; even after resort to those devices that to some degree overcome problems of proof, diffusion, and free riding, we can be confident that costs remain that are not accounted for. Is injunctive relief a partial solution to this problem? Since our purpose is to affect indirectly the incentives of the decision maker by making him pay the costs of his decision, why not simply accomplish the same effect directly by a court order enjoining the harmful activity — decreeing that it be stopped, or conducted in a less harmful way? The problem with this argument is that it overlooks a crucial purpose of cost internalization: to judge whether an activity is worth its costs once all those costs have been paid. A polluter will employ control techniques rather than pay damage costs if the former are less expensive than the latter; where the opposite is the case, however, he will continue to pollute and pay as long as he can do so and still make enough profit to stay in business — as long, in other words, as society sufficiently values his activities that they are worthwhile even with all costs internalized. Internalization by damage awards, then, is a means to test whether the social value of an activity outweighs its social costs. We have seen that at times this test is

only partial, or imperfect; but the trouble with the injunctive remedy is that it usually will provide no test at all. An activity might be enjoined that in fact was well worth all the costs it imposed. This is why the court in the *Boomer* case[13] denied injunctive relief: it feared that in granting an injunction it would do more harm than good.

Often injunctive relief would ask too much of a judge — to balance costs and benefits under conditions of high uncertainty, and in a context where his judgment, unlike a damage award, could not be tested and corrected by market forces if wrong. Consider the plight of the judge in *Diamond v. General Motors Corporation*,[24] a suit on behalf of all the inhabitants of Los Angeles County against 291 corporations allegedly responsible for air pollution in Los Angeles County. The court dismissed the suit, for the following, among other reasons:

It is readily apparent that the control of the emission of air pollutants is a highly complex problem. Our industrialized civilization is dependent upon energy derived from the oxidation of fossilized fuels. Many industrial processes involve the release directly or indirectly of volatile substances which can cause discomfort to others. Such unavoidable discomfort is one of the prices one pays for living in an industrialized civilization. The court does not have the facilities to undertake the balancing of the interests of the inhabitants of the Los Angeles Basin against the needs of productive industry in this same area. . . .[24]

So much for our survey of the citizen suit against *private* decision makers. The courts in this context can play a useful role, but our survey should suggest that they are too bound up by inflexible and unyielding limitations to be *of themselves* an effective means of intervention. Legislative programs are also necessary. We turn to them now.

Environmental Problems and Legislatures

We saw earlier that there are innumerable legislative programs, at the local, state, and federal level, dealing with virtually every aspect of environmental problems. Even if it were possible in the space available, it would be of little value to sketch the contours, much less trace the details, of all these programs. A more meaningful approach is to describe and appraise the three broad categories into which legislative intervention most generally falls: regulation, subsidization, and pricing. (Keep in mind during our discussion that at times a particular program will represent a mixture of these three approaches.) As was the case with the courts, we want to have some way of evaluating these three methods of legislative intervention. Two criteria seem important. First, the method should be fair. Second, it should aim to minimize the sum of all relevant costs to society (i.e., it should be efficient). The second criterion was discussed at length in Chapter 16, and so the constraint it imposes is only summarized in what follows. Then we can consider each of the three methods of intervention in turn.

CRITERIA FOR EVALUATING PROGRAMS

A legislative program should aim to minimize the *sum* of the costs of environmental degradation and the costs of avoiding environmental degradation, subject to a constraint of fairness. This test recognizes that while pollution, for example, imposes costs, *avoiding pollution does so also*. It says that we should not control pollution more than is worthwhile. The test is concerned with *all* relevant costs, even those that cannot be estimated in terms of dollars. On the degradation side, for example, it insists that account be taken of negative effects on health, peace of mind, scenic beauty, wildlife, the interests of future generations, and so forth. On the control (or avoidance) side, it insists on attention to the costs of a control program to polluters, the costs of administering the control program, such costs as inconvenience to members of society if they must change their behavior, and the costs of losing opportunities to deal with other important social problems if we choose to spend on environmental control. As noted, many of these costs — on both sides of the ledger — cannot be cast into monetary terms, but it is important to recognize that legislative bodies must nevertheless make judgments about how much control of environmental problems is worthwhile.

The constraint of fairness is equally ambiguous, but again it cannot be avoided. The legislature should consider whether a particular environmental program which appears worthwhile (efficient, or worth its costs) might nevertheless be undesirable because the resulting costs are distributed unfairly among members of society. Very clean water, for example, might mean increased unemployment among poor people; the efficient level of air quality might result in a situation where the well-off live in clean areas and the poor in polluted ones. (On the other hand, very clean air might once again mean that the poor will bear an unjust portion of the costs of achieving it.) In determining its environmental policy, the legislature must keep both efficiency and fairness in mind.

Quite obviously, it is no easy matter to define the "desirable" set of environmental objectives. In any event, once the objectives are determined (once the legislature decides, that is, how clean to make the water or air in various places), there are various means to attain them. These means are the three categories alluded to earlier: regulation systems, subsidy systems, and pricing systems. Using any one of these costs money, but the costs differ in amount and distribution from situation to situation. The problem of the legislature is to decide which means or mix of means to use. Consistent with our test, it should choose that means which has promise of reaching the legislatively set goal at least cost, subject to the constraint that the distribution of those costs among the members of society is considered fair.[25]

You should recognize that the objectives just described were implicit in the model we used for gauging the judicial role, and that the courts act more or less in accord with it. Behind the element of reasonableness employed by the courts, for example, is consideration of whether the benefits of a given activity (or level of activity) are worth the costs it generates. Behind the objective of cost internalization is a concern with judging the same question. Courts deny injunctions when to do otherwise would mean a net loss to society or result in an unfair loss of jobs to workers or support of education to students (through the payment of property taxes). Fairness and efficiency guide the actions of courts; they should also constrain the decisions of legislative bodies. With this in mind, let us consider the three categories of legislative intervention. Our

425

purpose will be to describe and illustrate each category and some of the different styles each can take, and — by way of appraisal — to briefly compare the different styles in any one category, and the different categories themselves, in terms of the objectives suggested above. In each case we shall treat the economic aspects briefly since they have been detailed in Chapter 16.

REGULATION

Regulation has been by far the predominant method of legislative intervention in the United States. The term implies a system of mandatory prescriptions and proscriptions enforced through sanctions such as fines, abatement orders, or a jail term. Within this general framework, however, one finds almost infinite variety. In the areas of water and air pollution, for example, regulatory programs at the local, state, and federal level commonly set emission or effluent limitations with which controlled sources must comply, or prescribe for these sources certain types of control techniques or production processes that must be used. Failure to comply might result in any of the sanctions mentioned above. Zoning is another typical regulatory approach: certain land uses (such as heavy manufacturing) can locate only in specified areas, and within those areas they may be required to comply with standards governing plant design and methods of construction and operation. State and federal controls on pesticides are another example. These prohibit application of certain chemicals (such as DDT) and provide that others may be manufactured and sold only after compliance with specified testing, registration, and labeling requirements; still other pesticides may be employed only by licensed applicators. Again, failure to comply will bring sanctions into play; a manufacturer, for example, may be prohibited from selling an improperly labeled pesticide.

A recent and rather novel illustration of the regulatory method is the National Environmental Policy Act of 1969 (NEPA).[4] NEPA is novel in that its direct effect is to control the activity of federal agencies (such as the dam building of the Corps of Engineers). In particular, the Act's well-known "impact statement" provisions require that federal agencies, with respect to any major actions or projects significantly affecting environmental quality,

prepare a "detailed statement" covering the impact of particular actions on the environment, the environmental costs which might be avoided, and alternative measures which might alter the cost–benefit equation. The apparent purpose of the "detailed statement" is to aid in the agencies' own decision making process and to advise other interested agencies and the public of the environmental consequences of planned federal action.[6]

Provisions similar to those of NEPA have been copied by a number of states; we shall discuss later some of the functions they perform. Note here that NEPA illustrates the regulatory approach in its provision of compulsory requirements backed up by coercive sanctions.

NEPA reflects one example — and as we said, a rather novel one — of the regulatory approach to environmental quality. Many more typical examples can be found on the federal and state level. As a concrete illustration, we will take one regulatory program that mixes federal and state control and discuss a few of its provisions and applications in terms of the criteria suggested at the outset of this section.

Under the Clean Air Act (as amended in 1970),[26] the federal government, among other things, prescribes minimum uniform national air quality standards for certain pollutants. These standards set the level of air quality that each state must meet; states may set more stringent standards if they wish, but for most pollutants the federal standard is so strict that very few, if any, will do so. The result is that for any given pollutant there is a uniform national standard. Given that states vary in terms of the size, physical characteristics, and values of their populations, in terms of meteorology and geography, economic base, and control capabilities, among other things, is it likely such uniform standards will minimize the sum of pollution costs and pollution avoidance costs? The answer appears to be, most assuredly not. Pollution costs vary from state to state, and so too do pollution avoidance (control) costs. It thus appears at this level of analysis that the uniform standards are inefficient: the appropriate (efficient) level of air quality for Iowa may well be too low for Montana and too high for California.

However, it might be argued that it would be so troublesome and expensive for the federal government to compile all relevant benefit–cost information for each state that the advantages (in terms of efficiency) of variable standards would be more than outweighed by the costs of realizing them. Thus, a program that is otherwise the most efficient imaginable might be so cumbersome and costly to administer as to be worthless. We can make a fairly sound guess, based on our intuitions about what the world is like, that uniform standards are unwise in economic terms, but without appropriate data we cannot be sure. The point is that Congress apparently never even inquired into the issue. Rather, it went forth on the basis of slogans like "Clean Air for Everyone" without thinking about what that slogan meant, whether its implicit expectations could be achieved, and whether in any event they would be worthwhile.

Furthermore, uniform national standards may not conform to the constraint of fairness. Perhaps what members of Congress had in mind when they talked of "clean air for everyone" is that it would be unfair if air quality for citizens of state *A* were not as high as that for citizens of state *B*. However, it may be much more expensive to attain a given level of air quality in State *A* than in state *B*; the result might be higher taxes for the citizens of state *A*, or fewer jobs (especially for the poor), or a diversion of resources from other objectives also important from the standpoint of fairness — an adequate level of welfare, a good health research program, and so forth. Fairness, like clean air, is not free, but in passing the Clean Air Act, Congress at times appeared to behave as though it were. Again, this does not suggest that the uniform standards of the Act are in the last analysis unfair, but rather that Congress, because it lacked articulated or even implicit tests for its decisions, did a poor job of focusing on the issue of fairness and weighing it against that of efficiency.

So much for the judgment about air quality standards. Whatever the standard, what means should be used to reach it? The question of means, again, should be considered in the framework of the cost-minimization-subject-to-fairness test.

For the most part, after the federal government promulgates the national air quality standards, it is up to the states to devise, subject to federal approval, an implementation plan that will achieve those standards. Take

the example of stationary sources (Chapter 9). It is the job of the states to prescribe for these sources emission levels low enough to meet the federal ambient air quality standards. The states have in essence two regulatory techniques available: *performance standards* and *specification standards*. Performance standards specify a maximum emission level (e.g., no more than X parts per million of a given pollutant during each hour of operation) but they *do not* direct how that level is to be reached. The regulated source can employ control equipment, change manufacturing processes, find a cleaner fuel, or use any other technique or combination of techniques consistent with achieving the established emission level. Specification standards go further; they require a particular type of control equipment, specify a particular fuel, and so on. Some control programs mix these approaches, promulgating performance standards for some sources, specification standards for others, or establishing an emission level (as in the performance approach) but requiring that it must be achieved by using one or more of some specified range of techniques (essentially like a specification program).

Which of these regulatory techniques is most consistent with our efficiency–fairness test? The answer depends very much upon the particular problem addressed. In general, the performance standard appears to be most efficient in that each controlled source has an incentive to achieve the established emission level at least cost to himself, and thus to society. No more will be spent on attaining the ambient standard than is necessary: control cost — an important component of the equation — will be minimized. With a specification program, on the other hand, the *regulator* rather than the regulated makes the decision, and often he will not be in the best position to know which techniques cost the least for a particular source. Moreover, his incentives often will differ, and bureaucratic inertia can take over: the regulator may develop the convenient and safe habit of approving only a familiar approach rather than taking the risk of a new technique which if more efficient is likely to bring no gain to him, but which in the case of failure will result in criticism from supervisors within the bureaucratic hierarchy. In short, performance programs tend in gen-

427

eral to give the strongest incentives to develop and employ least-cost solutions.

But analysis must reach deeper. In some circumstances the *administrative* costs of a performance program will be higher than in the case of the specification approach; more field inspectors, for example, might be needed if there is a very large number of controlled sources. In other instances a specification program will be most desirable simply because it is likely that the bureaucrat *does* know better than the controlled source. If an objective of control, for example, were to reduce emissions from backyard incinerators by 50 percent, it would probably be most sensible to prescribe the equipment to accomplish this. Most owners of backyard incinerators have little, if any, savvy about control technology and would be in no position to judge the least-cost approach to a performance standard. In addition, if there are a multitude of incinerators, it would be very expensive to check the output of each periodically. Under these circumstances it probably makes more sense from the standpoint of efficiency to require the installation of specified equipment, seal it once installed, and make random checks for violations.

The foregoing is hardly an exhaustive analysis of the problem. An important part of administrative costs, for example, has to do with enforcement once violations are discovered. Should fines be employed, and if so should they be civil or criminal? Should there be other criminal penalties, such as a mandatory jail term? Should abatement be used, rather than a fine or other penalty, so that offending sources are simply closed down? Again, the answer will depend very much on particular circumstances. We cannot analyze all the variables here; hopefully, the discussion has nevertheless suggested how the cost-minimization test should be employed. The same reasoning can be applied to any sort of environmental problem and the regulatory techniques available to deal with it.

But what about the fairness constraint? It seems seldom to come into play in the decision about which regulatory technique to use. It is difficult to see, for example, how fairness plays a part in deciding whether factory *A* should be controlled by a specification or performance standard. In the regulatory context, issues of fairness will most often arise with respect to deciding which sources should be controlled, and how much. If a control objective is to reduce automobile pollution by a certain amount, and if the poor usually own old cars while the better-off own new ones, considerations of fairness might suggest that new cars should be required to reduce emissions more than old ones.

The final two methods of legislative intervention — subsidization and pricing — have been explained in principle in Chapter 16, and in what follows we shall discuss only matters of application.

SUBSIDIZATION

In the context of environmental control, subsidies have exhibited few commendable properties. Nevertheless, they have found considerable use in the United States.

Subsidy programs can take two forms. The first and most common is that of tax incentives. Legislatures have provided, for example, that pollution sources may take the cost of investments in pollution control devices as a credit against their income tax liability, or depreciate these devcces on an accelerated basis, the result in either instance being lower taxes. The second form, seldom if ever used in the environmental field but nevertheless possible, is that of award payments — direct cash awards made to pollution sources if they achieve certain goals (e.g., if they reduce emission output by a specified degree).[27]

Subsidies tend to break down in practice. Unlike mandatory regulatory programs, subsidies in their pure state are voluntary. A source can control and receive the subsidy, or it can pollute and forego it. In its typical form —the tax incentive—the subsidy covers only a portion of a source's total control cost. Since pollution control generally represents a losing proposition for a source, government sharing of a *part* of the loss is hardly a reason for the company to voluntarily bear the remaining cost.

Direct cash awards could, at least in principle, avoid some of these shortcomings of tax incentives, but as a practical matter they would raise their own difficulties. For example, when a source claimed the subsidy as to certain of its alleged control investments, it would be hard for the government to determine what portion of those investments was, in fact, made to control pollution and what portion was made (and might have been made even without the subsidy) for other purposes, such as an expansion of productive capacity. A similar difficulty arises from the practical impossibility of preventing the recipient of the subsidy from pocketing the whole amount, even if greater than the cost of control. Note also that the cost of subsidies — whether tax incentives or award payments — is borne by taxpayers rather than pollution sources and the consumers of their products. Pollution costs, in other words, are not internalized, so there is no signal from the marketplace to encourage a shift away from the production and consumption of goods that generate high environmental costs.

Finally, tax incentives erode the tax base and escape the periodic review given direct expenditures by the legislature and the executive in the process of budget preparation. Nevertheless, subsidization has been used a good deal for pollution control. Part of the reason for this is probably that subsidies transfer wealth to industrial interests. If these industries have, as most believe, considerable influence over legislative bodies, it would follow that subsidy programs appear to be a "price" paid to industry for passage of pollution control regulations. This may be one reason why subsidies so commonly come in conjunction with regulatory programs. But one need not be so cynical in every instance. At times the legislative body might have judged that the impact of pollution control regulations would fall too heavily on employees (especially poor ones) of controlled sources, if no subsidy were paid. In such instances, subsidies combined with regulation might be the best combination of efficiency and fairness. Or take our earlier example of the control of pollution from old automobiles. Mandating the installation of control technology on such cars might fall too heavily on the poor without some form of subsidy added on. There are important roles for subsidy programs, from the perspectives both of efficiency and fairness. Still, one can reasonably suspect that on balance they have been poorly employed to date.

PRICING

Pricing systems are the conceptual opposite of subsidies, although they share with sub-sidies the characteristic of voluntary compliance. Pricing systems aim to reduce pollution by making the source pay if it chooses to pollute. The principle is clear: "If you charge a person for disposing of his wastes he will find ways to reduce the amount of wastes he disposes of, and . . . the more you charge him the stronger the incentive he will have to find some less damaging method of disposing of his waste."[28]

The operation and advantages of a pricing system in its most common form (the emission or effluent fee), have been explained in Chapter 16. Keep in mind the central points. We can be quite confident that each source would pollute and pay only to the extent that doing so would be less costly than any method to avoid paying (including going out of business). As long as sources try to minimize costs, each would act to keep its costs as low as possible. Moreover, there *is* a price such that under it the desired level of quality would be reached. Put these two observations together and you see that the desired level would be reached at the lowest cost to society. At least in theory, pricing has enormous advantages in terms of efficiency (i.e., in terms of minimizing pollution avoidance cost).

Of course, there are problems, some of them real and others imagined. Moreover, there is the constraint of fairness to consider. Let us look at these matters in turn.

One objection often made to the pricing principle is that sources will simply go on polluting, paying the established charge, and passing the charge on to consumers. This argument, however, has little substance. Even if firms could pass on the pollution charge, they would surely prefer to pass on a smaller rather than a larger one. Management knows that consumers tend to shift consumption away from goods as the goods go up in price, which means firms would lose more customers if they passed on higher prices. This same effect would encourage higher consumption of goods whose production involves little pollution and lower consumption of those involving relatively great pollution. Unlike subsidies, pricing systems internalize costs on producers and consumers, with the desired effect on the incentives of both groups.

But how simple is it to set the correct price — the one that would affect behavior in such a way that the net result would be achievement of the desired level of quality? The answer is that it is not simple but it is also not impossible. There *is* a schedule of charges that would yield the chosen result. To discover it requires market studies and some trial and error. The latter is hardly untypical of government intervention. Moreover, there is a pricing system that escapes the necessity of government undertaking the difficult job of determining the correct charges. Under it, the legislature or control authority would translate the desired quality level into a maximum annual number of pollution "rights" which anyone could buy at periodic auctions.[28] A polluter might decide, for example, that the cheapest approach for him would be to control up to a certain point, but to buy one "right" for each unit of pollution beyond that (roughly the same point at which he would begin to pay an emissions charge under the first system considered). Even conservation groups could bid at the auction and, perhaps, buy some rights and put them away, thus lowering pollution levels even more. (Recalling our discussion of the free rider, is it likely conservation groups would be very effective bidders at the auction?) The auction or "rights" model has the advantage of requiring no government decision about a price schedule; rather, only the simpler calculation of inferring the number of rights from the desired level of quality is necessary. The price is set by market forces in the course of the competitive bidding at the auction. Moreover, there would be no problem of passing on costs. The number of rights initially issued by the government would set the limit on the total amount of pollution. Once all those rights were used up, pollution would cease no matter how the costs of the rights were distributed.

There is no apparent reason to suppose that the administration costs of either system described above need be higher than those of regulatory or subsidy programs. A method is needed to measure emission output, to be sure; such information is necessary to determine the fee under the first price system or to assure that a source is not polluting beyond the number of rights it has purchased under the second. But measurement is also needed to enforce emission regulations or award payments. Even if administrative costs were

higher, the pricing approach would still be worthwhile as long as those higher costs were nevertheless lower than the benefits, in terms of efficiency, that we saw would be produced by employment of one of the pricing systems.

You might consider at this point why the regulatory approach cannot achieve these same efficiency benefits. The reason is quite simple. Pollution sources have different control costs — some can cut down pollution less expensively than others. In terms of saving society as much as possible (in terms, that is, of minimizing total control costs), the firms that can control pollution at relatively low cost should be required to reduce emissions by more than those with relatively high control cost. In other words, there should be a different emission regulation for each source. But to establish the correct regulation for each source would be almost impossible. Government would have to discover each firm's control cost, and it is not at all clear how it would get the information. If it asked each source, each would have an incentive to overstate costs, because that would result in a less stringent standard for it.

In sum, efficient pollution control regulations are possible in theory, but the real-world costs of imposing them in practice would be so high as to more than wipe out the gains from the efficient varying standards. With one uniform price per pollutant, on the other hand, each source would *automatically* adjust its output to the cost-minimizing point. Those with relatively low control costs would control more and pollute less; those with relatively high costs would control less and pay the pollution tax. There would be a different, and efficient, pollution output from each source — exactly the result that we wanted to achieve.[29]

Pricing systems, then, seem quite promising on efficiency grounds, but we must also consider whether they stand up sufficiently to the constraint of fairness. The issue is a complicated one, best approached on a case-by-case basis. A common objection is that pricing systems discriminate against the poor because they raise the prices of products. But this overlooks the fact that regulatory programs can have the same effect, and sometimes more so. (Compare, for example, an emissions tax on auto pollution, under which the poor person has the option of doing less driving, to the regulatory approach, under which each car must use the same expensive control devices whether the car is driven few or many miles each year. Moreover, subsidies could be combined with the emissions tax to ease the burden on the poor.)

It is hard to make final judgments about the practical merits of pricing systems because they have been used very little for control of environmental problems. This situation may soon change. Some states have already experimented with effluent fees to aid in controlling water pollution[30]; the Environmental Protection Agency is looking at emission taxes for the control of automobile pollution[31]; a variation of the rights system (gauged to gasoline rather than emissions) may be used to help control auto pollution in Los Angeles[32]; and the White House has suggested regulations combined with a sulfur tax to help control sulfur oxides pollution.[33] Pricing appears to be the approach of the future. It is not perfectly suited to every environmental problem, but it may be best suited to many.

Policing the Policemen

All of the legislative methods we have just considered are typically administered by bodies, administrative agencies, such as the Environmental Protection Agency on the federal level, or a state air pollution control authority, or a local water pollution control board. These agencies are, in a sense, the "policemen" of the environment. The problem is that at times the policemen do not do their job — they seem to favor polluters rather than control them. In this section we consider what might be the source of this problem, and then look at how the courts, by hearing citizen suits against the agencies, might be used to help assure that the agencies fulfill their obligations.

THE PROBLEM

With the growth of legislative and administrative control of social problems, and environmental problems are no exception, there has also come an increasing concern that legislative and administrative bodies are more responsive and sympathetic to the interests of those supposedly being controlled (usually in-

dustry) than to the public interest they are charged to advance. The concern is expressed in varying forms. At its most extreme it consists of allegations of virtual conspiracy — collusion among the governed and the government to advance the private interests of each at the expense of society's legitimate expectations.[34] ("The regulators are in bed with industry.") This has been often said of the Interstate Commerce Commission, the Federal Power Commission, the Atomic Energy Commission, state public utility commissions, and so on. There is little doubt that such collusive behavior occurs at times, although probably more rarely than appears to be believed. Rather, it is more likely to be true that outcomes which smack of conspiracy are most often simply results of the framework or structure in which legislative–administrative control goes forth. Take three examples. First, we saw that the costs of environmental degradation are widely diffused among members of the community, and that this, coupled with the problem of freeriding, tends to dilute the influence of the public in decision processes. Control of degradation, however, is an item of large importance to polluting industries. Moreover, the controlled class of polluters often consists of relatively few firms, and in any event the firms are usually organized through trade associations. Thus, the controlled class usually has a higher degree of cohesion, a more intense interest in influencing the course of intervention, and more funds, time, and expertise at its command than does the public at large.[35] And from this it is quite understandable that the controlled will generally have the most persuasive voice in the decision process — whether because it can buy votes, influence campaign contributions, or simply marshal the best and most convincing case.

Second, we saw from time to time how a legislative or administrative body needs various sorts of information in order best to do its job: information about the effects of environmental degradation, about the costs of control, about technologies that might be developed, and so forth. Often, quite obviously, the controlled will be the best (and sometimes the only) source of such information, and they strive to keep that information secret. They disclose relevant information to government only in return for "good relations" or a certain amount of "respect" for the controlled. Sometimes this can lead to abuse; the information machine is oiled a bit more than necessary, at a cost larger than the value of the information.

A third and final example is the common one of cross employment over time. Control agencies sometimes draw some of their employees from controlled firms, or — as is more typical — the opposite happens. A major motor company, for example, recently recruited a top executive who previously had been involved in California's automobile pollution control program. There is nothing particularly sinister in this phenomenon. Where better to find the expert you need? And the phenomenon need not be supposed to give bureaucrats incentives to treat the controlled favorably in the hope of a vice-presidency in the controlled firm at a later date, for we could imagine that the best strategy for the vice-presidential hopeful is to be tough while he sits on the agency: the firm will have to hire him to escape him. Still, whether sinister or not, cross employment is likely to lead to a sharing of values and views that is unhealthy if for no other reason than that it leads to close-mindedness and inhibited imagination.

THE CITIZEN SUIT AGAINST PUBLIC AUTHORITIES

Whether insidious conspiracy or innocent cooptation, there is good reason to believe that legislative and administrative decision processes will benefit from occasional airing and exposure to outside views. Someone should police the policeman. And this is just the role played by the citizen suit against *public* decision makers,[6] legislatures, and administrative agencies. Understand that the proper role for the courts in this context is rarely to substitute judicial for legislative and administrative judgment. Rather it is simply to review the legislative or administrative decision to assure that the rules were followed in the course of reaching that judgment.

REVIEW OF LEGISLATIVE DECISIONS

On the *legislative* level, this means rather limited judicial intervention, indeed. For the most part, courts cannot override legislative decisions *unless they find them inconsistent with the constitution*, and there is as yet no constitu-

tional right to a clean environment. And probably there should not be, for how would the judges decide the substance of such a right? We have seen how many value judgments and uncertainties are involved in issues about the "appropriate level" of such things as air or water quality, or about the most efficient or fair means to reach them. These are distinctly political questions, usually best settled through representative decision processes. Even leading advocates of expanded citizen suits in the environmental field urge caution about a constitutional right to a clean environment, lest judges become environmental czars.[36]

Still, there are more limited means by which judges can inquire into the propriety of legislative decisions about environmental quality. The most notable of these is the *public trust doctrine*. This doctrine, which has its origins in early Roman law, requires in essence that certain resources (parks, beaches, lakes, the air, etc.) are held by the government for the benefit of the general public and cannot be diverted by government action to environmentally unworthy uses, unless the diversion is of little consequence. It had long been the case that the public trust applied only to public waters; very recently, however, lawyers have urged, both in journal articles and in litigation, that it should apply to all environmental resources. The objective is to use the doctrine to require that legislatures and administrative agencies clearly justify governmental actions that interfere with environmental quality:

When a state holds a resource which is available for the free use of the general public, a court will look with considerable skepticism upon *any* governmental conduct which is calculated *either* to reallocate that resource to more restricted uses *or* to subject public uses to the self-interest of private parties.[37]

Michigan's Environmental Protection Act of 1970 has to some extent codified the public trust doctrine in that state.[38]

REVIEW OF ADMINISTRATIVE DECISIONS

The more active front with respect to expanded judicial policing of government decisions has had to do with *administrative agencies*. Here again, it is not the proper role of the court to substitute its judgment for that of the

432

administrative agency, or even so much to question the wisdom of the judgment (save if it is clearly arbitrary or capricious). Rather, the judge's task is the more limited one of assuring that the agency followed the rules — that it worked within the mandate given it by the legislature and on the basis of legislatively announced criteria, that it followed fair and appropriate procedures, and that it gave full consideration to all relevant views. Obviously, the courts here can play the very important role of assuring that agency decisions are well ventilated and sensitive to the public voice. And, especially in the last 10 years, they have done so; "they have succeeded in opening up the administrative process both at the hearing and the appellate stages to citizens and citizen groups. This is indeed the single most important development in recent administrative law."[39] There have been a number of facets to this development. Let us look at examples of three important ones.

An important facet has been judicial liberalization of the rules defining who it is that can come to court seeking review by the court of an agency decision. For many years the law of *standing*, as it is called, was narrow indeed: not any citizen could go into court to complain about the way an agency was doing its business. In order to have standing (the right) to sue, a citizen had to show that the agency's misdeeds had some quite direct impact, usually of a personal or economic nature, on that citizen. Thus, for a long while it was not sufficient to claim simply that the agency had acted unlawfully (e.g., by granting a permit to a polluter who was not in compliance with legislative standards), and to claim that the citizen wanted the decision reviewed because the quality of the environment would be impaired. The citizen needed a more direct stake than that. By the mid-1960s, however, standing doctrine began to broaden. The instrumental case in the environmental context was *Scenic Hudson Preservation Conference v. Federal Power Commission*.[40] The case arose when the Federal Power Commission approved the construction of a reservoir and pumping station on the Hudson River at Storm King Mountain in New York. Several conservation organizations, concerned only with the impact of the power facilities on the environmental quality of the Hudson River valley, brought a suit

claiming that the Commission had acted unlawfully in giving inadequate consideration to aesthetic and conservation factors. The Commission argued that the conservation organizations lacked standing, but the court held otherwise. It concluded that an economic or personal interest was not necessary to standing, as long as the suit was brought by individuals or groups who had, through their past activities and conduct, exhibited a special concern with the environment of the area affected by the agency decision (in this case, the Hudson River valley). After this holding, environmental groups such as the Sierra Club generally had standing to seek judicial review of agency decisions that might have an adverse impact on environmental quality. This does not mean the groups would win; to do so they would have to show that the agency decision (whether adverse to the environment or not) was not in accord with legislative standards. But at least now they had their foot in the door. Under the old standing rules, the court would not even have been able to consider the lawfulness of the agency action.

As sensible as the new law of standing seems, the U.S. Supreme Court appears to have taken a step backward by its decision in *Sierra Club v. Morton*.[41] The case arose in the following fashion. Walt Disney Productions sought permission from the Department of Interior to construct a ski resort at Mineral King, a remote, pristine area in the Sequoia National Forest in the Sierra Nevada Mountains. The Disney proposal contemplated a highway, rooms for over 1,500 people, a cog railway, and other facilities. The Sierra Club claimed that the development as proposed would violate relevant statutes and harm the environment of the area. The Club maintained that it had standing because in the past it had dedicated itself to preserving the quality of the environment, especially that of the Sierras. The Club did not claim that any of its members actually used the area. The Supreme Court held that without such use, the Club lacked standing. Its interest simply in preserving the area *for others* (and for future generations) was not enough. The holding is not particularly worrisome so far as Mineral King is concerned, because the Court permitted the Sierra Club to add the claim that some of its members *did* use the area in question and thus would be directly affected by adverse changes resulting from the Disney development. But the principle of the decision is still questionable. The right of a conservation group to sue should turn not on whether its members use the area in question, but on whether the group has a real interest in protecting the area for the public generally. The law of standing was developed by the courts to make sure that the person who brings suit has a sufficient interest that he will do a diligent job in presenting his case to the courts. The question, then, should be whether the group bringing suit has demonstrated an intense interest in the outcome, whether its members use the area or not.

Still, the *Sierra Club* ruling left intact some of the new law of standing: an individual (or group on behalf of its members) can establish standing by showing an impact on the environment that injures him (or the group members) because, for example, he uses the area. A direct *economic* impact is no longer necessary. This introduces a second important facet of recent development in administrative law: the *appearance of public interest groups* to assert the broadened standing rules that permit environmental interests to be considered. Such groups as the Sierra Club, the Environmental Defense Fund, and those sponsored by Ralph Nader have increased vastly in number and activity, directing most of their energies to suits seeking judicial inquiry into whether administrative agencies have followed the prescribed rules of the game in making decisions that have impact on environmental quality.

In the environmental field in particular, a third and very important facet is the *National Environmental Policy Act of 1969*[4] (and the state acts modeled after it). As we have seen, NEPA sets forth a number of procedural requirements with which agencies must comply in making decisions about projects and activities likely to impact on the environment. The requirements themselves are designed to make agency decisions more open and comprehensive. Public interest groups have had great success in convincing judges that in reaching particular conclusions, agencies failed to comply with some element of NEPA, especially the requirement to prepare and issue an environmental impact statement before approving a project that might significantly affect the environment. Many projects have been stopped as a result, not because the court found them unwise but because it found that the agency had not followed NEPA. For example, in *Wyoming Outdoor Coordinating Council v. Butz*,[42] the U.S. Forest Service sold to a private company the right to cut millions of board feet of timber in the Teton National Forest in Wyoming. The government had not prepared an environmental impact statement before the sale. A group of conservationists brought suit, claiming that the sale would significantly affect the environment and that, therefore, such a statement was required before any timber could be cleared. The court agreed and ordered that no timber could be cut until a statement was prepared. This does not mean that the timber sale will not ultimately go through. If the Forest Service prepares a statement, it might reveal that the sale could proceed in such a way as not to adversely affect the environment. The court might find, that is, that the Forest Service was not arbitrary in approving the sale. Nevertheless, the impact statement has at least two advantageous effects. First, it gives a court some basis for determining the reasonableness of the agency's decision. Second, and because of this, it forces such agencies as the Forest Service to plan carefully in order to minimize adverse environmental impact as much as possible. Thus, the result of NEPA's impact statement requirement is likely to be more thorough decisions and more attention by agencies to competing views. This is valuable even if the project (carefully thought out) is ultimately approved. Do not overlook, however, the fact that NEPA results in a much more protracted decision-making process than existed before, which can at times be unduly costly.

All these trends are healthy. They take account of the basic legitimacy of legislative and administrative decisions, but they also recognize the pressures that promise danger of distorted decisions; they give a role to the judiciary that it can manage without invading the proper realm of legislative–administrative authority; they help put balance into a somewhat skewed system, keeping the watchdogs not only honest but independent, open-minded, and supplied with relevant points of view. And perhaps this suggests a good note on which to summarize and conclude our survey of environmental law and its administration. Environmental problems are complex, and so are the ways in which the legal system can help resolve them. It is not sufficient simply to advocate a ban on all pollution, for this would overlook the fact that just as pollution imposes costs, so does its control. The problem is to find a proper balance. The balance cannot be achieved without intervention of the legal system, for without legal controls unfettered individual actions would lead to a level of environmental quality lower than that in the best interests of society. By the same token, it is not enough simply to observe problems with legislative and administrative decision making and, on that basis, urge that the courts should be entirely in charge of environmental control. The advice to "sue the bastards" makes sense in some contexts, but we have seen that often the courts are ill-suited to deal effectively with environmental problems. Because of the nature of those problems, citizens often cannot achieve access to the courts — the costs of a lawsuit to any one individual are higher than the costs of the problem to that person. Even if all the problems got before the courts, often the judiciary could not handle them, because of the complexity and broad reach of the issues. Generally, the courts can respond to the isolated instance of pollution that impacts on a few people; when recurring events affect large numbers, on the other hand, the courts suffer shortcomings.

But again, this is not to say that all environmental problems should be left exclusively to legislative and administrative control. This would overlook the fact that the courts *are* a good forum for some kinds of problems; it would also deny the courts the important function of policing legislative and administrative decisions. And even where primary responsibility is left to legislatures and administrative agencies, many issues remain to be resolved. Should the problem be regulated, and if so how? Should there be subsidies, or rather should we place charges on pollution? The answers to these questions obviously depend on the nature of the problem at hand — the number of pollution sources involved, the people affected, and so forth. To understand when and how to intervene requires in turn a subtle understanding of the nature of pollution problems and of the legal system, and how the various cogs of the system can be best fitted together and geared to the needs dictated by the problems.

It would be nice if matters were simpler than this but they are not; to ignore this reality is to be simplistic. If an environmentalist asks how he can best spend his efforts to improve environmental quality, there is no one answer to give. Sometimes the best response will be to support a lawsuit against a polluter, sometimes it will be to assist lobbying efforts to achieve more enlightened legislation or to contribute to public interest law firms. But in *each* instance the response will turn on a basic understanding of the issues at stake and the alternatives available to resolve them. The purpose of this survey has been to serve as a first step toward that understanding.

References

1. Chass, R. L., and Feldman, E. S. 1954. Tears for John Doe. *U. Southern Calif. Law Rev. 27:* 349.
2. 1611. 77 Eng. Rep. 816.
3. 33 United States Code §§407, 411, 413.
4. 42 United States Code §§4331–4335.
5. Heilbroner, R. 1970. The arithmetic of pollution. *The Economic Problem Newsletter,* Spring: 3.
6. Krier, J. E. 1971. *Environmental Law and Policy.* The Bobbs-Merrill Co., Inc. Indianapolis, Ind.
7. Prosser, W. 1964. *Torts.* West Publishing Co. St. Paul, Minn.
8. Reynolds Metals Company v. Yturbide. 1958. 258 F.2d 321.
9. McElwain v. Georgia-Pacific Corporation. 1966. 245 Ore. 247, 421 P.2d 957.
10. Baxter, W. F. 1968. The SST: from Watts to Harlem in two hours. *Stanford Law Rev. 21:* 1.
11. Handy v. General Motors, Inc. 1969. U.S. District Court, Central District of California.
12. Wis. Statutes, ch. 280, §280.02.
13. 1970. 26 N.Y.2d 219, 257 N.E.2d 870, 309 N.Y.S.2d 312.
14. Juergensmeyer, J. 1967. Control of air pollution through the assertion of private rights. *Duke Law J. 1967:* 1126.

15. Jordan v. United Verde Copper Co. 1925. 9 F.2d 144.
16. Krier, J. E. 1970. Environmental litigation and the burden of proof. In *Law and the Environment* (M. Baldwin, and J. K. Page, eds.). Walker & Company, New York.
17. Texas East. Trans. Corp. v. Wildlife Preserves, Inc. 1966. 48 N.J. 261, 225 A.2d 130.
18. Cramton, R. C., and Boyer, B. B. 1972. Citizen suits in the environmental field: peril or promise? *Ecol. Law Quart. 2:* 407.
19. Esposito, J. 1970. Air and water pollution: what to do while waiting for Washington. *Harvard Civil Rights–Civil Liberties Law Rev. 5:*32.
20. 1973. 42 U.S.L.W. 4087.
21. Wright, G. A. 1969. The cost-internalization case for class actions. *Stanford Law Rev. 21:* 383.
22. Krier, J. E. 1971. Environmental watchdogs: some lessons from a "study council." *Stanford Law Rev. 23:* 623.
23. H. R. 18,242, 91st Cong., 2d sess.
24. 1969. Superior Court of Los Angeles County.
25. Calabresi, G. 1970. *The Costs of Accidents*. Yale University Press, New Haven, Conn.
26. 42 United States Code §§1857–1858a.
27. Gerhardt, P. H. 1968. Incentives to air pollution control. *Law and Contemp. Prob. 33:* 358.
28. Dales, J. H. 1968. *Pollution, Property and Prices*. University of Toronto Press, Toronto.
29. Krier, J. E., and Montgomery, W. D. 1973. Resource allocation, information cost, and the form of government intervention. *Natural Res. J. 13:* 89.
30. Vermont Statutes, title 10, §912a.
31. 1971. *Federal Register 36:* 15487.
32. 1973. *Federal Register 38:* 2194.
33. Nixon, R. M. 1971. Program for a Better Environment, Message from the President of the United States. H.R. Doc. No. 92-46, 92nd Cong., 1st Sess.
34. Esposito, J. 1970. *Vanishing Air*. Grossman Publishers, New York.
35. Olson, M. 1971. *The Logic of Collective Action*. Schocken Books, Inc., New York.
36. Sax, J. L. 1971. *Defending the Environment*. Alfred A. Knopf, Inc., New York.
37. Sax. J. L. 1970. The public trust doctrine in natural resource law: effective judicial intervention. *Mich. Law Rev. 68:* 471.
38. Mich. Statutes §§14.528(201)–(207).
39. Jaffe, L. 1970. The administrative agency and environmental control. *Buff. Law Rev. 20:* 231.
40. 1965. 354 F.2d 608.
41. 1972. 40 U.S.L.W. 4397.
42. 1973. 484 F.2d 1244.

Further Reading

Baldwin, M., and Page, J. K. (eds.). 1970. *Law and the Environment*. Walker & Company, New York.

Brecher, J. J., and Nestle, M. E. 1970. *Environmental Law Handbook*. California Continuing Education of the Bar, Berkeley, Calif.

Sax, J. L. 1971. *Defending the Environment*. Alfred A. Knopf, Inc., New York.

Yannacone, V. J., and Cohen, B. S. 1971. *Environmental Rights and Remedies*. Lawyers Cooperative Publishing Co., Bancroft, Whitney & Co., Rochester, N.Y.

Aerial view of Tokyo, largest city in the world. Photo courtesy of Consulate
General of Japan, N.Y.

18

The Urban Growth Machine

Harvey L. Molotch *received his graduate training at the University of Chicago where he
was a Fellow at the Center for Urban Studies. He has completed studies of community
organizations in Chicago and California and has written extensively on mass media in America.
With a number of others, he recently completed a three-volume study,* Santa Barbara, The
Impacts of Growth. *He is currently Associate Professor of Sociology at the University of
California, Santa Barbara.*

Documenting the failures of American cities has become something of a full-time occupation for large numbers of sociologists, political scientists, economists, and educators.[1] Nothing seems to work: traffic is unbearable, violent crime is seemingly ever on the rise, housing is never adequate, the air is polluted, schoolchildren do not learn to read, the tax base is eroding, and the monster sprawls in all directions, engulfing more and more pristine countryside and productive farmland in its ugly grasp.

That's the bad news. The good news is that some of this is exaggeration. In actual fact, the amount of overcrowding in American cities has been continuously decreasing since World War II; the quality of housing has similarly been on the rise.[2] Some cities have declined economically, but others, at least their downtown cores, are quite healthy and have never really lost their economic vitality. And what critics of the city refer to as urban sprawl was the very mechanism through which large numbers of city dwellers came to enjoy the benefits and liabilities of suburban life: a playspace for the kids, the grandeur of a tract house separated by a side yard from the nearest neighbor, and a school bus instead of a streetcar for their children to go to school in. For a lot of people, that was seen as a great improvement.[3]

I thus argue for a balanced view — and one that takes into account the fact that any assessment of the city is ultimately based on one's own judgment as to what constitutes the good life. But when the balancing is done, we nevertheless are back at the conclusion that American cities are failures: they have not delivered the good life, and the way they continue to operate stops them from providing their residents with the amenities that the current state of technol-

ogy is quite capable of providing. Still further, I argue that American cities do not work as human settlements — as integrated "means of livelihood and ways of life"[4] — because they were not *meant* to work.

The shape and the development of American cities are not a response to some human vision of the good life. Unlike ancient Athens or Rome, contemporary Brasilia, or some of the new settlements of welfare-state Europe or China, nobody planned the American cities to achieve some larger social purpose. The grid streets of Chicago or New York, the scattered urban nuclei of Los Angeles, did not arise from some notion that within such a place the people would be happy or that a more perfect citizen would emerge. The American metropolis is not a deliberate social experiment or aesthetic expression. It is the coincidental outcome of the capitalist market process — where land is merely a commodity to be bought, sold, and exploited like any other. American cities have been merely arenas within which money could be made and the land of the city itself has become a part of the money-making process. Thus, urban spatial arrangements — who will live where, what will be made where, how far x will be from y — all of these socially crucial aspects of life have followed from the needs of individual entrepreneurs to make money.

The results are often irrational. Here is one illustration. Each day, there is a mass migration of low-skilled workers from the central cities where they live to the suburbs where they work. On the freeway, these workers pass a mass migration of high-skilled professionals who travel from the suburbs in which they live to the cities in which they work. Both groups pollute the air, endangering their collective health, and

438

both are forced to rely upon one of the most inefficient means of transportation available — autos and concrete highways, which squander resources.

How did this irrationality come about? Here is a rough explanation. Land in the central city has become increasingly expensive, so expensive that modern industry of the sort that requires expansive amounts of land cannot afford to remain there. Land in the countryside is cheap, and if the local government can be induced to grant the proper zoning, it becomes a good place to build a factory. Only businesses that operate with need for relatively little acreage (e.g., banks, corporate headquarters, department stores) can afford to be in the center of things. That is why the jobs that tend to be filled by low-income people often migrate to the country, and why jobs that are filled by high-income people tend to remain in the city.

Poor people, however, can only live in the city, where high-density zoning permits relatively low rents by stacking people high on a single plot of land. The subdividers in the country find the most profitability in the construction of luxury housing. Poor people's mobility to the country is also sometimes limited by restrictive zoning that artificially holds down the suburban housing supply, by racial discrimination that keeps out minority-group members even if they can afford the costs, and by a lack of mass transit in the country that makes living there a difficult matter for families who cannot possibly afford the requisite two cars.

Thus, it is that a number of decision makers, each following his own economic self-interest (sometimes with help from the law, as in exclusionary zoning), manages to generate an irrational situation. The whole thing could be avoided rather simply: before a factory is built, housing for workers could be required nearby. This is commonplace stuff in many parts of the world, but it has never caught on in America.

The Human Ecology of Urban Settlements

Without such overall planning, the deployment of human beings and the spatial network of their activities may thus lack any overall rationality, may be recklessly destructive of the physical resource base that supports the whole

enterprise, and may give rise to mean spirits and unhappy lives. This is the nature of the American urban human ecology, the ways in which people use local space to carry out their daily round of sustaining activity.

To begin to understand this uniquely human ecology, one must back up to some basic points. First, one must conceive of the urban settlement's analytic unit, a given parcel of land, as representing not just a physical resource but *a specific human interest*. That is, each landowner (or person who otherwise has some interest in the prospective use of a given piece of land) has in mind a certain future for that parcel which is linked somehow with his or her own well-being. If there is simple ownership, the relationship is straightforward: to the degree to which the land's profit potential is enhanced, one's own wealth is increased. The urban geography map is not just a display of spatial units, it is a mosaic of human interests.

This means that there is a struggle going on: each landowner is trying to attract the sorts of land users — builders, bankers, factories, or whatever — that will bring profit. Generally speaking, the more intensive the development, the higher the profitability of the land associated with that development. The rub comes from the fact that a finite amount of development is going to take place, making each landowner a competitor with every other in the continuous quest for a scarce resource: development. That means that, unlike the ecology of the natural world, where we envision a competition among species, or land users, struggling for a niche, in the human ecology, there is the additional struggle of land interests vying to attract the users.

Suppose that I own a piece of undeveloped land that I want to sell for the highest possible price. It is big enough for a factory. In my city there are two corporations currently looking for factory sites but many more than two pieces of available acreage. My task is to make my parcel more attractive than that of my competitors. I may invest in grading to make it perfectly level. I may clear it of shrubbery, or fill in a swamp, or do any number of things. If I do enough, I will be able to sell it and my gain will be my competitor's loss. My competitor tries the same thing and, all other things being equal, the ablest competitor combined with the best land will win. That is free enterprise among competing individual entrepreneurs.

Usually, however, all other things are not equal. Suppose that I have the ability to have the zoning of my land changed from agricultural to industrial. Under the former zoning category a new owner who intends to use the land for industry will have to risk the time and effort to achieve a zoning change. If I have good access to government myself, I can effect that zone change and make my land immediately more appealing and more valuable. There is money in it for me. Now that I have my zone change, I can try to use government in other ways: the state may be planning a new highway; if I can get it routed past my property, my property's access to employment centers, to markets, and to materials will be enhanced. Or maybe an airport is to be built nearby. Or perhaps a deep-water port. Or maybe a new campus of the state university. To the degree to which I can influence government to make these sorts of investments in places near my holdings, I can profit enormously. If my holdings are large enough, it will pay for me to take an active interest in politics and to invest the time and money necessary to have the right decisions made. My competitors will tend toward the same reasoning, and the competition among us will come to be partly reflected in the disputes among the various politicians who represent our somewhat diverse interests.

Now we see how government action and political power become an inextricable part of the human ecology. The situation is made even more complex by the fact that each landowner often acts not alone, but links up with others who are similarly situated. If I realize, for example, that the value of my own land is being depressed because an adjoining parcel of land is in poor condition, I get interested in the condition of the adjoining parcel as well as my own. Or, as another example of common interest, since a new highway would help the adjoining parcel as much as it would help my own, it would be rational to combine resources with that parcel's owner and attempt together to have the highway routed close to our common holdings. When a number of people perceive such a common interest, they have a *community*; when they are able to act together to enhance common profitability through the use of government, they have achieved political power.

The Locality as a Growth Machine

Because the process I have described is so common in American cities, city government has come to reflect the needs and interests of people who have vested interests in the development of land. Because campaign contributions are for them merely a sensible investment in government decisions that are crucial to their livelihood, they have come to be powerful in the affairs of local government in virtually every American city. I would argue that this management of the rewards deriving from growth and development is the very essence of local government as a political phenomenon. It is not the only function of government, but it is the key one. Among contemporary social scientists, Murray Edelman is almost alone in providing any real preparation for conceiving of government in such terms.[5] Edelman contrasts two kinds of politics. First, there is the "symbolic" politics that contain the "big issues" of public morality and the symbolic reforms that constitute the headlines and editorials of the daily press. Examples include debates over busing, welfare cheats, pornography, elections, and visits by foreign dignitaries. The other politics is the process through which government actions determine how goods and services are actually produced and distributed in the society. Which land will be rezoned? What products will be tariff-protected? Which firm will receive the lucrative cost-plus contract? Largely unpublicized, this is the politics that determines who, in material terms, gets what, where, and how.[6] This is the kind of politics that we are talking about at the local level; it is the politics of wealth distribution, and land is a crucial (but not the only) factor in this system.

Within any political jurisdiction there are competing land-interest groups scrapping among themselves for the scarce governmental resources that will lead to development. But there are also times and circumstances when at least some of the competing parties have a common interest. This common interest is in the development of the whole geographical area that they share in common. The whole political jurisdiction itself represents a shared common interest in the intense development of a shared set of parcels — all the land within the political unit. A rather large community of interests thus forms and is represented by a name that corresponds to the geography in question.

Such community can occur at any level — the Northside Business Club, the Baltimore Chamber of Commerce, the State of Illinois Development Council, the Southwestern States Regional Association, or the Republic of Costa Rica. The guiding consensus, one so commonly taken for granted that it is utterly uncontroversial, is that the area in question should develop — it should attract the economic base that will bring the population that will cause the entire economy to prosper. Land will be sold, builders will get contracts, washing machines and shoes will be bought, newspaper subscriptions will be taken, and bank interest will be paid. The city, or the state, or the region becomes a growth machine, an apparatus through which growth can be maximized.

This particular process of generating growth *for a particular community* differs somewhat from the goal of growth as it is ordinarily discussed by those addressing themselves to the larger issue of economic growth for the earth's people, the goal of generating more total productive wealth in the world. The kind of growth I am explaining is not aggregate growth, but always growth at the expense of someone else. It has nothing to do with bettering the lot of everyone; it merely involves a *transfer of wealth* from one party to another. It is a process of wealth *distribution*, not wealth *creation*.

It is a process in which people use government power to gain those resources that will enhance the growth potential of the community in question. Typically the governmental level where action is needed is at least one level higher than the community from which the activism springs. Thus, individual landowners aggregate to gain neighborhood services (e.g., a busline) from the city government, or a cluster of cities may aggregate to effectively influence the state government. In this process, the locality is striving to make these gains in competition with other localities that seek the same resources. Many cities in California competed for a new campus of the University of California, but only a few cities succeeded. Many regions competed to become the site of the Atomic Energy Commission's huge atom smasher at Weston, Illinois, but only the Chicago area got it. General Motors will invest in only so many new plants next year; Sheraton will create a limited number

of convention centers. Most localities want these facilities but the total amount of growth that can be supported at a given moment is finite; the existence of scarce resources for development means that government becomes the arena in which the different land-use interests compete for the public money that is crucial in determining where the growth goes. Localities thus compete with one another to gain the preconditions of growth, and they use governmental power, at all relevant levels, to achieve their goals. It is the main business of a locality to generate growth; the locality is a growth machine.

Of course, government decisions are not the only kinds of activities that affect land use and growth; some decisions made by private corporations also have major impact. For example, when a national corporation determines to locate a branch plant in a given locale, it will set the conditions for the surrounding land-use pattern. But even here, government actions are involved. Decisions about where to place a plant are made with reference to such issues as labor costs, tax rates, and costs of raw materials and transport to markets. It is governmental decisions (at one level or another) that determine the cost of access to markets and raw materials. This is especially true in the contemporary era of government subsidies for the development and exploitation of raw materials, the reliance on air transport and pipelines, and federal highway and railway programs. Similarly, through the imposition of regulations concerning pollution abatement, industrial zoning requirements, building codes, and employee safety standards, government decisions affect the cost of overhead expenses. Finally, government decisions will affect the costs of labor by determining unemployment rates, levels of welfare support, minimum-wage requirements, and the bounds within which labor organizing and labor-management bargaining must take place.[7] Government is thus a means to affect indirectly even the decisions in the private sphere that will influence growth patterns.

Who Supports the Growth Machine?

Within each community there is a group of people who provide the financial and spiritual energy to fuel their growth machine. A prominent number are local businessmen, particularly property owners and developers, who *need* local government in their daily routines to make money and secure their investments.[8] Their very livelihoods depend upon timely zoning decisions, strategic highway routings, and the maintenance of a continuous flow of new residents to the area — people to buy vacant land, subscribe to local newspapers, save and borrow at local financial institutions. There are also people whose business it is to serve those who make money from the land: the lawyers, management consultants, and realtors, who need to put themselves in situations where they can be most "useful" to those with land and property resources. Finally, there are those who, although not directly involved in land use, have their futures tied to local growth: for example, home-owned retail businesses may be in the position of having their own rates of growth tied to the growth of the metropolis as a whole; at least at the point at which the local market becomes saturated, one of the few possible avenues for business expansion is sometimes the expansion of the surrounding community itself.

The local business that seems to take prime responsibility for the sustenance of the growth machine is also the most important example of a business that has its interest anchored in the aggregate growth of the locality: the metropolitan newspaper. Increasingly, American cities are one-newspaper towns (or one newspaper-company towns), and the newspaper seems to be one kind of enterprise where expansion to other locales is difficult. The failure of the *New York Times* to expand operations in California is an important case in point. A paper's financial status (and, to a lesser extent, those of other media as well) tends to be wed to the size of the locality. The more the metropolis expands, the more ad lines can be sold on the basis of a larger and larger circulation base. The local newspaper thus tends to occupy an almost unique position: like many other local institutions, it has an interest in growth, but unlike most, it has no vested interest in the specific geographical pattern of that growth. That is, it does not ordinarily matter to a newspaper whether the additional population comes to reside on the north side or the south side, whether the money is made through a new convention center or a new

olive factory. The newspaper has no axe to grind, except of course the one axe that the whole community grinds together: growth. It is for this reason that the newspaper tends to achieve a statesmanlike status in the community and is deferred to as something other than a special interest by the special interests. These competing interests often use the publisher and editor as general community leader, as ombudsman and arbiter of internal bickering, and, at times, as an enlightened third party who can restrain the short-term profiteers in favor of more stable, long-term, and properly planned growth.

Thus it is that although newspaper editorialists have typically been in the forefront of expressing much sentiment in favor of "the ecology," they tend nevertheless to support growth-inducing investments for their regions. The *New York Times* likes office towers and additional industrial installations in the city even more than it loves the environment. The *Los Angeles Times* editorializes against narrow-minded profiteering at the expense of the environment but also favors the development of the SST because the plan would bring contracts to southern California. The papers do tend to support certain kinds of "good planning" principles, because such measures can help make for even more future growth. If the roads are not planned wide enough, the resulting congestion will eventually preclude the increasingly intense uses to which the land will hopefully be put. Planning for "sound growth" thus becomes the key environmental policy of the nation's local media and their statesmen allies. Such notions of "good planning" should not be confused with limited growth or conservation; they more typically represent the opposite sort of goal.

The media attempt to effect these goals not only through the kind of coverage they develop and editorials they write, but also through the kinds of candidates they support for local public office. The paper becomes the reformist influence, the "voice of the community" that exerts some influence over the competing subunits, especially the "fast-buck artists" among them. The papers are variously successful in their continuous battle with the particular special interests they may come to oppose periodically. The present point is thus not that the papers control the politics of the city, but merely that one of the sources of their special power is their commitment to growth per se, a goal that all important groups can rally round. That power is greater because the commitment is combined with appearance of "objectivity."

There are certain persons, ordinarily conceived of as members of the local elite, who have much less, if any, interest in local growth. For example, there are branch executives of corporations that are headquartered elsewhere who, although perhaps emotionally sympathetic with pro-growth outlooks, work for corporations that have no vested interest in the growth of the locality in question. Their indirect interest is perhaps in the existence of the growth ideology rather than growth itself. It is this ideology that helps make them revered people in the locality and that provides the rationale for the kind of local governmental policies that tend to generate relatively low operating costs. Nevertheless, this interest is not nearly as strong as the direct growth interests of developers, mortgage bankers, newspapers, and so on, and thus we find a tendency for such executives to play a lesser local role than the parochial home-grown executives whom they often replace as national corporations take over home-owned enterprises.[9]

Programs from the Growth Machine

The political power of the local interest groups that make up the growth machine is translated by city governments into policies appropriate to their needs. The consequence of this is that local government becomes primarily an extension of the local business class, rather than a mechanism that has as its mission the solving of public social and economic problems. The goal of the government as a growth machine is simple: growth. This leads to important work on a number of fronts: efforts must be expended to have higher-level governments provide the necessary resources that will induce development. It also means that local government policies "at home" need to be created and maintained such that industry will find the area attractive on as many grounds as possible.

As one illustration, a proper "business climate" means low taxes for potential industrial occupants. To lure industry, some localities provide tax breaks in the form of underassess-ment of business property or no-tax "honeymoon" periods for newly established industrial plants. More commonly, the possibility of "scaring off" industry causes taxes to be kept at a low rate, at least low relative to tax rates in other industrial societies. Similarly, a good business climate means permissive zoning to allow industry to have exactly the parcel of land it wants regardless of social or habitat consequences. It means weak pollution abatement laws and lax enforcement of the laws that do exist. It means a quiescent labor force that will work for relatively low wages and thus it means police action against militant labor organizing. It means the use of public expenditures to construct the airports, highways, and ports that will help make locating in the area more profitable. And, in some cases, it means the creation (sometimes with public money) of the "high" cultural institutions of prestige universities, symphonies, and culture palaces that will cater to the tastes or at least appeal to the pretensions of potential executive migrants. Needless to say, these "business conditions" affect not only the land-use decisions of entrepreneurs; they also go a long way toward determining the quality of life of the urban citizenry.

Once achieved, the fact of this proper business climate then becomes the basis through which the community advertises itself to prospective industrial users. In many issues of major business magazines one can find advertisements by localities, even whole countries, selling their virtues in just these terms. Mayors, governors, and presidents become "ambassadors" whose function it is to sell their locality to industrialists as the proper place for capital investment. Low taxes, a quiescent labor force, and sympathetic zoning and pollution control are key parts of the sales pitch. The growth machine also functions in other ways to sell the area. Government funds support boosterism of all sorts: subsidies to the Chamber of Commerce, advertisements in business magazines and travel publications, city-sponsored floats in the Rose Bowl parade, publicly subsidized convention centers, and support for professional ball teams carrying the locality name. All of these are advertising techniques: to boost locality visibility and thus, combined with other strategies, increase the chance of more development.

Selling Growth to Local Citizens

The selling of the city to outsiders has a counterpart in a series of efforts aimed at selling the idea of growth to the local residents, who will have to live with its consequences. The role of the local media has already been discussed at some length: they not only provide editorials that explicitly call for growth-inducing projects, they also sell growth by the very terminology in which they discuss civic betterment: "progress" is equated with growth; "greatness" is equated with growth, and the "health of the economy" is equated with growth.

But the media are aided by many other local institutions. The sports teams, in particular, are an extraordinary mechanism for instilling a spirit of civic jingoism. A stadium filled with thousands (joined by thousands more at home in front of television sets) screaming for Cleveland or Oakland (or whatever) would be a difficult scenario to otherwise fashion. This esprit can be drawn upon to inflict upon the local masses programs of growth (e.g., a convention center or airport expansion) not necessarily in their interests by making a gloss claim to the goal of a "greater Cleveland" or a "greater Oakland." Similarly, public school curricula, children's essay competitions, national spelling contests, beauty pageants, and soap box derbies help build an ideological base for local boosterism and acceptance of growth. My conception of the territorial bond among humans thus differs from those who conceive of it in terms of primordial instincts; I see this bond as socially organized and sustained, at least in part, by those who have a use for it.

The Problem of Jobs

But the key ideological prop for the growth machine, especially in terms of sustaining support from the middle and working classes for growth as a community goal, is the claim that growth "makes jobs." This claim is aggressively merchandised by developers, builders, and chambers of commerce and becomes a part of the "statesman" talk of editorialists and political officials. Such people do not speak of growth as useful for profits (which presumably is of greater interest to them); rather they speak of growth as necessary for making jobs.

But local growth does not *make* jobs; it only *distributes* jobs. The United States will next year see the construction of a certain number of new industrial plants, office units, and highways — regardless of *where* they are put. Similarly, a given number of automobiles, missiles, and lampshades will be manufactured, regardless of *where* they are manufactured. The number of jobs in the society, whether in the building trades or any other sector, will be determined by factors having to do with rates of investment return, federal decisions affecting the money supply, and other issues having very little to do with local decision making. All that a locality can accomplish is to attempt to guarantee that a certain proportion of newly created jobs will be in the locality in question. *Aggregate* employment is unaffected by the outcome of this competition among localities to "make" jobs.

A second fact is equally important: the labor force is essentially a single national pool, highly mobile, and capable of taking advantage of emerging employment opportunities at points geographically distant. As jobs develop in a fast-growing area, *the unemployed will be attracted from other areas to fill those developing vacancies and in sufficient additional numbers to create a continuous unemployed work force*. Just as local economic growth does not affect aggregate employment, it has very little long-term impact upon the *local* rates of unemployment. It may even have a *negative* effect, as is shown in the analysis in the next few paragraphs.

In the following analysis we look at unemployment rates in different localities as a function of the *population* growth in these areas. Now, what we would ideally like to analyze is unemployment as a function of the local *economic* growth rate, for example as a function of the rate of increase in total capital investment per capita. However, such data are simply not available. Instead, it has long been recognized that variation in population growth among localities reflects variation in economic growth. This is because much of the *variation* in local population growth reflects differences in population movements in and out of the area: localities with high economic growth rates attract population, those with low economic growth rates lose population.

For a sample of 115 middle-sized U.S. cities, it has been found that neither size of city nor rate of urban growth (between 1960 and 1970) correlates positively with lower rates of unemployment. In fact, what evidence there is points in the opposite direction: higher growth rates are associated with higher rates of unemployment.[10] Apparently, local expansion may lead to even more migration than that needed to keep unemployment constant. I have found that the rates of unemployment in the fastest-growing U.S. metropolitan areas are not different from the unemployment rate among U.S. metropolitan areas in general. Tables I and II list the 25 urban areas that grew fastest between 1950 and 1960, and between 1960 and 1970. The unemployment rates of those areas at the end of the respective decades are also listed. In both decades, half of the fastest-growing areas had unemployment rates above the national average for all metropolitan areas (i.e., there was no tendency for rapid growth to produce less joblessness). Just as striking are the comparisons of growth and unemployment rates for all metropolitan areas in California during the 1960–1966 period, a time of general boom in that state. Table III reveals that among all California metropolitan areas, there is virtually no correlation between 1960–1966 growth rates and the 1966 unemployment rate.

Table III is instructive also in that it reveals that although there is wide diversity of growth rates across metropolitan areas (1.7 to 8.7 percent), there is no comparable spread in the unemployment rates, which all cluster within the relatively narrow range of 4.3 to 6.5 percent. Consistent with our previous argument, I take this as evidence that the mobility of labor tends to flatten out variations in unemployment rates across metropolitan areas, regardless of widely diverging rates of local growth.

Still another way of confirming the thrust of this argument that growth does not make jobs is to compare areas in terms of the relationship between the expansion of the labor force over time and the associated unemployment rates. Labor-force expansion reflects both the economic expansion that produces it and the population growth that is its consequence. When examining all urbanized areas of the United States over the period 1960 to 1970, we find no relationship between labor-force expansion and unemployment rate. Like population growth itself, labor-force expansion does not have the effect of lowering the rate of unemployment.

Only in instances of severe declines in economic activity and consequent population outmigration, as in Appalachia, is there the consequence of depressed rates of employment following from a lack of population growth. But short of such chronic crisis, the clear implication is that local growth — the success of the growth machine in providing economic expansion and the increase in the labor force and population that follows — is no solution to the problem of local unemployment.

For the average worker in a fast-growing region, the effects are thus much the same as those in a stable area: there is a surplus of workers over jobs, generating continuous anxiety of unemployment and the effective depressant on wages that a surplus labor supply tends to exact. Indigenous workers may receive no particular benefit from the growth machine in terms of jobs; their "native" status gives them little claim over the "foreign" migrants to the additional jobs that may develop. Instead, they are merely members of the labor pool, and the degree of their insecurity is expressed in the local unemployment rate, just as is the case for the nonnative worker. Ironically, it is this very anxiety that often leads workers, or at least their union spokesman, to support enthusiastically their employers' preferred policies of growth. It is the case, of course, that an actual *decline* in local job opportunities, or growth not in proportion to natural increase, might require the hardship of mobility. But this price is not the same as, and is less severe than, the price of simple unemployment. It could also be easily ameliorated through a relocation subsidy for mobile workers, as is now commonly provided for high-salaried executives by the private corporation.

Workers' anxiety and its political consequences emerge from the larger fact that the United States is continuously a society of structural unemployment, with rates of unemployment conservatively estimated by the Department of Commerce at between 4 and 8 percent of that portion of the work force defined as active. There is thus a game of musical chairs being played at all times with workers circulating around the country, hoping to land in an empty chair at the moment the whistle is blown. Increasing the stock of jobs in any one place neither causes the whistle to be blown more frequently, nor increases the number of chairs relative to the number of players. The only way effectively to ameliorate this circumstance is either to create a full employment economy, a comprehensive system of drastically increased unemployment insurance, or some other method of breaking the connection between having a livelihood and the remote decisions of corporate and government functionaries. Until such a development occurs, the threat of unemployment acts to make workers at least passive

to land-use policies, taxation programs, and pollution nonenforcement schemes, which, in effect, represent income transfers from the general public to various sectors of the business class.[11] Thus, for a number of reasons, workers would follow their collective self-interest if they were consistently to use their political power not as part of the growth coalitions of the localities in which they are rooted, but as part of national movements that aim to provide full employment, income security, and tax, land-use, and environmental programs that benefit the vast majority of the population. They tend not to be doing this at present. Instead, they are, contrary to the interests of the rank and file, often a functioning part of the growth machine.

I Growth & Unemployment Rates of 25 Fastest-Growing Metropolitan Areas (SMSAS), 1960–1970

Metropolitan area	Rate of growth 1960–1970 (%)	Unemployment rate, 1970 (%)
1. Las Vegas, Nev.	115.2	5.2[a]
2. Anaheim–Santa Ana–Garden Grove, Cal.	101.8	5.4[a]
3. Oxnard–Ventura, Cal.	89.0	5.9[a]
4. Ft. Lauderdale–Hollywood, Fla.	85.7	3.4
5. San Jose, Cal.	65.8	5.8[a]
6. Colorado Springs, Col.	64.2	5.5[a]
7. Santa Barbara, Cal.	56.4	6.4[a]
8. West Palm Beach, Fla.	52.9	3.0
9. Nashua, N.H.	47.8	2.8
10. Huntsville, Ala.	48.3	4.4[a]
11. Columbia, Mo.	46.6	2.4
12. Phoenix, Ariz.	45.8	3.9
13. Danbury, Conn.	44.3	4.2
14. Fayetteville, N.C.	42.9	5.2[a]
15. Reno, Nev.	42.9	6.2[a]
16. San Bernardino–Riverside–Ontario, Cal.	41.2	5.9[a]
17. Houston, Tex.	40.0	3.0
18. Austin, Tex.	39.3	3.1
19. Dallas, Tex.	39.0	3.0
20. Santa Rosa, Cal.	39.0	7.3[a]
21. Talahassee, Fla.	38.8	3.0
22. Washington, D.C.	37.8	2.7
23. Atlanta, Ga.	36.7	3.0
24. Ann Arbor, Mich.	35.8	5.0[a]
25. Miami, Fla.	35.6	3.7
Total, United States	16.6	4.3

Growth does not affect unemployment rates.
[a]Denotes unemployment rate above SMSA national mean.
Source: 1970 U.S. Census, PC(1)-A1, U.S. Summary, "Number of Inhabitants," Table 32; "Summary of Economic Characteristics for Urbanized Areas, 1970"; "General Social and Economic Characteristics," Table 186.

The Problem of Natural Increase

Localities grow in population not only as a function of migration but also as a result of the fecundity of the existing population. Some mechanism is obviously needed to provide jobs and housing for such growth, either in the immediate area or at some distant location. There are ways of handling this without compounding the environmental and budgetary problems of existing settlements. First, there are some localities which, by many criteria, are not overpopulated. Their atmospheres are clean, water supplies are plentiful, and traffic conges-

tion is nonexistent. In fact, in certain places increased increments of population may bring benefits that spread the costs of existing road and sewer systems over a larger number of citizens, or increase the quality of public education by making rudimentary specialization (e.g., course electives in high school) possible. In California, for example, the great bulk of the population lives on a narrow coastal belt in the southern two-thirds of the state. Thus, a vast unpopulated region constitutes the northern one-third of the state—rich in natural resources including electric power, clean water, and breathable air. The option chosen in California, as evidenced by the State Water Project, is to

II Growth and Unemployment Rates of 25 Fastest-Growing Metropolitan Areas (SMSAs), 1950–1960

Metropolitan area	Rate of growth 1950–1960 (%)	Unemployment rate, 1960 (%)
1. Fort Lauderdale–Hollywood, Fla.	297.9	4.7
2. Anaheim–Santa Ana–Garden Grove, Cal.	225.6	4.6
3. Las Vegas, Nev.	163.0	6.7[a]
4. Midland, Tex.	162.6	4.9
5. Orlando, Fla.	124.6	5.1
6. San Jose, Cal.	121.1	7.0[a]
7. Odessa, Tex.	116.1	5.6[a]
8. Phoenix, Ariz.	100.0	4.7
9. West Palm Beach, Fla.	98.9	4.8
10. Colorado Springs, Col.	92.9	6.1[a]
11. Miami, Fla.	88.9	7.3[a]
12. Tampa–St. Petersburg, Fla.	88.8	5.1
13. Tucson, Ariz.	88.1	5.9[a]
14. Albuquerque, N.M.	80.0	4.5
15. San Bernardino–Riverside–Ontario, Cal.	79.3	6.7[a]
16. Sacramento, Cal.	74.0	6.1[a]
17. Albany, Ga.	73.5	4.4
18. Santa Barbara, Cal.	72.0	3.6
19. Amarillo, Tex.	71.6	3.3
20. Reno, Nev.	68.8	6.1[a]
21. Lawton, Okla.	64.6	5.5[a]
22. Lake Charles, La.	62.3	7.8[a]
23. El Paso, Tex.	61.1	6.4[a]
24. Pensacola, Fla.	54.9	5.3[a]
25. Lubbock, Tex.	54.7	3.9
Total, United States	18.5	5.2

Growth does not affect unemployment rates.

Note: Oxnard–Ventura and Middlesex County, N.J., although among the 25 fastest-growing SMSAs, were omitted because 1960 unemployment rates are not published for these SMSAs.

[a]Denotes unemployment rate above SMSA national mean.

Source: Table 33, Population Inside and Outside Central City or SMSAs in the U.S., 1940–1960, pp. 1–106 through 1–111, Vol. 1, P. 1, Census of Population, Bureau of Census; Table 154, Summary of Economic Characteristics, pp. 1–338 through 1–341, Vol. 1, P. 1, Census of Population, Bureau of Census.

move the water from the uncrowded north to the crowded, semiarid south, thus lowering the environmental qualities of both regions, and at a substantial long-term cost to the public budget. Other options were clearly present.

The point being made is that there are relatively unpopulated areas in this country which lack population not because they have "natural" problems of inaccessibility, ugliness, or lack of resources needed to support population. These areas are unpopulated because of the decisions made in the political economy to populate other areas instead. If the process were to be rendered more rational, the same investments in roads, airports, and factories could be made to effect a different outcome.

As a long-term problem, natural increase may well be on its way to oblivion. American birth rates have been steadily decreasing for the last several years and we are on the verge of a rate providing for zero population growth (Chapter 2). As that is achieved, a continuation of the present system of competing growth machines

will result in the proliferation of ghost towns and unused capital stocks as the price that is paid for the growth of the successful competing units. This will be an even more clearly preposterous situation than the current one, which is given to produce ghost towns only on occasion rather than as the inevitable consequence of growth of some localities at the expense of others.

Liabilities of the Growth Machine

Emerging trends are tending to weaken the growth machine. First and foremost is the growing suspicion among many citizens that growth brings with it more problems than the newspapers and businessmen are indicating. There are obvious problems of increased air and water pollution, traffic congestion, and overtaxing of natural amenities. These troubles become increasingly important and visible as increased

III   Growth and Unemployment Rates of All California Metropolitan Areas (SMSAs), 1960–1966

Metropolitan area	Rate of growth 1960–1966 (%)	Unemployment rate, 1966 (%)
Anaheim–Santa Ana–Garden Grove	6.5	4.3
Bakersfield	11.1	5.2
Fresno	12.3	6.5
Los Angeles–Long Beach	11.9	4.5
Modesto	Data not available	
Oxnard–Ventura	68.8	6.0
Sacramento	20.0	5.2
Salinas–Monterey	15.9	6.1
San Bernardino–Riverdale	27.9	6.2
San Diego	14.0	5.1
San Francisco–Oakland	11.1	4.4
San Jose	44.8	4.8
Santa Barbara	48.7	4.5
Santa Rosa	Data not available	
Stockton	12.5	6.3
Vallejo–Napa	20.6	4.4
California totals	23.29	5.25

Growth and unemployment rates are unrelated.
Source: Table 2, Estimates of the Population of SMSA's, July 1, 1966, and Components of Population Change Since April 1, 1960, p. 67, "Estimates of the Population of Counties and Metropolitan Areas, July 1, 1966, A Summary Report," Current Population Reports, Population Estimates and Projections, U.S. Department of Commerce, Bureau of the Census, Series P-25, No. 427, July 31, 1969; Table C-10, Civilian Labor Force, Employment and Unemployment, for Metropolitan Areas and Counties, 1958–69, *California Statistical Abstract, 1970,* State of California, Sacramento.

consumer income fulfills other material needs and as the natural cleansing capacities of the environment are increasingly overcome. Growth also introduces certain social pathologies, or at least makes existing pathologies more difficult to deal with. The larger the jurisdiction, for example, the more difficult it becomes to achieve a goal such as school integration without massive busing schemes. Smaller towns can have "naturally" integrated schools and other public facilities.

Another advantage of small-scale over large-scale settlements, one that is increasingly recognized by both urban planners and ordinary citizens, is the simple fact that growth frequently costs existing residents money. The costs of establishing new utilities, schools, and police services often are greater than the revenue contribution of the new residents. Under the current system most commonly used by localities in pricing public goods (e.g., water, sewer service, etc.), the dominant principle is that all users should share equally the costs of urban growth. For example, if a housing development is built 5 miles from any existing utility lines, the cost of running those utilities to serve an increment of, say, 100 new homes, will be shared equally by the existing residents and the new residents. The developer of these "leap-frog" projects thus creates a tremendous public cost, borne by all as a common expense. The contrasting principle would cause the developer to pay the *incremental costs* of utility expansion, a far more expensive proposition for him but one that would not subsidize growth where it is not economically sensible to have it. As another illustration, a locality may find itself using its full water capacity; any additional development may require the creation of an entire new water source at a vastly increased cost. Rather than having the developer pay the incremental costs of the new source, costs are spread evenly among all users, new and old. Here a city may grow far beyond what is its economically sensible limit.

In both these cases, the result of the current pricing system is to remove whatever rationality free-enterprise pricing mechanisms ordinarily provide. A developer has no incentive to build in areas proximate to existing utilities or to invest only in communities which have water supplies that are adequate. Present pricing systems thus mean that the state of the local resource base (both natural and man-made) tends to become irrelevant as a constraint on either the amount or the pattern of urban development.

Existing residents thus subsidize those who profit from the attracting of new residents. A 1970 study for the city of Palo Alto, California, indicated that it was actually cheaper for that city to acquire at full market value all the city's existing foothill open space rather than allow it to become an "addition" to the tax base through development.[12] It was computed that the cost to existing residents of urbanizing existing Palo Alto open space would be an average annual additional cost of $17 per capita, which amounts to approximately one-third more than the costs of acquiring the land and holding it open. A more detailed study in Santa Barbara, California, came up with similar results, showing that lowest city taxes would result from holding population levels constant.[10] Clearly, at least at certain population levels, and under certain topographical conditions, points of diminishing returns are crossed such that additional population increments lead to negative revenue gains.

Obviously, in addition to financial concerns, optimal population size is related to the sorts of values that are to be maximized[13]: it may indeed be necessary to sacrifice clean air to accumulate a population base large enough to support a major opera company. But the essential point remains that growth is certainly less of a financial advantage to "the taxpayer" than is conventionally depicted and that the values of most people would probably cause them to opt for the gains of relatively small cities compared to those made possible by large metropolises. It appears quite clear that under the conditions of affluence in the United States today, local growth is a financial and quality-of-life liability for the majority of local residents. Local growth is thus a wealth transfer from the local general public to a certain segment of the business community.

The Consequences of Reform

Because the city has traditionally been a growth machine, it has drawn a special sort of person into its politics. These people—whether acting on their own or on behalf of the constituency that financed their rise to power—are a certain subset of businessmen. In terms of the dynamics behind their existence as men of power, they come to politics not to save or destroy the environment, not to repress or liberate the blacks, not to eliminate civil liberties or enhance them. They may end up doing any or all of these things once having achieved formal authority, as an inadvertent consequence of being in a position to make decisions in other realms. But these sorts of symbolic positions are taken after the fact of having the resources for power; they are typically not the reason behind that power in the first place. Such people become "involved" in government, especially in the local party structure and fund raising, for reasons having to do with the land business and related processes of resource distribution. A few are "statesmen" who think in terms of the growth of the whole community rather than a more narrow geographical delimitation. But they are there to wheel and deal in order to effect resource distribution through local government —to bring wealth from outside and to distribute that wealth within. As a result of their position, and in part to develop the symbolic issues that will enable them to maintain a position of power, they get interested in welfare cheats, busing, crime in the streets, the price of meat or whatever. The "big issues" are thus substantially aftereffects of needing power for other purposes. That is not to say that these people don't "feel strongly" about these issues; they sometimes do. It is also the case that certain moral zealots and "concerned citizens" do go into politics to right wrongs, but the money and other supports that make them viable as politicians is largely amoral money.

Those who have ordinarily come to the forefront of local government have thus not been statistically representative or even representative of the higher social classes. The issues they introduce into public discourse are not representative either. They conform to Edelman's stipulation that the distributive issues, the matters that cause them to be there, are more or less deliberately removed from public debate.[14] The issues that are allowed into public debate derive from the world views of this most parochial group of entrepreneurs. It is their morality and agendas that dominate local politics.

There have been increasingly effective assaults on the growth machines in recent years.

Planning studies like the Palo Alto research[12] provide authoritative support for notions that have been around a long time: bigness does not seem to improve either the health of the public budget or the quality of life. Although it may be the economic arguments that appeal to the sentiments of many of the comfortable middle class who now oppose growth, there are many others who would sacrifice economic well-being, if it came to that, in order to halt locality growth. There are persons who value aesthetic and physical health over material goods and who, as individuals touched by the countercultural movements of the 1960s, give the antigrowth forces the energy and stamina of an authentic social movement. Many such participants were previously active in the peace and civil rights movements. They are now joined by middle-class professionals and workers who see their own life style in conflict with one aspect or another of the growth machine. Important in this group are employees and managers of organizations that are not dependent upon local expansion for profit, either directly or indirectly. In the Santa Barbara antigrowth movements, for example, a good deal of support is provided by executives from research and electronics firms, as well as branch managers of manufacturing corporations. Cosmopolitan in outlook and pecuniary interest, they use the local community only as a *setting* for life and work, rather than as an exploitable resource. Related to this constituency are certain very wealthy persons (particularly those whose wealth derives from the exploitation of nonlocal environments) who continue a tradition, with some modifications, of aristocratic conservationism.

These movements have, in some localities, begun to take local political power. The city of San Diego, the third largest metropolis in California, has elected a mayor who campaigned on an antideveloper platform. Although his politics seem to have more to do with controlling the *shape* of growth rather than its *size*, he has earned the enmity of the growth-oriented segment of the elite, who have claimed that he is out to "destroy" San Diego. In addition, antigrowth forces have had a pronounced effect upon local politics in such places as Oregon, Dade County, Florida, and many California communities, most notably in Marin and Santa Barbara counties. Similar movements are emerging in other sections of the country, such as Howard County in Maryland (which refused a major "patriotic" recreation development), and various areas of Vermont and New Hampshire. New Mexico has an "Underdevelopment Commission," which takes credit for scaring away new industrial plants.[15] Although most of the dramatic successes have been in communities of rather remarkable beauty, there is a good possibility that they will spread to other places as well because the basic facts are likely the same everywhere: the growth machine hurts the majority for the benefit of a minority.

The changes that the death of the growth machine will bring are obvious enough when applied to land-use policy. Local governments will establish holding capacities for their regions and then legislate, directly or indirectly, to limit population to those levels. The direction of any future development will tend to be planned to minimize deleterious environmental effects. A proposed development can be judged on its *social utility* rather than by some abstract reference to progress or jobs. The growth machine that has given American cities their present shape will be ended as the political and economic foundations of its existence are undermined. Industrial and business land users and their representatives will lose, at least to some extent, the effectiveness of their threat to locate elsewhere should public policies not be consistent with the profitability they desire. As the growth machine is destroyed everywhere, it will be the business interests who will increasingly be forced to make do with local policies, rather than the local populations having to make do with business wishes. New options for taxation, creative land-use programs, and the establishment of new forms of urban services will thus emerge as city government comes to resemble an agency that asks what it can do for its people rather than what it can do to attract more people. Any political change that could replace the land business as the key determinant of the local political dynamic would simultaneously weaken the power of one of the most reactionary political groups in the society. Political leaders, if drawn from a more representative base, would be expected not only to make land-use decisions in a new way but would likely take more progressive stands on other important issues as well. Thus, should such changes occur, there would likely be more progressive positions taken on civil rights, less harassment of welfare recipients, social "deviants," and other defenseless victims. Similarly, the public economy, the amount of total wealth spent for social services, might expand at the expense of the private economy, and for ends that are socially benign. The wealth of the society, so consistently held in the past by so few people, could at least partially begin to be shared more equally.[16]

It is noteworthy that in the places in which no-growth forces have established beachheads of power, their programs and policies have tended to be more liberal than those of their predecessors — on all issues, not just on the growth issue. In Colorado, for example, the environmentalist who led the successful fight against the Winter Olympics is the liberal Democratic legislator who also successfully sponsored abortion reform and other important liberal programs. The result of the no-growth takeover of localities may thus be a tendency for a general liberalization of local politics. To whatever degree local politics is the bedrock upon which the national political structure rests (and this is a highly debatable point), we can look forward to an increasing liberalization at the national level as well. It will then perhaps become possible to utilize national institutions to effect other policies which both solidify the death of the growth machine at the local level and which create new national priorities that are consistent with the new opportunities for urban civic life which may come to develop in the communities. Social life will make more sense and the nation's resources will less often be senselessly squandered. Proposed national programs involving vast public expenditures can be coolly evaluated for their social value in light of other priorities rather than mindlessly fought over to gain the jobs that they will supposedly create. Growth for growth's sake, at the aggregate level, will tend to be discouraged.

Finally, with the death of the growth machine at the local level, attention can be placed on the much more difficult (and important) problem of worldwide population growth and worldwide environmental destruction. This destruction of the world environment and the maintenance of the misery of half the world's peoples is perpetuated by the same type of growth machine as that which for so long has dominated the American local community. But it is more powerful, more intractable, and far less dependent upon local citizen sentiment (Chapter 19). It is the growth machine of the elites of the nation-states, the imperialism of those who control the corporate wealth of the world — not the developers and landowners of Cleveland, Boston, and Baltimore. These national elites consist of men who seem to be able to find no avenue for self- or class fulfillment other than in the continuous expansion of resource exploitation and material productivity. The elimination of their growth machine seems a truly awesome task. But if the general thrust of our analysis has been correct, it would be in the interests of most of the world's people to accomplish it.

References

1. A very small sampling of this work would include: Gordon, M., 1965, *Sick Cities*, Penguin Books, Inc., Baltimore, Md.; Weaver, R. C., 1965, *Dilemmas of Urban America*, Harvard University Press, Cambridge, Mass.; Gordon, L., 1971, *A City in Racial Crisis*, Brown Publishers, Detroit; Orleans, P., and Ellis, W. (eds.), 1971, *Race, Change and Urban Society*, Vol. 5 (Urban Affairs Annual Reviews), Sage Publications, Inc., Beverly Hills, Calif.

2. Downs, A. M. 1970. *Urban Problems and Prospects*. Rand McNally & Company, Chicago. p. 117.

3. Gans, H. 1967. *The Levittowners*. Pantheon Books, Inc., New York.

4. Goodman, P., and Goodman, P. 1960. *Communitas*. Random House, Inc. (Vintage Books), New York.

5. Edelman, M. 1964. *The Symbolic Uses of Politics*. University of Illinois Press, Urbana, Ill.

6. The terminology is drawn from Lasswell, H. 1936. *Politics: Who Gets What, When, How*. McGraw-Hill Book Company, New York.

7. See Piven, F. F., and Cloward, R. A. 1971. *Regulating the Poor*. Random House, Inc. (Vintage Books), New York.

8. See Spaulding, C. 1951. Occupational affiliations of councilmen in small cities. *Sociology and Social Research: 35*(3): 194–200.

9. Schulze, R. O. 1961. The bifurcation of power in a satellite city. In *Community Political Systems* (Janowitz, M., ed.). Macmillan Publishing Co., Inc., New York.

10. Appelbaum, R. P., et al. 1974. *Santa Barbara: The Impacts of Growth*. A Report to the City of Santa Barbara by the Santa Barbara Planning Task Force, Santa Barbara, Calif.; and Follett, R., The impacts of growth on middle class cities, unpublished Ph.D. dissertation, Department of Sociology, University of California, Santa Barbara, Calif.

11. See Whitt, J. 1972. The end of the road. Unpublished M.A. dissertation, Department of Sociology, University of California, Santa Barbara, Calif.

12. Livingston, L., and Blayney, J. A. 1971. Foothill environmental design study: open space vs. development. Final Report to the City of Palo Alto.

13. See Duncan, O. D. 1957. Optimum size of cities. In *Cities and Society* (P. Hatt and A. Reiss, Jr., eds.). Macmillan Publishing Co., Inc., New York.

14. See Schattschneider, E. E. 1960. *The Semisovereign People*. Holt, Rinehart and Winston, Inc., New York.

15. For a journalistic survey of these no-growth activities, see Cahn, R., Mr. Developer, someone is watching you. *Christian Science Monitor*, May 21, 1973, p. 9.

16. Evidence that the United States has seen no significant income redistribution in recent generations is contained in: Ackerman, F., Birnbaum, H., Wetzler, J., and Zimbalist, A., 1972, The extent of income inequality in the United States, in *The Capitalist System* (R. C. Edwards, M. Reich, and T. Weisskopf, eds.). Prentice-Hall, Inc., Englewood Cliffs, N.J., and in Kolko, G., 1971, Taxation and inequality, in *Political Economy: An Urban Perspective* (D. M. Gordon, ed.). D. C. Heath & Company, Lexington, Mass.

17. I have had the benefit of detailed comments from Richard Appelbaum, Norton Long, Gerald Suttles, Gaye Tuchman, and extraordinarily deft editorial advice from William Murdoch.

"All I want is just all the power there is."
From Herblock's State of the Union.

19

Environment:
Problems, Values and Politics

William W. Murdoch

Many environmental problems have arisen because of failure to predict all the consequences of technology. Although society can try to do better in the future, this will always be a danger. Individuals and governments are continually making decisions based on uncertainty, but it seems that there is a particularly high degree of uncertainty and complexity in the basic environmental issues, as well as a clash of social values and priorities. This chapter discusses the uncertainties first, then the complexity. It also explicitly recognizes that environmental action involves making value judgments, and discusses values and the political power needed to translate them into action.

UNCERTAINTIES IN RESOURCES

We do not know enough now about the demand for resources; and conditions in the future will change unpredictably. Man not only discovers but to some extent also invents resources, and predicting invention is difficult. Twenty-five years ago one could not have predicted the extent and consequences of the Green Revolution (Chapter 3). Because of that revolution, if everything goes well it should be possible to feed the population over the next 10 or 20 years at least as badly as they are fed now, and possibly a little better.

A similar uncertainty faces us in such resources as minerals. We will run out of presently known commercial reserves of certain important minerals at about the turn of the century (Chapter 4), but for some of them we will probably find substitutes or mine lower grades. Will substitutes be found for them all? Such an assumption is certainly implicit in our use of resources and is made explicitly by many economists.[1] Finding substitute resources is a particular example of one type of

general solution that has been proposed, the "technological fix." Whether or not the technological fix should be relied upon to solve our problems in the foreseeable future is a fundamental and arguable proposition which is not resolvable in an absolute way. (It can be ruled out as a *long-term* solution to indefinite expansion in a finite world.) The dilemma with respect to resources is exemplified by the following colloquy between two scientists, both of whom welcome "good" technology where it is appropriate. The discussion was part of evidence given before a Congressional committee.[2] Revelle has just commented that technology virtually assures our long-term survival and progress.

Cloud: First, I would like to comment that this is an act of faith, an act of faith which some of us don't share. . . .

Revelle: Which?

Cloud: The assumption that technology is going to find a way out.

Revelle: If it does not, we are sunk.

Cloud: We sure are.

Revelle: You must have this faith.

Cloud: No, you must not. Suppose you make a mistake; isn't it better to have made a mistake on the side of being sure that you were living within the limits of your resources?

Revelle: Expressed in terms of petroleum, you cannot live within the limits of your resources.

Cloud: Of course, we are already on the downgrade. By a miracle, a technological miracle, atomic energy comes along just in time to save our necks.

Revelle: I doubt that it is a miracle. It is an inevitable consequence of man's evolution.

Cloud: I think that is an act of faith which I don't share.

Faith is not generally a good basis for prediction. But equally, no one can categorically state that substitutes or alternative sources will *not* be found for a given resource at a given time in the future.

In the second place, a given resource is unlikely to run out everywhere at the same time. For example, mineral resources are distributed very unevenly throughout the world and nations are more or less dependent on their free flow across the face of the earth, as the Arabs' oil embargo demonstrated. All industrial nations, except possibly the USSR, are net importers of most of the minerals and ores that they use, and the United States will need to rely more and more heavily on foreign sources.[3] At the same time, world demand for industrial metals has been growing at 6 percent per year for a decade (i.e., three times faster than the population), and the underdeveloped countries in particular have enormous potential for increased demand. Since many important minerals occur largely in underdeveloped nations, it is not at all obvious that any nation can comfortably expect the free flow of minerals to continue indefinitely.

A third reason why the question of limits is difficult to answer is that each resource cannot be examined independently of other resources. Resources interact. Thus, to produce enough food on land we need a highly developed technology to produce machinery, fertilizers, and (for the present) pesticides. Each of these, in turn, is dependent upon a finite resource. Fertilizers have to be mined and manufactured. To produce all this material, vast industrial development, pressing on limited resources and economies, will be needed in the developing nations.

CATASTROPHES AND POLLUTION

Similar uncertainties arise in discussing environmental degradation as a limit to growth. If we suffer from pollution and environmental dislocation in a world of 4.0 billion people, of whom only 1 billion are industrialized, the imagination boggles at the specter of a population of 10 billion in which a higher proportion have evolved into the technological state — yet 10 billion is less than 50 years away at current growth rates. But how can we tell what will become technically or economically feasible or politically acceptable in dealing with pollution?

One can reasonably argue that there is already too much pollution and ecological dislocation — and that in itself should constitute a limit. But we also benefit from the pollution-producing activity, and might argue that the overall benefits outweigh the social costs.

The matter of costs and benefits is central in evaluating the risks of future environmental catastrophe. If continued growth implies a high probability of catastrophe, growth should stop. The crux of the matter is: How probable is the catastrophe and, less important, how large are the potential benefits of growth? The larger the potential catastrophe, the less probable it need be before we should give up the activity. We must not be misled by the fact that people do things that are irrational as judged by this kind of analysis; they climb mountains, for example. The actions taken by *individuals* at individual costs and for individual benefits must not be used as a yardstick for measuring *social* costs and benefits.

Where we can expect essentially infinite social costs and a relatively high probability of the occurrence of disaster, rational analysis is easy. But most situations are not this clear cut. However, we can probably place the detonation of large numbers of nuclear devices for digging canals and other purposes in the "clearly irrational" category. Similarly, nuclear reactor accidents seem to have very low probabilities of occurrence, but since the potential human cost is high, the additional economic cost of placing reactors underground probably is worthwhile. The economic costs of making fission power safe (Chapter 12) are also in this category of rational economic costs. Much more difficult is the question of pesticide use in the underdeveloped countries. And pesticides illustrate the additional problem that frequently we cannot measure the environmental costs. The social and economic benefits of pesticides, in the form of increased food for undernourished people, are obvious. But the environmental costs in the form of destabilized ecosystems and resistant pests are largely hidden and may not be measurable until a time in the future when the lag effects occur. The necessity to feed people is so great that here we probably need to continue to use pesticides and take a chance on catastrophe, but we must increase efforts to find alternatives. It may turn out, as pesticides become less efficient at controlling pests (Chapter 14), that we will cease to use them because the benefits have declined rather than because the environmental costs have risen.

An excellent illustration of uncertainty concerning catastrophe is provided by our potential effects on the climate; we do not even know if our activities, on balance, are heating it up or cooling it down (Chapter 15). Furthermore, the machinery of climate is made up of interacting parts (e.g., the relations among temperature, CO_2 in the atmosphere, and CO_2 in the oceans), and we cannot predict how they will feed back on one another as we tamper with one part. If we do cause changes in average temperature, it is not certain how this will affect climatic patterns. Will we have more drought in Africa, improve yields in tropical agriculture, reduce fishing yields in the North Atlantic?

Similar, though generally less serious, uncertainties face us in the area of noncatastrophic pollution. This is perhaps best illustrated in Table XIII of Chapter 5, which gives an overview of the environmental consequences of using various energy sources. This example is also useful in illustrating that the two problems — resources and pollution — are really inseparable, and indeed they are generally just different sides of the same coin. The table indicates some of the dilemmas that we face. Uncertainties about the safety of nuclear (fission) power plants and waste management make environmentalists hesitate to expand the amount of power we generate in this way, but we also do not really know if the technology of offshore oil drilling is dependable, and we certainly do not have adequate information about the biological consequences of oil spills. On the other hand, we are not sure if we can remove the sulfur from coal, although we know that coal is available in huge quantities. And how do we choose between strip mining for coal in Montana and oil spills along the coast?

The energy problem also illustrates uncertainties in resources: Can solar power be made economically feasible? Can the technological problems of fusion power be solved, and solved in time to replace fission power before we need to proliferate breeder fission reactors? Will Middle East oil be shut off again? It even provides us with an uncertain catastrophe: What is the chance that someday a group of terrorists will steal enough plutonium to make a bomb?

The presence of uncertainty, however, should not prevent us from making decisions. For example, no matter what energy sources we use in the future, energy conservation (particularly in the making of cities, buildings and life-styles that use less energy) will lessen environmental degradation.

Uncertainty is one aspect of the problem we face when it comes to internalizing *all* the costs, as discussed in Chapter 16. Suppose that a utility company wants to build a nuclear power plant near a particular town, and that we wish to follow Ruff's advice and make sure that all the costs are internalized. What are the costs? They obviously include all the usual capital and operating costs, and, in addition, they include environmental costs. Among activities leading to environmental costs are: (1) disposal or containment of radioactive waste, (2) radioactive emissions at the plant, (3) thermal pollution, (4) accident leading to the release of radioactive wastes, and (5) sabotage. Note that the last two events would have enormous costs but have a very low probability of occurring. One of the more depressing aspects of economics is that it *can* calculate such costs. For example, a human life is assigned a value (in earnings lost), and once property damage and the number of people killed or injured is estimated, total potential cost can be calculated. How are such costs, which could easily be many billions of dollars, to be internalized?

In principle, a utility company should buy insurance to cover such costs in the event that an accident should happen. Several problems arise here. First, no insurance company will take such risks; the government has to do the insuring (Chapter 12). Second, the company does not pay the true cost of such insurance (premiums), which would be very high. That cost is therefore not internalized, and perhaps this indicates that it is too high for us to pay; if the company cannot pay the premiums, it should not operate the plant. Third, a major problem in estimating the proper premium to be paid is the uncertainty associated with the likelihood that the accident will happen. This problem is tremendously difficult. Normally, we can put a probability on accidents happening, for instance from driving a car, because there is lots of information about how frequent car accidents are under various circumstances. But estimates of the probability of nuclear accidents vary widely because we have very little experience with them; such estimates are based largely on computer models, and to believe them we must believe that the modeler has been able to foresee all possible calamities. But our experience is that all the consequences of a new technology are *not* foreseen at the outset. So the estimates include a large amount of guesswork, and the guess varies according to whether or not one likes nuclear energy. The problem of uncertainty can thus seriously affect the most rational efforts to get a clean, safe environment.

COMPLEXITY: PEST CONTROL
AS AN EXAMPLE

Almost any environmental problem can be chosen to illustrate the complexity of the issues involved. I have chosen the example of reducing the use of pesticides in agriculture because it is a problem currently being intensively studied. The *technical* problems involved in this change can be severe, but I want to ignore that difficulty and look at other complexities that arise after all the technical problems have been solved. Even when we start with a small, narrowly focused problem, it quickly expands dramatically as we follow up the different issues involved.

Let us start with economics. A major economic problem is that of developing and marketing alternative pest management schemes. One such scheme is integrated control, which relies heavily on biological and cultural techniques and uses only narrow-spectrum chemicals or juvenile hormone. Pesticides are produced by large chemical companies that produce many other products as well. Before proceeding to develop a pesticide, these companies need to be assured that there is a large market. This depends on the size of the company and on the cost of developing and registering the compound (about $11 million on the average); and these costs would be the same for a company of any size. The surest way to have a large market is to sell a broad-spectrum biocide; but this is precisely a major fault with modern insecticides (Chapter 13).

Narrow-spectrum insecticides are better, but they are likely to have too small a market. One (though certainly not the only) reason that costs are so high is that the registration requirements set by the Environmental Protection Agency (EPA) are fairly strict, and recently have become more strict in response to pressure from environmental groups and lawsuits.

The registration requirements themselves, incidentally, may sometimes be a barrier to better pest control. A pesticide is defined as "any mixture of substances" that kills pests. Both juvenile hormone and ladybird beetles are mixtures of substances. Strictly speaking, therefore, both should undergo extensive tests for toxicity, environmental hazard, and so on (Chapter 8). Yet, it is clearly highly improbable that anyone will accidentally eat ladybird beetles, and if they do, they are unlikely to produce children with birth defects as a result! In fact, though, EPA has recently been discussing the question of whether ladybird beetles should be registered! This would make such biological control expensive.

In addition to development and registration costs, many new alternatives to pesticides face economic difficulties in marketing because of patenting and packaging problems. When a company changes an existing compound to produce a new one, it can patent that change, thus ensuring a return on its investment for a decade or more. But modern alternative pest control methods, such as integrated control, cannot be easily patented. They are less a set of particular compounds and more a collection of knowledge and behavior, both of which can be learned. Thus, a firm of pest management consultants might take on the control of pests on a farm, and the farmer might quickly learn how to sample, identify, and count the bugs and when to apply particular procedures. There is no guarantee that the consultant will ever get a return on his investment. One solution is to make sure that the consultant can keep learning new and improved techniques. This then requires that there be new research in the area (supported by the government?) and that there is an opportunity for the consultant to visit the research institute to keep up with developments. Even then, though, it may be necessary to ensure the consultant's job and income (by salary supports, or by nationalizing the pest consultant industry?).

Marketing pest control techniques also raises a series of behavioral and psychological problems. After all, in the end it is the farmer who has to be convinced to substitute pest management for regular applications of pesticides "by the calendar," and we do not want to wait for a catastrophe to convince him (Chapter 14). There are several problems here. Pest management often means sitting by, doing nothing (except monitoring), while the pest population increases toward the economic threshold. It may mean taking the advice of a new type of person, the pest consultant, rather than the known pesticide salesman, and it means trusting a set of procedures rather than a bag of pesticide, a bag that is clear and concrete evidence of a powerful technology. Indeed, there is good evidence that a strong motivation for the farmer to apply insecticides is in many cases *not* evidence of significant pest damage but a desire for *insurance*; the farmer often applies insecticides merely as a prophylactic (which will often actually stimulate pest problems). Thus, any new pest management scheme will have to devise a way of giving the farmer some assurance that the scheme will work. The obvious way to do this is through insurance schemes whereby the farmer is assured of a certain return on his crop, even if it is obliterated by pests, provided that he has followed the advice of the pest manager. But insurance companies have not shown much interest in insuring pest consultants for the large sums of money for which they might be sued by a farmer losing his crop, and perhaps, again, there is a need for government insurance.

A single step that would reduce insecticide use would be to cut the amount used on one crop: cotton. Because cotton accounts for about one-half of the insecticides used in agriculture, a solution to the insect pest problem in cotton would vastly reduce insecticide use. But the complexities previously discussed seem simple when compared with the baroque economic and political structures that surround the production of cotton and its use of pesticides. Cotton is unusual among crops in that there is a substitute, synthetic fiber. One way of reducing insecticide use is to plant less cotton, produce more synthetic fiber, and ex-

pand other crops that use less insecticide. (Of course, this raises the enormously complex issue of whether cotton or synthetic fibers are environmentally more damaging.) In the tangle of economic constraints that surround cotton production, the complex system of federal subsidies and payments is perhaps the most obvious. Cotton production is or has been subsidized by special loans, restrictions on acreage, import quotas, direct payments to farmers, and federal research and development. In some instances this results in cotton production in areas where it would otherwise not be economically feasible. For example, an area of 200,000 acres in the southern Rio Grande Valley in Texas is one of the heaviest users of insecticides, especially methyl parathion, and there are frequent reports of poisonings among field workers and spray-plane loaders. Yet in several years the only reason that the crop was grown by many of these farmers was the existence of the subsidies. Removal of subsidies would greatly lessen insecticide use because the alternative crops in the region use much less.

Finally, there are legal and social problems in getting better pest management. For example, because pests can move over large areas, the optimum practice may require all farmers, without exception, to plow down cotton stalks shortly after harvest. Should we rely on people's good sense and collective action, or should coercion be used? In northern Texas there are local districts that can enforce such practices, but this raises questions of people's freedom and imposes the costs of bureaucracy (Chapters 16 and 17).

Better pest management is likely to have important social consequences. Environmentalists are generally middle- and upper-class North Americans, and in their desire to improve the environment, they tend to forget social costs that fall heavily on the poor. The history of American farming has been the steady removal of the small farmer and farm laborer from the countryside to the city (Figure 1). Government programs have tended to accentuate the inequalities in income between rich (large) and poor (small) farmers. (The accentuation of income difference seems to be an almost universal consequence of public spending.[4]) The cotton subsidy program gives more money to large farmers than to small ones;

it is easier for rich farmers to get credit; and university research in agriculture has invariably produced labor-saving machinery (including herbicides) that reduces the opportunities for farm laborers and benefits large farmers, who can most efficiently use large machines, rather than machinery or products that improve the lot of the laborer or small farmer while still providing him with work. Rich farmers tend to be better educated and have higher status and have gained more from university extension services. Indeed, Gardner[5] has shown that over the long run, increased efforts in public research and in university and U.S. Department of Agriculture extension activities, have increased the inequality of farm incomes.

At the moment the social effects of the new pest management techniques are unclear, but unless government takes corrective measures, it seems likely that the new techniques will accentuate the inequality in farm incomes. University scientists, who want to spread the use of their techniques, find that richer farmers are better to deal with because the logistics are easier on a large farm and because if a large farmer is won over, it will have a good "demonstration" effect on other farmers. It may be difficult for many small farmers to get together to provide enough acreage for a pest consultant, and costs per acre are likely to be lower for larger farms. Unless we take steps to prevent it, the likelihood is that new techniques will be yet another mechanism that increases the difference between rich and poor in U.S. agriculture.

SOCIAL VALUES AND THE ENVIRONMENT

Environmental issues are enmeshed in our system of social values. Placing a pollution tax on gasoline is likely to help clear the air but to have a disproportionately harsh effect on the poor, because the poor spend almost all of their income on fixed expenses. Taxes on commonly consumed goods take proportionately more from a poor person's income than from a rich person's. And, since many of the poor live in city centers and have to travel to jobs in factories outside the city, the addition of such a pollution tax is not one they can avoid by more careful spending. We do need a cleaner environment, pollution taxes, and public spending. But we also need fairness and equity, which we can have by redistributing in-

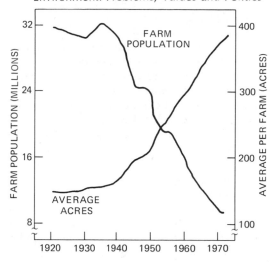

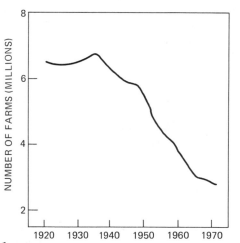

1 **Changes in Farms and Farmer Population in the United States.** While the farm population declined, the total population in the United States roughly doubled, from 106 million in 1920 to 210 million in 1973.

come — by making the poor richer at the expense of the rich and the middle class. Needless to say, this is a politically explosive issue but anyone who cares about the environment must confront it and weigh the social consequences of environmental action. I believe that no environmental action that is likely to increase inequalities should be pushed forward unless a method of redressing the inequity is also formulated and supported.

Some economists argue that the effort to help the poor by redistributing income will be

459

self-defeating because it will remove incentives, thereby slowing economic growth and making everyone poorer than they would have been if inequalities, and therefore incentives, were left as they are and the poor were pulled upward by a generally improved economy. There are several counterarguments. First, the experiment of redistribution has hardly been tried. Between 1935 and 1946 incomes in the United States became very slightly more equal (with no apparent effect on economic growth) and inequality has stayed consistent since then. Second, it is quite likely that poor people (and much of the rest of the population) would be economically and psychologically better off with improved social services (more public spending) and less inequity than with extra dollars to be spent in a milieu as socially derelict as our own. Third, from an environmental point of view, some types of economic growth (e.g., any that depend heavily on energy production) probably will have greater social costs than benefits. Finally, and perhaps most important, there is a fundamental question of values. Life should be something more than an economic footrace with greed the motivating force.

The United States, in particular, faces a difficult question of values in its relationship with the poor and hungry nations. Recently (1974) Lester Brown (the author of Chapter 3) argued that

There is little doubt but that a year from now we will see the largest food deficit of any region in history unfolding in Asia — a situation where political leaders in the more affluent countries, including the United States, may have to decide whether to throw up our hands and sort of cast Asia adrift . . . or go to consumers and ask the food equivalent of turning the thermostat down six degrees—that is, reducing the consumption of, say, livestock products in order to free many millions of tons of grain to move into Asia. Can we rely primarily on the marketplace to set the price and determine the distribution of so essential a commodity as food? And should Americans continue to consume as much fodder as they now do, most of those consuming more than we actually need?

World demand for food goes up by 30 million tons per year. In spite of 1973's good grain harvests, 1974's world grain reserves probably were below 20 days' worth compared with 95 days in 1961. In the face of these and other pressures, the price of wheat, corn, and soybeans doubled in the 20 months preceding July 1974. This, of course, is to the advantage of the U.S. farmer, who (along with Canada) produces a large fraction of the world's exportable food grains; one cropland acre of every six in the United States is now planted with soybeans, whose export now brings in more money than all of our high-technology exports, such as jet aircraft and computers. But the countries most in need of these grains have the greatest difficulty finding money to buy them. Should the United States let Asians starve in order to benefit its own balance of payments? On June 20, 1974, the San Francisco *Chronicle* noted that following a record crop, American farmers were holding massive amounts of wheat off the market "waiting for" (or creating) an increase in price, a strategy made possible by the removal of government controls as well as the disappearance of government-owned surplus wheat. "By mid July the American farmer could have under his control 90% of the world's supply of free wheat" said Charles Rhoades, executive director of the Oklahoma Wheat Commission.

Perhaps the most impressive recent evidence of the importance of social and political values as they affect environmental issues comes from China. The year 1949 can be looked upon as the start of an enormous experiment, admittedly without proper controls. In that year the Communist revolution was successful in China, and since then China's general development has been guided (and strengthened recently during the Cultural Revolution) by very explicit socialist notions of equality. India was faced at the same time with many of the same problems (poverty, large and rapidly increasing population, illiteracy, poor medical care, and a shortage of food) but chose development under the wing of Western capitalism, retaining a drastically unequal society. Both countries were about equally poor and both have remained poor (per capita annual income about $100); but there the similarity ends. India seems to fall further apart economically and socially with each passing week. The Chinese, by contrast, seem to have gone far to alleviating many of their problems. Recent observers[6] agree that the population is well fed and literate (which is extremely important in cutting the birth rate), and that medical care is remarkably adequate throughout the nation. The last point is especially significant and characteristic. China quite explicitly has designed its medical and health systems so that all citizens are assured of basic services. This has been done rather than developing a few excellent hospitals in the cities, hospitals that can give the rich good care but leave the majority of the people without medical help. These developments in China, together with an egalitarian economic system and a social system that respects the aged, have provided each person with a sense of security about the future. It seems that the birth rate and rate of population increase have declined dramatically, probably because poor people no longer feel the need to have many children to ensure that someone will be around to take care of them when they are old.

It is often claimed that poor countries with rapidly increasing populations need to be "developed" to break the vicious cycle of population increase. But China has shown that, in the absence of much economic growth but by adopting specific social values, the major environmental threat to poor countries — increasing population — can be defeated. No doubt China's culture and heritage differ in many ways from India's, and socialist revolution alone is not responsible for China's success, nor is its absence solely responsible for India's failure (e.g., India has tremendous problems with its diversity of languages), but the values pursued by the Chinese are clearly important. China has done it not only without aid, but despite obstruction from the West. True, the price paid by the Chinese is reduced political freedom; but it is not obvious that political freedom is of much value to a beggar starving to death on Calcutta's streets.

Equilibria and Optima

Although environmental problems are complicated, we must try to solve them; if we do nothing they will get worse, because of the momentum for growth in environmentally destructive activities. The simple principles of equilibria and optima can help us to make judgments in the face of the complex details that characterize environmental issues. First, it seems clear that continued growth of population, both in the United States and in the world, can only make environmental problems worse than they are now. In other words, the attainment of a nongrowing, equilibrium population should be a first-priority goal. Second, for both the United States and the world it seems clear than an *optimum* population — the best possible — is at most no larger than the present population. These two features are worth separating: a growing, nonequilibrium population prevents economic development in poor countries (Chapters 2 and 3), but a population might have stopped growing — that is, be at equilibrium — but still be too large (or too small) to be optimal. For the United States, zero population growth and a small equilibrium population should mean an increased chance for improving the general quality of our lives and the environment. For the poor and overpopulated nations of the world, zero population growth and lowered populations can mean avoiding such catastrophes as massive famine and the persistence of abject poverty.

SUSTAINABLE POPULATION LIMIT

Not everyone is convinced that zero population growth is a desirable goal, and indeed the town of Petaluma in California has found that it cannot easily legally achieve that goal by legislation. (Petaluma tried to establish an antigrowth ordinance that was rejected by the Court on the grounds that it interfered with the citizen's right to travel freely.) Even among those who agree that there is *some* limit to population size, there is a great range of estimates as to what that limit is (Chapter 2). One reason it is difficult to estimate the earth's carrying capacity is that the number of people the earth can support over an extended period of centuries is certainly less than the number it *might* support at any one time. It is this *sustainable* limit that concerns us. There is guarded optimism among agriculturalists that we can feed the 6 or 7 billion people who will be on earth at the turn of the century (Chapter 3). But while some believe that we can feed 20 billion or even a wildly optimistic 100 billion, there exists real doubt that the earth can sustain even 6 billion *over long periods*. Indeed, Hulett has estimated that, if we take account of all resources and of our effects on the environment, and if the total population came to have the current U.S. standard of living, the total sustainable world population would be only 1 billion, less than one-third of the current number.[7] It has been argued that considering simply the food supply, the sustainable capacity is only 1.5 billion.[8]

The argument here about the food supply is based on the fact that to feed 6 to 7 billion people we will have to do things to our food-producing ecosystems which are liable to impair their capacity to produce food in the future. Consider, for example, that our hopes for feeding more people in poor countries rest mainly on the new cereal grains. This means raising fertilizer applications in the developing countries by 10-fold and even up to 27-fold.[8] But many of the soils in the developing countries already have a very poor structure, and there is recent evidence that continued heavy application of fertilizers in even good soil may decrease its future productivity and its response to added fertilizers. So we might wind up with worse soils than we started with. Again, as increasing use is made of pesticides in poor countries, we may see the kind of burgeoning of pest problems described in Chapters 13 and 14. The resultant ecosystems may become so denuded of alternative control systems that yields may begin to decrease. Similar arguments can be developed for other aspects of the world's rapidly increasing food production.

These unhappy possibilities may or may not transpire. But the dangers are there and their likelihood is uncertain. The cautious procedure is to assume that they are highly probable. This means that we should place more emphasis on long-term solutions rather than on short-term "fixes" that may interfere with our chances of achieving a stable environment.

Probably no one knows what the long-term sustainable population is, but the doubling population and its greatly increased activities will place terrific and increasing strains on the environment over the next, say, 30 years. In trying to gauge the severity of the problem, a moderately conservative guess is that some large-scale environmental problems (such as agricultural failure in the poor countries) are likely to become severe and possibly even catastrophic within 30 years. (Some local problems such as urban air pollution are already severe.) A sober evaluation of these points and

of others made throughout this book indicates that in some respects we may be reducing the carrying capacity of the environment and suggests that we may be at, approaching, or exceeding the sustainable limit.

Discussion of maximum limits is illuminating and even necessary. However, the question of limits is to a great extent the wrong question insofar as it aims us toward the wrong goals. There is a more rational goal than living to the limits of our ecosystem, particularly when we do not know where the limits are and when we are likely to overshoot them in finding out.

THE CASE FOR OPTIMA

A more sensible course is to substitute optima for maxima in our goals. Although ecologists' concern over growth is frequently construed as a "negative" approach, it actually provides us with the stimulus to develop a positive approach in determining our relationship with our environment. A basic assumption is that, over a specified period of time and for defined conditions, there is an *optimum* population size — for the earth or any part of it — and that the optimum is smaller than the maximum size set by the physical and biological constraints of the environment. We can further define the optimum as one that allows long-term persistence of the population *in equilibrium* with its environment, equilibrium in the sense that resources and wastes are circulated to the extent needed for long-term survival and that the environment is kept as a diverse provider of our needs, including aesthetic needs. (A final refinement is that the optimum population can be said to have been reached when the addition of more members results in a lowering of the quality of life of those already present.) Estimating the optimum population size for any point in time, or even defining suitable techniques and criteria for doing so, is an extraordinarily important and difficult job. It is a job, too, which is not yet being tackled to any extent.

Similarly, "quality of life" is hard to define. It involves our physical and mental health and comfort, our material standard of living, social interactions, and other factors. But it surely must be higher when we have clean air and water, diverse surroundings, space for recreation and solitude, a rich culture willing to preserve manifestations of its history, and lively cities where social stress, congestion, and crime are at a minimum.

The goal of an optimum population seems eminently reasonable, and has become much more widely accepted in the last few years. Yet it is a puzzling, but I believe valid, observation that a nongrowing, equilibrium population is still not widely accepted as a worthwhile aim outside environmentalist circles, as is illustrated by several recent essays by social scientists.[9] Even if economic growth were our goal, most economists assure us that population growth per se is not necessary for economic growth. There is a great deal to be said in favor of a stable optimum population — in a nutshell, that it will be optimum with respect to the quality of life.

ECONOMIC GROWTH AND THE ENVIRONMENT

The question of growth versus equilibrium raises two major economic questions. The first is whether or not economic growth will soon be limited by dwindling resources.[10] There is no question that the growth of any physical quantity, such as the amount of a given mineral resource that is in use during a given period or the total of all such materials, is necessarily limited since we live in a finite world. What is less clear is when such a limit will be reached (or when the growth rate will be so small as to be negligible), and whether economic growth in terms of all goods and services need come to a stop within a period of even several centuries. Chauncey Starr[11] and others have noted that those who have claimed that growth must soon cease because of limited resources have ignored two facts: (1) that technological innovation (implying substitution, recycling, miniaturization, etc.) has increased roughly exponentially, and (2) that the price system in a sense "creates" resources where none existed previously (e.g., shale oil may now be a usable resource since oil is so expensive). There seems little doubt that economic growth is certainly possible for the foreseeable future (say 50 years), although there are widespread fears of a depression in the *immediate* future.[12]

The second question is: Can economic growth continue without a concommitant increase in environmental degradation, or should we stop economic growth in the rich countries (we cannot recommend that for poor nations) on environmental grounds? The benefits of economic growth hardly need to be stressed, and it is worth noting again that the poor in this country can hardly be expected to embrace nongrowth in their incomes with the same fervor as might many middle-class environmentalists. To answer these questions we must consider some of the features of economic growth, some of the environmental costs of that growth, and those aspects of growth that environmentalists might accept or reject.

Economic growth in a modern industrial society like the United States has been characterized most obviously by huge rates of increase in the use of energy and material resources, both of which lead to the major types of pollution discussed in earlier chapters. In the United States, as Holdren noted in Chapter 5, energy consumption has been increasing about 5 times as fast as the population, with all of the environmental consequences outlined in that chapter. The use of mineral resources has generally increased at about the same rate. For example, between 1905 and 1938 more metal was produced than had been produced in man's entire history before 1905, and metal consumption since 1939 is equal to the entire metal consumption before that period.[13] The world's industrial metal production has been growing at over 6 percent for 10 years. Real GNP in the United States tends to increase at about 3 to 4 percent per year, and we can expect a massive trend to industrialization in nations currently less developed. Therefore, an increase in industrial activity by at least a factor of 4 does not seem an unreasonable guess for the next 30 years. All of this means more mining and more waste production during manufacture and use of these products. As ores come from lower and lower grade rocks, each ton of ore will require the movement of more parent material and the use of more energy.

Perhaps less obvious is the change in the nature and distribution of environmental degradation that has accompanied growing affluence. Modern technology has produced thousands of exotic chemical compounds (Chapter 8), and their production has often been increasing faster than other components of industry. For example, the chemical industry in Europe doubled in the 8 years between 1953 and 1960.[14] During the same period the output of petrochemicals increased eightfold and plastics fourfold. Doubling times of 8 years give about a 16-fold increase in these products over the next 30 years! Second, there has been and continues to be a trend, in the United States and Europe at least, for industry to move to the coast, especially for industries that produce waste effluents and that will come under increasing pressure not to pollute inland rivers and lakes.

The special problem of growth in the coastal zone is even worse in estuarine areas, which are of great biological value (Chapter 1). One-third of the U.S. population now lives and works close to estuaries. The United States has lost to "progress" some 40 percent of its estuarine area.[15] As building land becomes scarcer and rivers and coastal areas are managed to prevent flooding and so on, estuaries must continue to dwindle unless some provision is made for them and the rivers that serve them. Since they act as a sink for pollution, they will inevitably feel the brunt of populational and industrial changes.

It is important to realize that potentially hazardous high growth rates and changes in technology are not restricted to industry, but occur also in agriculture. Thus, both in the rich and poor countries increases have occurred in two major polluters: fertilizers and pesticides. It is significant that the new strains of cereals being planted in the tropics require pesticides and fertilizers where they were not used before, or, if they were used, at least twice as much insecticide and four times as much fertilizer are needed. The great increase in use of these two types of products will take place in the tropics, especially in the Southern Hemisphere.

The increases are likely to be on a mammoth scale. Excluding China, about 70 million metric tons of fertilizer is used globally, almost all in the advanced countries, where about 80 percent of it is produced. Application rates are extremely low in the poor nations. For example, in 1967–1968 in India it was 8 kilograms per arable hectare, in Asia as a whole 15 kilograms, in Africa only 5 kilograms. By contrast, the rate was 219 kg in Britain and 70 kg in the United States. The rate of consumption has increased by about 10 percent over the past 10 years, and if this trend continues, we can expect consumption in 30 years to be about 10 times higher than it is now. Thus, it is likely to be as much as 700 million tons[16], perhaps even higher. Thus inland and inshore waters off the poorer countries of Asia and Africa are likely to have massive increases in "excess nutrients" and a consequent imbalance of nutrients (Chapter 10).

There will always be disagreement over the question: Has (or will) the "quality of life" or "social welfare" increased as per capita income has increased? Clearly, some aspects of growth benefit almost no one — for example, the increasing use of resources by the Department of Defense, including the manufacture of bombs and napalm used in Vietnam, all of which figured into per capita GNP. Indeed, Kenneth Boulding has suggested that GNP is altogether the wrong measure of economic welfare, which should be measured by the condition of people, the extent, quality, and complexity of their bodies and minds and the things in the system. In such a measure, GNP would be a *cost*, the cost of maintaining the system. However, we do not have a good measure of welfare, and economic growth in the past has certainly been correlated with an increase of many aspects of economic welfare.

The costs of economic growth are certainly not negligible, and with respect to environmental quality the weight of evidence probably favors the hypothesis that in recent years in the United States, as we have grown richer the quality of our shared environment has declined, partly owing also to increased population. One can try to measure this by indexes such as the clarity of the air, the dirtiness of the water we drink and swim in, the ratio of wilderness or National Park area to people (and the fact that visits to national parks may soon have to be rationed), the number of feet of clean beachfront per person, the time spent in nonvacation trips in one's car, the average noise level in the city, the increased probability of environmental disaster, and so on. A few horror stories do not measure the decline, but they illuminate it: for example, the frequent warning to children in Los Angeles not to run, skip, or jump on very smoggy days; smog in

Yosemite Valley; the fact that mothers' milk has contained so much DDT that it would be banned from interstate commerce.

The case is surely easy to make for the city dweller: smog, noise, traffic, mental stress, all the apparently necessary trappings of a healthy economy. Mishan[17] sees: "The spreading suburban wilderness, the near traffic paralysis, the mixture of pandemonium and desolation in the cities, a sense of spiritual despair scarcely concealed by the frantic pace of life" and notes that "such phenomena, not being readily quantifiable, and having no discernible impact on the gold reserves, are obviously not regarded as [economic] agenda."

It might be said in reply that people *choose* to live in such conditions, they want this kind of life; therefore, wealth has brought them increased amenities. Los Angeles, for example, is a rather unpleasant city, yet its population has been until recently one of the fastest growing in the United States. Clearly people "choose" to come to Los Angeles. But, in fact, people need jobs and they go where jobs are or where they think there is opportunity for economic success. Second, they have been led to believe that more money necessarily means a better life, and they go where there are highly paid jobs. That is, they make a choice based on misinformation. Madison Avenue has implied for years that our sexual attractiveness is directly related to our use of deodorants, toothpaste, and automobiles — and naturally no one wants to be unattractive. Finally, it is inaccurate to say that people have chosen to live in a city such as Los Angeles rather than some very different kind of city, since there exists virtually no choice. Almost all our cities have the same problems. Ringed by endless suburbs and suffocated by traffic smog, cities are fit for air-conditioned cars but not for people.

The matter of where people choose to live illustrates that, contrary to myth, increasing collective wealth does not necessarily increase individual choice. Mishan[17] observed that "As the carpet of increased choice is being unrolled before us by the foot, it is simultaneously being rolled up behind us by the yard." Consider, for example, the restriction of one's freedom to choose a pattern of living. Because of the impact of the car (that bastion of economic growth) it is essentially impossible for most people to live close to work, to walk there or cycle there, or to go by efficient public transport.

It would be foolish to blame all the problems of urban life on population and economic growth. Clearly, major reasons for these problems include the fact that successive national administrations (and the voters) have placed higher priorities on military excursions and the space race than on the welfare of the poor, in particular the poor black urban citizens, on educating people, or on the quality of life generally. Other reasons include our stress on private consumption rather than public spending and quite simply poor planning and administration. Nevertheless, the spread of disamenities that has blighted urban life is directly attributable to the headlong rush of production and growth. Part of the physical and administrative problems of cities is attributable to their high growth rate.

It is generally dangerous to assume that things will be the same in the future as they have been in the past, and most economists would claim that economic growth is compatible with enhanced environmental quality provided that environmental costs are internalized (Chapter 16) and that proper use is made of pollution technology. Ridker[18] has made this case clearly. He has pointed out that GNP is an artifact, an artificial measure reflecting in part the value people place on commodities rather than the amounts of commodities. It must be added, however, that in practice growing GNP *has* been associated with growth in the use of resources, space, energy, and so on, as we noted above. Ridker's question is: Can we continue to increase the rate at which we use "resources, space and environmental carrying capacity," at least over the next 30 years or so, without increasing pollution levels? The answer appears to be that, with very little slowing of economic growth, most pollutants levels could be *reduced* if the economy keeps growing, *provided an active pollution-abatement policy is actively pursued.*

A summary of Ridker's findings is given in Figure 2, which shows expected amounts of pollutants in the year 2000 under (1) different assumptions about population and economic

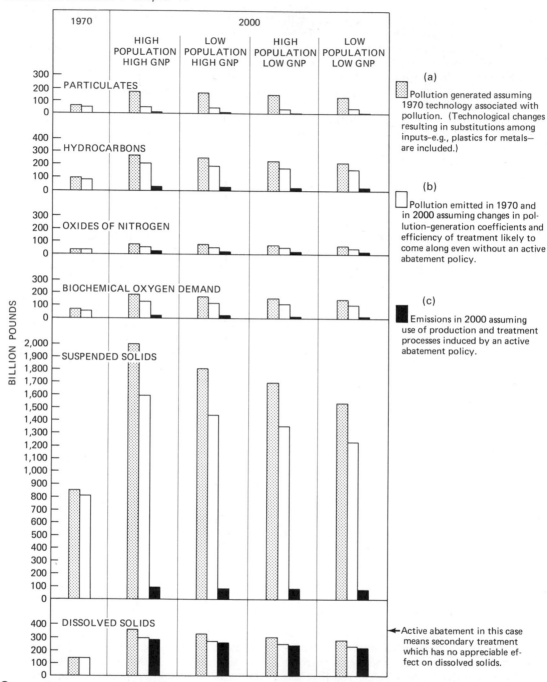

2 Pollution generated and emitted under different assumptions about population, GNP, and pollution technology. See the text for a discussion. *Source:* Reference 18.

growth rates (the columns in the figure), and (2) different assumptions concerning pollution-abatement policies (different types of bars in the diagram). The (a) bars indicate the amount of pollutants actually generated in 1970 and the amounts that would be generated in 2000 on the assumption that no significant changes in pollution technology occur after

466

1970 (although there would be changes in input, such as plastics for some metals). The (b) bars indicate actual pollution emissions in 1970, and for 2000 they indicate probable levels brought about by changes in technology that would come without the goad of an active abatement policy. The (c) bars indicate what would happen in 2000 if the standards recommended by the Environmental Protection Agency for 1976 were applied in 2000. Clearly, pollution levels can be greatly reduced.

Ridker then shows that the total cost of such abatement would be *less than 2 percent* of the GNP in 2000, compared with current costs of roughly 1 percent of GNP. Another way of viewing this is that each year we would have to give up only about one-tenth of 1 percent in annual growth in GNP, so that the growth rate would be, say, 3.9 percent per year rather than 4 percent. This is encouraging and is also evidence that there is no overall economic reason why environmentalists should not press for a strict pollution-abatement policy. However, as Ridker notes, this is a rather short-term projection and does not necessarily indicate that we can expect such a pattern to hold over, say, the following 50 years.

The analysis also omits discussion of two important points. First, it is easy to concentrate on pollution levels and ignore other aspects of environmental degradation. The most obvious of these is the obliteration of natural, seminatural, or agricultural ecosystems in the process of obtaining the raw materials for growth. Strip mining and open-pit mining, for example, lay waste vast tracts of land. Strip mining for coal alone has scarred about 2,500 square miles of land in the United States. The cost of restoring such land (if it can be restored) can be very high ($5,000 per acre[19]), and in some cases it may be impossible to "restore" the damaged natural community. Each year in the United States about 1 million acres of good farmland is taken out of production (there are roughly 350 million acres of cropland) and used for highways, shopping centers, and golf courses. Some of this is replaced, of course, by bringing more marginal land into production. But farming on marginal land generally is worse for the environment in terms of soil erosion, loss of wildlife habitat, and so on.

One of Ridker's comments illustrates how dangerous it is to reduce the problem of environmental degradation to one of pollution:

Another often-mentioned reason growth must eventually stop is environmental deterioration. In the last analysis, however, everything boils down to the availability of energy in useful forms, for with energy man can treat, move or protect himself from pollution and environmental deterioration. Even space can be increased with sufficient energy — by building upward and outward. A limitation on the use of energy could be the buildup of waste heat, but this depends on the source of energy and the efficiency with which it is utilized. To the extent that reliance is placed on geothermal heat and energy from the sun, a serious buildup of heat is not so certain.

Unfortunately, it is not true that massive amounts of energy can prevent environmental degradation or restore all ecosystems. How much energy is needed to reevolve an extinct raptorial bird species (Chapter 1)? Are we sure enough of our ability to reassemble a complete ecosystem (say, a West Coast estuary) that we think we can do it, given enough energy? I do not believe we can. Ridker's analysis also neglects the quality of the effluent. For example, although larger particles are easier to remove from a smokestack, there is evidence that the smaller particles are much more injurious, on a unit-weight basis, to the lungs; a pound of carcinogenic waste from the manufacture of some plastics (Chapter 8) or a pound of radioactive material is much worse than a pound of BOD (biochemical oxygen demand) or "suspended solids."

To analyze pollutants alone is an inadequate way of evaluating the environmental impact of future growth. It is very unlikely that energy use can be increased sharply without increasing environmental degradation. The various problems have been discussed generally in Chapters 5 and 12, but it is worth looking in detail at one area of the country where energy sources are likely to be heavily exploited over the next few decades: the Mountain and Western states. Here both coal and oil shale are abundant, and both can be obtained by strip mining, which is relatively cheap.

The Department of Interior has prepared an Environmental Impact Statement (EIS) for a prototype program for leasing to private industry tracts of land in Colorado, Wyoming, and Utah for the production of oil from oil shale. This EIS was reviewed by the Institute of Ecology, which has provided a description of the probable environmental hazards of such a program.[19] Taking Colorado as an example, the following problems are likely to arise, assuming that the "overburden" of soil and rock is stripped from above the shale:

1. The vegetation and associated wildlife will be destroyed over the mined area. No technology is now available for revegetating some of the old mined area, and establishing stable plant cover on spent shale is a particular problem, with no prospect of a solution.
2. Massive piles of overburden and spent shale would be dumped. Serious problems of erosion and leaching of salts from these piles to downstream areas would occur. Vast amounts of water would be needed, and leachings from spent shale together with huge quantities of highly saline waste "mine water" (even if the latter were treated, which would be very costly) would add a large burden of salts to the Colorado River. This would cause serious degradation of the water for downstream uses such as irrigation, and the Colorado already has severe problems in this respect (Chapter 6).
3. Contamination of groundwater.
4. Trace metals, which are highly toxic (Chapter 8), are likely to escape into the air and water.
5. Significant air pollution will arise from mining dust and from crushing and retorting the shale. The EIS admits that it will not be possible to meet SO_2 standards (Chapter 9), and other air pollution problems are likely to develop but were not adequately studied in the EIS.

Similar problems can arise in the strip mining of coal in these regions. Here the problem is aggravated by the need for huge quantities of water; indeed it simply may not be possible to provide enough water to mine and convert the coal in the area.[20] Furthermore, moving around great quantities of water in itself has a broad range of harmful environmental effects.

The question of the quality of the pollutants, raised earlier, illustrates the second point that I want to make about Ridker's analysis, which is

that it is based on the assumption of an improved technology. Even though we can certainly expect the technology to improve, this dependence on technology coupled with growth has some potential pitfalls. In the first place, it assumes that the proper political and economic decisions necessary to implement the technology will be made. But the recent energy crisis, during which the provisions of the 1970 Clean Air Act have been set aside or postponed, has shown us that, under the least pressure, government tends to scrap environmental standards in favor of providing enough energy to power industrial growth. Casting the problem in terms of technology obscures the social, political, cultural, and even moral dilemmas. Garrett Hardin has noted that there may be a whole class of "no technological solution" problems, such as stopping population expansion.

In the second place, technological "solutions" frequently embody unforeseen problems. The history of automobile exhaust devices has been a good example; reducing one set of pollutants tends to increase another (Chapter 9). The use of plastics has substituted some complex molecules that are severe health hazards for the more simple type of pollutants from the use of metals; "clean" nuclear energy is neither clean nor certainly safe (Chapters 5 and 12). The acid rain (Chapter 1) described by Likens and Bormann[21] is a very recent example of a solution that generates new problems. They have shown that strong acidity in the rain over areas in northeastern United States coincides with the introduction of new low-pollution fuels and particle-collecting equipment, because soot particles are no longer present to neutralize the SO_2 in the smoke. Moreover, the new, taller smoke stacks spread the pollution wider, changing "local soot problems into a regional acid-rain problem."

The application of technical "solutions" contains a more subtle danger. The technological complex — the society, structures, and machinery — is already a system composed mainly of positive feedbacks. The inherent danger of promised technological solutions to environmental problems is that they give the impression that the problem is being tackled and, in a society geared to growth, this allows the system to continue its headlong rush. In tinkering with the juggernaut of technology (which produced the problems), we are oiling its wheels. The current "antismog" devices on automobile exhausts are not the solution to the air pollution problem; the real solution is to restructure our transportation systems and our cities. Meanwhile the devices give the impression that something is being done and that we can go on using more and more automobiles. Technological solutions need to be applied within an overall systems approach that seeks remedies for the basic cause (untrameled growth) as well as for mere symptoms. The basic problem is to change our relationship to the environment and the opiate of purely technological solutions dulls our perception of this fundamental fact.

Many environmentalists (myself among them) have been overly simplistic in their contention that economic growth must stop because of environmental degradation. Although no analysis of the environmental consequences of economic growth has been complete, there is enough evidence that technological improvements and a vigorous government policy of abatement (essentially internalizing environmental costs) can actually reduce pollution levels while allowing economic growth. Extreme care needs to be taken to reduce the incidence of new problems arising from such solutions. On the other hand, the cost of preventing the destruction of ecosystems (or of restoring them) has not been well studied, and may impose much more severe restraints on growth than economists have considered tolerable. It may be impossible to grow economically without increasing this degradation, even though pollution levels can be reduced.

Many components of economic growth are unacceptable from an environmental point of view; environmentalists simply need to be more specific about which components concern them. For example, urban sprawl and new highway systems directly obliterate ecosystems and should in general be opposed. In any case, our cities would be better places to live without such expansion. Preventing the expansion of energy use, at least from the usual sources, must be a first priority. This means continuing to press for better conservation of energy and energy use. Sweden's high per capita income with a much lower per capita use of energy than the United States shows that we do not have to waste energy to be affluent. We should try to develop cleaner sources, such as solar energy, and to prevent growth that relies on the production of highly dispersive pollutants that disrupt ecosystems (see, e.g., Chapters 1 and 13). The concept of equilibrium is still useful, even if the economy continues to grow. Maintenance of biological diversity is a primary goal; in this sense we need to be at equilibrium with the environment even though we may be getting richer for some time in the future.

ECONOMIC VERSUS OTHER VALUES

Standard economic analysis assumes that we can place a monetary value on everything involved in transactions. This idea that everything is commensurable (translatable into a price tag) is basic to the notion of internalizing costs (Chapter 16), but some things cannot be valued in these terms, particularly irreplaceable environmental resources such as a species which cannot be reevolved once it is extinct. Economics provides no way of discounting far into the future (we cannot estimate the future value of an investment if the future is very far away); furthermore, there is no way to know if a given species will some day be valuable.

A more ambiguous case is the value placed upon the view from the rim of the Grand Canyon. Here, we might value it as the sum of all the payments people would be willing to make now, to save the view. But as we become more affluent, and as scenic beauty becomes rarer because of development, each person would be willing to pay more; and we still have the problem of evaluating its worth to future generations. These examples are more difficult than other exhaustible resources, such as oil, for which technology can foreseeably provide a substitute.

The existence of incommensurable things means that it is impossible to do a complete economic analysis of the effect of economic growth. When we have computed all the measurable costs and benefits, there will always be some things beyond measurement, some diseconomies (or perhaps even some benefits) that we cannot place on the economic scale to see if a given type of growth is "worthwhile."

The Environment, Politics, and Action

No matter how environmentally sound a proposed solution to an environmental problem may be, there is little hope of getting that solution implemented except through political channels. My own idealism was rudely shattered by a brief encounter with local politics in Santa Barbara. A developer wanted to build on some land close to the ocean, land that supported some interesting fauna. He had an environmental-impact report prepared and presented to the county supervisors. Local conservationists asked us (another ecologist and an ornithologist) to analyze the report, which turned out to be clearly inadequate and in error in several respects, especially with regard to the major fauna of interest. We presented testimony that showed that the report was in error, indeed incompetently done, and in fact the major biologist signing the report agreed with us in private that it was a very poor job. We should not have been, but were surprised when the supervisors essentially ignored our comments and prepared to give approval to the project. A few weeks later permission to develop was denied when it was discovered that the developer was allegedly involved in an attempt to bribe a supervisor. So much for a rational environmental policy!

In the early 1970s one particular Pogo cartoon was popular among environmentalists. Its message was that "we have met the enemy and he is us"; its implication was that our personal life styles and habits cause the environmental mess and that we can clean it up by each of us living a cleaner and perhaps simpler life. There is some truth to this, but it obscures important issues — for example that, for many people, riding to work in a polluting car is *not* a matter of choice but of necessity brought about by social, economic, and political forces.

Containing equal parts of truth, but equally less than the whole truth, are claims of various disciplines that *their* discipline, from ecology to engineering, provides the solution. Chapter 16, for example, provides a persuasive argument that the environmental problem is an economic problem — the internalizing of externalities. But, as most economists would agree, it is not a *purely* economic problem. A truly free market could internalize externalities via the pricing mechanism, but the political and economic reality is that we do not live in a free market. The U.S. economy is dominated by "corporate collectivism," in which heavy influence is exerted over some legislators and high government officers by some companies some of the time. (Surely all of them are not so influenced all of the time.) The legislators pass laws that affect the degree of internalization of social costs and that affect pollution standards, and government officers enforce these laws to a greater or lesser degree.

Improvement in environmental matters is not likely unless some powerful force in society is in favor of it, and it is extremely unlikely as long as powerful forces are against it. Real power still resides to a great degree in wealth and control of the means of production, that is, in the owners and executives of large corporations and financial institutions.[22] In 1962, the wealthiest 1 percent of the population owned 31 percent of the total wealth, which includes such things as land, homes, and automobiles, as well as businesses; even more striking, the top 1 percent owned 61 percent of the corporate stock.[23] Because of the heavy election expenses that stand between a candidate and power, wealthy campaign contributors exert enormous political influence. Attempts to reform the system have been spectacularly unsuccessful.

Questions of controlling pollution always bring us back to politics and power. For example, at the center of our environmental problems are our use of energy, the extent to which we have a range of options for energy sources, and how much control we have over these options. However, a result of the oil crisis begun in 1973 was that pollution standards for oil-burning plants were reduced and corporate earnings of oil companies were increased. Thus, both with respect to pollution and fairness we seem to have moved backward, and the average citizen, through his government, seems to have been powerless to prevent this.

The oil companies illustrate well the problem of the concentration of political power in the hands of a tiny fraction of the community and the difficulties that this places in the way of rational environmental planning by the elected representatives of the people. For example, when decisions had to be made about allocation of oil resources in the United States in late 1973, it turned out that the United States was the only Western industrial nation where the government had no independent way of finding out exactly what oil we had available. In something approaching a national crisis, the crucial information was in the hands of a few oil companies, and there was no way to force them to divulge or test the accuracy of that information for the national good. (The same problem applies, incidentally, with pollutants such as many of the pesticides; when only a few companies make them, it is impossible to get information on how much they make, so models of pesticide distribution in the environment often lack such basic data.)

The emergence and expansion of multinational corporations means that energy planning at an international level is also increasingly carried out by large companies. Again, during the oil crisis, oil shipments could be diverted from one nation to another (or one state to another) as the large oil corporations chose among options.

The problem is not merely that our oil reserves and policies are dictated by a mere handful of oil executives, but that the same companies are coming to dominate the development and exploitation of the alternative energy sources, a process that has been documented in detail by Ridgeway in his book *The Last Play*.[24] Ridgeway shows that starting as far back as the 1930s, oil companies have gradually taken control first of coal and now of uranium resources, so that much of the range of energy options is in their hands, both in the United States and abroad. By 1970 the oil industry essentially controlled the coal industry, with two of the three largest coal producers owned outright by oil companies and the largest single block of coal reserves in the country in the hands of the Jersey Standard Oil Company. (Incidentally, as demand for coal has been exceeding the supply, coal production on public lands, leased to such large companies, has been *declining;* that is, the oil companies control the reserves but largely do not develop them.) From 1965 to 1970, oil companies increased their share of coal production from 7 to 28 percent. In addition, oil companies control 45 percent of all U.S. uranium resources.

Ridgeway's book is depressing to read, as he describes rich coal and oil corporations expanding their profits and power while providing virtually nothing to the communities that contain the coal, almost ignoring the health and safety of the miners, and making a mockery of the notion that the government is able to regulate big business.

The amazing thing is that we have, through the federal government, paid the oil companies to wrest from us control over our energy options, a story clearly told recently by Rustow.[25] Oil depletion allowances have been a multibillion dollar gift from the taxpayer to the oil industry. In addition, all taxes paid to foreign oil-producing nations by the oil companies have been deductible, thus allowing the companies to avoid 50 percent of their U.S. taxes; the American taxpayer pays the oil taxes levied by foreign governments on U.S. oil corporations. Furthermore, import quotas on foreign oil (costing the taxpayer from $3 billion to $5 billion per year) have meant that we have followed a "drain America first" policy.

We are now in a position in which the average citizen seems to have lost all around. We have almost no control over which energy options we shall choose; the cost of power (gas and electricity) has skyrocketed along with corporate profits. In the midst of all the confusion, clean air standards have been lowered, the Alaska pipeline has been approved, and public subsidies have been proposed for private companies to explore for new sources of energy.

The response to these events has included a good deal of populist anger. The traditional response to such problems is government regulation, and in addition Congress has considered a new law on excess profits. The problem with these two solutions, as Ralph Nader has amply demonstrated, is that regulation has not worked; and as Harrington[26] points out, an excess profits tax is likely to fail:

The oil companies, given their privileged status in the Internal Revenue Code, are past masters at creative cost accountancy. Therefore, in the absence of structural change within the industry that would give the public the opportunity to monitor and contest the corporate definition of profits, I fear that the main beneficiaries of such a law would be the lawyers and accountants hired to evade it. This is particularly true in the case of multinational corporations which, with a little judicious juggling, can choose the country where they want to surface, or hide, their profits.

Harrington outlines a series of recent developments in the oil industry and the government's response to them, including the proposal to support private research and development with public money:

All of these things add up to a shocking capitalist indictment of late capitalism. Of the traditional rationalizations for profit — inventiveness, risk-taking, abstinence from consumption, the hazarding of private monies, and so on — literally nothing remains . . . private managers will set all the basic priorities while financing their preferences with consumers' and taxpayers' funds. Adam Smith would be appalled.

All this should not be confused with a "devil theory" of history. No claim is made that oil company executives are more evil than the rest of us. Certainly, the removal or replacement of these men would make no difference if the way that multinational corporations grow and evolve is left unchanged. Stockholders and boards of directors select executives whose policies increase the power and wealth of the corporation while avoiding indictments and antitrust actions. It is a self-perpetuating system, the survival of an odd sort of "fittest"; no giant corporation will suddenly become markedly more altruistic and community-minded, for fear that it will be devoured by its competitors.

And so we return to the problem of political power and control over policies that have enormous consequences for the environment. Somehow we must gain that power and control, and it is hard to see how anything short of restructuring the industry and placing control directly in the citizens' hands will work. That is, the corporations need to be dismantled and the industry nationalized so that its basic priorities are determined by democratic political processes.

Such a proposal, of course, is not a simple answer and it would be difficult to avoid some pitfalls of nationalization — that the same people would continue to run the industry for largely the same purposes as before, for exam-

ple. But specific steps can be taken to increase the likelihood of success. For example, Harrington has suggested: To start with, nationalize one existing company; establish a public review board and ombudsman to hear challenges from consumers and environmentalists; establish a TVA-type *public* corporation for research and development of energy reserves on public lands.

Nationalization of the oil industry, or any other corporation-dominated sector of the economy, is clearly not a panacea, nor will it necessarily stop the plundering of the environment. After all, the USSR seems to have been no kinder to its environment than we to ours. But democratic control over environmentally important decisions is a necessary first step, and this is apparently impossible when large corporations exist. Since corporate executives are not voted into their position and are not removable from it by the people at large (or even by the vast majority of their stockholders), and since they are relatively uncontrollable by the elected government, they are not responsible to the public over what they do with the public's environment. Yet they are able to shape that environment.

The crux of the matter is that large corporations not only have the power to pollute, they have the economic and political power to prevent, delay, and weaken regulatory legislation. They also have the power and connections to ensure that the regulatory agencies do not regulate as they ought to. In theory this can all be sorted out, but in practice the system is most likely to be self-perpetuating.[27,28] In the face of powerful evidence that poor automobile design kills people, Detroit successfully delayed and weakened car safety legislation for a decade.[31] Another example is the coal industry, which successfully combated legislation against pollution from sulfur oxide.[29]

Conceivably the leaders of private industry might someday become convinced of the worthiness of the cause of social welfare, including the value of a clean and stable environment. In this case they might be willing to use the best options because they have well-developed social consciences. This system has the same appeal as benevolent dictatorship. But all evidence points to the conclusion that the safety and welfare of the public are not a prime concern of the leaders of private industry, recent advertisements to the contrary notwithstanding. The auto industry responded so slowly to air pollution control that the Justice Department in January 1969 filed suit against four major auto manufacturers and their trade association on the grounds that they had conspired to delay the development and use of devices to control air pollution from automobiles. (The power of the companies is shown by the fact that the federal government later agreed to settle the suit by a consent decree that did not penalize the companies and sealed the grand jury records.)[29] Even more serious disregard of public safety has occurred in the nuclear power industry (Chapter 12).

The question of nationalization also raises the more general questions of property and ownership and their relationship to environmental quality. There has been a tendency to interpret Garrett Hardin's "Tragedy of the Commons" article[30] as evidence that private ownership will result in better husbandry of resources than will public ownership. But this is certainly not generally valid. In particular, private (or corporate) ownership of *exhaustible* resources, such as oil, coal, and minerals, has certainly not reduced the rate at which they are exploited; by contrast, international management of "publicly" owned fisheries, which are renewable resources, has frequently been successful in modern times, as has Peru's nationalization of the anchovy fishery (see Chapter 7).

Hardin's argument is that commonly owned property, like air and water, is abused for the precise reason that nobody owns it — it is a "commons." For example, it is not in the economic interest of an individual to spend money reducing his car exhaust fumes because he gains very little (the exhaust goes into a huge "commons") and he pays a lot. When all individuals operate like this, excessive air pollution is the result. Hardin's analogy of the commons is a useful one, especially in relation to *renewable* resources: if the grazing commons is collectively owned, each farmer will gain by putting one more of *his* sheep onto the commons and, if there are already enough sheep there, he will lose very little from the marginal overgrazing that results. However, since this is true for each farmer, there will tend to be severe overgrazing. Certainly one possible response to this is to divide up the commons and have each farmer own his own part or, what in fact happens, have a few farmers own the parts. There is an alternative however, which is to keep the land under collective ownership *and to do the same with the sheep*. The sheep and commons can then be managed collectively as a unit to maximize the common good.

One difficulty in husbanding important environmental resources seems to stem from the fact that in neither capitalist nor socialist systems is there any real sense on the part of the majority of people of *ownership* of these resources and, most importantly, of the industry that develops and exploits them. In fact, in socialist countries "as collectivist bureaucracy came more and more to exercise the prerogatives of owners [in the capitalist sense], collective ownership became formal and abstract."[31] It is this abstract aspect of public ownership that is a major weakness of nationalization (or socialism), a subject that has been very perceptively discussed by Erazim Kohák.[31]

Unless collective ownership implies a sense that the owned object is "mine" (or "ours"), rather than "theirs," it will not be treated with care. Yet in a large society, problems of scale make such "mineness" difficult to achieve. Such real tangible feelings of ownership are most easily achieved by working with the owned object, which may go far to explain the strong drive of many people in recent years to return to the land and care for a small part of the environment in a very tangible, manual way. As Kohák says, "the forms of ownership are liberating, sustaining the individual's freedom of identity and participation, if they express a lived relationship of belonging. They become alienating if they disrupt such belonging in the name of an abstract claim to possession." The problem that faces socialist public ownership of environmental resources and of the means of developing and producing them, a problem for which I can offer no all-purpose solution, is to ensure the participation of all the owners in major decisions. While that central problem of democratic control remains incompletely solved, the manipulation of these resources by a handful of the people, the owners of capital, is a poor alternative.

ENVIRONMENTAL ACTION

The first half of the 1970s has been marked by political apathy among students, in marked contrast to the political activism of the students of the 1960s. The activists apparently failed to achieve their goals, but they undoubtedly had a profound effect on government thinking and without doubt pulled America out of direct involvement in Vietnam much sooner than would otherwise have been the case. The environment is a much more nebulous and ill-defined cause, and it must be defended more by local holding actions than by a march on Washington. But it is no less important. Students have access to information on complex environmental issues and have the time and energy to ponder these issues, reach rational decisions about them, and then do something about it. From keeping an eye on the local city council to joining a Nader group or Environmental Defense Fund group, to getting environmentally conscious legislators elected, to designing a blueprint for public ownership of the power industry, there is no lack of things to do.

References

1. Barnett, H. J., and Morse, C. 1963. *Scarcity and Growth*. The Johns Hopkins Press (for Resources for the Future), Baltimore, Md. See also Reference 13.
2. Government Operations Committee. 1969. Effects of population growth on natural resources and the environment, p. 50. Hearings before a subcommittee of the Committee on Government Operations, House of Representatives.
3. Wade, N. 1974. Raw materials: U.S. grows more vulnerable to third world cartels. *Science 183:* 185–186.
4. Galbraith, K. 1967. *The New Industrial State*. Houghton Mifflin Company, Boston.
5. Gardner, B. L. 1969. Determinants of farm income inequality. *Amer. J. Agri. Econ. 51*.
6. Orleans, L. A. 1974. Progress in China. *Science 184:* 695–697. This is a review of six books dealing with medicine and public health in China.
7. Ehrlich, P. R., and Ehrlich, A. H. 1970. *Population, Resources, Environment: Issues in Human Ecology*. W. H. Freeman and Company, San Francisco.
8. Allaby, M. 1970. One jump ahead of Malthus. *The Ecologist 1:* 24–28.
9. Taylor, L. R. (ed.). 1970. *The Optimum Population for Britain*. Academic Press, Inc., New York.
10. Meadows, D. H., Meadows, D. L., Randers, J., and Behrens, W. W., III. 1972. *The Limits to Growth*. Universe Books, New York.
11. Starr, C., and Rudman, R. 1973. Parameters of technological growth. *Science 182:* 358–364.
12. Barraclough, G. 1974. The end of an era. *The New York Review of Books 21:* 14–20. This article reviews nine recent books concerned with the possibility of an imminent economic depression, and refers to many others.
13. Lovering, T. S. 1969. Mineral resources from the land. In *Resources and Man* (Preston Cloud, ed.). W. H. Freeman and Company, San Francisco, California.
14. World Health Organization. 1968. *Research into Environmental Pollution* (Tech. Rep. Ser. 406). WHO, Geneva.
15. Niering, W. A. 1970. The dilemma of the coastal wetlands: conflict of local, national and world priorities. In *The Environmental Crisis* (H. W. Helfrich, ed.). Yale University Press, New Haven, Conn.
16. Borgstrom, G. 1969. *Too Many*. Macmillan & Co. Ltd., London.
17. Mishan, E. J. 1969. *The Costs of Economic Growth*. Penguin Publishing Co. Ltd., Harmondsworth, England.
18. Ridker, R. G. 1973. To grow or not to grow: that's not the relevant question. *Science 182:* 1315–1318.
19. Environmental Impact Assessment Project of the Institute of Ecology. 1973. A Scientific and Policy Review of the Final Environmental Impact Statement for the Prototype Oil Shale Leasing Program of the Department of the Interior (K. Fletcher, and M. Baldwin, eds.) Government Printing Office, Washington, D.C.
20. National Academy of Sciences. 1973. Report of the Study Committee on the Potential for Rehabilitating Lands Surface Mined

for Coal in the Western United States. The Academy, Washington, D.C.

21. Likens, G. E., and Bormann, F. H. 1974. Acid rain: a serious regional environmental problem. *Science 184:* 1176–1179.

22. Domhoff, G. W. 1970. *The Higher Circles: The Governing Class in America*. Random House, Inc., New York.

23. Ackerman, F., et al. 1972. The extent of income inequality in the United States. In *The Capitalist System* (R. C. Edwards et al. eds.). Prentice-Hall, Inc., Englewood Cliffs, N.J.

24. Ridgeway, J. 1973. *The Last Play*. E. P. Dutton & Co., Inc., New York.

25. Rustow, D. A. 1974. Petroleum politics 1951–1974. *Dissent*, Spring: 144–153.

26. Harrington, M. 1974. The oil crisis — socialist answers. *Dissent*, Spring: 139–142.

27. The Nader Report. 1970. A series of separate reports on a number of administrative agencies, for example, J. S. Turner, *The Chemical Feast* (Study Group Report on the Food and Drug Administration). Grossman Publishers, New York.

28. Carter, L. J. 1968. Water pollution: officials goaded in raising quality standards. *Science 160:* 49–51.

29. Davies, J. C. 1970. *The Politics of Pollution*. Pegasus, Indianapolis, Ind.

30. Hardin, G. 1968. The tragedy of the commons. *Science 162:* 1243–1248.

31. Kohák, E. 1974. Possessing, owning, belonging. *Dissent*, Spring: 344–353.

Index

481

483

ABOUT THE BOOK

The text of this book is set in V.I.P. Palatino, a contemporary typeface designed by Herman Zapf, based on Renaissance type styles. The display type is set in Optima, a modern Sans Serif face distinguished by its combination of thick and thin elements. The book was designed by Andrew Zutis, set by David E. Seham Associates, printed and bound by The Book Press. The cover was printed by Algen Press Corp.